Engineering Superconductivity

Engineering Superconductivity

Editor:

Nelson Bolívar, Ph.D

ArclerPress

www.arclerpress.com

Engineering Superconductivity

Editor: Nelson Bolívar, Ph.D

708 3rd Avenue, 6th Floor
New York
NY 10017
United States of America
www.arclerpress.com

ISBN: 978-1-68094-398-6

Library of Congress Control Number: 2016951056

About the Editor

Nelson Bolívar, Ph.D

Nelson Bolivar has a PhD. in physics from the University of Lorraine in France finished in 2014. His expertise is in quantum systems and condensed matter. His interest includes spintronic devices and correspondences between general relativity and condensed matter. He is currently an associate professor at the Central University of Venezuela

Contents

List of Contributors

M P Das
Department of Theoretical Physics, Research School of Physics and Engineering, The Australian National University, Canberra ACT 2601, Australia

B J Wilson
Department of Theoretical Physics, Research School of Physics and Engineering, The Australian National University, Canberra ACT 2601, Australia

Xu Liu
National Lab for Superconductivity, Beijing National Laboratory for Condensed Matter Physics, Institute of Physics, Chinese Academy of Sciences, Beijing 100190, People's Republic of China

Lin Zhao
National Lab for Superconductivity, Beijing National Laboratory for Condensed Matter Physics, Institute of Physics, Chinese Academy of Sciences, Beijing 100190, People's Republic of China

Shaolong He
National Lab for Superconductivity, Beijing National Laboratory for Condensed Matter Physics, Institute of Physics, Chinese Academy of Sciences, Beijing 100190, People's Republic of China

Junfeng He
National Lab for Superconductivity, Beijing National Laboratory for Condensed Matter Physics, Institute of Physics, Chinese Academy of Sciences, Beijing 100190, People's Republic of China

Defa Liu
National Lab for Superconductivity, Beijing National Laboratory for Condensed Matter Physics, Institute of Physics, Chinese Academy of Sciences, Beijing 100190, People's Republic of China

Daixiang Mou
National Lab for Superconductivity, Beijing National Laboratory for Condensed Matter Physics, Institute of Physics, Chinese Academy of Sciences, Beijing 100190, People's Republic of China

Bing Shen
National Lab for Superconductivity, Beijing National Laboratory for Condensed Matter Physics, Institute of Physics, Chinese Academy of Sciences, Beijing 100190, People's Republic of China

Yong Hu
National Lab for Superconductivity, Beijing National Laboratory for Condensed Matter Physics, Institute of Physics, Chinese Academy of Sciences, Beijing 100190, People's Republic of China

Jianwei Huang
National Lab for Superconductivity, Beijing National Laboratory for Condensed Matter Physics, Institute of Physics, Chinese Academy of Sciences, Beijing 100190, People's Republic of China

X J Zhou
National Lab for Superconductivity, Beijing National Laboratory for Condensed Matter Physics, Institute of Physics, Chinese Academy of Sciences, Beijing 100190, People's Republic of China
Collaborative Innovation Center of Quantum Matter, Beijing, People's Republic of China

Kazumi Maki
Max-Planck Institute for the Physics of Complex Systems, Nöthnitzer Str. 38 01187 Dresden, Germany
Department of Physics and Astronomy, University of Southern California, Los Angeles, CA 90089-0484, USA

Hae Young Kee
Department of Physics, University of Toronto, Toronto, Ontario M5S 1A7, Canada

Yoshifumi Morita
Faculty of Engineering, Gunma University, Kiryu, Gunma 376-8515, Japan

Konrad Jerzy Kapcia
Electron States of Solids Division, Faculty of Physics, Adam Mickiewicz University in Poznan, Umultowska 85, ´ 61-614 Poznan, Poland

Jiangping Hu
Beijing National Laboratory for Condensed Matter Physics, Institute of PhysicsChinese Academy of Sciences
Collaborative Innovation Center of Quantum Matter

H. P. Roeser
Institute of Space Systems, University of Stuttgart, Pfaffenwaldring 31, 70569 Stuttgart, Germany

D. T. Haslam
Institute of Space Systems, University of Stuttgart, Pfaffenwaldring 31, 70569 Stuttgart, Germany

J. S. López
Institute of Space Systems, University of Stuttgart, Pfaffenwaldring 31, 70569 Stuttgart, Germany

M. Stepper
Institute of Space Systems, University of Stuttgart, Pfaffenwaldring 31, 70569 Stuttgart, Germany

F. von Schoenermark
German Aerospace Center, GSOC, Muenchnerstr. 20, 82234 Wessling, Germany

F. M. Huber
German Aerospace Center, GSOC, Muenchnerstr. 20, 82234 Wessling, Germany

A. S. Nikoghosyan
Department of Microwave and Telecommunication, Yerevan State University, Alex Manoogian 1, Yerevan, 0025, Armenia

LI HaiJie
Quantum Optoelectronics Laboratory, Southwest Jiaotong University, Chengdu 610031, China

WANG YiWen
Quantum Optoelectronics Laboratory, Southwest Jiaotong University, Chengdu 610031, China

WEI LianFu
Quantum Optoelectronics Laboratory, Southwest Jiaotong University, Chengdu 610031, China
State Key Laboratory of Optoelectronic Materials and Technologies, School of Physics and Engineering, Sun Yatsen University, Guangzhou 510275, China;

ZHOU PinJia
Quantum Optoelectronics Laboratory, Southwest Jiaotong University, Chengdu 610031, China

WEI Qiang
Quantum Optoelectronics Laboratory, Southwest Jiaotong University, Chengdu 610031, China

CAO ChunHai
Research Institute of Superconductor Electronics, Nanjing University, Nanjing 210093, China

FANG YuRong
Research Institute of Superconductor Electronics, Nanjing University, Nanjing 210093, China

YU Yang
Research Institute of Superconductor Electronics, Nanjing University, Nanjing 210093, China

WU PeiHeng
Research Institute of Superconductor Electronics, Nanjing University, Nanjing 210093, China

Yanlong Ding
Key Laboratory of Magnetic Levitation Technologies and Maglev Trains, Ministry of Education of China, and Superconductivity and New Energy R&D Center, Southwest Jiaotong University, Chengdu 610031, China

Shuiquan Deng
Key Laboratory of Magnetic Levitation Technologies and Maglev Trains, Ministry of Education of China, and Superconductivity and New Energy R&D Center, Southwest Jiaotong University, Chengdu 610031, China

Yong Zhao
Key Laboratory of Magnetic Levitation Technologies and Maglev Trains, Ministry of Education of China, and Superconductivity and New Energy R&D Center, Southwest Jiaotong University, Chengdu 610031, China

WANG YongTian
Key Laboratory of Condition Monitoring and Control for Power Plant Equipment of Ministry of Education, North China Electric Power University, Beijing 102206, China;
Institute of Physics, Chinese Academy of Sciences, Beijing 100190, China

LIU ZongDe
Key Laboratory of Condition Monitoring and Control for Power Plant Equipment of Ministry of Education, North
Suzhou Research Institute of North China Electric Power University, Suzhou 215123, China

WANG WeiHua
Institute of Physics, Chinese Academy of Sciences, Beijing 100190, China

ZHANG ShuLin
State Key Laboratory of Functional Materials for Informatics, Shanghai Institute of Microsystem and Information Technology, Chinese Academy of Sciences, Shanghai 200050, China;
Joint Research Laboratory on Superconductivity and Bioelectronics, Collaboration between CAS-Shanghai and FZJ, Shanghai 200050, China

ZHANG GuoFeng
State Key Laboratory of Functional Materials for Informatics, Shanghai Institute of Microsystem and Information Technology, Chinese Academy of Sciences, Shanghai 200050, China;
Joint Research Laboratory on Superconductivity and Bioelectronics, Collaboration between CAS-Shanghai and FZJ, Shanghai 200050, China

WANG YongLiang
State Key Laboratory of Functional Materials for Informatics, Shanghai Institute of Microsystem and Information Technology, Chinese Academy of Sciences, Shanghai 200050, China;
Joint Research Laboratory on Superconductivity and Bioelectronics, Collaboration between CAS-Shanghai and FZJ, Shanghai 200050, China

ZENG Jia
State Key Laboratory of Functional Materials for Informatics, Shanghai Institute of Microsystem and Information Technology, Chinese Academy of Sciences, Shanghai 200050, China;
Joint Research Laboratory on Superconductivity and Bioelectronics, Collaboration between CAS-Shanghai and FZJ, Shanghai 200050, China
Graduate University of the Chinese Academy of Sciences, Beijing 100049, China

QIU Yang
State Key Laboratory of Functional Materials for Informatics, Shanghai Institute of Microsystem and Information Technology, Chinese Academy of Sciences, Shanghai 200050, China;
Joint Research Laboratory on Superconductivity and Bioelectronics, Collaboration between CAS-Shanghai and FZJ, Shanghai 200050, China
Graduate University of the Chinese Academy of Sciences, Beijing 100049, China

LIU Ming
State Key Laboratory of Functional Materials for Informatics, Shanghai Institute of Microsystem and Information Technology, Chinese Academy of Sciences, Shanghai 200050, China;
Joint Research Laboratory on Superconductivity and Bioelectronics, Collaboration between CAS-Shanghai and FZJ, Shanghai 200050, China

KONG XiangYan
State Key Laboratory of Functional Materials for Informatics, Shanghai Institute of Microsystem and Information Technology, Chinese Academy of Sciences, Shanghai 200050, China;
Joint Research Laboratory on Superconductivity and Bioelectronics, Collaboration between CAS-Shanghai and FZJ, Shanghai 200050, China

XIE XiaoMing
State Key Laboratory of Functional Materials for Informatics, Shanghai Institute of Microsystem and Information Technology, Chinese Academy of Sciences, Shanghai 200050, China;

Zhao Liu
Key Laboratory of Magnetic Levitation Technologies and Maglev Trains, Ministry of Education of China, and Superconductivity and New Energy R&D Center, Southwest Jiaotong University, Chengdu 610031, People's Republic of China

Yongliang Chen
Key Laboratory of Magnetic Levitation Technologies and Maglev Trains, Ministry of Education of China, and Superconductivity and New Energy R&D Center, Southwest Jiaotong University, Chengdu 610031, People's Republic of China

Lupeng Du
Key Laboratory of Magnetic Levitation Technologies and Maglev Trains, Ministry of Education of China, and Superconductivity and New Energy R&D Center, Southwest Jiaotong University, Chengdu 610031, People's Republic of China

Pingyuan Li
Key Laboratory of Magnetic Levitation Technologies and Maglev Trains, Ministry of Education of China, and Superconductivity and New Energy R&D Center, Southwest Jiaotong University, Chengdu 610031, People's Republic of China

Yajing Cui
Key Laboratory of Magnetic Levitation Technologies and Maglev Trains, Ministry of Education of China, and Superconductivity and New Energy R&D Center, Southwest Jiaotong University, Chengdu 610031, People's Republic of China

Xifeng Pan
Western Superconducting Technologies Co., Ltd., Xi'an 710018, People's Republic of China

Guo Yan
Western Superconducting Technologies Co., Ltd., Xi'an 710018, People's Republic of China

Hideo Hosono
Frontier Research Center, Tokyo Institute of Technology, Yokohama 226-8503, Japan
Materials and Structures Laboratory, Tokyo Institute of Technology, Yokohama 226-8503, Japan
Materials Research Center for Element Strategy, Tokyo Institute of Technology, Yokohama 226-8503, Japan

Keiichi Tanabe
Superconductivity Research Laboratory, International Superconductivity Technology Center (ISTEC), 2- 11-19 Minowa-cho, Kohoku-ku, Yokohama, Kanagawa 223-0051, Japan

Eiji Takayama-Muromachi
National Institute for Materials Science, 1-2-1 Sengen, Tsukuba, Ibaraki 305-0047, Japan

Hiroshi Kageyama
Department of Energy and Hydrocarbon Chemistry, Graduate School of Engineering, Kyoto University, Nishikyo-ku, Kyoto 615-8510, Japan

Shoji Yamanaka
Department of Applied Chemistry, Graduate School of Engineering, Hiroshima University, HigashiHiroshima 739-8527, Japan

Hiroaki Kumakura
National Institute for Materials Science, 1-2-1 Sengen, Tsukuba, Ibaraki 305-0047, Japan

Minoru Nohara
Department of Physics, Okayama University, Okayama 700-8530, Japan

Hidenori Hiramatsu
Materials and Structures Laboratory, Tokyo Institute of Technology, Yokohama 226-8503, Japan
Materials Research Center for Element Strategy, Tokyo Institute of Technology, Yokohama 226-8503, Japan

Satoru Fujitsu
Materials Research Center for Element Strategy, Tokyo Institute of Technology, Yokohama 226-8503, Japan

Jing Wang
Department of Modern Physics, University of Science and Technology of China, Hefei, Anhui 230026, People's Republic of China

Guo-Zhu Liu
Department of Modern Physics, University of Science and Technology of China, Hefei, Anhui 230026, People's Republic of China
Max Planck Institut fur Physik komplexer Systeme, D-01187 Dresden, ¨ Germany

Konrad Kapcia
Electron States of Solids Division, Faculty of PhysicsAdam Mickiewicz University in Poznań

A. Bussmann-Holder
Max-Planck-Institut für Festkörperforschung, Heisenbergstr. 1, D-70569 Stuttgart, Germany

A. Simon
Max-Planck-Institut für Festkörperforschung, Heisenbergstr. 1, D-70569 Stuttgart, Germany

H. Keller
Physik Institut der Universität Zürich, Winterthurerstr. 190, CH-8057 Zürich, Switzerland

A. R. Bishop
Los Alamos National Laboratory, Los Alamos, NM87545, USA

H. R. Khan
Forschungsinstitut für Edelmetalle und Metallchemie, Schwäbisch Gmünd, West Germany

Ch. J. Raub
Forschungsinstitut für Edelmetalle und Metallchemie, Schwäbisch Gmünd, West Germany

H. R. Khan
Forschungsinstitute für Edelmetalle und Metallchemie, D-7070 Schwabisch Gmünd, Federal Republic of Germany

Daniel Homa
Virginia Polytechnic Institute and State University, 213 Holden Hall, Blacksburg, VA, 24060, USA

Yongxuan Liang
Virginia Polytechnic Institute and State University, 213 Holden Hall, Blacksburg, VA, 24060, USA

Gary Pickrell
Virginia Polytechnic Institute and State University, 213 Holden Hall, Blacksburg, VA, 24060, USA

M. Mumtaz
Materials Research Laboratory, Department of Physics, FBAS, International Islamic University (IIU), Islamabad 44000, Pakistan

Liaqat Ali
Materials Research Laboratory, Department of Physics, FBAS, International Islamic University (IIU), Islamabad 44000, Pakistan

Shoaib Azeem
Materials Research Laboratory, Department of Physics, FBAS, International Islamic University (IIU), Islamabad 44000, Pakistan

Saad Ullah
Materials Research Laboratory, Department of Physics, FBAS, International Islamic University (IIU), Islamabad 44000, Pakistan

G. Hussain
Materials Research Laboratory, Department of Physics, FBAS, International Islamic University (IIU), Islamabad 44000, Pakistan

M. W. Rabbani
Materials Research Laboratory, Department of Physics, FBAS, International Islamic University (IIU), Islamabad 44000, Pakistan

Abdul Jabbar
Department of Physics, Ghazi University, Dera Ghazi Khan 32200, Pakistan

K. Nadeem
Materials Research Laboratory, Department of Physics, FBAS, International Islamic University (IIU), Islamabad 44000, Pakistan

D. Shaltiel
Racah Institute of Physics, The Hebrew University of Jerusalem, Jerusalem, Israel

H.-A. Krug von Nidda
Experimental Physics V, Center for Electronic Correlations and Magnetism, University of Augsburg, 86135 Augsburg, Germany

B. Y. Shapiro
Physics Department, Bar-Ilan University, Ramat-Gan, Israel

B. Rosenstein
Department of Electrophysics, National Chiao Tung University, Hsinchu, Taiwan

A. Loidl
Experimental Physics V, Center for Electronic Correlations and Magnetism, University of Augsburg, 86135 Augsburg, Germany

B. Bogoslavsky
Racah Institute of Physics, The Hebrew University of Jerusalem, Jerusalem, Israel

I. Shapiro
Physics Department, Bar-Ilan University, Ramat-Gan, Israel

T. Fujii
Cryogenic Center, The University of Tokyo, Tokyo 113, Japan

T. Watanabe
School of Science and Technology, Hirosaki University, Hirosaki, Aomori 033-8561, Japan

T. Tamegai
Electrophysics Department, The University of Hongo, Tokyo 113-8656, Japan

Yujie Qin
Department of Computer and Information Engineering, Luoyang Institute of Science and Technology, Luoyang, Henan 471023, People's Republic of China

Baoluo Liu
Department of Computer and Information Engineering, Luoyang Institute of Science and Technology, Luoyang, Henan 471023, People's Republic of China

Preface

Superconductivity has been known for over a century. The liquefaction of Helium opened a whole new branch of experimental and theoretical work in low temperature physics. Since those days the discovery of High Temperature and Unconventional superconductors has been on the focus of a large amount of scientific research, with an exponentially growing spectra of applications for technologies.

Superconducting technology has the potential for future smart grid technologies. A combination of superconductor materials, electrical engineering, cryogenic insulation, cryogenics and cryostats constitutes some of the most recognized applications so far implemented. We engage some aspects of superconductivity using a theoretical and applied approaches, then several experimental set ups and results are included with potential applications. Hope this book can guide the reader into this promising field.. This book entitled Engineering Superconductivity covers some developments in these areas, making it essential for future applied superconductivity in electrical engineering and even in biomedicine. Content coverage of the book is as follows. First chapter gives an overview of recent advances in novel superconductors and second chapter presents a brief review on the investigation of the electronic structure and superconductivity of the FeSe superconductor and related systems. In third chapter, triplet superconductivity in a nutshell were studied, and fourth chapter discusses about superconductivity, charge orderings, magnetism, and their phase separations in the ground state of lattice models of superconductor with very short coherence length. Fifth chapter focuses on predicting unconventional high temperature superconductors in trigonal bipyramidal coordinations. Sixth chapter deals with electronic energy levels in high-temperature superconductors. Seventh chapter explores on experimental demonstrations of high-Q superconducting coplanar waveguide resonators. The main objective of eighth chapter is to study the electronic structure of Nb3Al/Nb3Sn using the state of the art first principle method. Ninth chapter deals with magnetoresistance oscillations in La-based metallic glass. Tenth chapter provides a novel superconducting quantum interference device for biomagnetic measurements. Eleventh chapter focuses on preparation of Nb3Al superconductor by powder metallurgy and effect of mechanical alloying on the phase formation. Twelfth chapter highlights on exploration of new superconductors and functional materials, and fabrication of superconducting tapes and wires of iron pnictides. In thirteenth chapter, competition between the superconductivity and nematic order in the high-Tc superconductor were studied. Fourteenth chapter presents interplay and competition between superconductivity and charge orderings in the zero-bandwidth limit of the extended Hubbard model with pair hopping and on-site attraction. Fifteenth chapter covers superconductivity in Fe and As based compounds. Sixteenth chapter deals with superconductivity of gold alloys and seventeenth chapter focuses on high-temperature superconducting fiber. Eighteenth chapter discusses about dielectric properties of (Zn) x /CuTl-1223 nanoparticle-superconductor composites. In nineteenth chapter experimental presentation of microwave absorption due to shaking of JV by AC magnetic field in Bi2212 and Bi2223 were presented. Last chapter deals with 3D-modeling numerical solution of guidance force of HTSC bulk above permanent magnetic guideway.

Editor
Nelson Bolívar, Ph.D

INTRODUCTION

Superconductivity is a phenomenon of exactly zero electrical resistance and expulsion of magnetic flux fields occurring in certain materials when cooled below a characteristic critical temperature. It was discovered by Dutch physicist Heike Kamerlingh Onnes on April 8, 1911 in Leiden.

Superconductivity, complete disappearance of electrical resistance in various solids when they are cooled below a characteristic temperature. This temperature, called the transition temperature, varies for different materials but generally is below 20 K (−253 °C).

The use of superconductors in magnets is limited by the fact that strong magnetic fields above a certain critical value, depending upon the material, cause a superconductor to revert to its normal, or non-superconducting, state, even though the material is kept well below the transition temperature.

Suggested uses for superconducting materials include medical magnetic-imaging devices, magnetic energy-storage systems, motors, generators, transformers, computer parts, and very sensitive devices for measuring magnetic fields, voltages, or currents. The main advantages of devices made from superconductors are low power dissipation, high-speed operation, and high sensitivity.

Thermal Properties of Superconductors

Superconductivity is a startling departure from the properties of normal (i.e., non-superconducting) conductors of electricity. In materials that are electric conductors, some of the electrons are not bound to individual atoms but are free to move through the material; their motion constitutes an electric current. In normal conductors these so-called conduction electrons are scattered by impurities, dislocations, grain boundaries, and lattice vibrations (phonons). In a superconductor, however, there is an ordering among the conduction electrons that prevents this scattering. Consequently, electric current can flow with no resistance at all. The ordering of the electrons, called Cooper pairing, involves the momenta of the electrons rather than their positions. The energy per electron that is associated with this ordering is extremely small, typically about one thousandth of the amount by which the energy per electron changes when a chemical reaction takes place. One reason that superconductivity remained unexplained for so long is the smallness of the energy changes that accompany the transition between normal and superconducting states. In fact, many incorrect theories of superconductivity were advanced before the BCS theory was proposed. For additional details on electric conduction in metals and the effects of temperature and other influences, see the article electricity.

Hundreds of materials are known to become superconducting at low temperatures. Twenty-seven of the chemical elements, all of them metals, are superconductors in their usual crystallographic forms at low temperatures and low (atmospheric) pressure. Among these are commonly known metals such as aluminum, tin, lead, and mercury and less common ones such as rhenium, lanthanum, and protactinium. In addition, 11 chemical elements that are metals, semimetals, or semiconductors are superconductors at low temperatures and high pressures. Among these are uranium, cerium, silicon, and selenium. Bismuth and five other elements, though not superconducting in their usual crystallographic form, can be made superconducting by preparing them in a highly disordered form, which is stable at extremely low temperatures. Superconductivity is not exhibited by any of the magnetic elements chromium, manganese, iron, cobalt, or nickel.

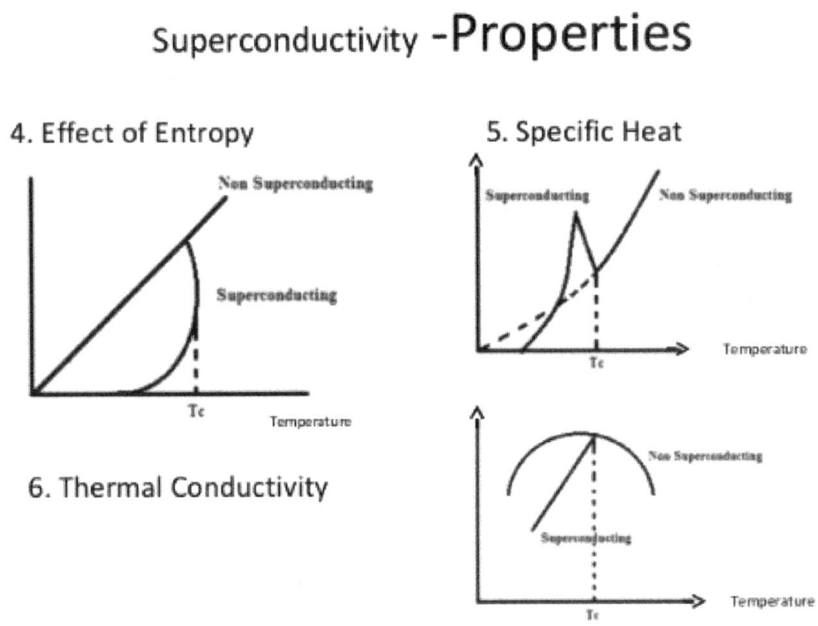

Most of the known superconductors are alloys or compounds. It is possible for a compound to be superconducting even if the chemical elements constituting it are not; examples are disilver fluoride (Ag_2F) and a compound of carbon and potassium (C_8K). Some semiconducting compounds, such as tin telluride (SnTe), become superconducting if they are properly doped with impurities.

Transition temperatures

The vast majority of the known superconductors have transition temperatures that lie between 1 K and 10 K. Of the chemical elements, tungsten has the lowest transition temperature, 0.015 K, and niobium the highest, 9.2 K. The transition temperature is usually very sensitive to the presence of magnetic impurities. A few parts per million of manganese in zinc, for example, lowers the transition temperature considerably.

Specific heat and thermal conductivity

The thermal properties of a superconductor can be compared with those of the same material at the same temperature in the normal state. (The material can be forced into the normal state at low temperature by a large enough magnetic field). When a small amount of heat is put into a system, some of the energy is used to increase the lattice vibrations (an amount that is the same for a system in the normal and in the superconducting state), and the remainder is used to increase the energy of the conduction electrons. The electronic specific heat (C_e) of the electrons is defined as the ratio of that portion of the heat used by the electrons to the rise in temperature of the system. The specific heat of the electrons in a superconductor varies with the absolute temperature (T) in the normal and in the superconducting state (as shown in Figure 1). The electronic specific heat in the superconducting state (designated C_{es}) is smaller than in the normal state (designated Cen) at low enough temperatures, but C_{es} becomes larger than Cen as the transition temperature T_c is approached, at which point it drops abruptly to Cen for the classic superconductors, although the curve has a cusp shape near T_c for the high-T_c superconductors. Precise measurements have indicated that, at temperatures considerably below the transition temperature, the logarithm of the electronic specific heat is inversely proportional to the temperature. This temperature dependence, together with the principles of statistical mechanics, strongly suggests that there is a gap in the distribution of energy levels available to the electrons in a superconductor, so that a minimum energy is required for the excitation of each electron from a state below the gap to a state above the gap. Some of the high-T_c superconductors provide an additional contribution to the specific heat, which is proportional to the temperature. This behavior indicates that there are electronic states lying at low energy; additional evidence of such states is obtained from optical properties and tunneling measurements.

The heat flow per unit area of a sample equals the product of the thermal conductivity (K) and the temperature gradient ΔT: $J_Q = -K\,\Delta T$, the minus sign indicating that heat always flows from a warmer to a colder region of a substance.

The thermal conductivity in the normal state (Kn) approaches the thermal conductivity in the superconducting state (Ks) as the temperature (T) approaches the transition temperature (T_c) for all materials, whether they are pure or impure. This suggests that the energy gap (Δ) for each electron approaches zero as the temperature (T) approaches the transition temperature (T_c). This would also account for the fact that the electronic specific heat in the superconducting state (C_{es}) is higher than in the normal state (C_{en}) near the transition temperature: as the temperature is raised toward the transition temperature (T_c), the energy gap in the superconducting state decreases, the number of thermally excited electrons increases, and this requires the absorption of heat.

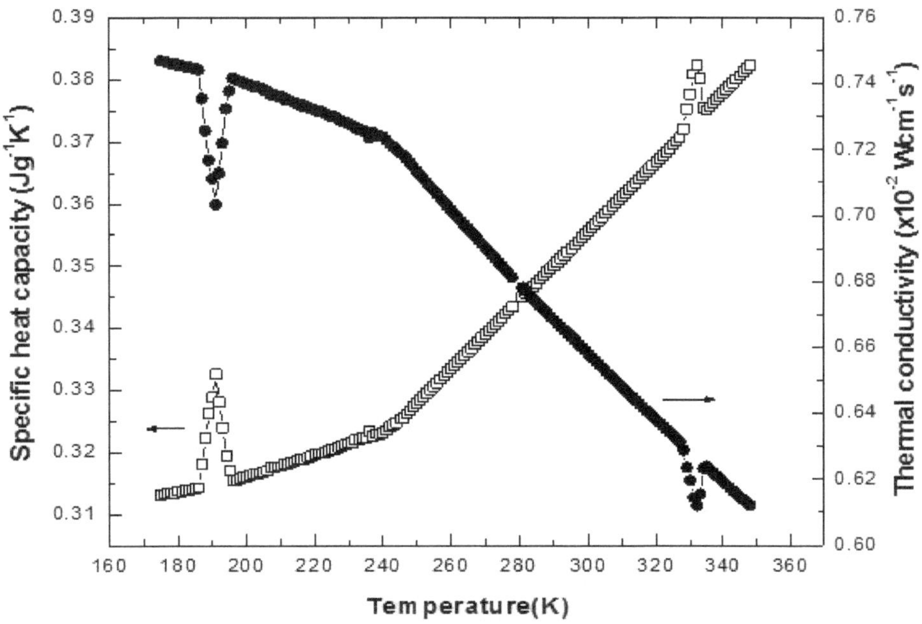

Energy gaps

As stated above, the thermal properties of superconductors indicate that there is a gap in the distribution of energy levels available to the electrons, and so a finite amount of energy, designated as delta (Δ), must be supplied to an electron to excite it. This energy is maximum (designated Δ0) at absolute zero and changes little with increase of temperature until the transition temperature is approached, where Δ decreases to zero, its value in the normal state. The BCS theory predicts an energy gap with just this type of temperature dependence.

According to the BCS theory, there is a type of electron pairing (electrons of opposite spin acting in unison) in the superconductor that is important in interpreting many superconducting phenomena. The electron pairs, called Cooper pairs, are broken up as the superconductor is heated. Each time a pair is broken, an amount of energy that is at least as much as the energy gap (Δ) must be supplied to each of the two electrons in the pair, so an energy at least twice as great (2Δ) must be supplied to the superconductor. The value of twice the energy gap at 0 K (which is 2Δ0) might be assumed to be higher when the transition temperature of the superconductor is higher. In fact, the BCS theory predicts a relation of this type—namely, that the energy supplied to the superconductor at absolute zero would be $2\Delta 0 = 3.53\ kT_c$, where k is Boltzmann's constant (1.38×10^{-23} joule per kelvin). In the high-T_c cuprate compounds, values of 2Δ0 range from approximately three to eight multiplied by kT_c.

The energy gap (Δ) can be measured most precisely in a tunneling experiment (a process in quantum mechanics that allows an electron to escape from a metal without acquiring the energy required along the way according to the laws of classical physics). In this experiment, a thin insulating junction is prepared between a superconductor and another metal, assumed here to be in the normal state. In this situation, electrons can quantum mechanically tunnel from the normal metal to the superconductor if they have sufficient energy. This energy can be supplied by applying a negative voltage (V) to the normal metal, with respect to the voltage of the superconductor.

Tunneling will occur if eV—the product of the electron charge, e (-1.60×10^{-19} coulomb), and the voltage—is at least as large as the energy gap Δ. The current flowing between the two sides of the junction is small up to a voltage equal to $V = \Delta/e$, but then it rises sharply. This provides an experimental determination of the energy gap (Δ). In describing this experiment it is assumed here that the tunneling electrons must get their energy from the applied voltage rather than from thermal excitation.

What is a superconductor?

Superconductors are materials that conduct electricity with no resistance. This means that, unlike the more familiar conductors such as copper or steel, a superconductor can carry a current indefinitely without losing any energy. They also have several other very important properties, such as the fact that no magnetic field can exist within a superconductor.

Superconductors already have drastically changed the world of medicine with the advent of MRI machines, which have meant a reduction in exploratory surgery. Power utilities, electronics companies, the military, transportation, and theoretical physics have all benefited strongly from the discovery of these materials.

A brief history of superconductors

The first discovery of a superconductive material took place in 1911 when a Dutch scientist named Heike Kammerlingh Onnes, who was also the first person to liquefy helium, and reached temperatures as low as 1.7 kelvin (K). In the 1960s, two unrelated discoveries made closely together ushered in a new era in which practical superconducting devices were developed and commercialized: one was the discovery of NbTi superconductor, which provided the first material for the practical manufacture of superconducting wire and shaped components; the second was the Josephson junction, which continues to provide the basis for a variety of unique electronic devices.

High-frequency electromagnetic properties

The foregoing descriptions have pertained to the behavior of superconductors in the absence of electromagnetic fields or in the presence of steady or slowly varying fields; the properties of superconductors in the presence of high-frequency electromagnetic fields, however, have also been studied.

The energy gap in a superconductor has a direct effect on the absorption of electromagnetic radiation. At low temperatures, at which a negligible fraction of the electrons are thermally excited to states above the gap, the superconductor can absorb energy only in a quantized amount that is at least twice the gap energy (at absolute zero, 2Δ0). In the absorption process, a photon (a quantum of electromagnetic energy) is absorbed, and a Cooper pair is broken; both electrons in the pair become excited. The photon's energy (E) is related to its frequency (ν) by the Planck relation, E = hν, in which *h* is Planck's constant (6.63×10^{-34} joule second). Hence the superconductor can absorb electromagnetic energy only for frequencies at least as large as $2\Delta_0/h$.

Magnetic-flux quantization

The laws of quantum mechanics dictate that electrons have wave properties and that the properties of an electron can be summed up in what is called a wave function. If several wave functions are in phase (i.e., act in unison), they are said to be coherent. The theory of superconductivity indicates that there is a single, coherent, quantum mechanical wave function that determines the behavior of all the superconducting electrons. As a consequence, a direct relationship can be shown to exist between the velocity of these electrons and the magnetic flux (Φ) enclosed within any closed path inside the superconductor. Indeed, inasmuch as the magnetic flux arises because of the motion of the electrons, the magnetic flux can be shown to be quantized; i.e., the intensity of this trapped flux can change only by units of Planck's constant divided by twice the electron charge.

When a magnetic field enters a type II superconductor (in an applied field between the lower and upper critical fields, H_{c1} and H_{c2}), it does so in the form of quantized fluxoids, each carrying one quantum of flux. These fluxoids tend to arrange themselves in regular patterns that have been detected by electron microscopy and by neutron diffraction. If a large enough current is passed through the superconductor, the fluxoids move. This motion leads to energy dissipation that can heat the superconductor and drive it into the normal state. The maximum current per unit area that a superconductor can carry without being forced into the normal state is called the critical current density (J_c). In making wire for superconducting high-field magnets, manufacturers try to fix the positions of the fluxoids by making the wire inhomogeneous in composition.

How does a Superconductor Work?

In order to understand how a superconductor works, it may be helpful to examine how a regular conductor works first. Certain materials such as water and metal allow electrons to flow through them fairly easily, like water through a garden hose. Other materials, such as wood and plastic do not allow electrons to flow through, so they are considered non-conducting. Trying to run electricity through them would be like trying to run water through a brick.

Even among the materials considered conductive, there can be vast differences in how much electricity can actually pass through. In electrical terms, this is called resistance. Almost all normal conductors of electricity have some resistance because they have atoms of their own, which block or absorb the electrons as they pass through the wire, water or other material. A little resistance may be useful to keep the electrical flow under control, but it can also be inefficient and wasteful.

A superconductor takes the idea of resistance and turns it on its head. A superconductor is generally composed of synthetic materials or metals such as lead or niobiumtitanium which already have a low atomic count. When these materials are frozen to nearly absolute zero, what atoms they do have grind to a near-halt. Without all of this atomic activity, electricity can flow through the material with practically no resistance. In practical terms, a computer processor or electric train track equipped with a superconductor would use very little electricity to perform its functions.

The most obvious problem with a superconductor is the temperature. There are few practical ways to supercool large supplies of superconductive material to the required transition point. Once a superconductor begins to warm up, the original atomic energy is restored and the material creates resistance again. The trick for creating a practical superconductor lies in finding a material which becomes superconductive at room temperature. So far, researchers have not discovered any metal or composite material which loses all of its electrical resistance at high temperatures.

To illustrate this problem, imagine a standard copper wire as a river of water. A group of electrons are in a boat trying to arrive at their destination upstream. The power of the water flowing downstream creates resistance, which makes the boat have to work even harder to get through the entire river. By the time the boat reaches its destination, many of the electron passengers are too weak to continue. This is what happens with a regular conductor -- the natural resistance causes a loss of power.

Now imagine if the river were completely frozen, and the electrons were in a sled. Since there would be no water flowing downstream, there would be no resistance. The sled would simply pass over the ice and deposit almost all of the electron passengers safely upstream. The electrons didn't change, but the river was altered by temperature to put up no resistance. Finding a way to freeze the river at a normal temperature is the ultimate goal of superconductor research.

What is the Josephson Effect?

The Josephson Effect is the passage of paired electrons through a thin, insulating dielectric barrier placed between two superconductors. A cooper pair of electrons passes through the insulating layer via a tunneling effect. There is no voltage drop while the current stays below a specific level, which is known as the critical current. Under constant, positive voltages, alternating currents as well as direct currents from the passage of electrons are maintained. The effect was predicted by theory in the early 1960s by Brian D. Josephson, and is used to take measurements of very low temperatures and in Josephson junction circuits that can rapidly switch signals to store data.

Electrons pass through an insulating film that is microscopically thin. The Josephson Effect can be controlled by applying a magnetic field which reduces the strength of a supercurrent across the barrier. Magnetic fields are blocked from entering the interior of the Josephson junction by fractional vortices. Current strength rises and falls at different points while the field strength is intensified, allowing for signal passage and switching to be controlled.

When the superconductors are exposed to direct current, electron pairs are passed through a barrier as electromagnetic waves are released, which results in the production of small quantities of light instead of heat. The Josephson Effect can also be applied to radio electronics used in extremely cold conditions, because a Josephson junction can work like an electromagnetic oscillation sensor. Circuits based on this junction can also store data, and can be manufactured into tight spaces because they are so efficient, so use in computers is possible.

The Josephson Effect occurs at very low temperatures, and is most efficient at temperatures close to zero degrees Kelvin (about -460&deg: F). Systems that use this effect can be loosely connected to measure magnetic fields. They can also generate low levels of power as part of generators that can be designed to be switched over many frequencies. How the Josephson Effect is used depends on an engineer's knowledge of quantum physics, and it is measured by using a variety of complex mathematical formulas.

Instruments that incorporate Josephson junctions use the Josephson Effect to make precise dimensional measurements, amplify electromagnetic signals, and drive fast computers. A Josephson tunnel junction switches signals faster than any other semiconductor switch. Such a system can operate at direct current or microwave frequencies, so superconductors can be used in many different metrology and computing applications.

CHAPTER 1

NOVEL SUPERCONDUCTIVITY: FROM BULK TO NANO SYSTEMS

M. P. Das and B. J. Wilson

Keywords: Multi-band superconductivity, Ginzburg-Landau theory, BCS theory, extended GL theory, nanosuperconductivity

INTRODUCTION (WITH A BRIEF HISTORY)

Superconductivity is an unanticipated phenomenon, which has a fascinating history in the realm of quantum science. The surprising fact that certain materials below a critical temperature completely lose their electrical resistance and at the same time repel a magnet, remained a mystery until a microscopic explanation became available in 1957 nearly five decades after its discovery. During these five decades there were many attempts to understand superconductivity both theoretically and experimentally. Two important phenomenological theories are mentioned here.

- A successful phenomenological theory [1] was proposed in 1935 by the London brothers, Fritz and Heinz. It is a macroscopically correct theory to treat electrodynamical aspects. They showed that the current density J in a superconductor is proportional to the magnetic vector potential A. The proportionality constant is $\sim \lambda^{-2}$ 2, where λ is the magnetic penetration depth, one of the characteristic lengths of a superconductor. It relates to the superfluid density ns, as $\lambda^{-2} \sim n_s$. The London brothers' theory both implies zero resistance and leads to a consistent explanation of the Meissner effect (repulsion of the magnetic field).
- In 1950 Ginzburg and Landau (GL) gave their celebrated phenomenological theory of superconductivity [2]. The GL theory, also known as Ψ theory, proposes an order parameter Ψ which encodes the nature of superconductivity through the symmetry of the system. To this day the GL theory remains a powerful practical tool that succinctly explains all the thermo- and electrodynamical properties. It also leads to a second fundamentally significant length ξ, the coherence length describing the spatial scale over which the order parameter varies.

In terms of these two characteristic lengths GL defined a quantity $\kappa = \lambda/\xi$. κ identifies the type of superconductivity (Type I or Type II) from the relation $\kappa < 1/\sqrt{2}$ or $> 1/\sqrt{2}$. This important relation signifies whether the surface free energy of a superconductor is positive or negative implying the magnetic field to be completely expelled from the superconductor (Meissner effect) or the magnetic field can be allowed in the superconductor as quantized vortices, respectively. For Type II superconductors, increasing the magnetic field causes the vortices to form a spatially ordered structure of vortex rods parallel to the magnetic field, known as an Abrikosov lattice [3]. Abrikosov's work was based on the solution of GL equations in presence of a magnetic field.

In 1950 two American groups observed the isotope effect that suggest that the transition temperature, T_c, changes inversely as the isotopic mass M, namely $T_c \sim M^{-a}$, where $a \sim 0.5$. This is strong evidence to invoke electron–phonon interaction in the theory of superconductivity as suggested by Fröhlich. In 1955 Bardeen and Pines [4] studied the nature of dielectric screening in a metal and there was a strong indication that, near the Fermi surface, the dynamics of the dielectric function can change the net interaction between two electrons to be effectively attractive. This happens when the dynamically screened Coulomb interaction is overcome by the electron–phonon interaction, which creates a local distortion of the positive-ion lattice in resonance with the electronic motion.

On this background Bardeen has a strong conviction that an electron pair near the Fermi surface can be bound by an attractive potential, which was confirmed cleverly by Cooper in 1956 [5]. He showed that a normal Fermi liquid in the presence of even the weakest attractive force close to the Fermi surface leads to an isolated bound state pair with a discrete negative energy relative to the Fermi energy. The

Citation: M P Das and B J Wilson, "Novel superconductivity: from bulk to nano systems," doi:10.1088/2043-6262/6/1/013001.

components of the pair retain their Fermi velocities, yet the pair as a quantized entity has zero total momentum and zero total spin. With no loss of time Bardeen, Cooper and Schrieffer [6] formulated what would happen when many such pair states near the Fermi surface undergo a fermionic condensation. By this a new ground state is generated separated from the normal ground state by an energy gap, D. This is the celebrated BCS theory, considered as the standard model for metallic superconductors. It successfully explains at least qualitatively a host of experimental findings and predicts many new superconducting properties. This theory explained almost all known properties of superconductors: excited states, specific heat, critical magnetic fields, ultrasonic attenuation, Meissner effect, penetration depth etc. Besides these, it predicted $\Delta(0)/T_c = 1.76$, the specific heat jump at T_c $\Delta C/C_N = 1.43$, the magnitude of the gap Δ, dependent on the strength of attractive interaction and many others.

Soon after the discovery Anderson [7] noted that the central character of superconductivity extends well beyond the domain of the BCS theory itself. He raised more fundamental issues such as broken gauge symmetry, which was further investigated by Nambu in 1960 [8]. Clarification of pairing mechanism and the simplification of its derivation was given in 1958 independently by Bogolyubov [9] and Valatin [10].

We noted above that the GL theory is our best phenomenological theory with many advantages for working out the real properties of inhomogeneous superconducting materials. This theory is also used in a variety of problems in other areas of physics, namely high energy physics, gauge theory, nonlinear optics and others.

GL Theory Derivable from BCS

In 1958 Gor'kov reformulated the BCS theory of superconductivity in the language of quantum field theory [11]. Soon after, Gor'kov made an important contribution to relate phenomenological GL theory with the microscopic BCS within some limitations. We present here briefly the equivalence.

GL Theory

In GL theory the free energy density, f of a superconductor near the normal to superconductive transition is expressed in terms of a complex order parameter (field) Ψ, which is nonzero below T_c

$$f_s = f_{n0} + \alpha|\Psi|^2 + \frac{\beta}{2}|\Psi|^4 + \frac{1}{2m^*} \times \left|\left(-i\hbar\nabla - \frac{2e}{c}A\right)\Psi\right|^2 + \frac{h^2}{8\pi}. \tag{1}$$

Here a and b are two phenomenological parameters. a is T-dependent and chosen in the form $\alpha = a(T - T_c)$. f_{n0} is the normal state free energy density, m* is the effective mass and e is the charge of an electron. A is the magnetic vector potential given by $\boldsymbol{h} = \nabla\times\boldsymbol{A}$, h being the magnetic field. By varying the free energy with respect to Y and A, one obtains two GL equations

$$\alpha\Psi + \beta|\Psi|^2\Psi + \frac{1}{2m^*}\left(-i\hbar\nabla - \frac{2e}{c}A\right)^2\Psi = 0, \tag{2}$$

$$j = \frac{c}{4\pi}\nabla\times\boldsymbol{h} = \frac{e\hbar}{m^*i}(\Psi^*\nabla\Psi - \Psi\nabla\Psi^*) - \frac{4e^2}{m^*c}|\Psi|^2\mathbf{A}. \tag{3}$$

Here j is the current density. Solution of these two equations (with appropriate boundary conditions) determines the order parameter and the current density.

For a homogeneous superconductor the first GL equation gives $|\Psi|^2 = -a(T - T_c)/\beta$ for $T < T_c$. It is a nontrivial solution typical for a second-order transition.

For non-zero magnetic field and non-zero gradient of Y, one can calculate the characteristic lengths x (coherence length) and l (penetration depth), as mentioned earlier. Here it is noted that the basis of the theory rests on two approximations correct around the critical temperature T_c. The order parameter $|\Psi|$ is small near T_c and $|\Psi| \sim \tau^{1/2}$ where $\tau = (1 - T/T_c)$. Despite these limitations there is a myth that the GL theory applies not only around T_c, rather it is useful for very low temperature. More in the next section.

BCS Theory

It is developed with an assumption that there is an attractive interaction between a pair of electrons by the mediation of a phonon in conventional metallic systems. The theory can be generalized such that any boson can mediate the interaction to produce a bound pair. The many body reduced Hamiltonian is written as

$$H = \sum_{\mathbf{k}\sigma}\xi_{\mathbf{k}} c^\dagger_{\mathbf{k}\sigma}c_{\mathbf{k}\sigma} + V_{int}, \tag{4}$$

$$V_{int} = \frac{1}{N}\sum_{\mathbf{kk}'} g_{\mathbf{kk}'} c^\dagger_{\mathbf{k}\uparrow}c^\dagger_{-\mathbf{k}\downarrow}c_{-\mathbf{k}'\downarrow}c_{\mathbf{k}'\uparrow}, \tag{5}$$

where ξ_k is the electron excitation energy, g_{kk} is the pairing interaction involving k and -k and k'and -k' with up and down spins, respectively. Using any one of the techniques (BCS variational method, Bogolyubov–Valatin canonical transformation, Bogolyubov–de Gennes approach or Gor'kov's thermodynamic Green functions) we obtain the BCS gap equation using the definition

$$\Delta_{\mathbf{k}} = -\frac{1}{N}\sum_{\mathbf{k}'} g_{\mathbf{kk}'}\langle c_{-\mathbf{k}'\downarrow}c_{\mathbf{k}'\uparrow}\rangle. \tag{6}$$

After going through some algebra [12–14], at T=0 one obtains the gap equation

$$\Delta_{\mathbf{k}} = -\frac{1}{N}\sum_{\mathbf{k}'} g_{\mathbf{kk}'}\frac{\Delta_{\mathbf{k}'}}{2\sqrt{\xi^2_{\mathbf{k}'} + |\Delta_{\mathbf{k}'}|^2}}. \tag{7}$$

Assuming $g_{\mathbf{kk}'} = -g_0$ for $|\xi_{\mathbf{k}}|, |\xi_{\mathbf{k}'}| < \omega_D$ (Debye cut-off), the k-independent gap equation is

$$\Delta_0 = \frac{g_0}{N}\sum_{\substack{\mathbf{k}' \\ |\xi_{\mathbf{k}'}|<\omega_D}}\frac{\Delta_0}{2\sqrt{\xi^2_{\mathbf{k}'} + |\Delta_0|^2}} \tag{8}$$

If the density of states near E_F is nearly constant within $\pm\omega_D$, the gap is given by

$$\Delta_0 = \frac{\omega_D}{\sinh\left(1/g_0 N(0)\right)}, \tag{9}$$

N(0) is the density of states at the Fermi energy. In the weak-coupling case $g_0N(0)$ being small, the gap is now

$$\Delta_0 = 2\omega_D \exp\left(1/g_0 N(0)\right). \tag{10}$$

At finite temperature, this equation is

$$\Delta(T) = g_0 N(0) \int_0^{\omega_D} d\xi \frac{\Delta(T)}{\sqrt{\xi^2 + |\Delta(T)|^2}} \times \tanh\left(\frac{\sqrt{\xi^2 + |\Delta(T)|^2}}{2T}\right). \tag{11}$$

Equivalence of GL with BCS near T_c

Following Gor'kov [13] one can obtain after some algebra the following equation

$$\left[\frac{1}{g_0 N(0)} - \ln\left(\frac{1.14\omega_D}{T}\right)\right]\Delta(T) + \frac{7\zeta(3)}{8}\frac{1}{\pi^2 T_c^2}\Delta^3(T) = 0. \tag{12}$$

Near Tc the coefficient of D(T) may be expanded as $g_0N(0)$ $(1\text{-}T/T_c)$, then in the uniform limit (A=0), the order parameter Y is proportional to D. The final results are

$$\psi = \Delta\sqrt{\frac{7\zeta(3)}{8}\frac{n}{\pi^2 T_c^2}}, \tag{13}$$

$$\alpha = -\frac{6\pi^2}{7\zeta(3)}\frac{T_c^2}{E_F}\left(1 - \frac{T}{T_c}\right), \tag{14}$$

$$\beta = -\frac{12\pi^2}{7\zeta(3)}\frac{\left(k_B T_c\right)^2}{E_F n}, \tag{15}$$

n is the electron density. Here we presented the order parameter Y proportional to the energy gap D of the BCS theory and the coefficients αb and b of GL theory in terms of physical parameters T_c, E_F and n. In view of these results the GL theory is almost complete in a limited sense, which is correct near T_c. We are reminded that $|\Psi| \sim \tau^{1/2}$, where $\tau = \left(1 - T/T_c\right)$. Here we have a clear understanding that the GL theory is just not phenomenological, rather it has a microscopic basis as discussed above.

Before we proceed further we give an outline of the rest of the paper. Novel superconductors, particularly the multiband (gap) ones are covered in section 2. We present GL theory, BCS theory and GL theory derivable from the BCS. In section 3 we discuss further expansion as mentioned above to include higher power of $\tau^{(2n+1)/2}$ $(n \geqslant 1)$ and their interesting consequences. We present some selected results on the gaps (both single and multiband materials) from our published calculations. In section 4 we discuss some current activity in nano-superconductors based on both BCS and GL theories. There have been many interesting reports on the novelty of nano-superconductors. This is a new area that is coming up vigorously. We present a balanced review of what confinement does to the smallness of nano-superconductors. Section 5 gives a brief summary and conclusions.

MULTIBAND SUPERCONDUCTORS

The electronic structure of real materials is much more complicated in comparison to the simple jellium model that was adopted in the original BCS theory. If we examine the band structure of superconducting materials, in most of them the Fermi energy passes through many bands and hence the Fermi surfaces are multi-sheeted with the same or different energy gap amplitudes. Evidence of two energy gaps are even obtained in high purity elemental superconductors, for example Nb, Ta and V [15], Nb doped $SrTiO_3$ [16]. More recently, multiband superconductivity involving 2 to 5 bands with more complicated Fermi surfaces are observed in a variety of novel superconductors, such as MgB_2 [17–19], RNi_2B_2C (R = Lu, Y) [20] and FeAs family of superconductors [21]. A variety of experiments are reported, namely magnetization, transport, specific heat, penetration depth, tunnelling, photo-electron spectroscopy, small angle neutron scattering etc, where multiple gaps are easily inferred.

Soon after the BCS theory appeared, Suhl et al [22] and Moskalenko [23] independently extended the one band BCS theory to two bands with overlapping energy bands at the Fermi surface. The two band theory was further discussed by Kondo [24], where the solutions are obtained by using Bogolyubov's variational method. Tilley [25] obtained these results by using Gor'kov's Green function technique. Additionally, Tilley derived GL equations for the two band case from the BCS method following the one band equivalence as mentioned above due to Gor'kov. The two-band model has been explored intensely starting with the work of Geilikmanet al [26]. Many other authors used this model in the context of high T_c oxides [27] and MgB_2[28–32]. Brandt and Das [33] have presented a comprehensive review of the latter topic.

Multiband GL Theory

Multiband GL theory can be seen as a straight forward generalization of the standard GL theory. The free energy density can be expanded as powers of multi-order parameters [34, 35]. Here one has total superconducting free energy composed of individual parts and mutual interaction terms arising out of various order parameters and gradient terms. Here we consider the case of two order parameters. Generalization to multi-order parameters is a straight forward case. The simplest inter-order parameter coupling is the internal Josephson coupling term that we consider below. The free energy density is written as

$$f_s = f_{n0} + \sum_{j=1,2} \alpha_j \left|\Psi_j\right|^2 + \frac{\beta_j}{2}\left|\Psi_j\right|^4 + \frac{1}{2m_j^*}\left|\left(-i\hbar\nabla - \frac{2e}{c}\boldsymbol{A}\right)\Psi_j\right|^2 - \Gamma\left(\Psi_1^*\Psi_2 + \Psi_2^*\Psi_1\right) + \frac{\boldsymbol{h}^2}{8\pi}. \tag{16}$$

Here the last but one term is the Josephson coupling term, with interaction strength Γ. Let us consider the applied field is zero and we are in the bulk at the temperature T close to T_c. By neglecting the $|\Psi_i|^4$ term, the free energy density (with reference to normal state) is given by

$$f = \alpha_1 |\Psi_1|^2 + \alpha_2 |\Psi_2|^2 - 2\Gamma |\Psi_1| |\Psi_2| \cos(\Delta\phi), \quad (17)$$

where $\Delta\phi$ is the phase difference between two condensates.

When $\Gamma = 0$ two bands are fully decoupled and two order parameters have two critical temperatures. For $\Gamma > 0$ the free energy is minimized with $\Delta\phi = 0$. For $\Gamma < 0$ the free energy is stable with $\Delta\phi = \pi$. This implies the condensation energy for two-band superconductivity is more stable than the two independent states. More on two-band nanosystems in section 4.f.i.

The GL equations for Ψ_i are obtained from the minimization of the free energy and similarly the equation for the current. The self-consistent solutions of these (i+1) equations describe the properties of multiband superconductors within the traditional GL theory (see [27–35]).

In the recent past it has been argued [28, 36–40] that the original GL theory is incomplete beyond the order parameter $\Psi_i \sim \tau^{1/2}$. Therefore, one has to incorporate the higher order terms $(\tau^{(2n+1)/2}$ for $n \geqslant 1)$ as a systematic expansion, see Fetter and Walecka [41]. For the derivation of this result one has to expand the order parameter equation in powers of Y and its spatial gradients. Details of this theory are in our recent paper [40].

Multiband BCS Theory

One band theory is generalized to a multiband [22, 23] situation. For simplicity we consider a two-band case. Multiband generalization is straightforward

$$H = \sum_{ik\sigma} \xi_{i\mathbf{k}} c^{\dagger}_{i\mathbf{k}\sigma} c_{i\mathbf{k}\sigma} + \frac{1}{\Omega} \sum_{ij\mathbf{k}\mathbf{k}'} g_{ij} c^{\dagger}_{i\mathbf{k}\uparrow} c^{\dagger}_{i-\mathbf{k}\downarrow} c_{j-\mathbf{k}'\downarrow} c_{j\mathbf{k}'\uparrow}, \quad (18)$$

here i, j stand for the band indices 1, 2. g_{ij} is the intra- and interband couplings, and W is the volume.

Using the same variational principle as in the one-band case, the gap equation is

$$\Delta_i = \frac{1}{\Omega} \sum_{j\mathbf{k}} g_{ij} \frac{\Delta_j}{2E_{j\mathbf{k}}} \tanh\left(\frac{E_{j\mathbf{k}}}{2k_B T}\right), \quad (19)$$

where $E_{j\mathbf{k}} = \sqrt{\xi_{j\mathbf{k}}^2 + \Delta_{\mathbf{k}}{}^2}$.

Explicitly, two gaps are given by

$$\Delta_1 = g_{11} A_1 \Delta_1 + g_{12} A_2 \Delta_2, \quad (20)$$

$$\Delta_2 = g_{21} A_1 \Delta_1 + g_{22} A_2 \Delta_2, \quad (21)$$

where

$$A_i = \frac{1}{\Omega} \sum_{\mathbf{k}} \frac{1}{2E_{i\mathbf{k}}} \tanh\left(\frac{E_{i\mathbf{k}}}{2T}\right), \quad (22)$$

g_{ij} are interaction matrix involving diagonal g_{ii} and off-diagonal g_{ij} parameters related to bands I and j. The equation for T_c as largest $\Delta \to 0$, $T_c = 1.13\omega_D \exp(1/\Lambda)$, where Λ is the largest eigenvalue of $\lambda_{ij} = g_{ij} N_j(0)$

$$\Lambda = \frac{1}{2}\left[\lambda_{11} + \lambda_{22} + \sqrt{(\lambda_{11} - \lambda_{22})^2 + 4\lambda_{12}\lambda_{21}}\right]. \quad (23)$$

T_c of the two band system is enhanced by the second term of the equation below

$$T_c \approx T_{1c}\left[1 + \frac{\lambda_{12}\lambda_{21}}{\lambda_{11}^2(\lambda_{11} - \lambda_{22})} + O(\lambda_{12}^4)\right]. \quad (24)$$

T_{1c} is the critical temperature for band 1.

GL DERIVED FROM BCS

As we have done in the one-band case (uniform without gradient and A=0), we follow the similar procedure in the two-band case. Self-consistency condition through third order in Δ_i gives the same equation as in the one-band case. Near T_c the coefficient of Δ_i (T) may be expanded to lowest order in t. Now by comparing the first GL equation, the values of α_i and β_i can be obtained. Ψ_i is proportional to Δ_i. Results are [40]

$$\alpha_i = \frac{N_i(0)}{N_T(0)} \frac{6\pi^2 (k_B T_c)^2}{7\zeta(3)\varepsilon_F}\left(1 - \frac{T}{T_c} + \mathscr{A} - \frac{g_{jj}}{(g_{11}g_{22} - g_{12}g_{21})N_i(0)}\right), \quad (25)$$

$$\beta_i = \frac{N_i(0)}{N_T(0)} \frac{6\pi^2 (k_B T_c)^2}{7\zeta(3) n\epsilon_F}, \quad (26)$$

$$\mathscr{A} = \log\left(\frac{2\hbar\omega_D e^{\gamma}}{\pi k_B T_c}\right). \quad (27)$$

BEYOND THE STANDARD GL THEORY

This is an important development in recent works on the GL theory. It is an extended version beyond the traditional GL theory [2]. Traditional GLT includes terms beyond $\Psi \propto \tau^{1/2}$ which are incomplete. One is inclined to extend it to complete the terms $\propto \tau^{3/2}$ as done in [37–40] by which one improves the gap and other properties to lower temperatures. To accomplish the extended theory $\Psi \propto \tau^{3/2}$ and beyond, one follows Gor'kov (see also Fetter and Walecka [41]) by expanding the self-consistent gap equation in powers of the order parameter and its spatial gradients to high orders. Depending on the desired precision the expansion terminates at some order. Details of this theory can be found in [40]. We present here some results. In figure 1(a) the energy gap Δ_1/T_c for one-band superconductor is shown as a function of temperature T in the BCS and GL theory. The black dots are BCS results. The solid red line overshoots the BCS in the traditional GL theory, which refers to $\tau^{1/2}$ in the expansion $\tau^{(2n+1)/2}$terms, which means n=0. For $n \geqslant 1$ are shown as higher order corrections in the extended GL. The inset is a close-up of the region near $\tau = 1$. It is clear that the traditional GL result is unreliable away from T_c. In figure 1(b) the magnitude of individual terms is shown on a Log plot. More in the figure caption. (reference [40]).

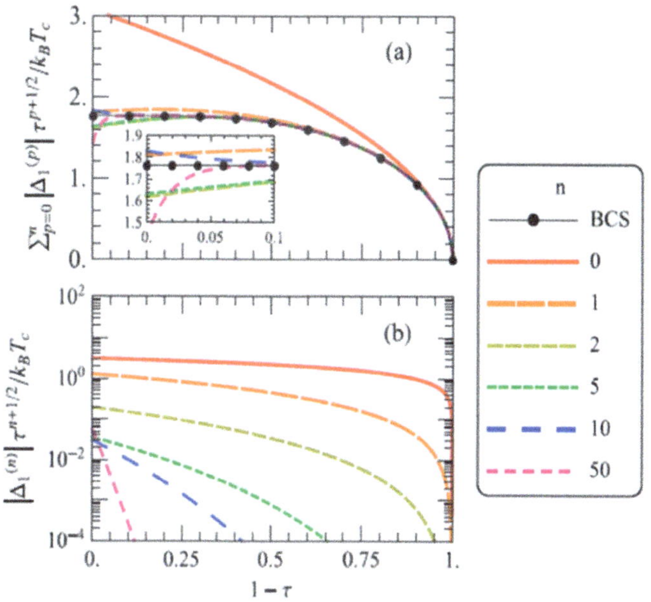

Figure 1. (a) The extended GL expansion is compared to a numerical full BCS result. The extended GL converges to the true solution in the region t<1 and for moderate t it converges quickly to that of the BCS. Inset: a close up of the region near t=1. There are singularities in the BCS function infinitesimally close to t=1, which prevent extended GL from converging at this point. (b) The magnitude of individual terms in the extended GL expansion are shown on a Log plot. The magnitude of higher terms decays quickly except near the point t=1, where it remains finite. It shows the expansion is converging in the region t<1 (from [40]).

Now we discuss the two-band situation. Firstly, we plot the BCS solution for both the gaps with a range of values of inter-band couplings g_{ij}. The values of g_{ij} chosen are 0.001, 0.01, 0.1 and 0.55. Other parameters are in the figure caption. In figure 2(a) the energy gap for the first band is shown as a function of T/T_c = 1-t, where the inter-band coupling has a weak effect on the first gap. Nothing spectacular is seen in the first gap results. In (b) the second gap is shown for the same couplings and their effect is drastic for weaker coupling. For g_{ij} = 0 the second critical temperature T_{2c} is ~0.33 in the scale of T/T_c. The second gap 0 at this T_{2c}. As the couplings increase the second band has induced superconductivity from the first band and it results in the plots shown. At higher couplings the plot is reminiscent of a one-band BCS result.

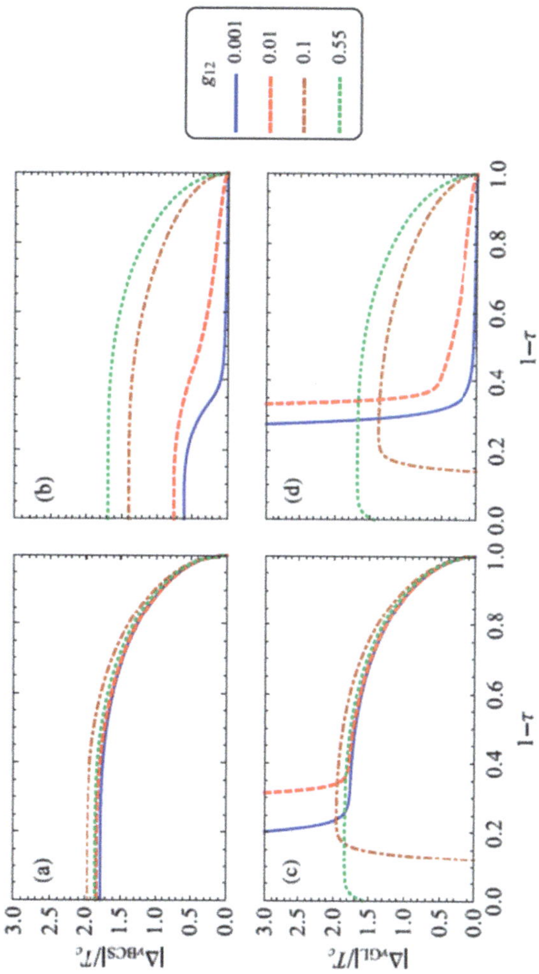

Figure 2. (a) and (b) are BCS gaps for first and second bands. They are compared with high expansion (n=50) in the extended GL theory in (c) and (d), respectively. The parameters used here are g_{11} = 0.6, g_{22} = 0.5, $N_1(0) = N_2(0)$ = 0.3 and ω_D = 0.09 . (From [40]).

In figures 2(c) and (d) we show order parameters 1 and 2 calculated by using the extended GL formalism using the same parameters. For 1 - t ³ 0.3the behavior of the GL plots is similar to that of BCS. However, for $1 - \tau \leqslant 0.3$ the behavior of the order parameters is drastically different in comparison to BCS in (a) and (b), respectively. The point where the plots begin to disagree is close to the location of $T_{2c} \sim 0.33T_c$. Komendova et al [42] have argued that there is a possibility of hidden criticality near T_{2c}, which becomes critical in the limit of vanishing inter-band coupling. This feature is likely to be associated with the anomalous behavior of the gap near this T_{2c}. Although the BCS solution of the first band showed only a weak perturbation with the inter-band coupling, the nonconvergent behavior seen in the GL solution of the second band affects the dominant band drastically at lower temperature.

Now we consider a range of real materials $FeSe_{0.94}$ [43], OsB_2 [44], LiFeAs[45] and MgB_2 [46]. The parameters used for these calculations are representative of the behavior of the materials, though the exact choice is not claimed to be unique. For each set of parameters we plot the BCS and extended GL gaps with the terms in the GL calculations retained to order t $^{(2n+1)/2}$ with n=50. These results are shown in figure 3. For details see [40].

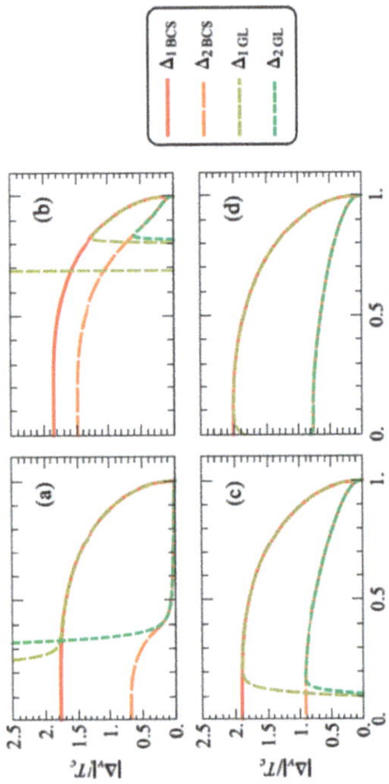

Figure 3. Extended GL calculations compared to full multiband BCS results. The materials are (a) $FeSe_{0.94}$, (b) OsB_2, (c) LiFeAs and (d) MgB_2.(From [40]).

Similar to two-band results we also calculate for three-band materials using both extended GL and BCS theories (figure 4). The legend shows the BCS by black dots and other symbols are for extended GL with different powers of t as indicated before. $N_i(0) = 0.3$ for all three bands, $g_{11} = 0.6$, $g_{22} = 0.55$, $g_{33} = 0.4$, $g_{12} = g_{21} = 0.1$ and four columns are for $g_{13} = g_{23} = g_{31} = g_{32}$ for the values 0.001, 0.01, 0.1 and 0.5, respectively.

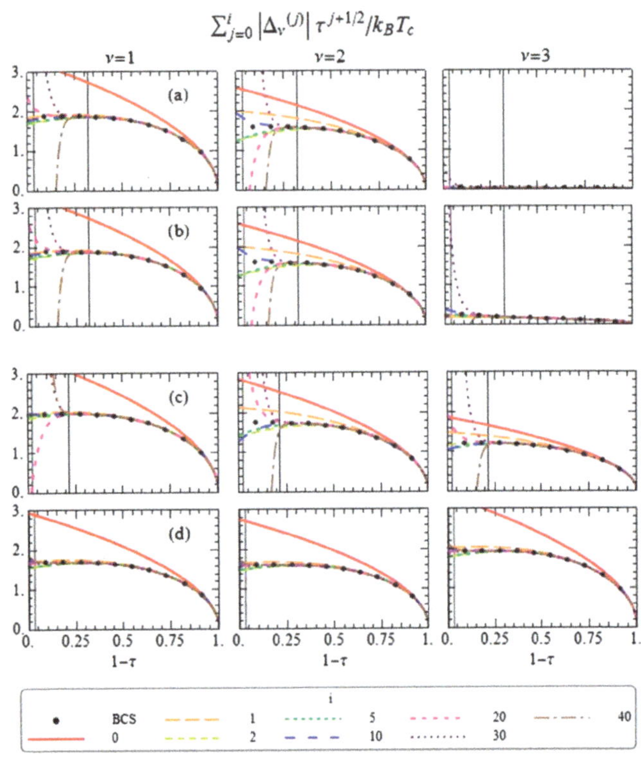

Figure 4. Extended GL and BCS results compared for three bands. (B J Wilson and M P Das unpublished).

To summarize this part we have clearly demonstrated the importance of $\tau^{(2n+1)/2}$ expansion for large n multi band superconductors. This point emphasizes the weaker validity of the GL theory for lower temperatures for some materials and especially for applications with small inter-band couplings.

Now we address the next part when the material is nano in size, yet a superconductor.

FROM BULK TO NANO OR 'NANO IN BULK' IN A SUPERCONDUCTOR

As a definition, 'nanoscience' is the study of materials whose physical size is on the nanometre scale (in the range of 1–1000 nm). While 'nano' means precisely small, 'meso' is a broader term, being intermediate between the microscopic (molecular) and macroscopic (bulk) scale. In practice the 'mesoscopic' regime partly overlaps the 'nanoscopic' description [47].

For a superconductor it will not be inappropriate to argue 'nano in bulk' for the following reasons. We have said before that there are two characteristic length scales in a superconductor, coherence length(s) x and penetration length l. Although a superconductor as described above may be a bulk material and superconductivity is a macroscopic quantum phenomenon, these particular and relevant length scales are nanometric in dimension. Much of interesting physics relies on these length scales when one studies the surfaces of bulk superconductors or vortices of superconductors in the presence of a magnetic field. These length scales are also meaningful for a material, whose physical size is nanometric subject to some conditions such as parity of number of electrons and fluctuations. In addition to these, the current developments in experimental nanotechnology have made nanoelectronics a very attractive subject [48, 49]. We present here some basic issues of physics including theories and experiments.

Kubo Gap

In 1962, Kubo [50] proposed a theory for metallic particles, which are small in size unlike the bulk. For such systems it is only quantum mechanics that accounts for observed physical properties. In the jargon of the field, 'quantum confinement' means that the de Broglie wavelength of the particles is comparable to the size of the system that contains them. Small size implies strong quantum confinement effects. Bulk extended matter, when sufficiently curtailed in one or more of its dimensions, will behave as a quasi-two dimensional (two-DEG), a quasi-one dimensional (quantum wire) or a quasi-zero dimensional (quantum dot) system. The price to be paid for this is to lose the long-range extendedness of the wave functions. The energy of motion becomes quantized as the wave function is confined. Subject to confinement, the continuum of the density of states becomes discrete. However, the energy gaps between the states are tiny, in the meV range, if the particle size is nanometric.

Kubo recognised the inherent statistical nature of the problem. He argued the energy spectrum depends on the precise geometry of the system, even though the average energy gaps may be same. Energy spectrum in many respects leads to anomalies in the physico-chemical properties of small metallic particles at low temperatures. In brief, the Kubo gap depends on the spectrum. This gap $d_K \sim 1/D\ (E_F)$, inversely proportional to the electronic density of states at the Fermi level. Kubo gap is essentially the energy required to remove an electron from the small metal particle. See earlier reviews by Halperin [51] and von Delft and Ralph [52]. In comparison to the thermal energy, if d_K £ T, the material is a metal and otherwise it is an insulator. Our further discussion is about only metallic nanoparticles.

Anderson Criterion

Following the BCS theory (meant for clean superconductors), Anderson [53] presented a theory for dirty superconductors, where the elastic scattering strength is large compared to the energy gap. Independently similar results are obtained by Abrikosov and Gor'kov [54], see also [55] for more detailed calculations. The scattering mean free path does not change the Gor'kov's anomalous Green function, which is the same as the energy gap.

Anderson's idea emphasizes that often in the presence of some disorder, a superconducting state can be more stable and less likely to be suppressed [56]. This loss of superconductivity has been explored not only in nanoparticles and in nanowires, but also in more complex systems exhibiting many exciting and novel effects. For a small number of particles, as mentioned before, the allowed energy levels are quantized, discrete and the mean energy level spacing d_K (the Kubo gap) becomes bigger for the smallness of the sample. When the superconducting energy gap D approaches d_K, superconductivity is suppressed at this point by the Anderson's criterion [57]. In this limit D/ d_K L< 1, there are no available energy levels correlated by the pairing interaction. More precisely, it has been argued [58] that superconductivity will vanish for d_K greater than a critical value d_{KC} = 3.56D for even N and d_{KC} = D/4 for odd N. Presently Kubo-Anderson criteria are utilized prolifically to predict superconductivity in nanosystems [59].

Although Anderson's criterion remained unexplored for some time, from the mid-nineties in view of many experimental investigations on nanoparticles, theoretical activities began vigorously based on the BCS reduced Hamiltonian. While the BCS theory is used for macoscopic systems, we are reminded of the Richardson solution [60], originally developed for finite nucleus. The latter approach is based on the canonical ensemble, where the number of particles is fixed. This is in contrast with the BCS theory, where grand canonical ensemble is used to deal with the macroscopic limit. While the BCS theory provides a mean-field solution, Richardson's solution is exact [61–63]. More on this in the next sub-section.

Richardson solution depends on d_K near the Fermi level E_F. The property of Cooper pairing depends on the 'parity' of number of electrons N. For odd N, one electron is unpaired and carries an additional energy Δ_P = D. This feature is experimentally detected [57, 64, 65] exploring the Coulomb blocked phenomenon and theoretically analyzed by Matveev and Larkin [66].

Theories

During the past two decades there have been a lot of activities in this area of finite size nanosuperconductors. We give here a brief review. Most of the theoretical work reported here is carried out with the BCS theory applied to a finite number of particles appropriate for nanosystems. For thin films Blatt and Thompson [66] showed the energy gap is an oscillatory as a function of film thickness passing through resonances, where the period of oscillation is the Fermi wavelength. The gap increases with a decrease of the thickness. The enhancement of superconductivity by the size effect was also discussed by Parmenter [67]. With the decrease of size the gap increases, where the weak coupling limit goes to the strong coupling case (D(0)/T_c varies from 3.528 to 4). Thermodynamical properties, particularly specific heat, spin susceptibility of small grains are investigated in [68] by using both GL and BCS theories. Since GL is unable to deal with quasi-particles, a static approximation within the BCS was used, where contributions due to quasi particles are incorporated explicitly.

GL Theory for Nanosuperconductors

In recent years, the traditional GL theory has been applied to meso/nanosuperconductors. As mentioned earlier, it is important to apply the correct boundary conditions if the superconductivity is occurring in a small volume. On the boundary the order parameter has to vanish. On application of a magnetic field above a certain critical field a vortex can be created. The authors of this work [69] have applied traditional 3-dimensional GL theory in a parallelepiped cell containing a small mesoscopic disk at the center. Vortex state solutions are obtained by minimizing the free energy numerically by the method of simulated annealing. See reference [69] for details of calculations. The result has been shown in figure 5 as a simple example. Clearly we notice the vanishing of order parameter near the edges of the disk and at the center where the magnetic field has penetrated.

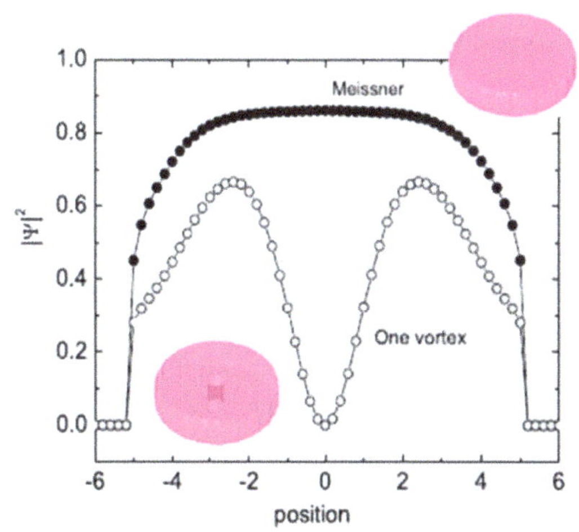

Figure 5. Density of superconducting electrons in a disk for no vortex (Meissner state) with black filled dots and a single vortex with unfilled dots, occurring at the center of the disk. (See [69]).

By using the same traditional GL theory Peeters and co-workers have studied mesoscopic superconductivity for vortex phase diagram [70]. Free energies of different vortex configurations are calculated by increasing both the disk thickness and magnetic field. The authors showed a smooth transition from a multi-vortex state to a giant vortex state (figure 6). With increment of magnetic field with steps of 0.01 H_{c2} the superconductor goes from a Meissner state to vortex states as shown in the figure 6. This picture is reminiscent of formation of vortices in rotating ^{4}He superfluid and atomic

Bose–Einstein condensate [71]. In this sub-field of mesoscopic superconductors there has been plenty of theoretical and experimental activities. Some of the references are given here [72]. We also point out more recently available extended GL theory for nano/meso superconductors in the later section 4.f.ii.

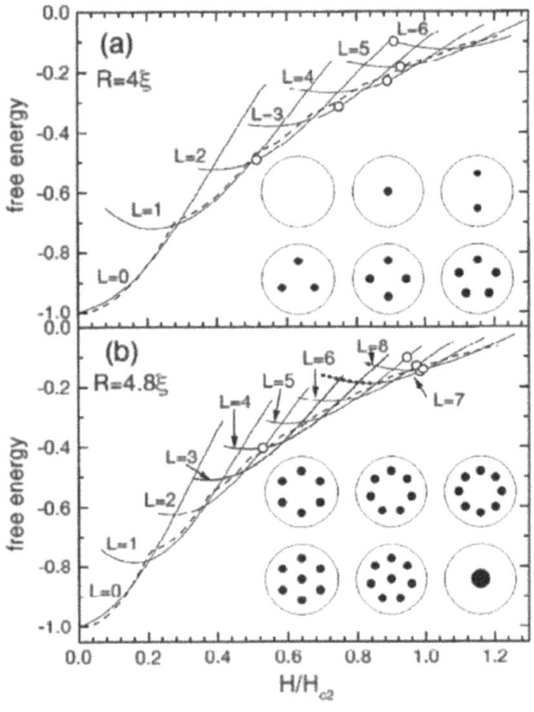

Figure 6. Free energy configurations with vorticity L for two disks of radius R = 4x in (a) and R = 4.8x in (b), x being the coherence length, is shown as a function of magnetic field H scaled by the second critical field H_{c2}. For $L > 7$ states are metastable, hence a giant vortex is formed for large L. (See [70]).

Reduced BCS Hamiltonian for Nanosuperconductors

We mentioned earlier that the BCS model for bulk superconductor deals with grand canonical ensemble. For small grains it is inappropriate for use due to the fluctuations in smallness of N. This requires a treatment by canonical ensemble (see details in [58]). In metallic condensed systems with large N pairing occurs due to virtual exchange of phonons between the conduction electrons. In atomic nucleus with a fixed number of fermions (small N) pairing appears due to short-range nature of nuclear interaction and it includes contributions from the singlet-S and triplet-P channels. In both cases the pairs are correlated in the time-reversed states. A canonical treatment of BCS states are applied to nuclear physics soon after the former's appearance. In 1963 Richardson [60] provided an exact solution in a simplified form of the BCS, for application to nuclear physics with some more papers followed soon. Despite this, the condensed matter community did not pay much attention until the superconductivity in small metallic grains was reported in the mid-nineties [57]. Currently Richardson model is an attractive picture for nanosuperconductors.

Richardson's reduced Hamiltonian is given by

$$H = \sum_{j=0,\sigma}^{N-1} \varepsilon_j c^\dagger_{j\sigma} c_{j\sigma} - \lambda d \sum_{j,j'=0}^{N-1} c^\dagger_{j\uparrow} c^\dagger_{j\downarrow} c_{j'\downarrow} c_{j'\uparrow}. \tag{28}$$

Here $c_{j\sigma}\left(c^\dagger_{j\sigma}\right)$ destroys (creates) electrons in free time-reversed states $|j,\sigma\rangle$ with discrete uniformly spaced degenerate eigenenergies $\varepsilon_j = jd + \varepsilon_0.\ d$. d is the (Kubo)gap of the discrete spectrum, l is dimensionless pairing strength. In BCS model g = λd. The interaction g scatters only time-reversed electron pairs within the Debye cutoff frequency w_D of the energy shell around the Fermi energy ε_F. l is related to the parameters bulk gap D and the cutoff frequency w_D via the bulk gap equation $1/\lambda = \omega_D/\Delta$. In this model $N = 2\omega_D/d$.

We notice here an interesting point that the nanostructure boundary condition is entering in a subtle way through the discrete spectrum. Reference [62] considers finite size corrections to the Richardson model, where the low energy spectrum of the problem is included as an expansion (1/N) in the inverse of the number of electrons in the grain.

Going back to the original BCS the ground state is written as

$$|\,\mathrm{BCS} \propto \exp\left[\sum_j \frac{v_j}{u_j} c^\dagger_{j\uparrow} c^\dagger_{j\downarrow}\right] |0\rangle, \tag{29}$$

where u_j and v_j are variational parameters of the original BCS theory. These two parameters are obtained as BCS solutions from a variational minimization of the reduced Hamiltonian.

$$u_j^2 = \frac{1}{2}\left(1 + \frac{\varepsilon_j - \mu}{E_j}\right), \tag{30}$$

$$v_j^2 = \frac{1}{2}\left(1 - \frac{\varepsilon_j - \mu}{E_j}\right), \tag{31}$$

with the quasi-particle excitation energy given by $E_j = \sqrt{\varepsilon_j^2 + |\Delta|^2}$. D is the energy gap that we have discussed earlier. In the grand canonical ensemble this BCS ground state is asymptotically exact for $N \to \infty$. For finite N it requires a canonical ensemble, for which a projected BCS (PBCS) ansatz is used, which is given by

$$|\,\mathrm{PBCS} \propto \left(\sum_j \frac{v_j}{u_j} c^\dagger_{j\uparrow} c^\dagger_{j\downarrow}\right)^N |\,0 \tag{32}$$

It is an 'N conserving theory' [61]. Apart from BCS and PBCS ansatz, there are other numerical approaches, namely Lanzcos method [74], perturbative renormalization group method [65] and density matrix renormalization group technique [58]. These methods have been extensively used for nanosuperconductivity. See reviews [52, 73].

For the novel physics for finite N, we have size, shape and shell effects, effects of fluctuations and odd–even parity [75–79]. Randomness of eigenvalue distributions raises new question of statistics of ensembles in the Dyson–Wigner sense. These features are in the presence of pairing correlations unlike in single-particle physics of disordered systems [80]. Thermodynamic properties of small superconducting grains are studied in detail in [81] by analytical and numerical methods. Here the authors have studied in the limit of $d/\Delta > 1$ and the crossover of superconducting to normal grains.

An interesting point of all this important work is to emphasize how a crossover of bulk to a nanosystem is possible. Although Anderson›s criterion has been in the literature for over five decades, the ground breaking experiments of Tinkham and co-workers [57] brought new light on ultra-thin nanosuperconductivity. Without going through details, a list (incomplete) gives an idea of experimental activities [59].

On the other hand, by using the BCS theory appropriate for nanosystems, namely Richardson model, it opened a new panorama in this novel area of superconductivity.

A illustrations we present here some results of calculations. In figure 7 (upper panel) calculations of condensation energy given by the difference of ground state energy of the pairing Hamiltonian and the energy of the filled Fermi state, $E_b^c = E_b^{GS} - FS\left|H_{BCS}\right|FS$. The condensation energy is scaled by the bulk superconducting gap. They are shown for even (b = 0) and odd (b = 1) parity state in the PBCS and exact calculations. The parity effect in terms of the gap parameter is given by the difference of ground state energy of an odd grain and the mean energy of the neighboring even grains by adding or removing an electron. This gives the Matveev–Larkin (ML) gap by the expression [65] $\Delta_{ML} = E_1(N) - 1/2\left[E_0(N+1) + E_0(N-1)\right]$. The lower panel of figure 7 gives the ML gap scaled to the bulk gap as a function of grain size by the PBCS (dashed line) and exact (continuous line) methods. One can see the asymptotic limits of both PBCS and exact results. It is also interesting to note that the Anderson's criterion $d/\Delta \sim 1$ is satisfied. For details see [73].

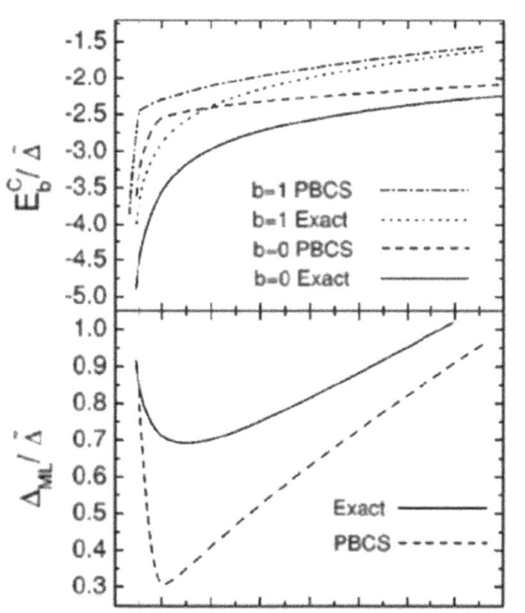

Figure 7. Upper panel shows condensation energy for (b=0) and odd (b=1) grains calculated using PBCS and exact wave functions as a function of grain size. The lower panel gives the Matveev–Larkin gap in PBCS and exact calculations (from [73]).

Fluctuation effects are calculated in [62]. Figure 8 shows the condensation energies for even and odd N both for uniformly spaced and randomly spaced levels (discussions in [62, 80 and 73]).

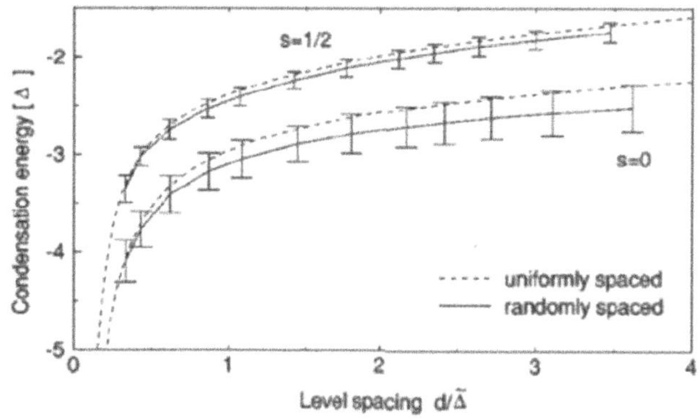

Figure 8. Even and odd condensation energies are shown for uniform equally spaced level (dashed line) and ensemble-averaged energies for randomly spaced level (solid line). Vertical bars show fluctuations from [62] reported in [73].

Nano Superconductivity in Multiband Systems

Earlier, in section 2, we have discussed multiband theories in the BCS and GL models. Particularly, two-band systems were analyzed in some detail. From the material perspective Nb, Ta and V are the elemental metals which have two-band superconductivity [15]. Superconducting studies have been made in Nb-doped $SrTiO_3$ in bulk [16]. After the discovery of two-band superconductivity in MgB_2 [82] and Fe pnictides [83], important work continued to be done in multiband materials (see these two reviews [84, 85]).

In the area of nanosuperconductivity a number of studies have been reported. Quantum size effect (QSE) modulated superconductivity has been observed in nanograins of Pb on Si [86]. QSE has been studied in Pb, In and V [87–89], where nanosuperconductivity seems to obey the Anderson criterion. Authors of [90] have investigated size dependent T_c, coherence length, upper critical field H_{c2}, and irreversible field H_{irr}, in Nb. Reduction of the gap in ultra-thin films of Pb on Si by variation of thickness is reported in [91]. Magnetization in Pb nanoparticles has been studied in the range of 4–1000 nm [92]. The Meissner effect is clearly seen for particle sizes >30 nm, consistent with the Anderson criterion. Multilayer ultra-thin Pb films (5–15 monolayers) are the subject of a study of scanning tunnelling spectroscopy [93], where layer dependent superconducting gap and T_c are reported. Superlattices of artificially engineered pnictide superconductors are fabricated for new interfacial phenomena and many device applications [94].

While experimental progress on multiband superconductivity is heading forward, there are some developments on the theoretical front. The possibility of superconductivity in doped $SrTiO_3$ in films and interfaces is reported [95] by using a two-dimensional two-band model. Shape resonances and shell effects are investigated for multiband superconductors in thin film form [96]. In this work the one-band model of Blatt and Thompson [66] has been extended for two band superconductors relevant to MgB_2 systems. From this work we show in figure 9 the calculated T_c

/ T_c (Bulk) value as the film thickness varies. The implication of this calculation is that the pattern of the shape resonances is more regular as the offset e_{op} decreases. (Note, this offset is between the two σ and π bands.) In the one- band limit the shape resonances (blue color) have larger amplitudes compared to the two-band case (black color). Therefore, the size effect is suppressed by the multiband structure. For details see [96].

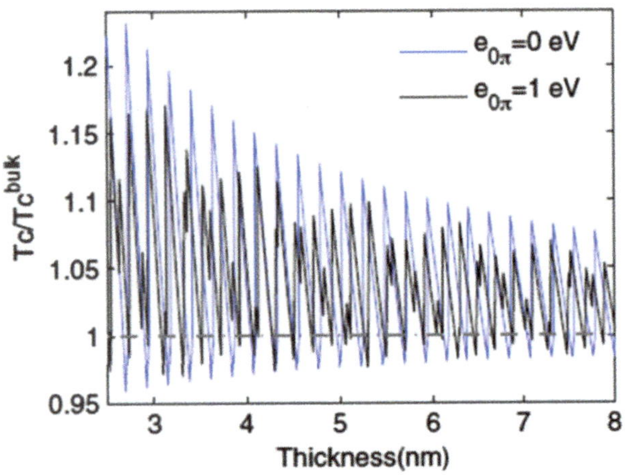

Figure 9. T_c/T_c(Bulk for MgB_2 = 38 K) as a function of film thickness in nm for one-band (in blue colour) and two-band (black colour) limits. (See [96]).

Multiband Nano Superconductivity by BCS-Richardson Model

In a long paper, Kruchinin and Nagao (KN) [97] considered a two-band superconductivity model (for MgB_2) by extending Richardson's exact model. We describe this approach in some detail. They considered the two-band Hamiltonian as in the bulk for the nanograin.

$$H = \sum_{ik\sigma} \xi_{ik} c^{\dagger}_{ik\sigma} c_{ik\sigma} + \frac{1}{\Omega} \sum_{ijkk'} g_{ij} c^{\dagger}_{ik\uparrow} c^{\dagger}_{i-k\downarrow} c_{j-k'\downarrow} c_{jk'\uparrow}. \tag{33}$$

Unlike in the bulk, ks are now discrete state indices, js—band indices. g is the interaction matrix with attraction for the intra-band ($g_{ii} < 0$) and repulsion ($g_{ij} > 0$) for the inter-band couplings. Since the states are discrete, the sums in the Hamiltonian are over the set I of N_{1I} states corresponding to band 1 with fixed width $2\omega_{1D}$ and the set J of N_{2I} states for band 2, respectively. For simplicity one can assume Debye energies ω_{1D} and ω_{2D} are the same and equal to ω_D. Then $N_{1I}/N_{2I} = \rho_1/\rho_2$, where ρ_1 and ρ_2 are density of states for respective bands. As in the Richardson's model the interaction energies are $g_{11} = d_1\lambda_{11}$ and $g_{22} = d_2\lambda_{22}$, $d_1 = 2\omega_D/N_{1I}$ and $d_2 = 2\omega_D N_{2I}$. The last two are the mean energy level spacings. λ_1 and λ_2 are dimensionless parameters for two sub-bands. The inter-band coupling is written as $g_{12} = \lambda_{12}\sqrt{d_1 d_2}$ The ratio $\rho_1/\rho_2 = d_2/d_1$.

KN [98] employed the functional integral approach, commonly used in the problems of pairing correlations of a finite number of fermions [99, 100] With this technique the coupled gap equations at T=0 are obtained (for bulk) as $\Delta_i = \omega_D/\sinh(1/\eta_i)$ with i = 1, 2.

$$\frac{1}{\eta_1} = \frac{\lambda_{22} + \alpha_{\pm}[\eta_1, \eta_2]\lambda_{12}}{\lambda_{11}\lambda_{22} - \lambda_{12}^2}, \tag{34}$$

$$\frac{1}{\eta_2} = \frac{\lambda_{11} + \alpha_{\pm}^{-1}[\eta_1, \eta_2]\lambda_{12}}{\lambda_{11}\lambda_{22} - \lambda_{12}^2}. \tag{35}$$

The phase of the gap is $\alpha_{\pm}[\eta_1, \eta_2] = \pm\sinh(1/\eta_2)/\sinh(1/\eta_1)$. Here there are two cases, (i) the phases can be same and (ii) the phases are opposite. When the inter-band coupling λ_{12} is large, $\Delta_1 = -\Delta_2$ and for $\lambda_{12} = 0$, bulk gaps are independent two gaps. These results we have obtained earlier by the GL method (section 2.2) for the bulk two-band superconductivity. KN calculates the condensate energy E_c for the two-band nanosuperconductor. As conventionally defined it is the difference of ground state energies of normal and superconducting states including the interaction term. The final expression for the total condensation energy E_c is given by

$$E_c = E_c(N_1, \lambda_{11}) + E_c(N_2, \lambda_{22}) - \Delta E_c \times (\Delta_1, \Delta_2, d_1, d_2, \lambda_{11}, \lambda_{22}, \lambda_{12}), \tag{36}$$

where the first two terms on the right hand side are condensation energy of two independent sub-systems and the last term is the contribution of inter-band contribution given by

$$\Delta E_c = \frac{\lambda_{12}^2}{\lambda_{11}\lambda_{22} - \lambda_{12}^2}\left[\frac{\Delta_1^2}{d_1\lambda_{11}} + \frac{\Delta_2^2}{d_2\lambda_{22}} + \frac{2(\Delta_1^*\Delta_2 + \Delta_1\Delta_2^*)}{\lambda_{12}\sqrt{d_1 d_2}}\right]. \tag{37}$$

The quantity ΔE_c increases or decreases depending on the value of the phase difference. $(\Delta_1^*\Delta_2 + \Delta_1\Delta_2^*) = 2(|\Delta_1||\Delta_2|\cos(\Delta\phi))$. If its value is > 0, it leads to an instability, whereas its value < 0, makes E_c more stable. This we have discussed in section 2.1 under the multiband GL theory.

Among other things KN discusses critical level spacing, the parity gap, Kondo effect and application to quantum computing for this model.

Multiband Nano Superconductivity by Extended GL Model

This is a new paradigm of nanofilms involving multiband superconductors. Quantum confinement and fluctuations are important in quasi-2D systems. Therefore, it will be alluring to implement the extended GL method as worked out in detail for multiband bulk system in [39, 40]. Shanenko et al [100] have made a preliminary attempt in this realm. They have included terms of order $\tau^{3/2}$ in the extended GL equation. Yet the validity of GL equations remains t † 1.

EPILOGUE

In this review we presented a bird's eye view of recent advances in novel superconductors. We gave a pedagogic brief history of superconducting phenomenon by presenting both theories and experiments. We begin with the one-band/gap superconductors. Phenomenological theory of GL and microscopic theory of

BCS are outlined with an important landmark result. Gor'kov showed that the GL theory was derived from the BCS theory, particularly its correctness nearT_c. There is a technical point that for the equivalence of both the theories, GL theory relies on the expansion of $\tau^{1/2}$ ($t = 1 - T/T_c$) in the lowest order.

In view of possible occurrence of superconductivity in s-d band metals, multi-band generalization of GL and BCS was done by a number of workers. An important parameter appearing in these theories is the inter-band couplings, which show some interesting consequences such as, time reversed broken symmetry state, kink solutions and frustration. In the multiband case we showed the equivalence of GL and BCS models near the T_c. Going beyond the standard GL theory we made expansion of the order parameter to higher order in $t^{(2n+1)/2}$. We showed some new results in the calculations of the order parameter compared to the BCS values.

After having stated these background materials we discussed superconductivity going from 'bulk to nano'. A criterion predicted by Anderson that superconductivity can survive in a small grain of material above a certain critical size is the main issue highlighted in the later part of the paper. In this context we discussed the BCS theory for finite size systems, what has been studied in great detail in nuclear physics for pairing of nucleons. An exactly solvable model due to Richardson for finite number of particles is discussed with many important results reviewed.

Physics being an experimental science, a large number and variety of experiments have been done recently on nanosuperconductivity. We have given an ample amount of information of this development. Arising out of this necessity of understanding the experiments, some highlights of theory are presented. With these prospects we are hopeful that there will be more investigations in this fast growing area.

ACKNOWLEDGMENTS

MPD is grateful to Professor Nguyen Van Hieu for the kind invitation and warm hospitality that is immensely enjoyable coming to Ha Long City for the IWAMSN-2014.

REFERENCES

1. London F and London H 1935 Proc. R. Soc. A 149 71
2. Ginzburg V L and Landau L D 1950 Zh. Eksp. Teor. Fiz. 20 1064 (English translation) Landau L D 1965 Collected Papers of L D Landau ed D ter Haar (Oxford: Pergamon Press) p 546
3. Abrikosov A A 1957 Zh. Eksp. Teor. Fiz. 32 1442 Abrikosov A A 1957 Sov. Phys. JETP 5 1174 (Engl. transl.)
4. Bardeen J and Pines D 1955 Phys. Rev. 99 1140
5. Cooper L 1956 Phys. Rev. 104 1189
6. Bardeen J, Cooper L and Schrieffer J R 1957 Phys. Rev. 108 1175
7. Anderson P W 1958 Phys. Rev. 110 827
8. Nambu Y 1960 Phys. Rev. 117 648
9. Bogolyubov N N 1958 Sov. Phys. JETP 7 41
10. Valatin J 1958 Il Nuovo Cimento VII 843
11. Gor'kov L P 1958 Sov. Phys. JETP 7 505
12. Abrikosov A A, Gor'kov L P and Dzyaloshinski I E 1963 Methods of Quantum Field Theory in Statistical Physics (Englewood Cliffs, NJ: Prentice Hall) chapter 7
13. Gor'kov L P 1959 Sov. Phys. JETP 9 1364
14. Ketterson J B and Song S N 1999 Superconductivity (Cambridge: Cambridge University Press)
15. Shen L Y L, Senozan N M and Phillips N E 1965 Phys. Rev. Lett. 14 1025
16. Binnig G, Baratoff A, Hoenig H E and Bednorz J G 1980 Phys. Rev. Lett. 45 1352
17. Ivarone M, Karapetrov G, Koshelev A, Kwok E W K, Crabtree G W, Hinks D G, Kang W N, Choi E-M, Kim H J and Lee S I 2002 Phys. Rev. Lett. 89 187002
18. Eskildsen M R, Kugler M, Tanaka S, Jun J, Kazakov S M, Karpinski J and Fischer Ø 2002 Phys. Rev. Lett. 89 187003
19. Xi X X 2008 Rep. Prog. Phys. 71 116501
20. Shulga S V, Drechsler S-L, Fuchs G, Müller K-H, Winzer W, Heinecke M and Krug K 1998 Phys. Rev. Lett. 80 1730
21. Zehetmayer M 2013 Supercond. Sci. Technol. 26 043991
22. Suhl H, Matthias B T and Walker L R 1959 Phys. Rev. Lett. 3 552
23. Moskalenko V A 1959 Fiz. Metal. Metalloved. 8 503
24. Kondo J 1963 Prog. Theor. Phys. 29 1
25. Tilley D R 1964 Proc. Phys. Soc. 84 573
26. Geilikman B T, Zaitsev R O and Kresin V Z 1967 Sov. PhysSolid State 9 642
27. Lee D H and Ihm J 1987 Solid State Commun. 62 811
28. Zhitomirsky M E and Dao V-H 2004 Phys. Rev. B 69 054508
29. Golubov A A and Koshelev A E 2003 Phys. Rev. B 68 104503
30. Mazin I I and Schmalian J 1995 Physica C 469 614
31. Kogan V G, Martin C and Prozorov R 2009 Phys. Rev. B 80 014507
32. Gurevich A 2003 Phys. Rev. B 67 184515
33. Brandt E H and Das M P 2011 J. Supercond. Novel Magn. 24 57
34. Das M P, He H-X and Choy T C 1988 Inter. J. Mod. Phys. B 2 1513
35. Askerzade I N 2006 Physics: Uspekhi 49 1003
36. Kogan V G and Schmalian J 2011 Phys. Rev. B 83 054515
37. Shanenko A A, Milesevic M V, Peeters F M and Vagov A V 2011 Phys. Rev. Lett. 106 047005
38. Vagov A V, Shanenko A A, Milesevic M V, Axt V M and Peeters F 2012 Phys. Rev. B 85 014502
39. Vagov A V, Shanenko A A, Milesevic M V, Axt V M and Peeters F 2012 Phys. Rev. B 86 144514
40. Wilson B J and Das M P 2014 J. Phys.: Condens. Matter 26 325701
41. Fetter A L and Walecka J D 1971 Quantum Theory of ManyParticle Systems (New York: McGraw Hill)

42. Komendova L, Chen Y, Shanenko A A, Milesovich M V and Peeters F M 2001 Phys. Rev. Lett. 108 207002
43. Khasanov R, Bendele M, Amato A, Konder K, Keller H, Klaus H-H, Luetkens H and Pomjakushina E 2010 Phys. Rev. Lett. 104 087004
44. Singh Y, Martin C, Budko S L, Ellen A, Prozorov R and Johnston D C 2010 Phys. Rev. B 82 144532
45. Kim H, Tanatar M A, Song Y J, Kwon Y S and Prozorov R 2011 Phys. Rev B 83 100502
46. Golubov A A, Kortus J, Dologov O V, Jepson O, Kong Y, Anderson O K, Gibson B J, Anh K and Kremer R K 2002 J. Phys.: Condens. Matter 14 1353
47. Das M P 2010 Adv. Nat. Sci.: Nanosci. Nanotechnol. 1 043001
48. Moshchalkov V, Woedenweber R and Lang W (ed) 2010 Nanoscience and Engineering in Superconductivity (Berlin: Springer) Morelle M, Chibotaru L F, Ceulemans A, Carballeira C and Moshchalkov V V 2005 Proc. SPIE Strongly Correlated Electron Materials: Physics and Nanoengineering ed I Bozovic and D Pavuna 59321B
49. Bonca J and Kruchinin S (ed) 2010 Physical Properties of Nanosystems (NATO Science for Peace and Security Series B) (Berlin: Springer)
50. Kubo R 1962 J Phys. Soc. Jap. 17 975
51. Halperin W P 1986 Rev. Mod. Phys. 58 533
52. von Delft J and Ralph D C 2001 Phys. Rep. 345 61
53. Anderson P W 1959 J Phys. Chem. Solids 11 26
54. Abrikosov A A and Gor'kov L P 1958 JETP 35 1558 Abrikosov A A and Gor'kov L P 1960 JETP 39 1781
55. Tsuneto T 1962 Prog. Theo. Phys. 28 857
56. Ma M and Lee P A 1985 Phys. Rev. B 32 5658 Altschuler E L and Lades M 1999 arXiv:cond-mat/9901257
57. Ralph D C, Black C T and Tinkham M 1995 Phys. Rev. Lett. 74 3241 Black C T, Ralph D C and Tinkham M 1996 Phys. Rev. Lett. 76 688 Ralph D C, Black C T and Tinkham M 1997 Phys. Rev. Lett. 78 4087
58. Dukelsky J and Sierra G 2000 Phys. Rev. B 61 12302
59. Yang C-C, Huang W-L, Lin Y-H, Weng C-Y, Mo Z-Y and Chen Y-Y 2011 IEEE Trans. Magnetics 47 3535 Bose S, Raychaudhuri P, Banerji R and Ayyub P 2006 Phys. Rev. B 74 224502 Li W H, Yang C C, Tsao F C and Lee K C 2003 Phys. Rev. B 68 184507 Li W H, Yang C C, Tsao F C, Wu S Y, Huang P J, Chung M K and Yao Y D 2005 Phys. Rev. B 72 214516 Guo Y et al 2004 Science 306 1915
60. Richardson R W 1963 Phys. Lett. 3 277
61. Braun F and von Delft J 1998 Phys. Rev. Lett. 81 4712
62. Yuzbashyan E A, Baytin A A and Altshuler B L 2005 Phys. Rev. B 71 094505
63. Pogosov W V, Lin N and Misko V R 2013 Eur. Phys. J. B 86 235
64. Tuominen M T, Hergenrother J M, Tighe T S and Tinkham M 1992 Phys. Rev. Lett. 69 1997
65. Matveev K A and Larkin A I 1997 Phys. Rev. Lett. 78 3749
66. Blatt J M and Thomson C J 1963 Phys. Lev. Lett. 10 332
67. Parmenter R 1968 Phys. Rev. 166 392
68. Mühlschlegel B, Scalapino D G and Denten R 1972 Phys. Rev. B 6 1767
69. Romaguerra A R de C, Doria M M and Peeters F M 2007 Physica C 460–462 1232
70. Schweigert V A, Peeters F M and Deo P S 1998 Phys. Rev. Lett. 81 2783
71. Tsubota M 2006 J. Phys.: Conf. Ser. 31 88
72. Grigorieva I V, Escoffier E, Richardson J, Vinnikov L Y, Dubonos S and Oboznov V 2006 Phys. Rev. Lett. 96 077005 Grigorieva I V, Escoffier E, Misko V R, Baelus B J, Peeters F M, Vinnikov L Y and Dubonos S V 2007 Phys. Rev. Lett. 99 14703 Budzin A I and Brosn J P 1998 Phys. Lett. A 196 267 Palacios J J 1998 Phys. Rev. B 58 R5948 Baelus B J, Cabral L R E and Peeters F M 2004 Phys. Rev. B 69 064506 Fink H J and Presson A G 1966 Phys. Rev. 151 219 Moshchalkov V V, Qui X G and Bruyndoncx V 1997 Phys. Rev. B 55 11793 Kanda A, Baelus B J, Shimizu N, Tadano K, Peeters F M, Kadpwaki K and Ootuka Y 2006 Physica C 445–448 253 Moshchalkov V V, Gielen L, Strunk C, Jonckheere R, Qiu X, Van Haesendonck C and Bruynseraede Y 1995 Nature 373 319
73. Dukelsky J, Pittel S and Sierra G 2004 Rev. Mod. Phys. 76 643
74. Mastellone A, Falci G and Fazio R 1998 Phys. Lev. Lett. 80 4542
75. Garcia-Garcia A M, Urbina J D, Yuzbashiyan E A, Richter K and Alschuler B L 2011 Phys. Rev. 83 014510
76. Romero-Bermudez A and Garcia-Garcia A M 2014 Phys. Rev. 89 024510
77. Gladilin V N, Fomin V M and Devreese J T 2002 Solid State Commun. 121 519
78. Krezin V and Ovchinnikov Y N 2006 Phys. Rev. 74 024514
79. Smith R A and Ambegaokar V 1996 Phys. Rev. Lett. 77 4962
80. Sierra G, Dukelsky J, Dussel G G, von Delft J and Braun F 2000 Phys. Rev. B 61 R11890
81. Schechter M, Imry Y, Levinson Y and von Delft J 2001 Phys. Rev. B 63 214518
82. Nagamatsu J, Nakagawa N, Muranaka T, Zenitani Y and Akimitsu J 2001 Nature 410 63
83. Kamihara Y, Watanabe T, Hirano M and Hosono H 2008 Am J. Chem. Soc. 130 3296
84. Xi X X 2008 Rep. Prog. Phys. 71 116501
85. Zehetmeyer M 2013 Supercond. Sci. Technol. 26 043001
86. Guo Y et al 2004 Science 306 1915
87. Li W H, Yang C C, Tsao F C and Lee K C 2003 Phys. Rev. B 68 184507
88. Li W H, Yang C C, Tsao F C, Wu S Y, Huang P J, Chung M K and Yao Y D 2005 Phys. Rev. B 72 214516
89. Yang C-C, Huang W-L, Lin Y-H, Weng C-Y, Mo Z-Y and Chen Y-Y 2011 IEEE Trans. Magn. 47 3535
90. Bose S, Raychaudhuri P, Bannerji R and Ayyub P 2006 Phys. Rev. B 74 224502

91. Bron C, Hong I-P, Patthey F, Sklyadneva Y I, Heid R, Echenique P M, Bohnen K P, Chulkov E V and Schneider W-D 2009 Phys. Rev. Lett. 102 207002

92. Reich S, Leitus G, Popovitz-Biro R and Schechter M 2003 Phys. Rev. Lett. 91 147001

93. Eom D, Qin S, Chou M-Y and Shih C K 2006 Phys. Rev. Lett. 96 027005

94. Lee S et al 2013 Nat. Mater. 12 392

95. Fernandes R M, Haraldsen J T, Woelfle P and Balatsky A V 2013 Phys. Rev. B 87 014510

96. Romero-Bermudez A and Garcia-Garcia A M 2014 Phys. Rev. B 89 024510

97. Kruchinin S P and Nagao H 2012 Intern. J. Mod. Phys. B 26 1230013

98. Popov V N 1987 Functional Integrals and Collective Excitations (Cambridge: Cambridge University Press)

99. Mühlschlegel B 1962 J. Math. Phys. 3 522

100. Shanenko A A, Orlova N V, Vagov A, Milosevic M V, Axt V M and Peeters F M 2013 EPL 102 27003

CHAPTER 2

ELECTRONIC STRUCTURE AND SUPERCONDUCTIVITY OF FESE-RELATED SUPERCONDUCTORS

Xu Liu, Lin Zhao, Shaolong He, Junfeng He, Defa Liu, Daixiang Mou, Bing Shen, Yong Hu,Jianwei Huang and X J Zhou

Keywords: FeSe films, ARPES, electronic structure, iron-based superconductors

INTRODUCTION

Iron-based superconductors discovered in 2008 [1] represent the second class of high temperature superconductors after the discovery of the first class of high-T_c cuprate superconductors in 1986 [2]. The superconducting transition temperature (T_c) has reached ~55 K [3–6], which is beyond the generally-believed McMillan limit of the conventional superconductors. Indications of even higher T_c have emerged in single-layer FeSe films [7–10]. Since the discovery of the cuprate superconductors, understanding the high temperature superconductivity mechanism remains a prominent and challenging task facing the condensed matter physics community. The discovery of iron-based superconductors provides an opportunity to compare and contrast with the cuprate superconductors that may help to uncover the secret to high temperature superconductivity. Great progress has been made in materials preparation, experimental investigation and theoretical understanding of iron-based superconductors [11–20].

So far, several families of iron-based superconductors have been discovered and can be mainly categorized into '11' [21], '111' [22], '122' [23] and '1111' [1, 3–5] systems according to their crystal structure (figure 1) [12]. Similar to cuprate superconductors, iron-based superconductors are quasi-two-dimensional in their crystal structure. The FePn (Pn = As or Se) layer is an essential building block that is believed to be responsible for the superconductivity in iron-based superconductors. Different from the cuprate superconductors where the CuO_2 plane is basically co-planar, the FePn (Pn = As or Se) unit consists of three layers with the central Fe layer sandwiched in between two adjacent Pn layers. This results in the doubling of the unit cell in the iron-based superconductors and the folding of the corresponding electronic structure (figure 2). Most significantly, different from cuprate superconductors where the low-energy physics is mainly dominated by the single Cu $d_{x^2-y^2}$ orbital, in iron-based superconductors, all the five Fe 3d orbitals participate in the low-energy electronic structures [26–29]. Generally, there are multiple bands crossing the Fermi level that form hole-like Fermi surface sheets near the Brillouin zone center and electron-like Fermi surface sheets near the zone corner [30–37]. The multiple-orbital nature (figure 2) [149, 150] plays an important role in understanding the physical properties and superconductivity mechanism in iron-based superconductors.

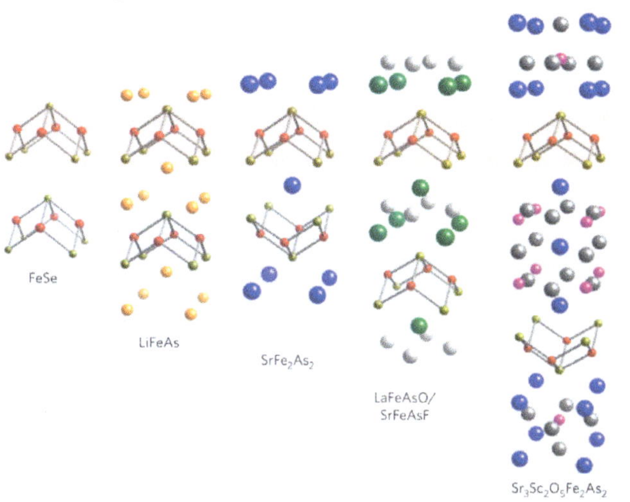

Figure 1. Several major classes of iron-based superconductors. All of these iron-based superconductors contain FeSe or

Citation: Xu Liu, Lin Zhao, Shaolong He, Junfeng He, Defa Liu, Daixiang Mou, Bing Shen, Yong Hu, Jianwei Huang and X J Zhou, "Electronic structure and superconductivity of FeSe-related superconductors," J. Phys.: Condens. Matter 27 (2015) 183201 (22pp) doi:10.1088/0953-8984/27/18/183201.

FeAs building blocks (the iron ions are shown in red and the pnictogen/chalcogen anions are shown in gold) that are essential for the occurrence of superconductivity. Reprinted from [116].

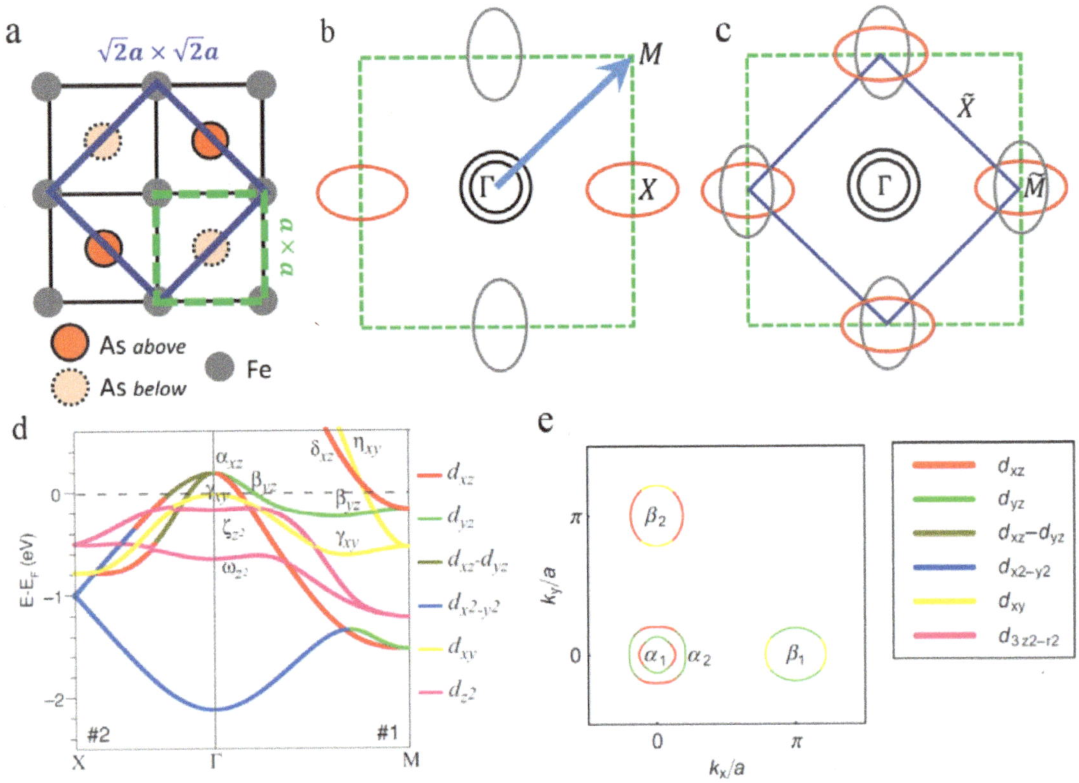

Figure 2. Brillouin zones of iron-based superconductors containing one iron and two irons in a unit cell and the orbital-related electronic structure. (a) FeAs lattice indicating As above and below the Fe plane. Dashed green and solid blue squares represent 1- and 2-Fe unit cells, respectively. (b) Schematic two-dimensional Fermi surface in the 1-Fe Brillouin zone whose boundaries are indicated by a green dashed square. The arrow indicates a folding wave vector to convert from the 1-Fe zone to the 2-Fe zone. (c) Fermi sheets in the folded Brillouin zone whose boundaries are now shown by a solid blue square. In this case, the size of the first Brillouin zone is half of that of the 1-Fe case (dashed green squeare). Reprinted from [16]. (d) A typical orbital character of the bands in the iron pnictides as shown in [149, 150]. (e) The corresponding Fermi surface sheets of the five-band model for the undoped compound in [150].

Among all the iron-based superconductors, the FeSe superconductor and its related systems have gained particular attention due to its simple crystal structure and peculiar electronic and physical properties. The FeSe superconductor has a simple crystal structure consisting of the FeSe layer that is the essential building block in iron-based superconductors. Such a simple structure is ideal for theoretical and experimental study of the superconductivity mechanism. The bulk FeSe superconductor has a superconducting transition temperature (T_c) of ~8 K [21] and can be dramatically enhanced to 37 K at high pressure [38]. On the other hand, its cousin system, FeTe, possesses a unique antiferromagnetic ground state but is non-superconducting [39]. Substitution of Se by Te in the FeSe superconductor results in an enhancement of T_c up to 14.5 K and superconductivity can persist over a large composition range in the Fe(Se,Te) system [40, 41]. Intercalation of the FeSe superconductor leads to the discovery of the $AxFe_{2-}ySe_2$(A = K, Cs and Tl) system that exhibits a T_c higher than 30 K [24, 42, 43] and unique electronic structure of the superconducting phase [17, 44, 45]. The latest report of possible high temperature superconductivity in the single-layer FeSe/$SrTiO_3$ films with a T_c above 65 K has generated much excitement in the community [7–10]. This pioneering work opens a door for interface superconductivity to explore for high T_c superconductors [7]. The distinct electronic structure and superconducting gap, layer-dependent behavior and insulator–superconductor transition of the FeSe/$SrTiO_3$ films provide critical information in understanding the superconductivity mechanism of iron-based superconductors.

In this paper, we will present a brief review on the investigation of the electronic structure and superconductivity of the FeSe superconductor and related systems. We will put particular focus on the FeSe film, which is an exciting and fast-growing field. The paper is organized as follows. In section 2, bulk FeSe and related materials are first introduced. In section 3, we will discuss the electronic properties of the FeSe films, including Fermi surface, band structure, gap symmetry and the evolution of electron structure with annealing. The implications and theoretical understandings are presented in section 4. In section 5, we end with further issues to be investigated and a future perspective.

ELECTRONIC STRUCTURE OF BULK FE (SE, TE) SYSTEMS AND INTERCALATED FESE

Bulk Fe (Se,Te)

The bulk FeSe superconductor with a T_c of 8 K was first discovered in 2008 in a tetragonal phase β-FeSe with PbO structure at ambient pressure (figure 3(a)) [21]. It has been found that excess Fe is inevitable to stabilize the crystal structure of $Fe_{1+}\delta Se$ and the superconductivity is very sensitive to its stoichiometry (δ) [46]. The FeSe superconductor exhibits a dramatic pressure dependence (figure 3(c)). In particular, its T_c can be enhanced to 36.7 K under high pressure [38]. At ambient pressure, the FeSe superconductor undergoes a structural transition from tetragonal to orthorhombic around 90 K but without a magnetic transition [46, 47]. This is distinct from many other iron-based parent compounds where the structural transition is usually accompanied by a magnetic phase transition [48–54]. However, under high pressure, static magnetic order is observed in superconducting FeSe [55]. Short range spin fluctuation has also been reported in FeSe superconductors that is enhanced under high pressure [56].

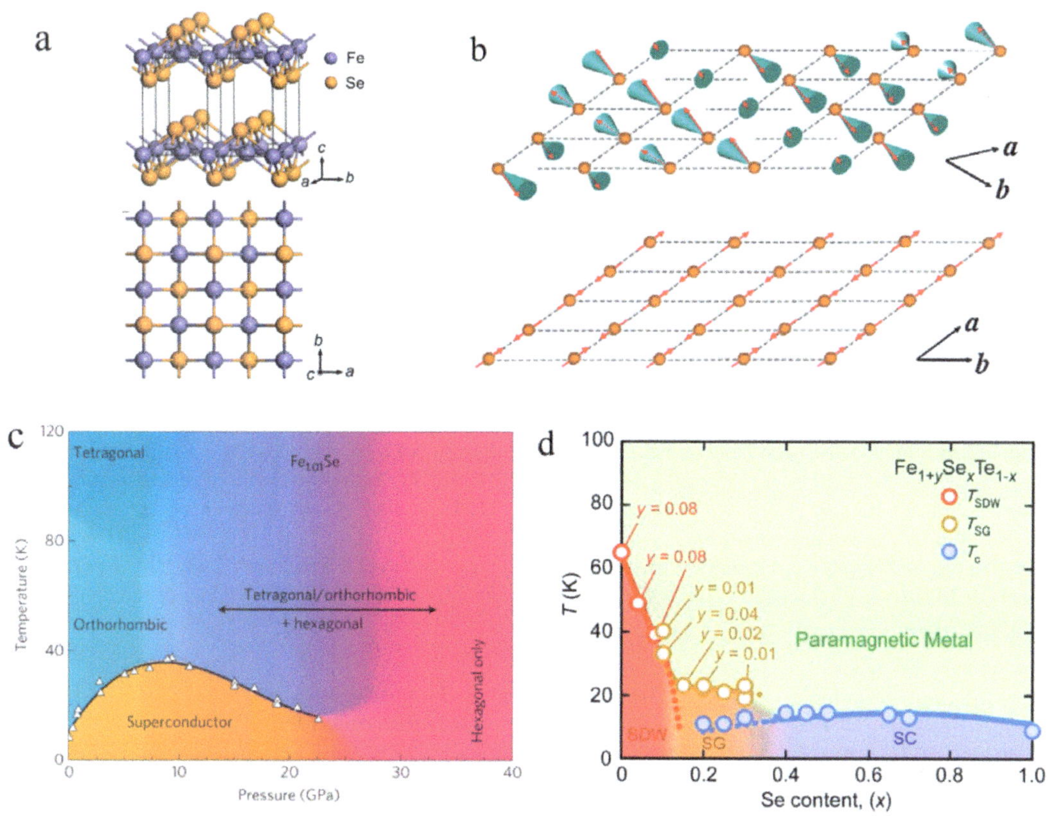

Figure 3. Crystal structure, magnetic structure and phase diagram of bulk $Fe_{1.01}Se$ and the phase diagram of the $Fe_{1+}ySexTe_{1-}x$ system. (a) Schematic crystal structure of β-FeSe. Four unit cells are shown to reveal the layered structure. Reprinted from [21]. (b) Magnetic structures of α-FeTe (upper panel) and $BaFe_2As_2$ (lower panel), which are shown in the primitive Fe square lattice. Reprinted from [39]. (c) Phase diagram of $Fe_{1.01}Se$ as a function of pressure. Reprinted from [38]. (d) Phase diagram of $Fe_{1+}ySexTe_{1-}x$ with y ~ 0 as a function of x and T. Reprinted from [41].

As Se is gradually replaced by Te in FeSe, superconductivity with a maximum T_c around 14.5 K can be observed in a large composition range (x) in the $Fe_{1+}\delta Se_{1-}xTex$ system (figure 3(d)) [40,41]. Again, the excess Fe content (δ) has a big effect on the physical properties and superconductivity in this case [46]. The excess Fe is found to be located at the interstitial sites [39]. $Fe_{1+}\delta Te$ is present at the other end of the $Fe_{1+}\delta(Se_{1-}xTex)$ phase diagram (figure 3(d)). It undergoes a phase transition between 60–75 K [40, 57] and no superconductivity is observed in bulk $Fe_{1+}\delta Te$. The signature of superconductivity is reported in FeTe films but is likely due to the incorporation of oxygens [58–60]. At low temperature, $Fe_{1+}\delta Te$ shows an antiferromagnetic order of bi-collinear structure (figure 3(b), upper panel) that is distinct from the usual collinear magnetic structure observed in many other parent compounds like $BaFe_2As_2$ (figure 3(b), lower panel) [39, 61].

The band structure and Fermi surface of the bulk FeSe superconductor (figures 4(a) and (b)) from the band structure calculations show similar behaviors to other iron-based superconductors, i.e. the low energy electronic states originate mainly from the iron 3d orbitals and there are two hole-like Fermi surface sheets at the zone center and two intersecting electron-like Fermi surface sheets around the zone corner [62]. Direct ARPES measurements on the FeSe superconductors have been hindered due to the difficulty in obtaining FeSe single crystals. With the latest progress on growing high-quality FeSe single crystals [63–66], several ARPES measurements on the FeSe superconductor have become available [32, 67, 68].

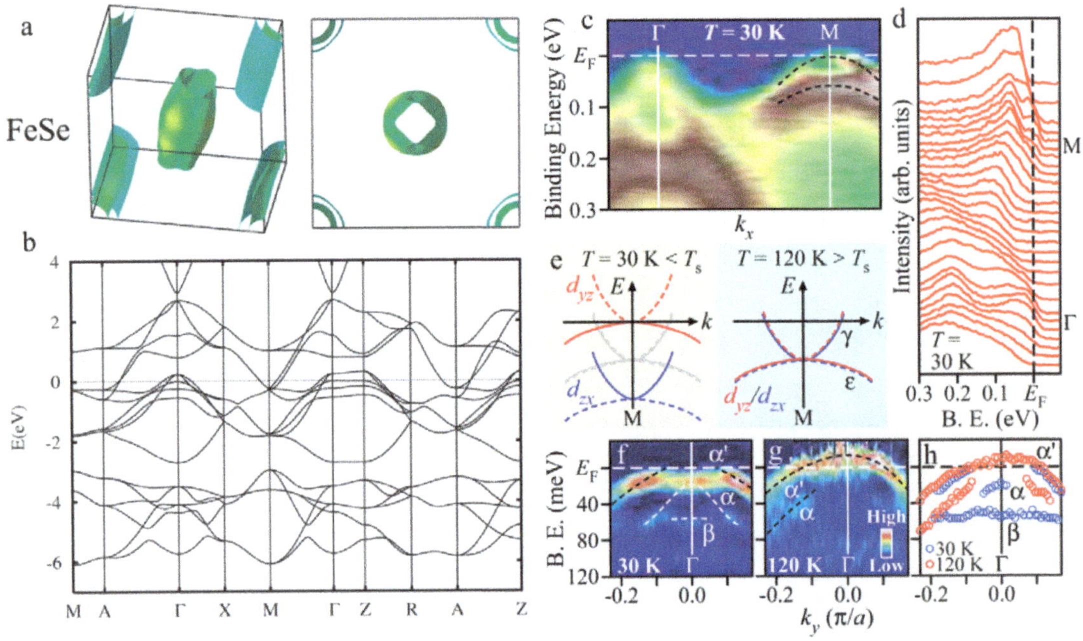

Figure 4. Fermi surface and band structure of the bulk FeSe from band structure calculations and angle-resolved photoemission measurements. (a) and (b) LDA calculated Fermi surface and band structure of bulk FeSe from [62]. (c) and (d) ARPES intensity and corresponding EDCs, respectively, along the Γ-M cut at T = 30 K. (e) Schematic band diagram around the M point below/above the structural transition temperature T_s. Red and blue curves indicate the dyz and dzx orbitals, respectively. Solid and dashed curves represent the band dispersion along the (0, 0)–(π, 0) and (0, 0)–(0,π) directions (longer Fe–Fe and shorter Fe–Fe directions) of the untwined crystal, respectively. (f) and (g) Comparison of the second-derivative plot of the near-E_F ARPES intensity around the Γ point between T = 30 and 120 K. (h) Experimental band dispersion around the Γ point atT = 30 K (blue circles) and 120 K (red circles), extracted by tracing the peak maxima of the EDCs divided by the Fermi-Dirac function. Reprinted from [67].

The ARPES results on FeSe single crystals reported so far give a basically consistent picture [32,67, 68], as exemplified in figure 4. First, at low temperature below the structural transition (~90 K), the Γ point is dominated by two hole-like bands (figure 4(f)), while near the M point, an electron-like band crossing the Fermi level and a hole-like band at higher binding energy are observed. Second, the measured electronic structures show a significant difference from the band structure calculations and the bands undergo a pronounced orbital-dependent renormalization. Third, there is an obvious change in the electronic structure across the structural phase transition [67, 68]. An energy splitting of bands up to 50 meV around the M point is observed below the structural transition. This is interpreted as the splitting of the dyzand dxz orbital bands (figure 4(e)), indicating the orbital ordering. In particular, the Fermi energy for both the hole-like bands near the zone center and the electron-like band near M is rather small. These are consistent with the quantum oscillation measurements [69, 70] and STM measurement [71] of the FeSe crystals. The comparable energy scale of the Fermi energy, superconducting gap and Zeeman energy in the FeSe superconductor also provides a window to look into the BCS-BEC crossover regime [71].

FeTe [40, 57], as a cousin of FeSe, shows quite different behaviors. It is non-superconducting and undergoes a structural/magnetic transition around T_{MS} = 70 K [40, 57] (figure 5(e)). It is semiconductor-like above the transition while it becomes metallic below the transition (figure5(e)). Initial ARPES measurements reveal a pair of nearly electron–hole compensated Fermi pockets, strong Fermi velocity renormalization and an absence of a spin-density-wave gap [72]. The measured electronic structure is qualitatively similar to other iron pnictides although the antiferromagnetic structure in FeTe is different and rotated by 45° when compared with that in other iron pnictide parent compounds. This is not consistent with the Fermi surface nesting picture for the formation of spin-density waves (SDW) below the transition temperature. Further work studied the Fermi surface and band structure of FeTe both in the paramagnetic state and the SDW state [73]. As shown in figure 5, above the magnetic transition temperature, the spectral weight mainly concentrates near the Γ point and M point (figure 5(a)) (as noted by [74], there may be a 45° rotation on the Brillouin zone definition in [73]). The Γ point is dominated by two hole-like bands while there are two electron-like bands near the M point. Upon decreasing the temperature below T_{MS}, the spectral weight distribution experiences an obvious change. The spectral weight around M and X points is suppressed, which is interpreted as the spectral weight transfer from low binding energy to high binding energy [73]. Detailed temperature-dependent ARPES measurements on FeTe reveal a coherence quasiparticle peak near the Fermi level below the transition, accompanied by a hump structure located at higher binding energy and a dip in between these two features (figures 5(h) and (i)) [73, 74]. Such a 'peak–dip–hump' structure bears a strong resemblance to that observed in manganites and heavily underdoped cuprates that can be explained in terms of polaron formation [74]. The hump feature is interpreted as being from incoherent excitation of electrons strongly coupled to bosons while the quasiparticle feature is related to coherent polaron motion. In this picture, the coherent polaron motion at low temperature may explain the metallic transport in the spin-density-wave state [74].

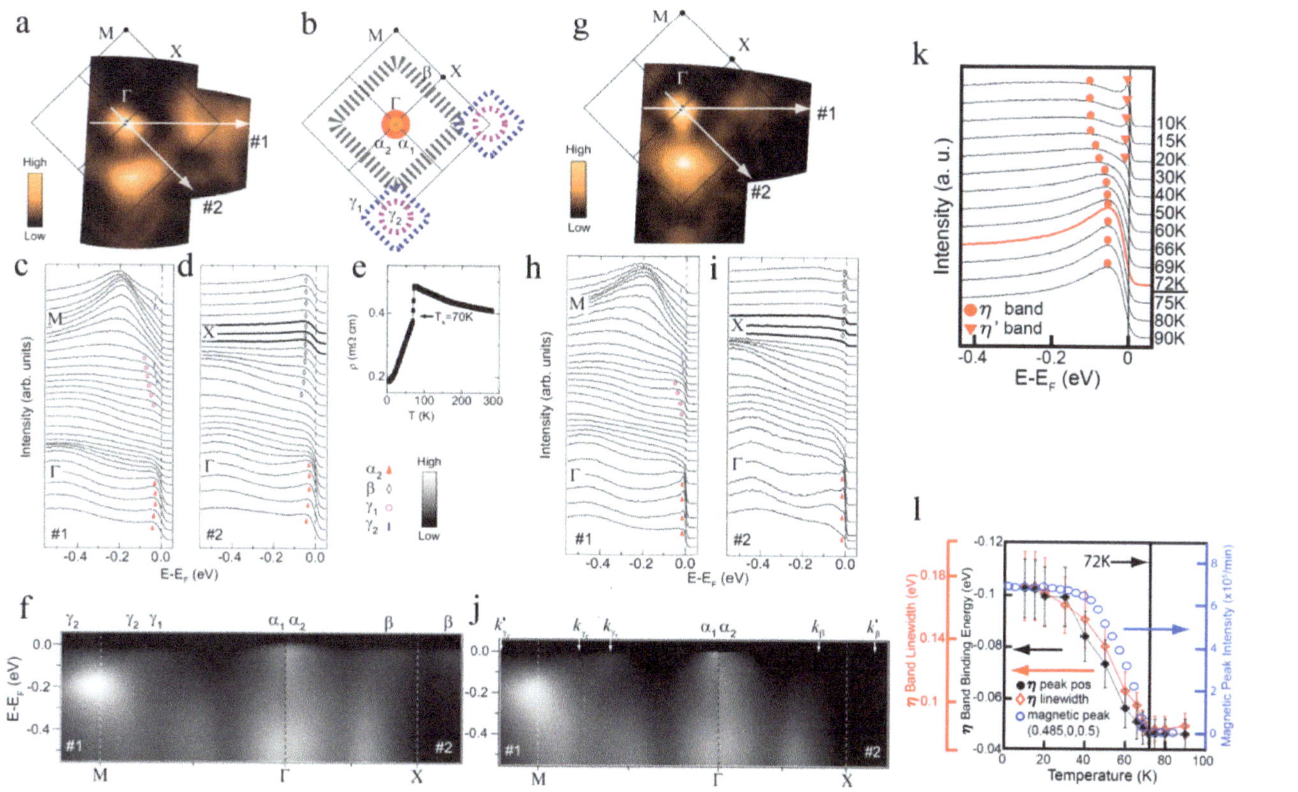

Figure 5. Fermi surface and band structure of FeTe. (a)–(f) Electron structure of $Fe_{1.06}Te$ in the paramagnetic state above the magnetic transition. (a) Fermi surface of $Fe_{1.06}Te$ measured at 135 K. (b) The spectral weight distribution around E_F. (c) and (d) The EDCs along cuts 1 and 2 labelled in (a), respectively. (e) The temperature dependence of the resistivity for $Fe_{1.06}Te$. (f) The photoemission intensities along the M-Γ-X high-symmetry line in the paramagnetic state. (g)–(j) Electron structure of $Fe_{1.06}Te$ in the magnetic state. (g) Fermi surface of $Fe_{1.06}Te$ measured at 15 K. (h) and (i) The EDCs along cut 1 and cut 2 labelled in (g), respectively. (j) The photoemission intensities along the M-Γ-X high-symmetry lines in the magnetic state. Reprinted from [73]. (k) Plot of the EDCs at the M point at various temperatures. (l) Plot of the M band binding energy and linewidth at the M point together with the (0.5, 0, 0.5) antiferromagnetic Bragg peak intensity versus temperature. Reprinted from [74].

Extensive ARPES measurements have been carried out on the Fe(Se,Te) system to investigate its electronic structure and superconducting gap [30, 31, 75–80]. Typical results are summarized in figure 6. There are three hole-like bands around the Γ point with two bands crossing the Fermi level (labeled as β and γ) and the third band (labeled as α) barely touching the Fermi level, while there are two electron-like bands around the M point in Fe(Se,Te) (figures 6(c)–(e)) [30]. The measured Fermi surface is sketched in figure 6(f). Taking advantage of the photoemission matrix element effect by performing ARPES measurements under different polarization geometry, the α band is assigned as from dxz/dyz orbitals, the β band is mainly from dxz, dyz and some from dxy orbitals and the γ band is assigned as from $d_{x^2-y^2}$ orbital [30]. Direct comparison between measurements and band structure calculations indicates a strong orbital-selective renormalization in the normal state [80]. Such strong renormalizaiton is consistent with the recent theoretical calculations to show that the 11 system exhibits a stronger correlation effect in iron-based superconductors [81].

In the optimally-doped $FeTe_{0.55}Se_{0.45}$ superconductor, the measured superconducting gap is nearly isotropic both near the zone center and the zone corner (figure 6(g)) [75]. Further high-resolution measurements indicate that the superconducting gap around the Fermi surface near Γ is anisotropic, which can be represented by $\cos(4\varphi)$ modulation (figure 6(h)) [76]. These ARPES results are consistent with the STM measurement on Fe(Te,Se) that points to an unconventional s-wave ($S_\pm$-wave symmetry) superconducting gap [82]. In Fe(Se,Te) superconductors, like in the FeSe superconductor case [71], the Fermi energy is comparable to the superconducting gap. This system thus provides an opportunity to explore the BCS-BEC crossover phenomenon in a superconducting material [31, 77].

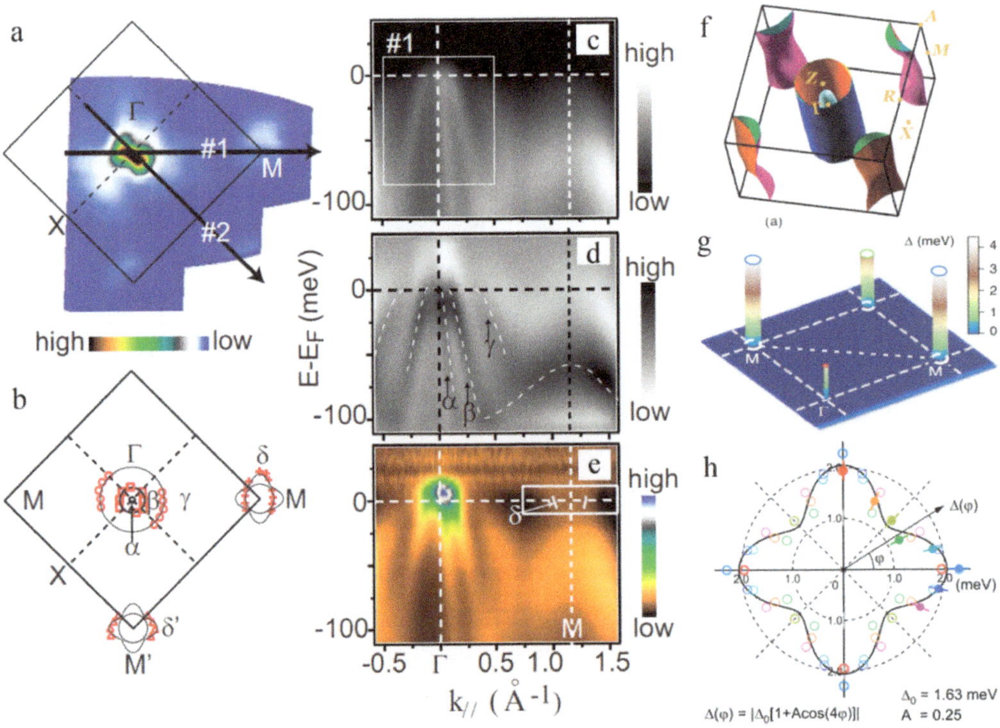

Figure 6. Fermi surface, band structure and superconducting gap of Fe(Te,Se). (a) Fermi surface of $Fe_{1.04}Te_{0.66}Se_{0.34}$. (b) The Fermi surfaces are constructed based on the measured Fermi crossings. (c) The photoemission intensity along the cut 1 in the Γ-M direction and (d) its second derivative with respect to energy. (e) The data in panel (c) are re-plotted after dividing the angle integrated energy distribution curve. (f) The calculated Fermi surface of $Fe_{1.04}Te_{0.66}Se_{0.34}$. Reprinted from [30]. (g) Three-dimensional representation of the superconducting gap with the Fermi surface topology of $FeTe_{0.55}Se_{0.45}$. Reprinted from [75]. (h) Fermi surface-angle dependence of superconducting gap size of $FeTe_{0.6}Se_{0.4}$. Reprinted from [76].

Intercalated FeSe System

The attempt to intercalate FeSe with potassium (K) resulted in a successful preparation of the $KxFe_{2}ySe_{2}$ superconductor with a T_c higher than 30 K [24]. Later on, it was found that FeSe can be intercalated with many other intercalates including Cs, Rb, Tl, Li, Na, Sr, Ca, Yb, Eu and even molecules [25, 43, 83–85, 42]. This system is complicated, with the coexistence of different phases. For reviews on this topic, one may refer to [17, 20].

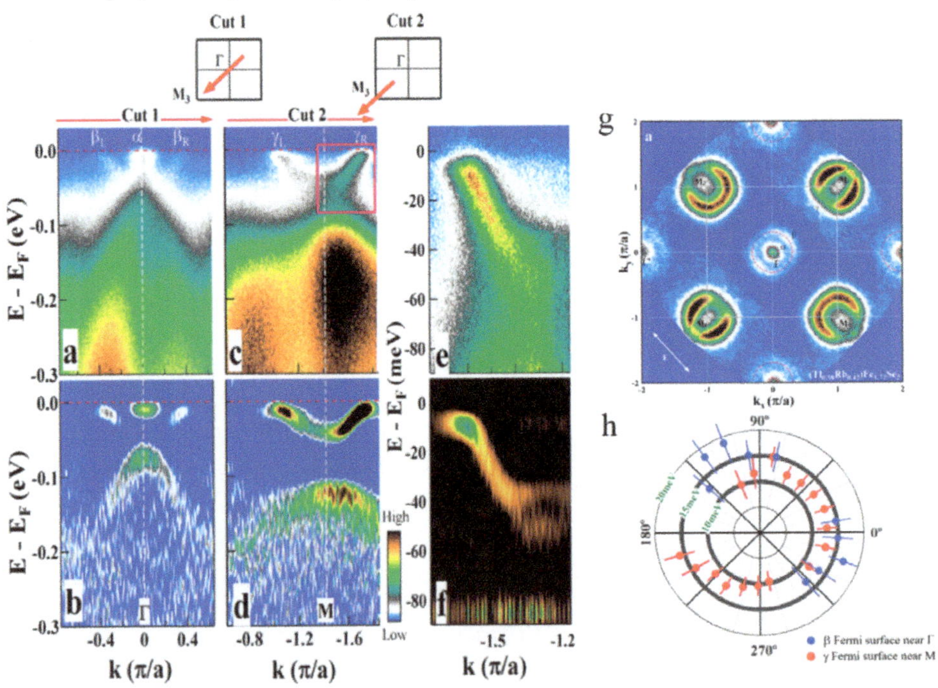

Figure 7. Band structure, Fermi surface and superconducting gap of $AxFe_{2-}ySe_{2}$. Band structure and photoemission spectra of $AxFe_{2-}ySe_{2}$ superconductors measured along two high symmetry cuts. Cut locations are illustrated in the top-left and top-right

inserts. (a) and (c) Measured band structure of $(Tl_{0.58}Rb_{0.42})Fe_{1.72}Se_2$ along cut 1 and cut 2, respectively. (b) and (d) Their corresponding EDC second derivative images. (e) and (f) Fine measurement of the band structure in the red square of (c) and its corresponding EDC second derivative images. (g) Fermi surface of $(Tl_{0.58}Rb_{0.42})Fe_{1.72}Se_2$. (h) Momentum-dependent superconducting gap size for the β Fermi surface sheet around Γ and the γ Fermi surface sheet around M, respectively. Reprinted from [17, 87].

For the completeness of the review, here we present the main ARPES results on the $AxFe_{2\_}ySe_2$(A = K, Cs, (Tl, K), (Tl, Rb)) system [17, 44, 86–92]. Although the intercalates can vary, the electronic structure of the superconducting phase in the $AxFe_{2\_}ySe_2$ system turned out to be similar. As shown in figure 7, the measured Fermi surface consists of two electron-like Fermi surface sheets (α and β in figure 7(g)) around Γ and one electron pocket around M (γ in figure7(g)) that includes two nearly degenerate Fermi surface sheets. The bands around Γ (figures7(a) and (b)) are dominated by two electron bands α and β. The bottom of the α band just touches the Fermi level while the origin of the β band remains unclear [87]. Around the M point, a clear electron band is observed in the vicinity of the Fermi level and a hole-like band with its top about 100 meV below the Fermi level (figures 7(c) and (d)). More detailed measurements indicate that there are two electron bands with similar Fermi momenta but different band bottoms around M (figures 7(e) and (f)) [17].

It was proposed that the interband scattering between the hole-like bands near Γ and electron-like bands near M gives rise to electron pairing and superconductivity in iron-based superconductors [93–99]. The distinct Fermi surface topology in the $AxFe_{2_}ySe_2$superconductor presents a serious challenge to this picture since there is no hole-like Fermi surface present near the Γ point. An alternative scattering mechanism between the electron pockets was proposed [93, 100–102]. In this case, the interelectron pocket scattering tends to produce $d_{x^2-y^2}$ gap symmetry. A nearly isotropic superconducting gap is observed around the γ electron pocket at the M point (figure 7(h)) [17, 44, 86–88]. In particular, the electron Fermi pockets around the zone center provide a good opportunity to distinguish among various gap symmetries. A nearly isotropic superconducting gap is observed on the small α electron pocket [91, 92] and a nearly isotropic superconducting gap was also observed around the larger βelectron pocket around Γ [17]. These results are not consistent with the d-wave gap symmetry in the $AxFe_{2\_}ySe_2$ system with only electron Fermi surface.

ELECTRONIC STRUCTURE AND SUPERCONDUCTIVITY OF FESE FILMS

Electronic Structure and Superconductivity of FeSe Films Grown on Graphene Substrate

High-quality crystalline FeSe films can be grown on graphene substrate by a molecular beam epitaxy (MBE) method [103]. The graphene substrate used in this case is double-layer graphene formed on the SiC(0 0 0 1) surface and the FeSe films grow as islands on the graphene layer. The grown FeSe films are along the (0 0 1) direction. One unit cell consists of one Fe layer and two Se layers (one above the Fe layer and the other below) (see figure 1) (loosely speaking, we will call this one unit cell of FeSe as single-layer in the following). The interaction between the FeSe film and the underlying graphene substrate is very weak so the lattice constant of the FeSe film is the same as the bulk FeSe (3.76 Å). Also because of this weak interaction, the FeSe films form domains with different orientations within the (0 0 1) plane. In this case, several STM works have been done on the FeSe/graphene films [104–106], but no ARPES measurement can be done due to the multi-domain nature of the films.

The superconductivity of the FeSe/graphene films behaves in a usual manner, i.e. its superconducting transition temperature decreases when the films get thinner (figure 8) [103]. Figure 8(a) shows a series of normalized tunneling conductance spectra on an 8-layer FeSe/graphene film. A clear coherence peak associated with superconductivity is observed and the gap size (Δ) can be obtained by measuring the distance between the two peaks (2Δ). It is 2.1 meV at 3 K for the 8-layer FeSe/graphene film with a corresponding superconducting temperature (T_c) at 8 K. In comparison, bilayer FeSe/graphene film shows a lower T_c that is only 3.7 K (figure 8(b)). The superconducting gap and T_c are determined for the FeSe/graphene films with different layers, as shown in figure 8(c) which plots the variation of T_c with the film thickness. T_c is found to be linearly proportional to the inverse of the film thickness 1/d. This is consistent with the usual expectation that T_c of the ultra-thin films decreases with the decrease of the film thickness: $T_c(d) = T_{c0}\ (1 - d/d_c)$ with T_{c0} being the bulk transition temperature andd_c being the critical thickness where superconductivity emerges [107]. Similar behaviors are also observed in ultra-thin Pb [108] and ultra-thin $YBa_2Cu_3O_{7\_}\delta$ [109] films. Note that no superconductivity is observed in single-layer FeSe/graphene film above 2.2 K. The extrapolatedT_{c0} from the layer-dependence of T_c is ~9.3 K, which is consistent with the T_c of the bulk FeSe superconductor [103].

In an atomically flat and defect-free (0 0 1) surface of the FeSe/graphene film (about 30 unit cells thick) with large terraces (figure 9(a)), the topmost layer is Se-terminated (figure 9(b)) [104]. Scanning tunneling spectroscopy (STS) probes the quasiparticle density of states and measures the superconducting gap at the Fermi energy (E_F) (figure 9(c)). The maximum of the superconducting gap is measured as Δ_0 = 2.2 meV. The most striking feature of the spectra at 0.4 K is the V-shaped dI/dV curve and the linear dependence of the quasiparticle density of states on energy near E_F. This feature is similar to that found in the d-wave cuprate superconductors with gap nodes [110] and in contrast to the U-shaped dI/dV curve observed in bulk Fe(Se,Te) superconductor [82]. This V-shaped feature in the FeSe/graphene film explicitly indicates the existence of line nodes in the superconducting gap function [104]. Further Fermi surface topology examination (figure 9(d)) suggests that the line nodes may be related to the component of the extended $s_\pm$-wave form $\Delta_2(\cos kx + \cos ky)$ that gives rise to nodes near M points. The reason why the Cooper pairing is nodal in FeSe/graphene films but nodeless in bulk Fe(Se,Te) remains a theoretical challenge although it is suggested that the pnictogen height may introduce a switch between the nodal and nodeless pairings [111].

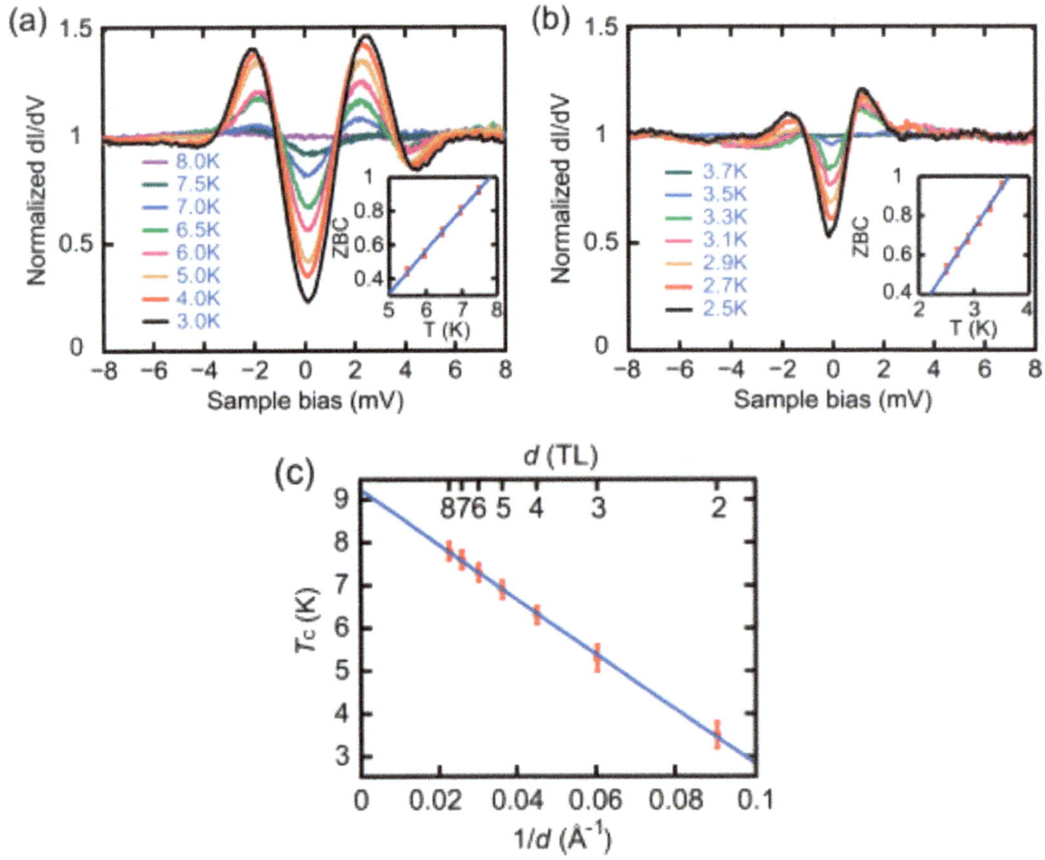

Figure 8. Normalized tunneling conductance spectra and T_c as a function of the film thickness in the FeSe/graphene films. Normalized tunneling conductance spectra on (a) 8-layer and (b) 2-layer FeSe/graphene films measured at different temperatures. (c) Superconducting transition temperature T_c versus the inverse of the film thickness d. Reprinted from [103].

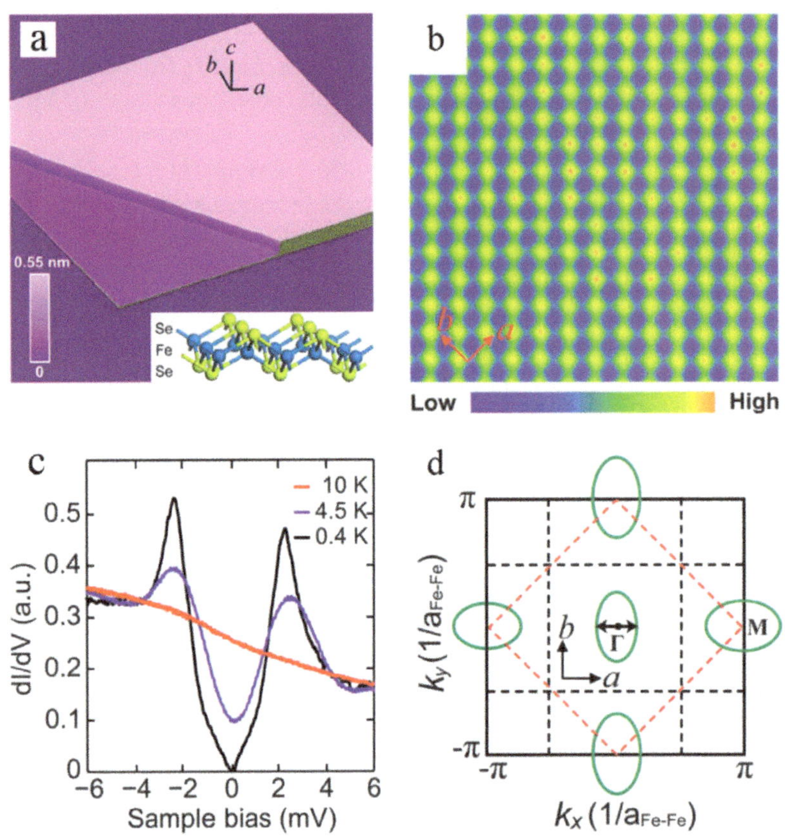

Figure 9. STM/STS measurements on an as-grown FeSe/graphene films about 30 layers thick. (a) Topographic image of the FeSe/graphene film. The step height is 5.5 Å. (b) Atomic-resolution STM topography of the FeSe/graphene film. The bright

spots correspond to the Se atoms in the top layer. (a) and (b) correspond to either of the Fe–Fe bond directions. (c) Temperature dependence of differential conductance spectra at different temperatures. (d) Schematic of the unfolded Brillouin zone and the Fermi surface (green ellipses). The nodal lines for cos kx cos ky and (cos kx+cos ky) gap functions are indicated by black and red dashed lines, respectively. Reprinted from [104].

In the FeSe/graphene films, the electron pairing with twofold symmetry, instead of four-fold symmetry, was demonstrated by direct imaging of quasiparticle excitations in the vicinity of magnetic vortex cores, Fe adatoms and Se vacancies [104]. The twofold pairing symmetry was further supported by the observation of striped electronic nanostructures in the slightly Se-doped samples. The anisotropy was explained in terms of the orbital-dependent reconstruction of the electronic structure in FeSe. It was also discovered that, in the FeSe/graphene films, superconductivity near the twin boundaries is suppressed, which also provides a measure of the superconducting coherence length of the order of 5.1 nm [105]. Such a superconductivity suppression is linked to the increased Se height in the vicinity of twin boundaries [105]. STM/STS also revealed signatures of a bosonic mode in the local quasiparticle density of states of the superconducting FeSe/graphene films [106].

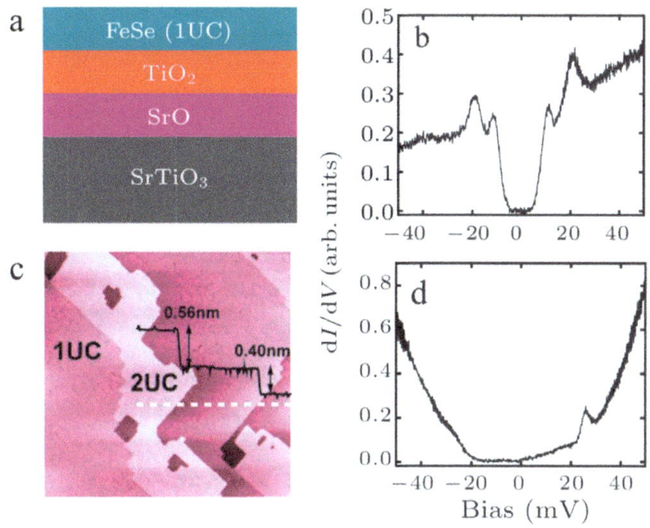

Figure 10. Discovery of possible high temperature superconductivity in single-layer FeSe/$SrTiO_3$ film. (a) Schematic structure (side-view) of the FeSe films on the $SrTiO_3$substrate along the c-axis. (b) Tunneling spectrum taken on the single-layer FeSe/$SrTiO_3$film at 4.2 K revealing the appearance of a superconducting gap. Four pronounced coherence peaks appear at ±20.1 mV and ±9 mV, respectively. (c) The STM image of the FeSe film with both single-layer (1UC) and double-layer (2UC) FeSe/$SrTiO_3$ films. (d) Tunneling spectrum taken on the double-layer FeSe/$SrTiO_3$ films. Reprinted from [7].

Electronic Structure and Superconductivity of FeSe Films Grown on $SrTiO_3$ and Related Substrates

FeSe films grown on the $SrTiO_3$ substrate exhibit surprising and different behaviors from those grown on graphene substrate [112]. When a single-layer FeSe film was grown on a $SrTiO_3$ (0 0 1) substrate by a MBE method (figure 10), four pronounced peaks were identified at ±20.1 mV and ±9 mV in the tunneling spectrum at 4.2 K that are symmetric with respect to the Fermi level (figure 10(b)) [7]. The gap at 20.1 meV, when taken as a superconducting gap, would correspond to a superconducting transition temperature near 80 K if one assumes a similar ratio between the superconducting gap and T_c as in the FeSe/graphene films [104]. The temperature dependence of the gap, the observation of a magnetic field-induced vortex state and the transport measurement all indicate that the single-layer FeSe/$SrTiO_3$ film is likely superconducting [104]. This pioneering work has generated great excitement in the community and many experimental and theoretical works will follow. On the other hand, it was found that the two-layer FeSe/$SrTiO_3$ films, grown under the same condition, show a semiconducting behavior (figure 10(d)) [7].

The immediate ARPES measurements on a superconducting single-layer FeSe/$SrTiO_3$ film reveal a simple and distinct electronic structure [8]. The observed Fermi surface (figure 11(a)) consists of only an electron pocket near the zone corner M without any Fermi surface crossing near the zone center Γ. This Fermi surface topology is the simplest among all the observed Fermi surfaces in iron-based superconductors (figure 11); it is also dramatically different from the calculated Fermi surface of bulk FeSe (figure 11(d)).

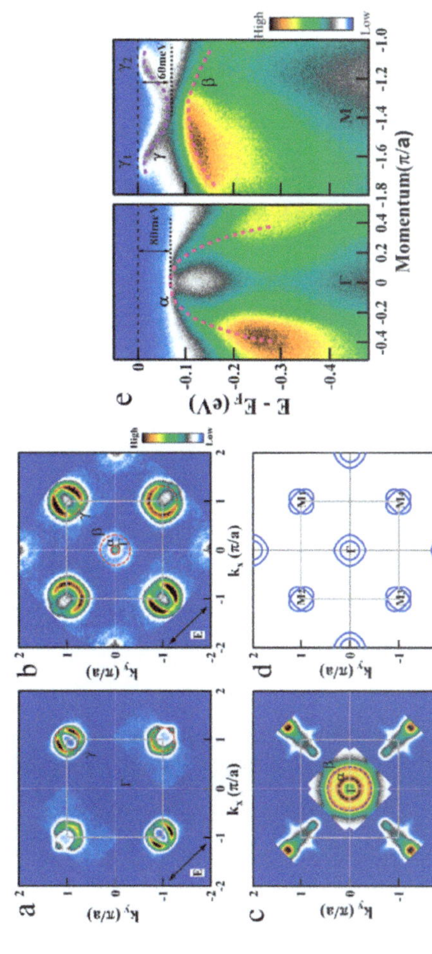

Figure 11. Distinct Fermi surface and band structures of the superconducting single-layer FeSe/$SrTiO_3$ film. (a) Fermi surface mapping of the single-layer FeSe/$SrTiO_3$ film measured at 20 K. (b) Fermi-surface mapping of (Tl,Rb)xFe_2_ySe_2 superconductor (T_c = 32 K) [87]. (c) Fermi-surface mapping of $Ba_{0.6}K_{0.4}Fe_2As_2$ superconductor (T_c = 35

K) [34]. (d) Fermi surface of β-FeSe by band-structure calculations for kz = 0 (blue thick lines) [80]. For convenience, the four equivalent M points are labelled as M1(π, π), M2(−π, π), M3(−π, −π) and M4(π, −π). (e) Band structure along the cut crossing the Γ point (left panel) and along the cut crossing the M3 point (right panel). The pink dashed line in the left panel shows schematically a hole-like band near the Γ point with its top at 80 meV below the Fermi level. The purple dashed line in the right panel shows schematically an electron-like band with its bottom at 60 meV below the Fermi level. Reprinted from [8].

Considering two electron pockets around the M point due to two Fe sites in a unit cell [113, 114], the carrier concentration of this particular sample can be estimated by the area of the electron pocket; it corresponds to ~0.10 electrons per Fe. The band structure near the zone center (left panel of figure 11(e)) is dominated by a hole-like band with its top about 80 meV below the Fermi level. The band structure across the zone corner (M point) consists of an electron-like band with its bottom about 60 meV below the Fermi level. An effective mass of ~$3m_e$ can be estimated for this particular electron-like band [8].

ARPES also provides an alternative approach to examine whether the single-layer FeSe/$SrTiO_3$is superconducting or not and its superconducting transition temperature by directly measuring the possible superconducting gap opening and its temperature dependence [8]. The temperature-dependent photoemission spectra are shown in figure 12(a). Sharp peaks develop at low temperature and become gradually broadened with increasing temperature. The symmetrized photoemission spectra are shown in figure 12(b) in order to visually inspect possible gap openings and remove the effect of Fermi distribution function near the Fermi level. There is a clear gap opening at low temperature. The gap size decreases with increasing temperature and closes at 50–55 K. The variation of the gap size as a function of temperature (figure 12(e)) follows a standard BCS form. Momentum-dependent measurements (figures12(c) and (d)) indicate that the gap is nearly isotropic without any sign of zero gap (node) around the Fermi surface. Since the single-layer FeSe film is highly two-dimensional, it avoids complications from the three-dimensional Fermi surface. In this case, it is definitive to claim that the gap along the electron-like Fermi surface near M is s-wave like without nodes. The near-E_F band back-bending at low temperature (figure 12(g), right panel), the sharp coherence peaks at low temperature (figure 12(a)), the BCS-form-like temperature dependence of the gap (figure 12(e)) and the momentum dependence of the gap (figure 12(c)) provide strong evidence to indicate that the gap observed here is a superconducting gap.

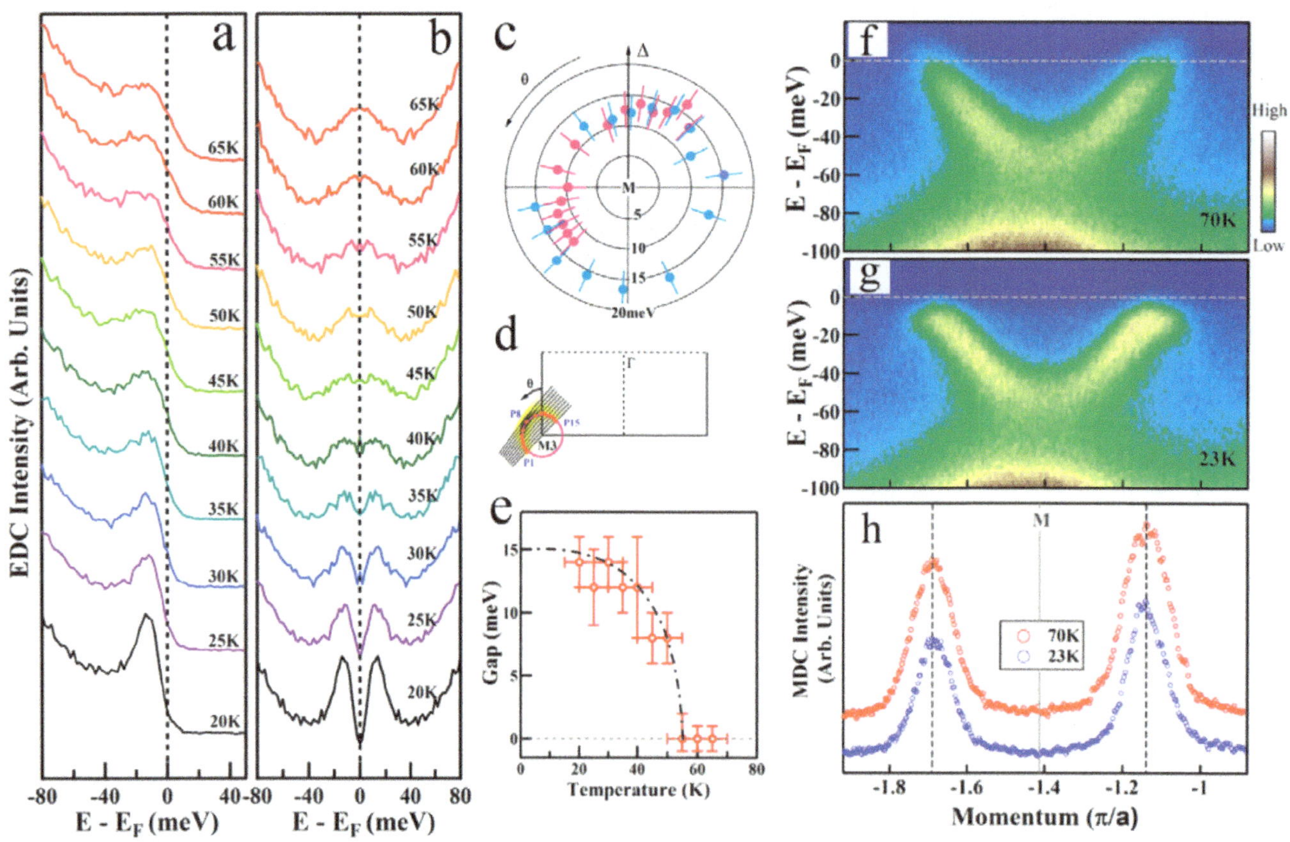

Figure 12. Temperature and momentum dependence of the superconducting gap in the superconducting single-layer FeSe/$SrTiO_3$ film. (a) Photoemission spectra (EDCs) at the Fermi crossings of the electron-like γ Fermi surface near M and their corresponding symmetrized spectra (b) measured at different temperatures. (c) Momentum dependence of the superconducting gap along the γ Fermi surface. (d) The Fermi surface mapping near M3 and the corresponding Fermi crossings. The violet circle represents the Fermi surface. The red solid circles represent measured Fermi momenta that are labeled as P1 to P15. (e) Temperature dependence of the superconducting gap. Reprinted from [8]. (f) and (g) Band structure along Γ-M cut measured at 70 and 23 K, respectively. The corresponding MDCs (momentum distribution curves) at the Fermi level for the two measurements are shown in (h). The two MDC peaks show little change in their positions above and below the gap opening temperature. Reprinted from [9].

In the electronic phase diagram of both the high temperature cuprate superconductors and the iron-based superconductors [115, 116], it is clear that the physical properties depend strongly on the carrier concentration, and superconductivity can be optimized at a particular carrier concentration. It is tempting to ask whether one can establish a similar phase diagram for the single-layer FeSe/$SrTiO_3$ system and whether it is possible to optimize its superconductivity. Such an idea is motivated by the MBE growth process where the single-layer FeSe/$SrTiO_3$ film is first grown at a relatively low temperature which is non-superconducting; then it becomes superconducting only after the film is annealed in vacuum at a relatively high temperature [7,117, 118]. To keep track of how the evolution from the as-grown non-superconducting single-layer FeSe/$SrTiO_3$ film to the vacuum-annealed superconducting film occurs, extensive ARPES measurements were carried out on the as-grown single-layer FeSe/$SrTiO_3$ film that was vacuum-annealed in situ by many sequences at different temperatures and different times [9]. The main results are summarized in figure 13. For the as-grown non-superconducting single-layer FeSe/$SrTiO_3$ film, its electronic structure (figures 13(e)–(h)) is distinct from that of the superconducting films (figures 13(a)–(d)). In this case, the underlying Fermi surface shows four strong spots around the M point (figure 13(e)). A hole-like band is present near the Γ point but its top is closer to the Fermi level compared to that in the superconducting film. A pronounced hole-like band is observed along the Γ-M direction near the M point that is totally different from the electron-like band in the superconducting film. The distinct electronic structures between the as-grown non-superconducting films and the superconducting films indicate there are two different phases present in the single-layer FeSe/$SrTiO_3$ films (figures 13(a)–(h)); for convenience, the phase with the electronic structure similar to the as-grown films is called the 'N' phase while the one with the electronic structure similar to the superconducting FeSe films is called the 'S' phase [9]. It turned out that the evolution proceeds in three stages during the vacuum-annealing process [9]. For the as-grown non-superconducting film, after initial first-stage mild annealing, it stays in the pure N phase (left pink region in figure 13(i)). Further annealing leads to the emergence of the S phase which increases with the annealing process at the expense of the N phase. This gives rise to the second stage of the mixed phase region (middle light blue region in figure 13(i)). After extensive annealing at sufficiently high temperature and long time, the film can convert to pure S phase in the final third stage (right bright green region in figure 13(i)).

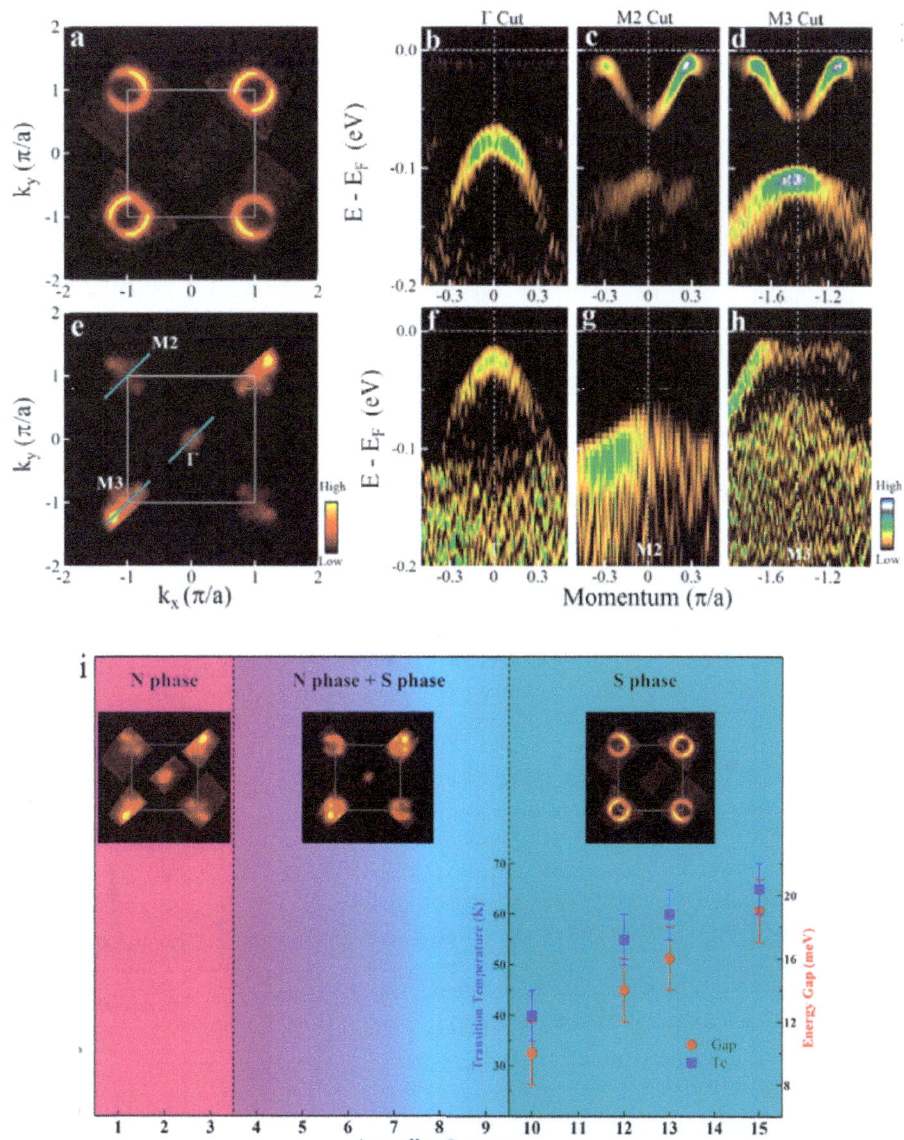

Figure 13. Electronic structure of the S phase and the N phase and phase diagram of the single-layer FeSe/$SrTiO_3$ film with vacuum annealing. (a) Fermi surface of the S phase of the single-layer FeSe/$SrTiO_3$ film. (b)–(d) Corresponding band structure of the S phase along the Γ, M2 and M3 cuts, respectively. (e) Fermi surface of the N phase of the single-layer FeSe/$SrTiO_3$ film. (f)–(h) Corresponding band structure of

the annealed single-layer $FeSe/SrTiO_3$ film along the Γ, M2 and M3 cuts, respectively. The band structures shown in (b)–(d) and (f)–(h) are second derivatives of the original band with respect to energy. (i) Schematic phase diagram of the single-layer FeSe film during the annealing process. The inset shows the Fermi surface corresponding to each stage. Reprinted from [9].

The annealing process also made it possible to tune the superconductivity of the single-layer $FeSe/SrTiO_3$ film by varying the carrier concentration [9]. The carrier concentration of the S phase during the annealing process can be obtained by calculating the area of the electron-like Fermi surface near M. It has been shown that during the vacuum annealing process, the S phase gets more and more electron-doped, with the electron concentration increasing from 0.07 to 0.12 electrons per Fe [9]. The corresponding gap measurements indicate that the gap size and the gap closing temperature increase with the annealing process (figure 13(i)). A gap of ~19 meV and a gap closing temperature of ~65 K were achieved for the annealed single-layer $FeSe/SrTiO_3$ film (figure 13(i)) [9]. This provides the first electronic evidence of superconductivity at ~65 K in the single-layer $FeSe/SrTiO_3$ films that was also confirmed by other groups [10, 114]. Such a gap increase with vacuum annealing was also observed in the STM/STS measurements [117].

Since the S phase appears in the early stage of vacuum annealing of the single-layer $FeSe/SrTiO_3$films which coexists with the N phase [9], it is intriguing to ask whether the S phase becomes superconducting as long as it appears. To address this issue, the evolution of the electronic structure and the energy gap as a function of the carrier concentration is studied for the S phase of the single-layer FeSe films [119]. It was found that, at low temperature, there is a gap opening along the underlying electron-like Fermi surface near M at low carrier concentration, as seen from the symmetrized photoemission spectra around the Fermi surface (figure 14(b)). As the carrier concentration of the S phase increases, the gap size decreases and reaches zero at a critical carrier concentration around 0.089 electrons/Fe (figures 14(a) and (b)). Further increase of the carrier concentration results in another gap opening at low temperature. From the appearance of the coherence peak and the gap closing with increasing temperature (figure14(d)), this high carrier concentration gap is apparently the superconducting gap we have discussed before. The low carrier concentration gap shows quite different behaviors from the high carrier concentration superconducting gap. The broad photoemission peaks, weak temperature dependence of the gap (figure 14(c)) and its relatively large size compared to the superconducting gap are similar to those observed in insulating heavily-underdoped cuprates [120]. We believe the low carrier concentration energy gap is consistent with an insulating gap. Therefore, there is an insulator–superconductor transition across the critical carrier concentration of 0.089 in the S phase of the single-layer $FeSe/SrTiO_3$ films [119]. It was noted that such a transition exhibits many similar behaviors to those found in the heavily underdoped cuprates [120]. At present, the origin of this insulator–superconductor transition remains to be understood, which may be associated with the two-dimensionality that enhances electron localization or correlation. In a multi-orbital system like the iron-based compounds, the carrier density-induced Mott transition may be realized in an orbital-selective fashion [121]. Such an orbital-selective Mott transition has been examined in $AxFe_{2-y}Se_2$ superconductor [122]. In the S phase of single-layer $FeSe/SrTiO_3$ films, it is possible that the same mechanism is in operation for the observed carrier concentration-induced insulator–superconductor transition [119].

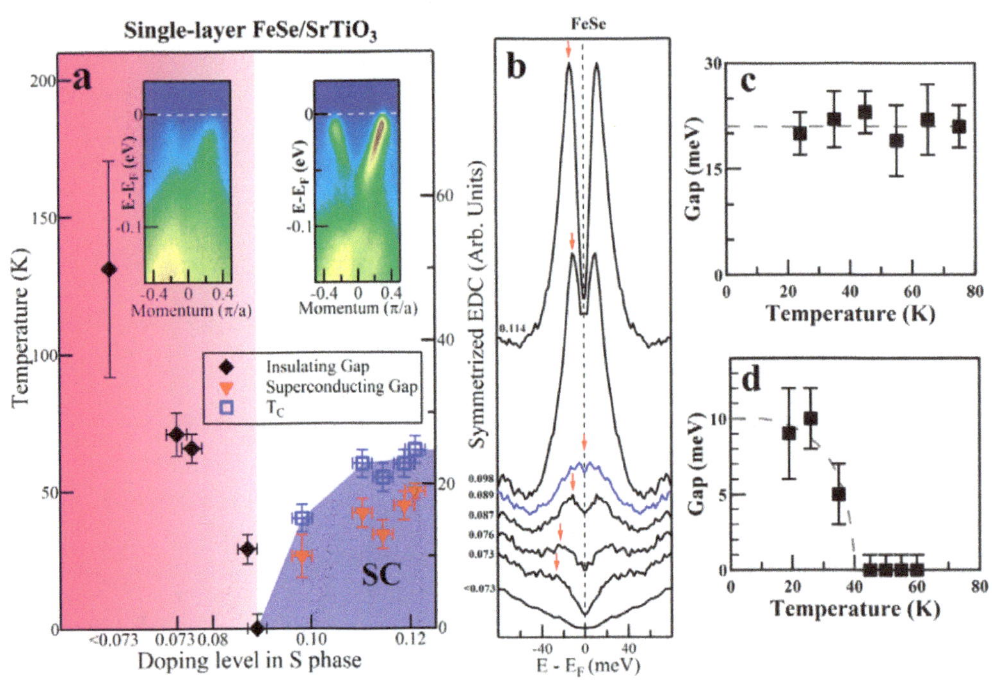

Figure 14. Evidence of an insulator–superconductor transition in the single-layer $FeSe/SrTiO_3$ film. (a) Phase diagram of the S phase in the single-layer $FeSe/SrTiO_3$ film that shows the decrease of the insulating energy gap (black solid diamond) with increasing doping on the low doping side and the increase of the superconducting gap (red solid triangle) and the corresponding superconducting transtion temperature T_c(blue empty square) with increasing doping on the high doping side. There is an insulator–superconductor transition near ~0.09 doping level. (b) Doping-evolution of the symmetrized EDCs at a Fermi momentum k_F along the cut crossing the M2 point. (c) and (d) The variation of the energy gap at different temperatures at low doping level and high doping level, respectively. Reprinted from [119].

The electronic structure and superconductivity of the $FeSe/SrTiO_3$ films show drastically different layer-dependence from that of the FeSe/graphene films (figure 8) [103], as well as FeSe films grown on the MgO substrate [123]. For $FeSe/SrTiO_3$ films, while single-layer $FeSe/SrTiO_3$ film can become superconducting, STM/STS does not find the signature of superconductivity in the second and multiple-layer $FeSe/SrTiO_3$ films regardless of how the samples are annealed [7,118]. Such a dramatic difference between the single-layer, double-layer and multiple-layer $FeSe/SrTiO_3$ films is surprising if one considers the usual layer-dependence found in FeSe/graphene films [103]. To address this issue, a comparative investigation between single-layer and double-layer $FeSe/SrTiO_3$ films is carried out by performing a systematic angle-resolved photoemission study on the samples annealed in vacuum [124]. The as-prepared double-layer $FeSe/SrTiO_3$ film exhibits an electronic structure that is similar to the N phase of the single-layer $FeSe/SrTiO_3$ film (figure 15). The double-layer $FeSe/SrTiO_3$ film shows a quite different doping behavior from the single-layer film. It is hard to get doped and it remains in the semiconducting/insulating state even under an extensive annealing condition. However, after sufficient vacuum annealing, the double-layer $FeSe/SrTiO_3$ film may follow a similar doping trend as in the single-layer film although superconductivity has not been realized. Such a behaviour is understood as originating from the much reduced doping efficiency in the bottom FeSe layer of the double-layer $FeSe/SrTiO_3$ film from the FeSe-$SrTiO_3$ interface when the dominant doping process comes from electron transfer from the $SrTiO_3$ surface [124].

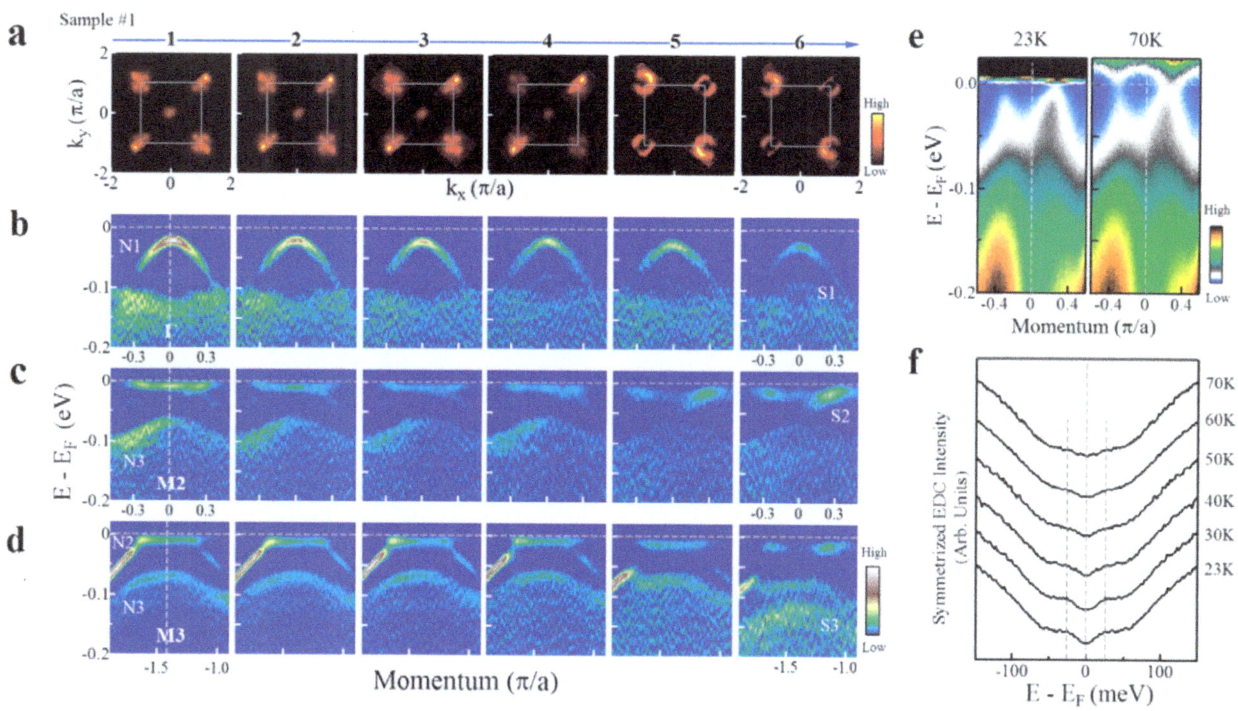

Figure 15. Fermi surface and band structure evolution of the double-layer $FeSe/SrTiO_3$ film annealed at a constant annealing temperature of 350 °C in vacuum for different times. (a) Fermi surface and band structure evolution of the double-layer $FeSe/SrTiO_3$ film annealed at a constant annealing temperature of 350 °C in vacuum for different times. The sequences 1–6 correspond to accumulative times of 15, 30.5, 46.5, 66.5, 87.5 and 92.5 h, respectively. (a) Fermi surface evolution as a function of the annealing time. (b)–(d) Band structure evolution with annealing time for the momentum cuts across Γ(b), M2 (c) and M3 (d). (e) Band structure of the annealed double-layer $FeSe/SrTiO_3$ film (corresponding to the sequence 6) measured at 23 K (left panel) and 70 K (right panel). The photoemission images are divided by the corresponding Fermi distribution function to highlight opening or closing of an energy gap. (f) Corresponding symmetrized EDCs on the Fermi momentum measured at different temperatures. Reprinted from [124].

Detailed ARPES studies on the layer-dependence of the electronic structure for the $FeSe/SrTiO_3$ films have been reported in [10, 114] and typical results are shown in figure 16 [10]. It is found that the electronic structure of the $FeSe/SrTiO_3$ films becomes similar when the number of layers is above two. In this case, the Fermi surface is quite different from the S phase of the superconducting single-layer $FeSe/SrTiO_3$ film in that the Fermi surface of the multi-layer $FeSe/SrTiO_3$ films consists of four strong spots near the M point and there are bands crossing the Fermi level around the Γ point (figure 16(a)). It was also found that the multi-layer $FeSe/SrTiO_3$ films experience a transition at a high temperature where the electronic structure exhibits a change above and below the transition temperature [10]. The transition temperature increases as the number of FeSe layers decreases (figure 16(b)). Such a transition is attributed to the spin-density-wave formation and the layer-dependent transition temperature is associated with the strain difference in samples with different layers [10].

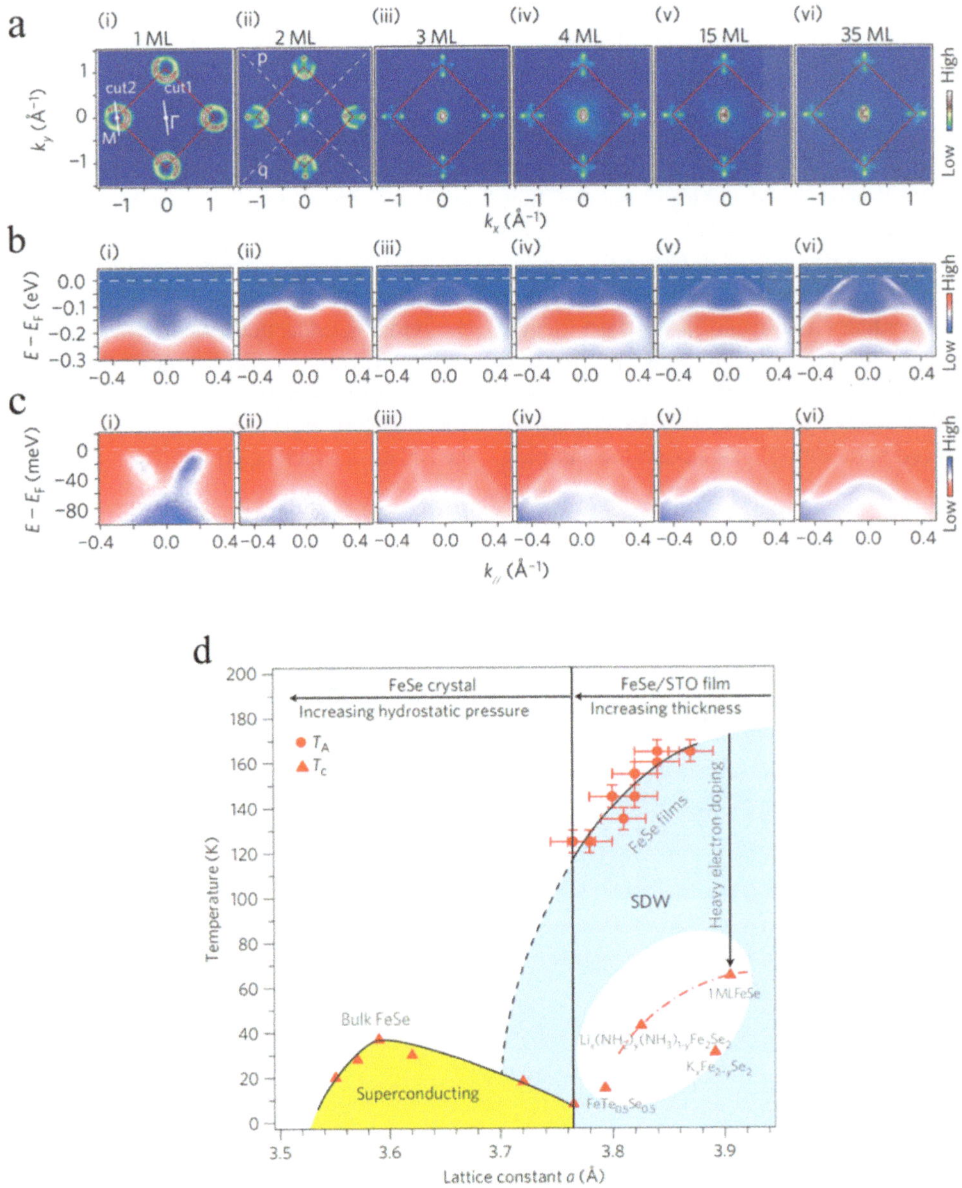

Figure 16. Layer-dependent Fermi surface and band structure in the FeSe/$SrTiO_3$ films.(a) The thickness dependence of the Fermi surface as represented by the photoemission intensity map at the Fermi energy at 30 K. (b) and (c) The thickness dependence of band structure along the cut crossing the Γ point and M point respectively. (d) Phase diagram of the FeSe-related systems. Values of T_c and T_A for FeSe are plotted against the lattice constant. The dashed line represents the extrapolated values of T_A, suggesting the existence of spin-density-wave (SDW) order in bulk FeSe under pressure. Values of T_c for other iron selenides are also plotted in the elliptical region. Reprinted from [10].

We note that the N phase of the single-layer FeSe/$SrTiO_3$ film is not observed in [10, 114]. Also there are discrepancies in the electronic structure of the double-layer FeSe/$SrTiO_3$ films between different groups [10, 114, 124]. The difference can be attributed to the different preparation conditions used and possible mixing of the single-layer and double-layer FeSe/$SrTiO_3$ films. In our case, we have clearly observed a pure N phase in both single-layer and double-layer FeSe/$SrTiO_3$ films [9, 124]. In fact, the electronic structure of the N phase in the single-layer and double-layer FeSe/$SrTiO_3$ films is rather similar to that found in multi-layer films [10, 114] except for some qualitative band position difference. The electronic structure of the N phase in the single-layer and double-layer FeSe/$SrTiO_3$ films also experiences a transition at high temperature (figure 17), the transition temperature in single-layer FeSe/$SrTiO_3$ is the highest among all the FeSe/$SrTiO_3$ films and basically follows the trend found for multi-layer FeSe/$SrTiO_3$ films [10]. Such a transition is reminiscent of the electronic structure change in the parent compound $BaFe_2As_2$ across the structural/magnetic transition temperature (~140 K for $BaFe_2As_2$) (figures 17(e) and (f)) [35]. In bulk FeSe, there is a structural transition near 90 K [46]. Recently, an electronic structure change was also reported in the ARPES study of bulk FeSe crystals across this structural transtion [67, 68]. Since there is no static magnetic transition found in bulk FeSe, whether the electronic structure transition in the FeSe/$SrTiO_3$ films is caused by spin-density-wave formation needs further investigation and direct magnetic measurements.

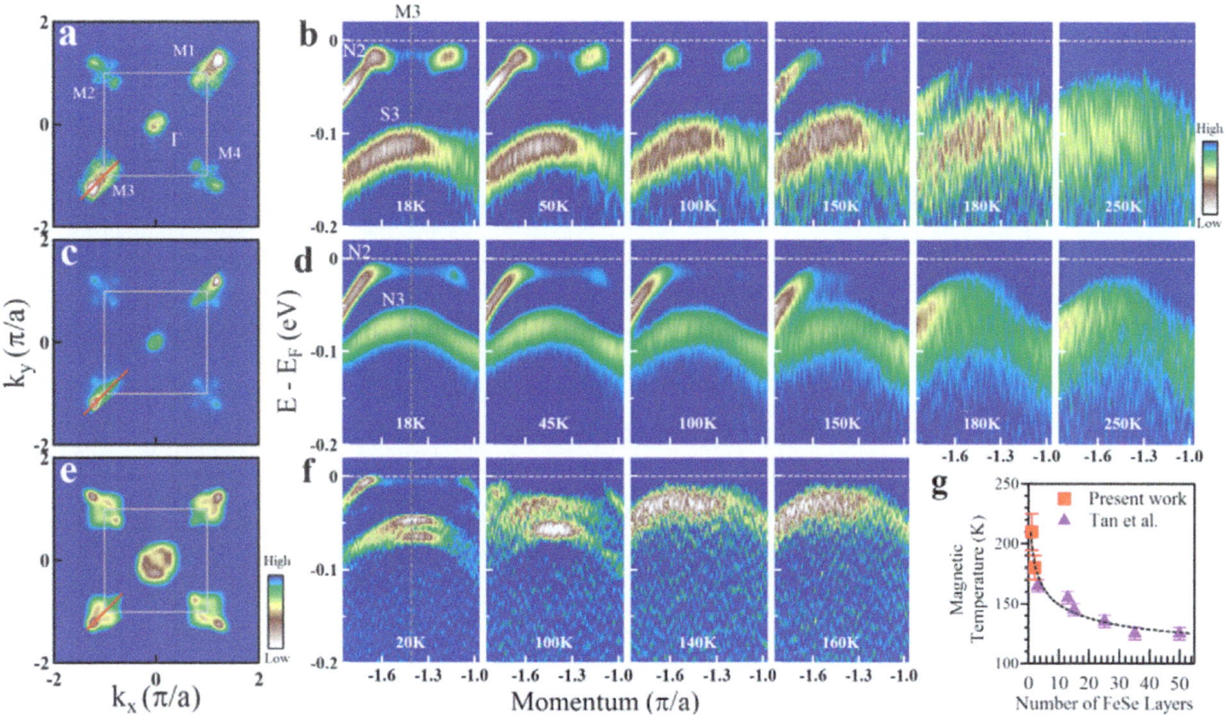

Figure 17. Temperature dependence of the band structure of the N phase in the single-layer and double-layer FeSe/$SrTiO_3$ films. (a), (c) and (e) are the Fermi surface of single-layer FeSe/$SrTiO_3$ film, double-layer FeSe/$SrTiO_3$ film and $BaFe_2As_2$, respectively. (b), (d) and (f) Temperature dependence of the band structure measured along the M3 cut for the single-layer FeSe/$SrTiO_3$ film, double-layer FeSe/$SrTiO_3$ film and $BaFe_2As_2$, respectively. (g) The transition temperature as a function of the number of FeSe layers in the FeSe/$SrTiO_3$ films. Here the transition temperature for the single-layer and double-layer FeSe films (red squares) is determined from the temperature dependence of the band structure where the hole-like bands begin to disappear. The temperature for other films (violet triangles) with more FeSe layers is taken from [10]. Reprinted from [124].

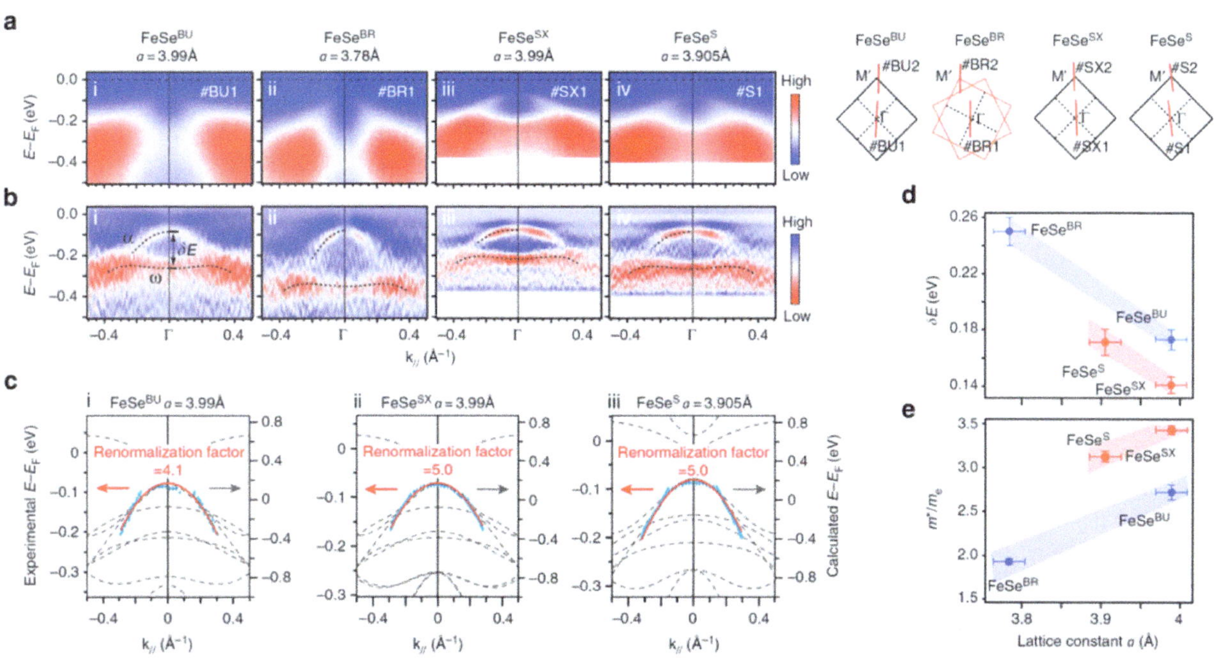

Figure 18. Band structure of the single-layer FeSe films grown on various substrates. (a), (b) The photoemission intensity along #BU1, #BR1, #SX1 and #S1 across Γ and the corresponding second derivative with respect to energy to highlight the dispersions for $FeSe^{BU}$, $FeSe^{BR}$, $FeSe^{SX}$ and $FeSe^{S}$, respectively. For the definition of different substrates, refer to [127]. (c) Comparison of the dispersion of band α from ARPES data and the DFT calculated band structures along the Γ-M' direction. (d) The energy separations between the α and ω bands at Γ versus a. (e) Effective mass m^* of band α as a function of in-plane lattice constant a, where m_e is the free-electron mass. Reprinted from [127].

The dramatic difference of superconductivity between the FeSe/$SrTiO_3$ films and the FeSe/graphene films points to the key role played by the growth substrate. Furthermore, as the superconductivity of the FeSe superconductor is found to be sensitive to pressure in the bulk form [38] or strain in the thin film form [125, 126], it is natural to ask whether the strain may play the key role in enhancing superconductivity in the FeSe/$SrTiO_3$ films. As the bulk FeSe has a lattice constant of 3.76 Å and $SrTiO_3$ has a lattice constant of 3.90 Å, when FeSe films are epitaxially grown on the $SrTiO_3$ substrate, the FeSe films will experience a tensile stress from the underlying $SrTiO_3$ substrate [7]. To check on the effect of strain on the superconductivity of the FeSe films, different substrates with disparate lattice constants have been used to grow FeSe films [113, 127]. With the substrate varying from $SrTiO_3$ to $SrTiO_3$/$KTaO_3$ ($SrTiO_3$ buffer-layers grown on $KTaO_3$ substrate as the substrate for FeSe films), to $BaTiO_3$/$KTaO_3$ ($BaTiO_3$buffer layers grown on $KTaO_3$ substrate as the substrate for FeSe films), there are changes in the corresponding electronic structures of the FeSe films (figures 18(a) and (b)), but the gap closing temperature shows little change within the range of (70 ± 5) K [113, 127]. This seems to indicate that the tensile stress on the FeSe films exerted from the underlying substrate does not affect superconductivity of the FeSe films significantly.

The latest episode sees the observation of a replica band in the band structure of the superconducting single-layer FeSe/$SrTiO_3$ films (figure 19) [114]. As shown in figure 19, the main hole-like band near the Γ point (band d in figure 19(b)), the electron-like band (band a in figure 19(c)) and hole-like band (band b in figure 19(c)) all produce weak 'replica' bands that are ~100 meV below the original bands. Such replica bands persist above the gap opening superconducting temperature (figures 19(d)–(g)). The replica bands appear weak in other measurements [8–10] possibly because of the photoemission matrix element effect caused by different measurement conditions. These replica bands have been understood as due to the presence of bosonic modes, most probably optical phonons in $SrTiO_3$, that couple to the FeSe electrons with only a small momentum transfer [114]. It was also suggested that such interfacial coupling could assist superconductivity in most channels and this coupling is responsible for the T_c enhancement in the single-layer FeSe/$SrTiO_3$ films [114].

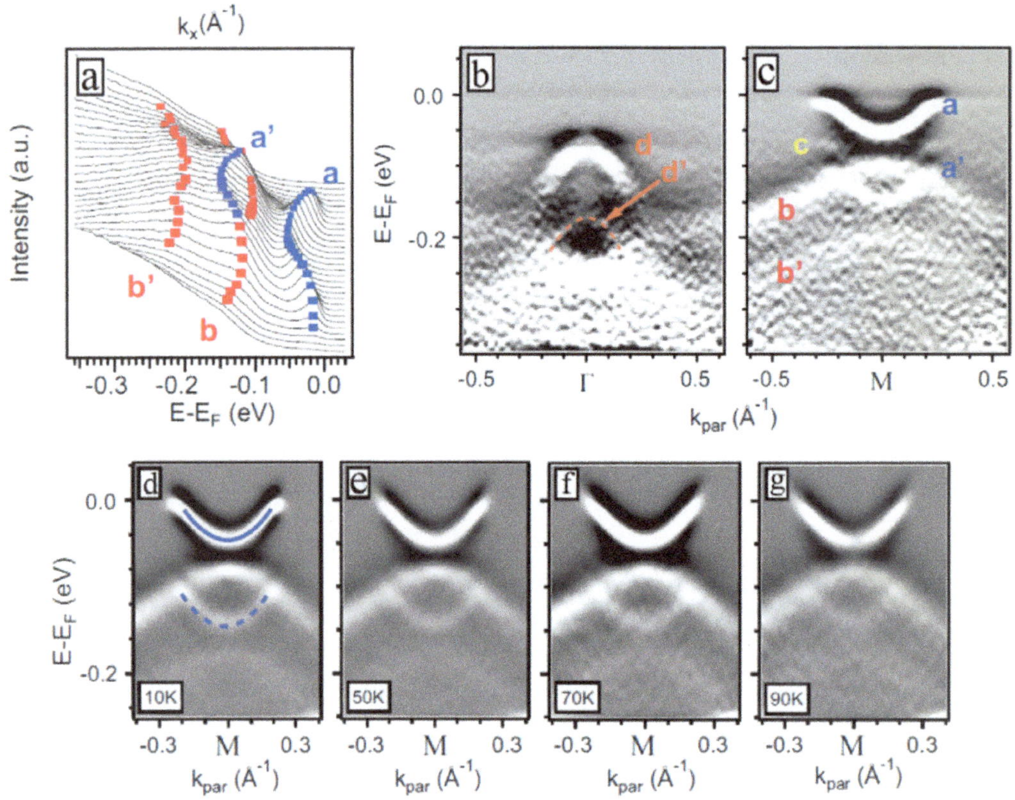

Figure 19. Replica bands observed in the single-layer FeSe/$SrTiO_3$ films and their temperature dependence. (a) EDCs at M shown as a waterfall plot, with markers indicating band peaks. (b) and (c) Second derivatives in energy of the high symmetry cuts from Γ cut and M cut. An additional weaker replica, labelled c, can now be seen at M in (c), sitting 50 meV below a, and at the Γ point in (b) we see the hole band and a corresponding replica, labelled d and d', respectively. (d)–(g) Temperature dependence of the replica bands, which persist at temperatures higher than the gap-opening temperature. Reprinted from [114].

Transport and Magnetic Measurements of FeSe/$SrTiO_3$ Films

Although ARPES and STM/STS have observed gap opening and the behaviors of the gap are consistent with the superconducting gap, such as the sharp coherence peak, the BCS form of the gap size as a function of temperature and the particle–hole symmetry around the Fermi surface [7–10], direct evidence from the transport and magnetic measurements is necessary to establish whether the FeSe/$SrTiO_3$ films are truly superconducting or not. Such measurements on ultra-thin films are challenging because of the complication of the sample deterioration in air, $SrTiO_3$substrate conductance and very weak magnetic signal. Initial transport measurement on a five-layer FeSe/$SrTiO_3$ film covered by a 20 nm thick amorphous silicon protection layer saw a transition with an onset temperature of ~53 K (figure 20(a)) [7]. Later on, the single-layer FeSe/$SrTiO_3$ film was better protected by covering with a 10-layer FeTe film followed by a 30 nm thick amorphous silicon layer so that the sample could be stable at ambient condition for a relatively long time. The resistance as a function of temperature clearly reveals the appearance of superconductivity. The resistance begins to drop at 54.5 K and reaches zero at 23.5 K with an onset T_c of 40.2 K (figure 20(b)) [128]. Magnetic measurement on the same sample reveals a diamagnetic response with an abrupt change of both in-phase and out-of-phase signals at 21 K (figure 20(d)) that is consistent with the zero-resistance critical temperature from the transport measurement (figure 20(b)). This has unambiguously demonstrated that the single-layer $FeSe/SrTiO_3$ film under measurement is superconducting although the measured T_c is not as high as that expected from the STM/ARPES measurements. This is understandable because it is known that the superconducting transition temperature of the $FeSe/SrTiO_3$ films is sensitive to the carrier concentration [9]. Also the protection layer may affect the superconductivity of the single-layer $FeSe/SrTiO_3$ film; in particular, FeTe is known to be magnetic at low temperature, which may suppress the superconductivity of the adjacent FeSe film. Superconductivity is also reported in double-layer $FeSe/SrTiO_3$ films from the transport measurements [129]. Hints of higher transition temperatures are spotted in some $FeSe/SrTiO_3$ films [130] that need further investigations.

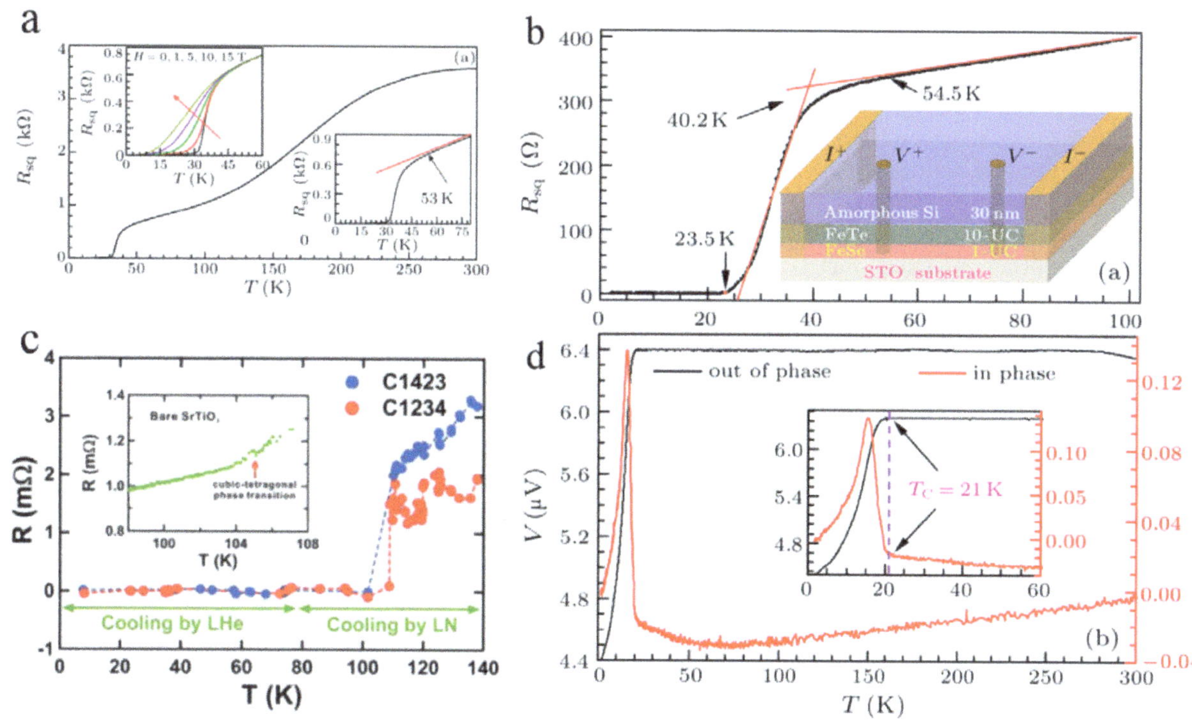

Figure 20. Transport and magnetic measurements of the superconductivity in the $FeSe/SrTiO_3$ films. (a) Temperature dependence of square resistivity (Rsq) of a 5-layer $FeSe/SrTiO_3$ film from 0 to 300 K. Upper inset: Rsq–T curves at various magnetic fields along the c-axis. Lower inset: the Rsq–T curve from 0 to 80 K [7]. (b) The temperature dependence of resistance under zero field, showing $T_c^{onset} = 40.2K$ and $T_c^{zero} = 23.5K$. Inset: a schematic structure for the transport measurements in the heterostructure of 30 nm amorphous Si/(10-layer)FeTe/(1-layer)$FeSe/SrTiO_3$ [128]. (c) Temperature dependence of the resistance in situ measured on a single-layer $FeSe/SrTiO_3$ film [131]. The inset shows the temperature dependence of resistance taken on a bare $SrTiO_3$ surface. (d) The diamagnetic response measured by a two-coil mutual inductance system. Inset: the data near the superconducting transition in an amplified view showing the formation of diamagnetic screening at 21 K [128].

It is ideal to carry out in situ transport and magnetic measurements on the $FeSe/SrTiO_3$ films in order to remove the effect of sample damage and protection layers. The in situ transport measurements have been performed lately on the single-layer $FeSe/SrTiO_3$ film [131]. Surprisingly, the measured zero-resistance temperature reaches as high as 109 K (figure 20(c)), which is well above the $T_c \sim 80$ K expected from the 20.1 meV gap from the STM/STS measurements [7] and the $T_c \sim 65$ K from ARPES measurements [9, 10]. Further measurements are needed to reproduce and confirm this exciting result.

IMPLICATIONS AND THEORETICAL UNDERSTANDING

Implications

The discovery of high temperature superconductivity in single-layer $FeSe/SrTiO_3$ films has important implications in understanding the superconductivity mechanism and the pairing symmetry in iron-based superconductors. In most iron-based superconductors, the Fermi surface consists of hole-like Fermi surface sheets near the Brillouin zone center (Γ point) and electron-like Fermi surface sheets near the zone corner M(π, π) point (figure 2) [132]. It has been proposed that the electron scattering between the hole-like bands around Γ and electron-like bands near M is responsible for electron pairing, and superconductivity of the iron-based superconductors has a dominant s± superconducting order parameter [93–99]. The absence of the Fermi surface near Γ in the superconducting single-layer $FeSe/SrTiO_3$ film [8], together with the fact that only electron-like Fermi surface sheets are observed in the intercalated $AxFe_{2-y}Se_2$ superconductors [44, 86, 87], rules out the necessity of the electron scattering between the hole-like bands near Γ and electron-like bands near M in this ‹Fermi nesting› picture.

The observation of the nearly isotropic superconducting gap around the electron-like Fermi surface near the M point (figure 12(c)) [8] also provides crucial information in understanding the pairing symmetry of iron-based superconductors. With the absence of a hole-like Fermi surface near Γ, electrons can only scatter across the Fermi surface sheets between M points, which is predicted to result in a d-wave superconducting gap [93, 100–102]. The dxy-like superconducting order parameter is obviously ruled out because it will produce zero gap on the electron-like Fermi surface near M. With an alternative $d_{x^2-y^2}$ -like superconducting order parameter that can avoid the direct gap nodes, it is argued that two Fermi surface sheets around a given M point with opposite phases can also give rise to gap nodes around the Fermi surface [133]. In this case, it remains inconsistent with the observation of a nodeless gap on the electron-like Fermi surface near the M point. The superconducting pairing symmetry in the single-layer $FeSe/SrTiO_3$ films and the $AxFe_{2-y}Se_2$ system remains puzzling and requires further experimental and theoretical efforts.

The different behaviors between the FeSe/graphene and $FeSe/SrTiO_3$ films clearly indicate the critical role played by the interface in giving rise to the high temperature superconductivity. The discovery of high temperature superconductivity in the single-layer $FeSe/SrTiO_3$ films represents the first clear case of interface-enhanced superconductivity. Interface (and surface) superconductivity has a long history which is expected to increase superconducting temperature [134]. Superconductivity has been reported in various interface systems, including those between two insulators [135], between one insulator and one metal [136] and between one metal and one semiconductor [137]. However, in all these cases, the observed T_c in the interface is not obviously higher than the maximum T_c of its individual constituents. In the single-layer $FeSe/SrTiO_3$ films, its superconducting transition temperature is much higher than the maximum T_c of either the FeSe system or the $SrTiO_3$ system, thus establishing a convincing case for interface superconductivity. This discovery proves the concept of interface-enhancement of superconductivity and opens a door to further explore for new superconductors with higher T_c.

Theoretical Understanding

$FeSe/SrTiO_3$ films have attracted much theoretical attention, in regard to understanding their unique electronic structure, carrier doping mechanism and the possible origin of high temperature superconductivity [138–146]. By using first-principles calculations, the atomic and electronic structures of the single-layer and double-layer $FeSe/SrTiO_3$ films were studied [138]. The band structure of free single-layer FeSe film shown (figure 21(a)) is similar to that of bulk FeSe (figure 4) [62]. It is found that both single-layer and double-layer FeSe films on the $SrTiO_3$substrate behave like a slightly doped semiconductor with a collinear antiferromagnetic order on the Fe ions and a Dirac cone-like band around the Γ point (figure 21(b)) [138]. These results differ from the measured results, for both the N phase and S phase of the single-layer and double-layer $FeSe/SrTiO_3$ films [8–10, 124]. It was further shown that the $SrTiO_3$ substrate-induced tensile strain tends to stabilize the collinear antiferromagnetic state in FeSe thin film by enhancing the next-nearest-neighbor super-exchange antiferromagnetic interaction bridged through the Se atoms [142]. The calculated Fermi surface, considering the collinear antiferromagnetic order and electron doping, can be consistent with experimental observations that the hole-like Fermi surface sheets near Γ disappear [142]. In the presence of constrained magnetization, it was also shown that the observed Fermi surface can be described by the checkerboard antiferromagnetic spin pattern [140, 141]. In this case, it was found that the system has a considerable charge transfer from the $SrTiO_3$ substrate to the FeSe layer and so has a self-constructed electric field. The disappearance of the Fermi surface around the Γ point can also be explained by the antiferromagnetic checkerboard phase with charge doping and electric field effects [141]. These results show considerable consistency with the behavior of the N phase of the $FeSe/SrTiO_3$ film [9, 124]. So far, most theoretical calculations involve antiferromagnetic order and overall show some qualitative agreement with the experimental observations. However, we note that the ARPES results are only suggestive that the N phase of the $FeSe/SrTiO_3$ films may be magnetic [9] or the low temperature form of the multi-layer $FeSe/SrTiO_3$ films may be in the spin-density-wave state [10]. Further direct evidence is needed to pin down their magnetic nature.

It is clear from ARPES measurements that the superconducting S phase of the single-layer $FeSe/SrTiO_3$ films is electron-doped [8–10, 114]. The origin of the electron doping may have two possibilities: oxygen vacancy formation on the $SrTiO_3$ surface that gives rise to electron doping and the formation of Se vacancies in the FeSe film [8]. Further ARPES results substantiate the picture of oxygen vacancy formation on the $SrTiO_3$ surface [10, 124]. The role of the oxygen vacancy on the top layer of the $SrTiO_3$ substrate was also investigated by first principles calculations [145]. These show that the oxygen vacancies strongly bind the FeSe layer to the substrate, giving rise to a (2 × 1) reconstruction, and also serve as the source of electron doping. Theoretical investigation on the doping effect of Se vacancies in the $FeSe/SrTiO_3$ films leads to a somehow counter-intuitive result that, in terms of the Fe-3d bands, Se vacancies behave like hole dopants rather than electron dopants as usually expected [144]. Such results are considered to exclude Se vacancies as the origin of the large electron pockets measured by ARPES [144].

The most prominent issue concerning the superconductivity of single-layer $FeSe/SrTiO_3$ films is on the origin of their dramatic T_c enhancement. The discovery of high temperature superconductivity in single-layer $FeSe/SrTiO_3$ films was motivated by the idea that strong electron–phonon coupling can enhance superconductivity [7]. Subsequent study on the effects of screening due to the $SrTiO_3$ ferroelectric phonons on the Cooper pairing in FeSe shows that it can significantly enhance the energy scale of Cooper pairing and even change the pairing symmetry [139]. It was shown that the role of the

$SrTiO_3$ substrate in increasing the electron–phonon coupling and the resulting higher T_c is two-fold. First, the interaction of the FeSe and TiO_2 terminated face of $SrTiO_3$ prevents the single-layer FeSe from undergoing a shearing-type (orthorhombic) structural phase transition. Second, the substrate allows an antiferromagnetic ground state of FeSe which opens certain electron–phonon coupling channels within the single layer that are prevented by symmetry in the non-magnetic phase [143]. The replica bands observed in the superconducting single-layer FeSe/$SrTiO_3$ films provide experimental evidence on the coupling of electrons in the FeSe layer with the $SrTiO_3$ phonons [114]. Further theoretical analysis suggests that such interfacial electron–phonon coupling assists superconductivity in most channels and is responsible for raising the superconducting temperature in single-layer FeSe/$SrTiO_3$ films [114].

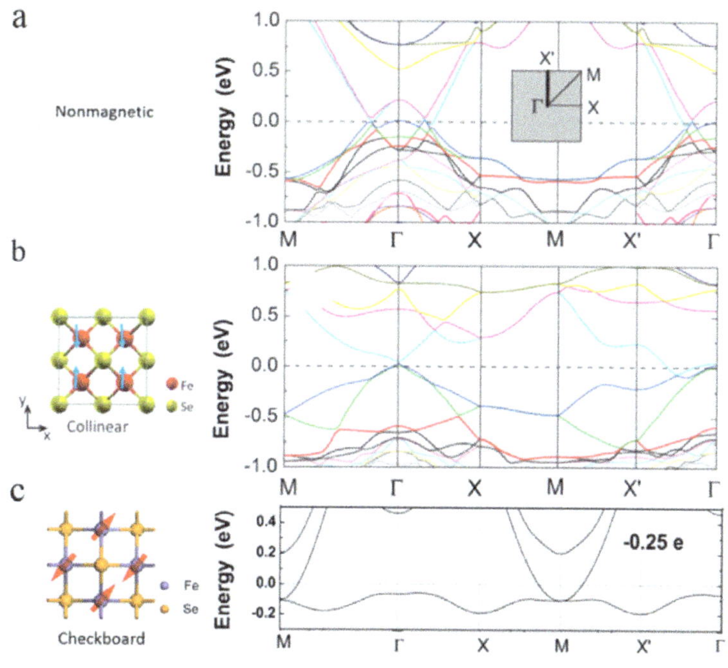

Figure 21. Band structure calculations of single-layer FeSe film with different magnetic orders. (a) and (b) The calculated band structure of single-layer FeSe/$SrTiO_3$ film in nonmagnetic state and collinear antiferromagnetic state respectively [138]. (c) The calculated band structure of single-layer FeSe film in checkerboard antiferromagnetic state [141].

SUMMARY AND PERSPECTIVE

In this brief review, we have shown that FeSe superconductors and related systems have exhibited rich and fascinating phenomena and physical properties. Since the report of possible high temperature superconductivity in single-layer FeSe/$SrTiO_3$ films [7], great progress has been made in understanding related issues both experimentally and theoretically. Some questions are answered and more questions are asked. Since this is a burgeoning and fast-evolving field, it is natural that many issues remain to be addressed. Here we briefly list some further issues that are either particular to FeSe/$SrTiO_3$ systems or more general to iron-based superconductors.

Interface Superconductivity?

So far many experiments point to interface superconductivity in single-layer FeSe/$SrTiO_3$ films. More decisive evidence is needed to pin down the nature of the interface superconductivity. What new effects can be expected from a system with interface superconductivity?

Doping Mechanism?

It has been shown that during the vacuum annealing process, the single-layer FeSe/$SrTiO_3$ films can be electron-doped. It has also been proposed that oxygen vacancies on the $SrTiO_3$ surface may play the role of electron doping [8–10]. It remains to be discovered how to directly prove the role of oxygen vacancies in electron-doping. Recent STM measurements on the vacuum annealed single-layer FeSe/$SrTiO_3$ film found that [117] there are some extra Se adatoms on the surface of the as-prepared FeSe film and the sample is insulating. In the initial stage of the annealing process, these extra Se adatoms are gradually removed. Further annealing leads to the formation of Se vacancies. What is the role of these Se adtoms and Se vacancies on the doping of FeSe films?

Role of Electron–Phonon Coupling?

There is experimental evidence to show the existence of electron–phonon coupling in FeSe/$SrTiO_3$ films [114]. There are theoretical arguments that such electron–phonon coupling can enhance superconductivity [7, 114, 139, 143]. However, more experimental evidence is needed to prove that such an electron–phonon coupling is the cause of the T_c enhancement in the single-layer FeSe/$SrTiO_3$ films.

Pairing Mechanism?

The observation of the distinct Fermi surface topology in the superconducting FeSe/$SrTiO_3$films [8] has ruled out the possibility of the Fermi surface nesting picture in giving rise to superconductivity in the single-layer FeSe/$SrTiO_3$ films. With only electron pockets present near the M point, what is the pairing mechanism then?

Pairing Order Parameter?

With only electron pockets present near M points in the single-layer $FeSe/SrTiO_3$ films [8–10,114], it is predicted that d-wave pairing may dominate [93, 100–102], which is not consistent with the experimental results of a nearly isotropic nodeless superconducting gap [8]. What can be the superconducting order parameter for the superconducting single-layer $FeSe/SrTiO_3$ films?

Theoretical Understanding of Electron Structure

The unique electronic structure of the N phase and S phase of the single-layer $FeSe/SrTiO_3$ film and the multi-layer $FeSe/SrTiO_3$ films poses a challenge to band structure calculations and theoretical understandings. More effort is needed to understand the electronic structure by including the effect of the substrate, vacancy and disorder, electron correlation, the spin-orbit coupling and so on.

Inconsistency of Superconductivity Measurements

There are inconsistencies among different measurements on the superconducting properties of $FeSe/SrTiO_3$ films. First, transport measurements on single-layer $FeSe/SrTiO_3$ films give different T_c, varying from 23.5 K (zero resistance [128]) to 109 K [131]. Second, there is an obvious difference of T_c on single-layer $FeSe/SrTiO_3$ films between transport measurements [7,128, 129, 131] and STM/ARPES measurements [7–10]. Third, for multi-layer $FeSe/SrTiO_3$ films, STM/STS and ARPES indicate they are not superconducting [7, 124]. However, transport measurements show signs of superconductivity in double-layer $FeSe/SrTiO_3$ film [129] and five-layer $FeSe/SrTiO_3$ film [7]. How to reconcile these differences?

Nature of the N Phase and S Phase in Single-Layer $FeSe/SrTiO_3$ Films

With vacuum annealing, two different phases (N phase and S phase) with distinct electronic structures are clearly observed [9]. It remains unclear about the real difference between the N phase and S phase, i.e. whether the electronic structure difference is caused by the difference in crystal structure, composition or the carrier concentration. It is also not clear how the N phase and S phase are distributed in real space.

Magnetic State?

ARPES has provided suggestions that the N phase of the single-layer $FeSe/SrTiO_3$ film may be magnetic [9] and the low-temperature form of the multi-layer $FeSe/SrTiO_3$ films may be in the spin-density-wave state [10]. Direct evidence is needed to pin down their magnetic nature.

Further Enhancement of Tc?

The superconducting transition temperature of $FeSe/SrTiO_3$ film is sensitive to carrier concentration [9]. T_c has not reached the maximum in the doping range studied; more electron doping may result in even higher T_c [9]. For double-layer or multi-layer $FeSe/SrTiO_3$ films, it has been found that the absence of superconductivity is related to insufficient electron doping [124]. It is interesting to explore, if sufficient electron doping is provided, whether double-layer or multi-layer $FeSe/SrTiO_3$ films can become superconducting or exhibit even higher T_c than the single-layer case. So far, $SrTiO_3$ and related $BaTiO_3$ substrates are proven to be magical in producing high temperature superconductivity in single-layer FeSe films. Are there other substrates present that can produce a similar effect and even higher T_c?

The surprising discovery of high temperature superconductivity in $FeSe/SrTiO_3$ films has provided a new platform for investigating interface superconductivity, the superconductivity mechanism of iron-based superconductors and opened a new path in exploring for new high temperature superconductors. Theoretical calculations suggest robust topological phases may be realized in single-layer $FeSe/SrTiO_3$ films, which is intertwined with superconductivity [146]. The ultra-thin nature of $FeSe/SrTiO_3$ films makes them ideal for making heterostructures [147]; the topological insulator/superconductor heterostructure has been proposed as a prototype to detect Majorana fermions [148]. The ideal two-dimensional character, dichotomy between single-layer and double-layer films and high T_c superconductivity open a door for potential device fabrication and applications. We have already witnessed surprises brought by single-layer FeSe films; we can expect more surprises to come in the near future.

ACKNOWLEDGMENTS

We thank W Zhang, Y-B Ou, F Li, C Tang, Q-Y Wang, Z Li, L Wang, X Chen, X Ma, Q Xue, J Hu, Q M Si and D Lee for collaborations and helpful discussions. XJZ thanks financial support from the NSFC (11190022,11334010 and 11374335) the MOST of China (973 program No: 2011CB921703 and 2011CBA00110) the Strategic Priority Research Program (B) of the Chinese Academy of Sciences (Grant No. XDB07020300).

REFERENCES

1. Kamihara Y et al 2008 J. Am. Chem. Soc. 130 3296
2. Bednorz J G et al 1986 Z. Phys. B 64 189
3. Chen X H et al 2008 Nature 453 761
4. Chen G F et al 2008 Phys. Rev. Lett. 100 247002
5. Ren Z A et al 2008 Chin. Phys. Lett. 25 2215
6. Wang C et al 2008 Europhys. Lett. 83 67006
7. Wang Q Y et al 2012 Chin. Phys. Lett. 29 037402
8. Liu D F et al 2012 Nat. Commun. 3 931
9. He S L et al 2013 Nat. Mater. 12 605
10. Tan S Y et al 2013 Nat. Mater. 12 6340
11. Ishida K et al 2009 J. Phys. Soc. Japan 78 062001
12. Paglinone J et al 2010 Nat. Phys. 6 645
13. Johnston D C et al 2010 Adv. Phys. 59 803
14. Stewart G R et al 2011 Rev. Mod. Phys. 83 1589
15. Wang F et al 2011 Science 332 200
16. Hirschfeld P J et al 2011 Rep. Prog. Phys. 74 124508
17. Mou D X et al 2011 Front. Phys. 6 410
18. Dai P C et al 2013 Nat. Phys. 8 709
19. Chubukov A V et al 2012 Annu. Rev. Condens. Matter.

Phys. 3 13
20. Dagotto E et al 2013 Rev. Mod. Phys. 85 849
21. Hsu F C et al 2008 Proc. Natl Acad. Sci. USA 105 14262
22. Wang X C et al 2008 Solid State Commun. 148 538
23. Rotter M et al 2008 Phys. Rev. Lett. 101 107006
24. Guo J et al 2010 Phys. Rev. B 82 180520
25. Fang M et al 2011 Europhys. Lett. 94 27009
26. Ma F J et al 2010 Front. Phys. China 5 150
27. Yildirim T 2008 Phys. Rev. Lett. 101 057010
28. Ma F J et al 2008 Phys. Rev. B 78 224517
29. Ma F J et al 2009 Phys. Rev. Lett. 102 177003
30. Chen F et al 2010 Phys. Rev. B 81 014526
31. Lubashevsky Y et al 2012 Nat. Phys. 8 309
32. Maletz J et al 2014 Phys. Rev. B 89 220506
33. Kordyuk A A 2011 Phys. Rev. B 83 134513
34. Zhao L et al 2008 Chin. Phys. Lett. 25 4402
35. Liu G D et al 2009 Phys. Rev. B 80 134519
36. Liu H Y et al 2008 Phys. Rev. B 78 184514
37. Liu H Y et al 2010 Phys. Rev. Lett. 105 027001
38. Medvedev S et al 2009 Nat. Mater. 8 630
39. Bao W et al 2009 Phys. Rev. Lett. 102 247001
40. Dong C et al 2011 Phys. Rev. B 84 224506
41. Katayama N et al 2010 J. Phys. Soc. Japan 79 113702
42. Wang H D et al 2011 Europhys. Lett. 93 47004
43. Ying T P et al 2012 Sci. Rep. 2 426
44. Zhang Y et al 2011 Nat. Mater. 10 273
45. Tian Q et al 2011 Phys. Rev. Lett. 106 187001
46. McQueen T M et al 2009 Phys. Rev. Lett. 103 057002
47. Mizuguchi Y et al 2010 Physica C 470 S338
48. de la Cruz C et al 2008 Nature 453 899
49. Zhao J et al 2008 Nat. Mater. 7 953
50. Chen Y et al 2008 Phys. Rev. B 78 064515
51. Zhao J et al 2008 Phys. Rev. B 78 132504
52. Zhao J et al 2008 Phys. Rev. B 78 140504
53. Goldman A I et al 2008 Phys. Rev. B 78 100506
54. Huang Q et al 2008 Phys. Rev. Lett. 101 257003
55. Bendele M et al 2014 J. Supercond. Nov. Magn. 27 965
56. Imai T et al 2009 Phys. Rev. Lett. 102 177005
57. Chen G F et al 2009 Phys. Rev. B 79 140509
58. Han Y et al 2010 Phys. Rev. Lett. 104 017003
59. Nie Y et al 2010 Phys. Rev. B 82 020508
60. Zheng M et al 2013 arXiv:1301.4696
61. Li S L et al 2009 Phys. Rev. B 79 054503
62. Subedi A 2008 Phys. Rev. B 78 134514
63. Lin J Y et al 2011 Phys. Rev. B 84 220507
64. Boehmer A E 2013 Phys. Rev. B 87 180505
65. Huynh K K et al 2014 Phys. Rev. B 90 144516
66. Ma M W et al 2014 Supercond. Sci. Technol. 27 122001
67. Nakayama K et al 2014 Phys. Rev. Lett. 113 237001
68. Shimojima T et al 2014 Phys. Rev. B 90 121111
69. Terashima T et al 2014 Phys. Rev. B 90 144517
70. Audouard A et al 2014 arXiv:1409.6003
71. Kasahara S et al 2014 Proc. Natl Acad. Sci. 111 16309
72. Xia Y et al 2009 Phys. Rev. Lett. 103 037002
73. Zhang Y et al 2010 Phys. Rev. B 82 165113
74. Liu Z K et al 2013 Phys. Rev. Lett. 110 037003
75. Miao H et al 2012 Phys. Rev. B 85 094506
76. Okazaki K et al 2012 Phys. Rev. Lett. 109 237011
77. Okazaki K et al 2014 Sci. Rep. 4 4109
78. Ieki E et al 2014 Phys. Rev. B 89 140506
79. Nakayama K et al 2010 Phys. Rev. Lett. 105 197001
80. Tamai A et al 2010 Phys. Rev. Lett. 104 097002
81. Yin Z P et al 2011 Nat. Mater. 10 932
82. Hanaguri T et al 2010 Science 328 474
83. A Krzton-Maziopa et al 2011 J. Phys.: Condens. Matter 23 052203
84. Li C-H et al 2011 Phys. Rev. B 83 184521
85. Burrard-Lucas M et al 2013 Nat. Mater. 12 15
86. Qian T et al 2011 Phys. Rev. Lett. 106 187001
87. Mou D X et al 2011 Phys. Rev. Lett. 106 107001
88. Zhao L et al 2011 Phys. Rev. B 83 140508
89. Wang X et al 2011 Europhys. Lett. 93 57001
90. Chen F et al 2011 Phys. Rev. X 1 021020
91. Xu M et al 2012 Phys. Rev. B 85 220504
92. Wang X P et al 2012 Europhys. Lett. 99 67001
93. Kuroki K et al 2008 Phys. Rev. Lett. 101 087004
94. Mazin I I et al 2008 Phys. Rev. Lett. 101 057003
95. Wang F et al 2009 Phys. Rev. Lett. 102 047005
96. Chubukov A V et al 2008 Phys. Rev. B 78 134512
97. Stanev V et al 2008 Phys. Rev. B 78 184509
98. Wang F et al 2009 Europhys. Lett. 85 37005
99. Mazin I I and Johannes M D 2009 Nat. Phys. 5 141
100. Mailer T A et al 2011 Phys. Rev. B 83 100515
101. Wang F et al 2011 Europhys. Lett. 93 57003
102. Das T et al 2011 Phys. Rev. B 84 014521
103. Song C L et al 2011 Phys. Rev. B 84 020503
104. Song C L et al 2011 Science 332 1410
105. Song C L et al 2012 Phys. Rev. Lett. 109 137004
106. Song C L et al 2014 Phys. Rev. Lett. 112 057002
107. Simonin J 1986 Phys. Rev. B 33 7830
108. Ozer M M et al 2006 Nat. Phys. 2 173
109. Tang W H et al 2000 Supercond. Sci. Technol. 13 580
110. Fischer O et al 2007 Rev. Mod. Phys. 79 353
111. Kuroki K et al 2009 Phys. Rev. B 79 224511
112. Wang L L et al 2013 Chin. Phys. B 22 086801
113. Peng R et al 2014 Phys. Rev. Lett. 112 107001
114. Lee J J et al 2014 Nature 515 245
115. Lee P A, Nagaosa N and Wen X F 2006 Rev. Mod. Phys. 78 17
116. Paglione J and Greene R L 2010 Nat. Phys. 6 645

117. Zhang W H et al 2014 Phys. Rev. B 89 060506
118. Li Z et al 2014 J. Phys.: Condens. Matter 26 265002
119. He J F et al 2014 Proc. Natl Acad. Sci. 111 18501
120. Peng Y Y et al 2013 Nat. Commun. 4 2459
121. Yu R and Si Q M 2013 Phys. Rev. Lett. 110 146402
122. Yi M et al 2013 Phys. Rev. Lett. 110 067003
123. Schneider R et al 2012 Phys. Rev. Lett. 108 257003
124. Liu X et al 2014 Nat. Commun. 5 5047
125. Nie Y F et al 2009 Appl. Phys. Lett. 94 242505
126. Nebashima F et al 2013 Appl. Phys. Lett. 103 172602
127. Peng R et al 2014 Nat. Commu. 5 5044
128. Zhang W H et al 2014 Chin. Phys. Lett. 31 017401
129. Sun Y et al 2014 Sci. Rep. 4 6040
130. Deng L Z et al 2013 arXiv:1311.6459
131. Ge J F et al 2015 Nat. Mater. 14 285
132. Singh D J and Du M-H 2008 Phys. Rev. Lett. 100 237003
133. Mazin I 2011 Phys. Rev. B 84 024529
134. Ginzburg V L 1964 Phys. Lett. 13 101
135. Reyren N et al 2007 Science 317 1196
136. Gozar A et al 2008 Nature 455 782
137. Zhang T et al 2010 Nat. Phys. 6 104
138. Liu K et al 2012 Phys. Rev. B 85 235123
139. Xiang Y Y et al 2012 Phys. Rev. B 86 134508
140. Bazhirov T and Cohen M L 2013 J. Phys.: Condens. Matter 25 105506
141. Zheng F W et al 2013 Sci. Rep. 3 2213
142. Cao H Y et al 2014 Phys. Rev. B 89 014501
143. Ye Q Q et al 2014 Phys. Rev. B 88 205130
144. Berlijn T et al 2014 Phys. Rev. B 89 020501
145. Bang J et al 2013 Phys. Rev. B 87 220503
146. Hao N N and Hu J P 2014 Phys. Rev. X 4 031053
147. Wang Y L et al 2012 J. Phys.: Condens. Matter 24 475604
148. Fu L and Kane C L 2008 Phys. Rev. Lett. 100 096407
149. Zhang Y et al 2011 Phys. Rev. B 83 054510
150. Graser S et al 2009 New J. Phys. 11 025016

CHAPTER 3

TRIPLET SUPERCONDUCTIVITY IN A NUTSHELL

Kazumi Maki, Hae Young Kee, Yoshifumi Morita

Keywords: Unconventional superconductivity, Half quantum vortex

INTRODUCTION

The first triplet superconductors have been discovered in 1980 by Jerome et al. [1] in Bechgaard salts $(TMTSF)_2PF_6$ and $(TMTSF)_2ClO_4$ [2, 3]. Since then many triplet superconductors have been discovered, which is concisely reviewed by Sigrist and Ueda [4]. Although it has been noticed that all of these triplet superconductors are nodal (i.e. non s-wave), the gap symmetry has not been addressed seriously until around 1994. We now have a proposal on the gap symmetry of UPt_3, Sr_2RuO_4 and $PrOs_4Sb_{12}$ as shown in Fig. 1 [5, 6]. Here we understand that these triplet superconductors are characterized by the equal spin pairing (ESP) and with extremely small spin-orbit coupling energy $E_{so}\sim10^{-3}\Delta$, where is the superconducting energy gap [7]. Therefore the superconducting order parameter is characterized by $\hat{l}$ (the chiral vector) and $\hat{d}$ (the spin vector) as in superfluid ^{3}He-A, except for the gap symmetry [8, 9].

On the other hand, the discovery of a non-centrosymmetric superconductor $CePt_3Si$ by Bauer et al. in 2004 [10] opened a new door. According to Anderson's argument [11], such a triplet superconductor cannot exist. In order to clarify the role of the parity-breaking term (e.g. the Rashba term [12]), Frigeri et al. [13, 14] considered a model with a Rashba term. They found, first of all, the Fermi surface is split into the one for up-spin and the another for downspin, as shown in Fig. 2. Here we have introduced the Rashba Hamiltonian $\mathcal{H}_R = \alpha k_z \sigma_z$, different from the one used by Frigeri et al. [13, 14]. Secondly, they found the strong admixture of the singlet component. Later, many noncentrosymmetric superconductors have been discovered in $CeIrSi_3$, $CeRhSi_3$, UIr and Li_2Pt_3B [15–18]. Also the spinorbit coupling energy E_{so} or the Rashba term is found invariably extremely large [15, 18], i.e. $E_{so}\sim10^3$ K in these systems. Recently in [19], we have proposed that the triplet superconductors in non-centrosymmetric crystals should be like the one in ^{3}He-A1, in contrast to the previous proposals. The superconductivity occupies the Fermi surface associated with one spin component (say, up-spin) while the other Fermi surface remains in the normal state. With this assumption, we can describe many unusual experimental results of CePt3Si.

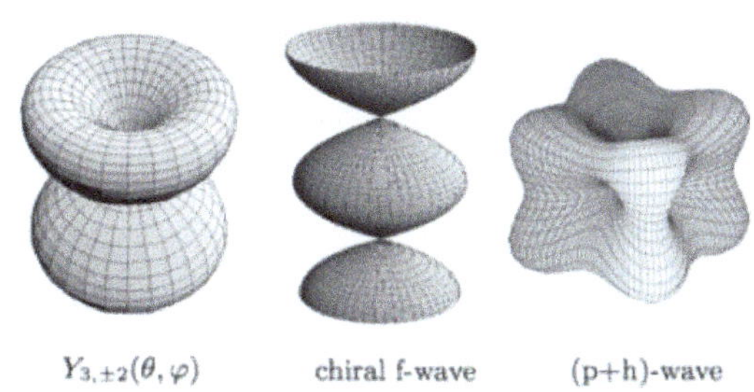

Figure 1: Our proposal on gap functions of Y3, ±2(θ,φ) for UPt_3 and $CePt_3Si$, chiral f-wave for Sr_2RuO_4, (p + h)-wave for $PrOs_4Sb_{12}$.

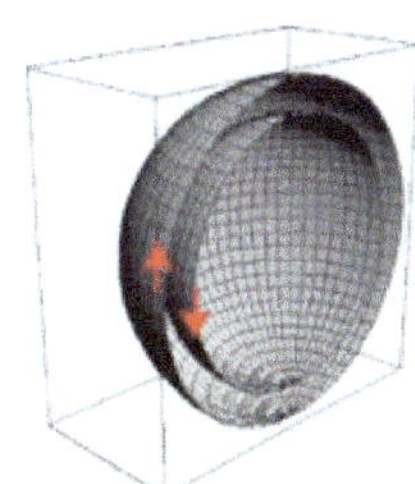

Figure 2: The split Fermi surface for each spin due to the parity-breaking term.

THE ZERO MODE AND THE MAJORANA FERMION

In this section, we shall summarize the bound-state spectrum around an Abrikosov's vortex in triplet superconductors. In both type-A and type-A1 triplet superconductors, we conclude that vortices harbor the zero mode which behaves as the Majorana fermion.

Citation: Maki, K., Kee, H.Y. & Morita, Y., "Triplet Superconductivity in a Nutshell," J Supercond Nov Magn (2009) 22: 71. doi:10.1007/s10948-008-0363-7.

First of all, we note that ^l (the chiral vector) of many triplet superconductors are fixed parallel to the crystal $\hat{c}$ -axis. Secondly, in the absence of a magnetic field, $\hat{d} \| \hat{l}$ is due to the spin-orbit coupling. However, when the magnetic field $\vec{H}$ (parallel to $\hat{l}$) is applied in the vicinity of $H_{c2}(T)$ (the upper critical field), $\hat{d} \perp \hat{l}$ is realized. In this particular situation, Ivanov [20, 21] have shown that the BdG equations for the triplet superconductor decouple into the one for the up-spin component and the other for the down-spin component. Then these BdG equation have the same structure as the one for s-wave superconductor first written down in [22, 23]. Therefore, first let us consider a single vortex.

Following Caroli, de Gennes and Matricon [22, 23], the bound state energy formula is given as follows (see, for detail [24–26])

$$\epsilon_n = n\omega_0 \quad \text{with } n = 0, \pm 1, \pm 2, \pm 3, \ldots, \tag{1}$$

where $\omega_0 \simeq \Delta (p_F \xi)^{-1} [\ln(\frac{2}{\delta})]^{-1}$ (δ is the cut-off parameter) and $\delta \simeq \sqrt{\Gamma/\Delta}$ (Γ is the quasiparticle scattering rate in the normal state) in the very clean system. So ω0 is essentially the same as in s-wave superconductors except for the logcorrection due to the nodal structure. Note that the boundstate spectrum is very different from the one for the singlet superconductors (s-wave, d-wave etc.). For the singlet superconductors, $\epsilon_n = (n + \frac{1}{2})\omega_0$ with n = 0,±1,±2,±3,... with a gap $\omega_0/2 \sim \Delta^2/E_F$. On the other hand, in triplet superconductors, there is the zero mode or the Majorana fermion [9, 27, 28] with zero energy (right on the Fermi surface). The Majorana fermion wavefunction associated with the zero mode is given by φ

$$\phi_0(r, \hat{k}) \sim \left[\operatorname{sech}(r/\xi)\right]^{C|f|} \tag{2}$$

and C = ξv^{-1}~1. These wavefunctions should be accessible at low temperatures.

HALF-QUANTUM VORTICES

Half-quantum vortices are generated in type-A triplet superconductors, which correspond to the $\hat{d}$ -soliton or the $\hat{d}$ -domain wall [24–26]. Since $\hat{l}$ ^l-vector is fixed parallel to the crystalline $\hat{c}$ -axis, we limit ourselves to the texture free energy associated with the $\hat{d}$ vector. Especially for UPt_3, Sr_2RuO_4 and $PrOs_4Sb_{12}$ in a magnetic field parallel to $\hat{l}$, the texture free energy simplifies as [29]

$$\mathcal{F} = \frac{1}{2}\chi_N C^2 \int dx\, dy \Big[K(\nabla\Phi)^2 + \sum_{i,j} |\partial_i \hat{d}_j|^2 + \xi_D^{1-2}(\hat{d}_x^2 + \hat{d}_y^2) \Big], \tag{3}$$

where χ_N and C are the spin susceptibility and the spinwave velocity, and $\xi_D^{1-2} = \xi_D^{-2} - \xi_H^{-2}$. Here

$$K(t) = \frac{\rho_s(t)}{\rho_{sp}(t)} = \frac{1 + \frac{F_1}{3}}{1 + \frac{F_1^a}{3}} \frac{1 + \frac{F_1^a}{3}(1 - \rho_s^0(t))}{1 + \frac{F_1}{3}(1 - \rho_s^0(t))}, \tag{4}$$

where t = T/T_c, ρ_s(t), ρ_{sp}(t) and ρ^0_s (t) are the super- fluid density, the spin superfluid density and the bare superfluid density. In $PrOs_4Sb_{12}$, $\rho_s^0(t)$ is well approximated by $\rho_s^0(t)$ = 1−0.460572t −0.53428t1.729 [30]. Experimentally, m∗/m = 50 in $PrOs_4Sb_{12}$ and, if we ignore Fa 1 ,

$$K(t) = \left[1 - \left(1 - \frac{m}{m*}\right)\rho_s^0(t)\right]^{-1}. \tag{5}$$

Here we note that the temperature dependence of K(t) in UPt_3 and Sr_2RuO_4 is very similar to that of $PrOs_4Sb_{12}$ [24, 25].

Here we shall compare the free energy of an Abrikosov's vortex and a bound pair of half-quantum vortices by the $\hat{d}$ -soliton. For this purpose, we consider the free energy of an Abrikosov's vortex in a unit circular cell with radius a which encloses a flux quantum. Then a is given by

$$a = \left(\frac{\phi_0}{\pi H}\right)^{1/2}, \tag{6}$$

where H is the field strength. Note also a similar relation due to de Gennes:

$$H_{c2}(t) = \frac{\phi_0}{2\pi\xi^2(t)}, \tag{7}$$

where ξ(t) is the coherence distance. This indicates, for $H < H_{c2}(t)$, $a > \sqrt{2}\xi(t)$. The free energy for an Abrikosov's vortices is readily given by

$$\mathcal{F}_A = \pi\chi_N C^2 K \ln(a/\xi). \tag{8}$$

For a bound pair of HQVs separated by a distance R and inside the unit cell,

$$\mathcal{F}_{BP} = \frac{\pi}{2}\chi_N C^2 \Big\{ K \ln\Big(\frac{a + \sqrt{a^2 - R^2/4}}{2\xi}\Big) + \ln(a/\xi) - \frac{R}{4a}\sin^{-1}\Big(\frac{R}{\sqrt{a^2 - R^2/4}}\Big) + \frac{1}{4}\Big(\frac{R}{\xi_D'}\Big)^2 \ln\Big(\frac{2\xi_D'}{a}\frac{K+1}{\sqrt{2K+1}}\Big)\Big\}. \tag{9}$$

Also, for K(t)1 $K(t) \gg 1$, this equation is optimized at R/2a = $\sqrt{2K+1}/K+1$ and [31]

$$\mathcal{F}_{BP} = \frac{\pi}{2}\chi_N C^2 \Big\{ K \ln\Big(\frac{a}{2\xi}\frac{2K+1}{K+1}\Big) + \ln\Big(\frac{a}{\xi}\Big) - \frac{1}{2}\frac{\sqrt{2K+1}}{K+1}\sin^{-1}\Big(\frac{2\sqrt{2K+1}}{K+1}\Big) + \frac{2K+1}{(K+1)^2}\Big(\frac{a}{\xi_D'}\Big)^2 \ln\Big(\frac{2\xi_D'}{a}\frac{K+1}{\sqrt{2K+1}}\Big)\Big\}. \tag{11}$$

Now from FA = F_{BP}, we find that

$$H^*(t)/H_{c2}(t) = \left[2^{-(K-2)}(2K+1)^{(2K+1)}(K+1)^{-2(K+1)}\right]^{-1/(K-1)}, \quad (11)$$

where H*(t) is the phase boundary below which HQVs are stable. For $PrOs_4Sb_{12}$, we show the stability region of HQVs in Fig. 3 [26], which is close to the phase boundary determined experimentally by Izawa et al. [32]. According to the our theoretical picture, the B phase discovered in [32] is the vortex state filled with HQVs. Also we have got a similar result for the B phase in UPt_3 [24, 25]. On the other hand, as to Sr_2RuO_4, there is no clear hallmark of such a phase transition. However, we speculate that the anomalous angle-dependent magneto-specific heat found by Deguchi et al. [33] below 300 mK can be due to HQVs.

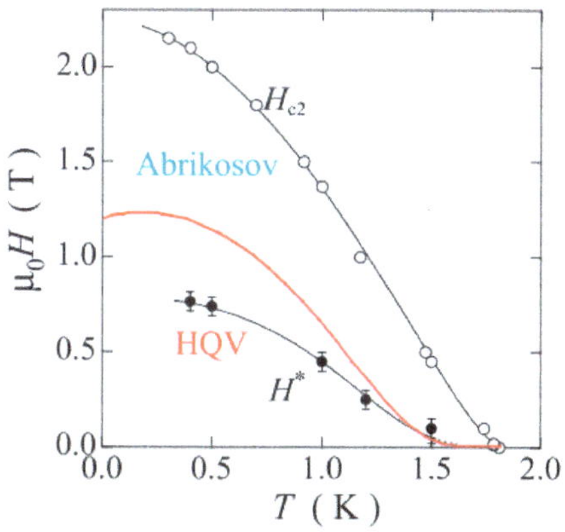

Figure 3: For PrOs4Sb12, the stability region of HQV (inside the line without circles) is shown, based on our theoretical result [26]. It is close to the phase (whose boundary is given by black circles) determined experimentally by Izawa et al. [32]

Finally, it is to be noted that we can access the HQVs by scanning tunneling microscopy (STM) [34], micromagnetometry [35] and small angle neutron scattering (SANS) at ultra-low temperature.

CONCLUDING REMARKS

We have proposed a new paradigm for triplet superconductivity.

- Most of triplet superconductors, which were discovered before 2004, belong to type-A. Their topological structure is very similar to superfluid 3He-A. At low temperatures, a vortex splits into a pair of half-quantum vortices (HQVs).
- A new class of superconductors in non-centrosymmetric crystals, which were discovered after 2004, belong to type-A1. The topological structure is the same as of superfluid ^{3}He-A1.
- All these triplet superconductors have the zero mode/ Majorana fermion attached to a vortex, the implications of which should be explored further.

ACKNOWLEDGEMENTS

We thank Ana Celia Mota and Coneliu Florian Miclea for very informative discussions on the ongoing experiments on the flux pinning in $PrOs_4Sb_{12}$ and $CePt_3Si$. We thank also Balazs Dora, Stephan Haas and Masaru Kato for enlightening discussions. Y.M. is supported by a Grant-in-Aid for Scientific Research from the Ministry of Education, Culture, Sports, Science and Technology of Japan.

REFERENCES

1. Jerome, D., Mazaud, A., Ribault, M., Bechgaard, K.: J. Phys. Lett. 41, L95 (1980)
2. Takigawa, M., Yasuoka, H., Saito, G.: J. Phys. Soc. Jpn. 56, 873 (1987)
3. Lee, I.J., Brown, S.E., Naughton, M.J.: J. Phys. Soc. Jpn. 75, 051011 (2006)
4. Sigrist, M., Ueda, K.: Rev. Mod. Phys. 63, 239 (1991)
5. Won, H., Haas, S., Parker, D., Telang, S., Vanyolos, A., Maki, K.: In: Lectures on the Physics of Highly Correlated Electron Systems IX. AIP Conference Proceedings, vol. 789. Melville, Brooklyn (2005)
6. Won, H., Morita, Y., Maki, K.: Phys. Status Solidi B 244, 4371 (2007)
7. Maki, K., Haas, S., Parker, D., Won, H.: In: Topology in Condensed Phases. World Scientific, Singapore (2006)
8. Vollhardt, D., Wölfle, P.: The Superfluid Phases in Helium Three. Taylor & Francis, London (1990)
9. Volovik, G.E.: The Universe in a Helium Droplet. Clarendon, Oxford (2003)
10. Bauer, E., Hilscher, G., Michor, H., Paul, C., Scheidt, E.W., Gribanov, A., Seropegin, Yu., Noël, H., Sigrist, M., Rogl, P.: Phys. Rev. Lett. 92, 027003 (2004)
11. Anderson, P.W.: Phys. Rev. B 30, 4000 (1994)
12. Rashba, E.J.: Sov. Phys. Solid State 2, 1109 (1960)
13. Frigeri, P.A., Agterberg, D.F., Koga, A., Sigrist, M.: Phys. Rev. Lett. 92, 097001 (2004)
14. Frigeri, P.A., Agterberg, D.F., Sigrist, M.: New J. Phys. 6, 115 (2004)
15. Settai, R., Takeuchi, T., Onuki, Y.: J. Phys. Soc. Jpn. 76, 051003 (2007)
16. Kimura, N., Muro, Y., Aoki, H.: J. Phys. Soc. Jpn. 76, 051010 (2007)
17. Kobayashi, T.C., Hori, A., Fukushima, S., Hidaka, H., Kotegawa, H., Akazawa, T., Takeda, K., Ohishi, Y., Yamamoto, E.: J. Phys. Soc. Jpn. 76, 051007 (2007)
18. Nishiyama, M., Inada, Y., Zheng, G.-Q.: Phys. Rev. Lett. 98, 047002 (2007)
19. Morita, Y., Maki, K.: Preprint (2008)
20. Ivanov, D.A.: In: Vortices in Unconventional Superconductors and Superfluids. Springer, Berlin (2002)
21. Ivanov, D.A.: Phys. Rev. Lett. 40, 268 (2001)
22. Caroli, C., de Gennes, P.G., Matricon, J.: Phys. Lett. 9, 310 (1964)
23. de Gennes, P.G.: The Superconductivity of Metals and Alloys. Benjamin, New York (1966). Reprinted in Percus Book, Reading (1999)
24. Kee, H.Y., Maki, K.: Preprint (2008)
25. Kee, H.Y., Raghavan, A., Maki, K.: Preprint (2007)

26. Morita, Y., Maki, K.: Preprint (2008)
27. Majorana, E.: Nuovo Cim. 14, 171 (1937)
28. Penrose, R.: The Road to Reality. Vintage Books, New York (2007)
29. Kee, H.Y., Kim, Y.B., Maki, K.: Phys. Rev. B 62, R9275 (2000)
30. Parker, D., Maki, K., Haas, S.: Eur. Phys. J B 49, 77 (2006)
31. Kee, H.Y., Maki, K.: Europhys. Lett. 80, 46003 (2007)
32. Izawa, K., Nakajima, Y., Goryo, J., Matsuda, Y., Onuki, S., Settai, R., Thalmeier, P., Maki, K.: Phys. Rev. Lett. 90, 117001 (2003)
33. Deguchi, K., Mao, Z.Q., Maeno, Y.: J. Phys. Soc. Jpn. 73, 1313 (2004)
34. Fischer, Ø., Kugler, M., Maggio-Aprile, I., Berthod, C., Renner, C.: Rev. Mod. Phys. 79, 353 (2007)
35. Tsuei, C.C., Kirtley, J.R.: Rev. Mod. Phys. 72, 969 (2000)

CHAPTER 4

SUPERCONDUCTIVITY, CHARGE ORDERINGS AND MAGNETISM

Konrad Jerzy Kapcia

Keywords: Extended Hubbard model, Phase separation, Superconductivity, Magnetism, Charge ordering, Pair hopping, Phase diagrams

INTRODUCTION

The interplay and competition between superconductivity (SC) and other electron orderings, such as charge orderings (CO) or various magnetic (M) orderings is currently under intense investigations in broad range of materials including cuprates, bismutates, iron-pnictides, organic conductors and heavy-fermion systems (for review, see e.g. refs. [1–14] and references therein). Moreover, the phase separation (PS) has been evidenced (by various microscopy techniques such as STM, MFM, etc.) in some of the materials from the groups mentioned previously, especially in iron-pnictides (PS with M), bismutates (PS with CO) and cuprates.

In this paper, we study a simplified model which can pertain to that problem. The effective model considered has a form of single-band extended Hubbard model with pair hopping at the atomic limit (t=0) [4–17]:

$$\begin{aligned}\hat{H} = {} & U\sum_i \hat{n}_{i\uparrow}\hat{n}_{i\downarrow} - \mu\sum_i \hat{n}_i \\ & -I\sum_{\langle i,j\rangle}\left(\hat{\rho}_i^+\hat{\rho}_j^- + \hat{\rho}_j^+\hat{\rho}_i^-\right) + \frac{W}{2}\sum_{\langle i,j\rangle}\hat{n}_i\hat{n}_j \\ & -2J^z\sum_{\langle i,j\rangle}\hat{s}_i^z\hat{s}_j^z - J^{xy}\sum_{\langle i,j\rangle}\left(\hat{s}_i^+\hat{s}_j^- + \hat{s}_j^+\hat{s}_i^-\right), \qquad (1)\end{aligned}$$

where $\hat{n}_{i\sigma} = \hat{c}_{i\sigma}^+\hat{c}_{i\sigma}$, $\hat{n}_i = \sum_\sigma \hat{n}_{i\sigma}$, $\hat{\rho}_i^+ = (\hat{\rho}_i^-)^\dagger = \hat{c}_{i\uparrow}^+\hat{c}_{i\downarrow}^+$, $\hat{s}_i^z = (1/2)(\hat{n}_{i\uparrow} - \hat{n}_{i\downarrow})$ and $\hat{s}_i^+ = (\hat{s}_i^-)^\dagger = \hat{c}_{i\uparrow}^+\hat{c}_{i\downarrow}$. $\hat{c}_{i\sigma}^+$ and $\hat{c}_{i\sigma}$ denote the creation and annihilation operators, respectively, of an electron with spin σ (σ = ↑,↓) at site i, which fulfil standard anticommutation relations for fermionic operators. μ is the chemical potential. The nearest-neighbors interactions U, I, W, $J^z$and J^{xy} are effective model parameters and are assumed to include all the possible renormalizations and contributions like those coming from the strong electron-phonon coupling or from the coupling between electrons and other electronic subsystems in solid or chemical complexes.

The analysis of the ground state of model (1) has been performed by means of a variational approach (VA) in the grand canonical ensemble [6–14], which treats on-site U term exactly and the intersite interactions within the mean-field approximation (MFA). The MFA is a rigorous treatment of the intersite terms in the limit of infinite dimensions d→+∞ (or large coordination number z); thus, the result obtained in this paper for model (1) are also exact in the limit d→+∞. This approach allows us to calculate the grand canonical potential per site $\omega = \langle\hat{H}\rangle/L$ and the (free) energy per site $E = \langle\hat{H} + \mu\sum_i \hat{n}\rangle/L = \omega + \mu n$ (at T=0, L is a number of lattice sites). $\langle \widehat{A} \rangle$ is an average value of operator $\widehat{A}$.

Let us introduce the following mean-field order parameters:

$$n_{\vec{q}} = \tfrac{1}{N}\sum_i \exp\left(i\vec{q}\cdot\vec{r}_i\right)\langle\hat{n}_i\rangle, \qquad (2)$$

$$\Delta_{\vec{q}} = \tfrac{1}{N}\sum_i \exp\left(i\vec{q}\cdot\vec{r}_i\right)\langle\hat{\rho}_i^-\rangle, \qquad (3)$$

$$m_{\vec{q}}^\alpha = \tfrac{1}{N}\sum_i \exp\left(i\vec{q}\cdot\vec{r}_i\right)\langle\hat{s}_i^\alpha\rangle, \qquad (4)$$

where $\vec{r}_i$ determines the location of i-site and α=z,± $(m_{\vec{q}}^+ = (m_{\vec{q}}^-)^*)$. We restrict ourselves only to a case of two sublattice orderings on the alternate lattices, i.e. $\vec{q} = \vec{0}, \vec{Q}$, where $\vec{Q}$ is a half of the smallest reciprocal lattice vector. $n \equiv n_{\vec{0}}$ is an electron concentration in the system. In the charge-ordered (CO) phase $n_{\vec{Q}} \neq 0$. Model (1) exhibits a symmetry between I>0 (SC with s-pairing, $\Delta_{\vec{0}} \neq 0$) and I<0 (SC with η-pairing, $\Delta_{\vec{Q}} \neq 0$) cases. Similar symmetry occurs for spin orderings in both direction (α=z,±) between ferro- (with $m_{\vec{0}}^\alpha \neq 0$) and antiferro- (with mαQ⃗ ≠0mQ→α≠0) magnetic

Citation: Konrad Jerzy Kapcia, "Superconductivity, Charge Orderings, Magnetism", doi:10.1007/s10948-014-2906-4.

(M) phases. Notice that for general case of both J^{xy} and J^z non-zero that symmetry is valid only for VA results. The boundary between M phases with orderings in z-direction and xy-direction is for $|J^{xy}/J^z|=1$ (in VA). Thus, we define $|J|=\max\{|J^z|,|J^{xy}|\}$. Both symmetries in the above-mentioned discussion are for model (1) on the alternate lattices (i.e. lattices, which consist of two interpenetrating lattices). In the presence of finite single electron hopping $t\neq0$, both of the above-mentioned symmetries are broken in the general case [18–30], but the detailed discussion of effects of $t\neq0$ is beyond the scope of this publication. Moreover, because of the particle-hole symmetry of model (1) the phase diagrams obtained are symmetric with respect to $\bar{\mu} = 0$ ($\bar{\mu} = \mu - U/2 - W_0$, $W_0=zW$) as well as n=1. The equations for energies of phases and states for $n\geq1$ can be obtained using the relation $\bar{E}(2-n) = E(n) + (U + 2W_0)(1-n)$, where

E(n) is the energy for $n\leq1$ ((10)–(15) and (17)) and $\bar{E}(n)$ is the energy for $n\geq1$. For order parameters, the relation $\bar{\psi}(2-n) = -\psi^*(n)$ is fulfilled ($n\neq1$), where $\psi = \Delta_{\vec{q}}, m^{\alpha}_{\vec{q}}, n_{\vec{Q}}$ ($n_{\vec{Q}} = n^*_{\vec{Q}}$, $(m^z_{\vec{q}})^* = m^z_{\vec{q}}$).

The PS state (macroscopic separation) is a state in which two domains with different electron concentrations (n_+ and n_-) exist in the system (coexistence of two homogeneous phases). The energies of the PS states are calculated from minimization of the expression

$$E_{PS}(n_+, n_-) = mE_+(n_+) + (1-m)E_-(n_-), \quad (5)$$

where $E_\pm(n_\pm)$ are values of the energy of two separating phases at $n_\pm$ corresponding to the lowest homogeneous solution for a given phase, m is the fraction of the system with charge density n_+, and $mn_+ + (1-m)n_- = n$ [8, 31–34].

In this paper, we focus on the ground state of model (1) in two cases: (i) J=0 and $U\leq0$ – Section2.1; as well as (ii) W=0 – Section 2.2. Names of transition orders used are consistent with our earlier works [7–14, 33–36]. In particular, a transition between a homogeneous phase and the PS state is symbolically named as a "third-order" transition.

THE GROUND STATE RESULTS (*T*=0)

Superconductivity and Charge Orderings (*J*=0 and *U*≤0)

For any $U\leq0$ and $J^z, J^{xy}=0$ the ground state phase diagrams of model (1) are shown in Figure 1. Figure 1a is a result of comparison of the grand canonical potentials of particular phases, which are as following:

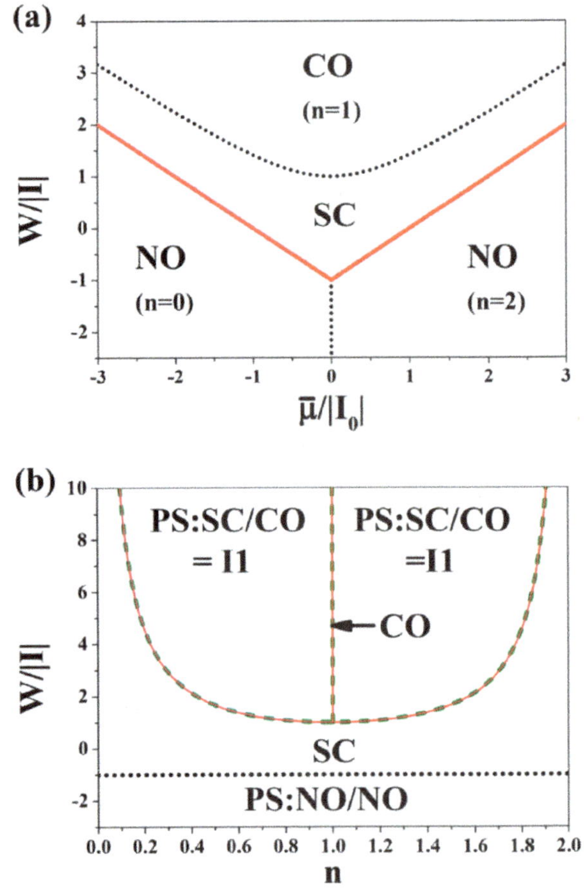

Figure 1: Ground state diagram for $U\leq0$ and J=0 as a function of μ (**a**) and as a function of n (**b**).Dotted, solid and dashed lines denote first-order, second-order and "third-order" transitions, respectively.

$$\omega^a_{NO} = 0, \qquad (n=0) \quad (6)$$

$$\omega^b_{NO} = -2\bar{\mu}, \qquad (n=2) \quad (7)$$

$$\omega_{CO} = -\bar{\mu} - W_0, \qquad (n=1,\ n_{\vec{Q}}=1) \quad (8)$$

$$\omega_{SC} = -\frac{(\bar{\mu} + |I_0| + W_0)^2}{2(|I_0| + W_0)}, \quad \left(n = 1 + \frac{\bar{\mu}}{|I_0| + W_0}\right) \quad (9)$$

where $I_0=zI$ and $W_0=zW$. The second-order **NO–SC** phase boundary is determined by $W_0/I_0 = |\bar{\mu}/I_0| - 1$ (n and $\Delta_{\vec{q}}$ are continuous). The first-order **SC–CO** transition occurs for $(W_0/I_0)^2 = (\bar{\mu}/I_0)^2 + 1$, at which $\Delta_{\vec{q}}$ and $n_{\vec{Q}}$ are discontinuous. n changes (discontinuously, except $\bar{\mu}=0$) from $n_{SC} = 1 \mp \sqrt{(W_0 - |I_0|)/(W_0 + |I_0|)}$ in the SC phase (for $n \lesseqgtr 1$, respectively) to $n_{CO}=1$ in the CO phase. A first-order **NO–NO** transition occurs for $\bar{\mu}=0$ and $W/|I|<-1$.

As a function of n, the energies of the particular homogeneous phases are as follows (for $n\leq1$):

$$E_{NO} = \tfrac{1}{2}(U + W_0 n)\, n, \quad (10)$$

$$E_{SC} = \tfrac{1}{2}\left[(U + 2W_0)n - (|I_0| + W_0)n(2-n)\right], \quad (11)$$

$$E_{CO} = \tfrac{1}{2}Un, \quad (n_{\vec{Q}} = n) \quad (12)$$

$$E_{I1} = E_{SC} + -\tfrac{1}{2}\left[\sqrt{W_0 + |I_0|}(1-n) - \sqrt{W_0 - |I_0|}\right]^2, \quad (13)$$

where we also consider the mixed phase (I1) in which three order parameters: $n_{\vec{Q}}$, $\Delta_{\vec{0}}$ and $\Delta_{\vec{Q}}$ are non-zero (microscopic coexistence of SC and CO). In the SC phase, $|\Delta|^2_{\vec{0}/\vec{Q}} = \frac{1}{4}n(2-n)$, $|\Delta|^2_{\vec{Q}/\vec{0}} = 0$ (for I≷0I≷0, respectively), and $n_{\vec{Q}}$ =0. For the I1 phase (in the range of its definiteness and n≤1) one obtains: $|\Delta|^2_{\vec{0}/\vec{Q}} = \frac{1}{4}\left\{n(2-n) - \left[1 + (n-1)\sqrt{\frac{W_0+|I_0|}{W_0-|I_0|}}\right]\right\}$, $|\Delta|^2_{\vec{Q}/\vec{0}} = \frac{1}{4}(n-1)\left[1 - n - \sqrt{\frac{W_0-|I_0|}{W_0+|I_0|}}\right]$ (for I≷0I≷0, respectively), and $n^2_{\vec{Q}} = 1 + (n-1)^2 + (n-1)\frac{2W_0}{\sqrt{W_0^2-I_0^2}}$. In this phase $\bar{\mu}_{I1} = \mp\sqrt{W_0^2 - I_0^2}$ (for $n \lessgtr 1$ (for n≶1n≶1, respectively), what implies that the I1 phase can occur only at the SC–CO boundary in Figure 1a. Notice that in the I1 phase the $\bar{\mu}$-dependences of n, $n_{\vec{Q}}$, $\Delta_{\vec{0}}$ and $\Delta_{\vec{Q}}$ are indefinite, whereas in the SC phase, it can be easily obtained by using (9) (i.e. n=$n(\bar{\mu})$).

The first-order boundaries (for fixed $\bar{\mu}$) can be associated with the existence of the PS states in defined ranges of n (for fixed n). In our model, two such states can occur: (i) PS:SC/CO with concentrations in domains: nSC=1$\mp\sqrt{(W_0 - |I_0|)/(W_0 + |I_0|)}$ (n≶1n≶1, respectively) and n_{CO}=1 as well as (ii) PS:NO/NO with concentrations in domains: $n^a_{NO} = 0$ and $n^b_{NO} = 2$. Using (5) and (10)–(12) one obtains (for n≤1):

$$E_{PS:SC/CO} = \frac{1}{2}Un + (1-n)\left[\sqrt{W_0^2 - I_0^2} - W_0\right], \tag{14}$$

$$E_{PS:NO/NO} = \left(\frac{1}{2}U + W_0\right)n. \tag{15}$$

Comparing energies (10)–(15), we obtain the diagram shown in Figure 1b. The I1 phase and the PS:SC/CO state are degenerated in the whole ranges of their occurrences, but at T>0, this degeneration is removed and only the PS:SC/CO state exists [13]. The SC–I1 and I1–CO transitions are second-order, whereas the SC–PS:SC/CO and PS:SC/CO–CO transitions are "third-order". At the PS:NO/NO–SC transition, the sizes of domains change discontinuously as well as the occurring orderings change; thus, we classified this transition as first-order rather than "third-order" (cf. the definitions of phase boundaries in [33–36]). Notice that in all above-mentioned phases and states, there are no single occupied sites $(\sum_i \langle|\hat{s}^z_i|\rangle/L = 0$ and $m^{\alpha}_{\vec{q}} = 0)$ and all electrons are locally paired. The results obtained are in agreement with those for U→−∞U→−∞ (VA, fixed n) [32, 37]. In that limit, the results derived by RPA approach for d=1,2,3 lattices [4, 38, 39] are consistent with Figure 1b; but for these dimensions, the critical concentration n_c≠0 exist that for 0<n<n_c only the SC phase exists (for any W/|I|>0, here: n, n_c<1). The PS:SC/CO state for W>0 has not been considered in [4, 37–39].

Superconductivity and Magnetism (*W*=0)

For W=0 and T=0, the phase diagrams of model (1) are shown in Figure 2. On the diagram, as a function of $\bar{\mu}$ (shown in Figure 2a) the NO (with n=0 or n=2), SC, and M phases occur, whose grand potentials are given by (6), (7), (9) (for W_0=0), and

$$\omega_M = -\bar{\mu} - U/2 - |J_0|/2 \qquad (n = 1,\ m^{\alpha}_{\vec{q}} = 1/2), \tag{16}$$

respectively, ($|J_0|$=z|J|). The first-order SC–M boundary is located at $(U + |J_0|)/|I_0| = (\bar{\mu}/I_0)^2 + 1$. The SC–NO transitions at $|\bar{\mu}/I_0| = 1$ are second order.

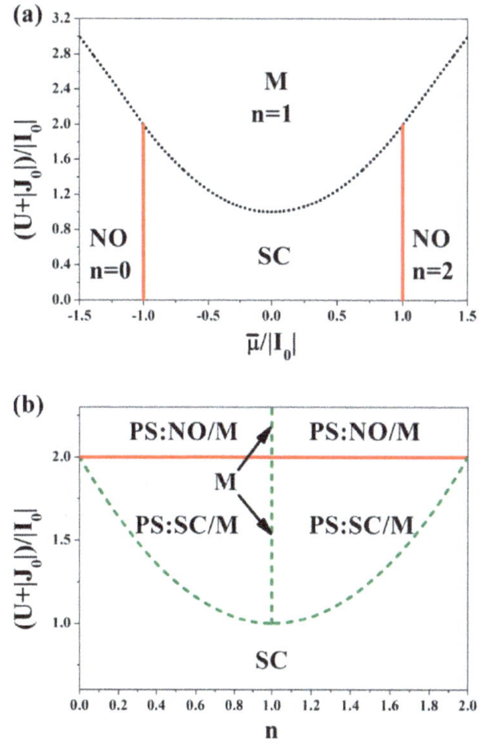

Figure 2: Ground state diagram for W=0 and J≠0 as a function of μ (**a**) and as a function of n (**b**). Denotations are the same as in Figure 1.

Similarly, as in the previous section, we derived the energies of particular phases and possible PS states for fixed n. For the M phase, we obtain (n≤1):

$$E_M = -\frac{1}{2}|J_0|n^2, \qquad \left(m^{\alpha}_{\vec{q}} = n/2\right) \tag{17}$$

whereas the energies: (i) $E_{PS:SC/M}$ of the PS:SC/M state with concentrations in domains: $\sqrt{(U + |J_0|)/|I_0| - 1}$ (for n≶1n≶1) and n_M=1; (ii) $E_{PS:NO/M}$ of the PS:NO/M state with concentrations in domains: $n^a_{NO} = 0\ (n < 1)$ or $n^b_{NO} = 2\ (n < 1)$ and n_M=1 as well as E_{SC} for the SC phase can be derived from (5) and (10), (11) (with W_0=0) and (17).

The resulting diagram as a function of n is shown in Figure 2b. Notice that in the M phase, μ_M=−$|J_0|$n and $\frac{\partial\mu}{\partial n} < 0$ for n<1. Thus, the M phase occurs only for n=1 and (U+$|J_0|$)/$|I_0|$>1 (cf. refs. [12, 40–42]). The mixed phase (I2) in which order parameters $\Delta_{\vec{q}}$ and $m^{\alpha}_{\vec{q}}$ are non-zero (microscopic coexistence of SC and M) does not exist in any range of $\bar{\mu}$ and n. The PS:SC/M–PS:NO/M is second-order transition between two PS states, because in the SC domain, $\Delta_{\vec{q}}$ goes continuously to zero, whereas in the other domain (M), $m^{\alpha}_{\vec{q}} \neq 0$ and it changes continuously [33–36]. The PS:SC/M–M and PS:NO/M–M transitions are "third-order" ones. For J=0 the M phase reduces into the NO″ phase with $\omega_{NO'} = -\bar{\mu} - U/2$ (n=1) and ENO′(n)=0 (for n<1) [8–11], and the PS states on Figure 2b change into PS″:SC/NO″ and PS″:NO/NO″ states, whose energies can be derived from (5) analogously. The energies of this NO″phase and the PS″:NO/NO″ state are equal at T=0, but at T>0 the PS″ does not occur. In such a case, the PS″:SC/NO″-PS″:NO/NO″ is a second-order transition between two PS states [33–36], whereas the PS″:SC/NO″–NO″ transition is "third-order".

SUMMARY AND FINAL COMMENTS

The analysis of the model considered shows that the superconductivity can coexists with magnetism only in the state with phase separation (PS:SC/M for $1<(U+|J_0|)/|I_0|<2$, W=0), whereas superconductivity and charge orderings—in the state with phase separation (PS:SC/CO) as well as in homogeneous mixed I1 phase (for U<0, $W/|I_0|>1$ and $J^z,J^{xy}=0$). These last mentioned two states are degenerated at T=0. This degeneration can be removed in the presence of long range (at least between next-nearest neighbors) interactions [32]. The mixed homogeneous phases I2 (coexistence of SC and M orderings) does not appear on the phase diagrams.

Let us discuss how the properties of the system in the PS states evolve with n. If the SC fraction of the system is rather large (as it occurs near the boundary with the homogeneous SC phase, i.e. for $n \gtrsim n_{SC}$), for example, the measured penetration depth will remain constant with increasing n (since $n_- = n_{SC}$ is independent of n) [8, 31, 32]. On the other hand, near half-filling ($n \approx 1$) the SC fraction is strongly reduced and there will be only diluted SC domains (with $n_{SC}<1$) in a insulating or semiconducting CO/M background (with $n_{CO/F}=1$). When the SC domains do not percolate, one should observe a partial Meissner effect but without zero resistance. On the contrary, in the homogeneous mixed phases (I1, I2) behaviour of the SC characteristics would be totally different, because the whole system exhibits superconductivity and all of its SC properties (in particular, a penetration depth) would change with n, but with a full Meissner effect and zero resistance [37–39].

Notice that the results of Section 2.1 together with results of [13, 32, 37, 43–46] give the full picture of behavior of model (1) for U≤0, W≠0, and $J^z,J^{xy}=0$ within VA approach. Cases of U>0 as well as $J^z,J^{xy}\neq 0$ are left to future works. Some analyses concerning the interplay between Wand J^z,J^{xy} interactions for I=0 have been performed in [5, 47–49].

ACKNOWLEDGMENTS

The author is indebted to Professor Stanisław Robaszkiewicz and Professor Roman Micnas for their very fruitful discussions during this work and careful reading of the manuscript. The work has been financed by National Science Center (NCN, Poland) as a research project in the years 2011–2013, under Grant No. DEC-2011/01/N/ST3/00413. We thank the European Commission and the Ministry of Science and Higher Education (Poland) for the partial financial support from the European Social Fund—Operational Programme "Human Capital"–POKL.04.01.01-00-133/09-00—"Proinnowacyjne kształcenie, kompetentna kadra, absolwenci przyszłości". We also thanks NCN for the doctoral scholarship No. DEC-2013/08/T/ST3/00012 in years 2013-2015 and the Foundation of Adam Mickiewicz University in Poznań for the support from its scholarship programme.

REFERENCES

1. Johnston, D.C.: The puzzle of high temperature superconductivity in layered iron pnictides and chalcogenides. Adv. Phys. 59, 803 (2010). doi:10.1080/00018732.2010.513480
2. Dagotto, E., Hotta, T., Moreo, A.: Colossal magnetoresistant materials: the key role of phase separation. Phys. Rep. 334, 1 (2001). doi:10.1016/S0370-1573(00)00121-6
3. Micnas, R., Robaszkiewicz, S., Bussman-Holder, A.: Two-component scenarios for nonconventional (exotic) superconductors. Struct. Bond. 114, 13 (2005). doi:10.1007/b101016
4. Micnas, R., Ranninger, J., Robaszkiewicz, S.: Superconductivity in narrow-band systems with local nonretarded attractive interactions. Rev. Mod. Phys. 62, 113 (1990). doi:10.1088/0022-3719/21/6/009
5. Robaszkiewicz, S.: Magnetism and charge ordering in the localized state. Phys. Status Solidi (b) 70, K51 (1975). doi: 10.1002/pssb.2220700156
6. Robaszkiewicz, S.: Pseudospin models of superconductors with very short coherence length. Acta Phys. Pol. A 85, 117 (1994)
7. Robaszkiewicz, S., Pawłowski, G.: Effects of finite pair binding energy in a model of a superconductor with local electron pairing. Physica C 210, 61 (1993). doi: 10.1016/0921-4534(93)90009-F
8. Kapcia, K., Robaszkiewicz, S., Micnas, R.: Phase separation in a lattice model of a superconductor with pair hopping. J. Phys.: Condens. Matter 24, 215601 (2012). doi:10.1088/0953-8984/24/21/215601ADS
9. Kapcia, K., Robaszkiewicz, S.: The magnetic field induced phase separation in a model of a superconductor with local electron pairing. J. Phys.: Condens. Matter 25, 065603 (2013). doi: 10.1088/0953-8984/25/6/065603ADS
10. Kapcia, K.: Metastability and phase separation in a simple model of a superconductor with extremely short coherence length. J. Supercond. Nov. Magn. 27, 913 (2014). doi:10.1007/s10948-013-2409-8
11. Kapcia, K.J.: Superconductivity, metastability and magnetic field induced phase separation in the atomic limit of the Penson-Kolb-Hubbard model. Acta Phys. Pol. A 126, A-53 (2014). doi:10.12693/APhysPolA.126.A-53
12. Kapcia, K.: Some properties of the model of a superconductor with pair hopping and magnetic interactions at half-filling. Acta Phys. Pol. A 121, 733 (2012)
13. Kapcia, K.: Interplay and competition between superconductivity and charge orderings in the zero-bandwidth limit of the extended Hubbard model with pair hopping and on-site attraction. J. Supercond. Nov. Magn. 26, 2647 (2013). 10.1007/s10948-013-2152-1
14. Kapcia, K.J.: Uporządkowania elektronowe i ich separacje w rozszerzonych modelach Hubbarda (in Polish). PhD thesis, Adam Mickiewicz University in Poznań, Poznań (2014). Accessed 6 Dec 2014. http://hdl.handle.net/10593/10857

15. Bari, R.A.: Superconductivity, ferroelectricity, and the Mott insulator. Phys. Rev. B 7, 2128 (1973). doi:10.1103/PhysRevB.7.2128
16. Ho, W.-C., Barry, J.H.: Tricritical behavior and phase diagrams of the Bari model for Mott insulator-superconductor transitions. Phys. Rev. B 16, 3172 (1977). doi:10.1103/PhysRevB.16.3172
17. Wiecko, C., Allub, R.: Superconductivity and random disorder in the zero-bandwidth limit. Phys. Rev. B 35, 2041 (1987). doi:10.1103/PhysRevB.35.2041
18. Robaszkiewicz, S.: The charge-ordered state in an extended Hubbard model. Phys. Status Solidi (b) 59, K63 (1973). doi:10.1002/pssb.2220590155
19. Micnas, R., Ranninger, J., Robaszkiewicz, S.: An extended Hubbard model with inter-site attraction in two dimensions and high-Tc superconductivity. J. Phys. C: Solid State Phys.21, L145 (1988). doi:10.1088/0022-3719/21/6/009
20. Hui, A., Doniach, S.: Penson-Kolb-Hubbard model: a study of the competition between single-particle and pair hopping in one dimension. Phys. Rev. B 48, 2063 (1993). doi:10.1103/PhysRevB.48.2063
21. Robaszkiewicz, S., Bułka, B.: Superconductivity in the Hubbard model with pair hopping. Phys. Rev. B 59, 6430 (1999). doi:10.1103/PhysRevB.59.6430
22. Ptok, A., Maśka, M.M., Mierzejewski, M.: The Fulde-Ferrell-Larkin-Ovchinnikov phase in the presence of pair hopping interaction. J. Phys.: Condens. Matter 21, 295601 (2009). doi:10.1088/0953-8984/21/29/295601
23. Ptok, A., Maśka, M.M., Mierzejewski, M.: Coexistence of superconductivity and incommensurate magnetic order. Phys. Rev. B 84, 094526 (2011). doi:10.1103/PhysRevB.84.094526
24. Czart, W.R., Robaszkiewicz, S.: Thermodynamic and electromagnetic properties of the Penson-Kolb model. Phys. Rev. B 64, 104511 (2001). doi:10.1103/PhysRevB.64.104511
25. Robaszkiewicz, S., Czart, W.R.: Superconductivity in the two-dimensional extended Hubbard model with pair-hopping interaction. Acta Phys. Pol. B 32, 3267 (2001) ADS
26. Czart, W.R., Robaszkiewicz, S.: Superconducting properties of the η-pairing state in the Penson-Kolb-Hubbard model. Acta Phys. Pol. A 106, 709 (2004)ADS
27. Japaridze, G.I., Kampf, A.P., Sekania, M., Kakashvili, P., Brune, Ph.: η-pairing superconductivity in the Hubbard chain with pair hopping. Phys. Rev. B 65, 014518 (2001). doi:10.1103/PhysRevB.65.014518
28. Dziurzik, C., Japaridze, G.I., Schadschneider, A., Zittartz, J.: Triplet superconductivity vs. easy-plane ferromagnetism in a 1D itinerant electron system with transverse spin anisotropy. Eur. Phys. J. B 37, 453 (2004). doi:10.1140/epjb/e2004-00081-5
29. Czart, W.R., Robaszkiewicz, S.: Properties of the extended Hubbard model with transverse (XY-type) spin-exchange interaction. Phys. Status Solidi (b) 243, 151 (2006). doi:10.1002/pssb.200562502
30. Czart, W.R., Robaszkiewicz, S.: Electron orderings of half-filled extended Hubbard models with spin-and charge-exchange interaction. Acta. Phys. Pol. A 109, 577 (2006)
31. Bąk, M.: Mixed phase and bound states in the phase diagram of the extended Hubbard model. Acta Phys. Pol. A 106, 637 (2004)
32. Robaszkiewicz, S., Pawłowski, G.: Superconductivity, charge orderings and phase separations in systems with local electron pairing. Acta Phys. Pol. A 90, 569 (1996)
33. Kapcia, K., Robaszkiewicz, S.: The effects of the next-nearest-neighbour density-density interaction in the atomic limit of the extended Hubbard model. J. Phys.: Condens. Matter23, 105601 (2011). doi:10.1088/0953-8984/23/10/105601ADS
34. Kapcia, K., Robaszkiewicz, S.: The effects of the next-nearest-neighbour density-density interaction in the atomic limit of the extended Hubbard model. J. Phys.: Condens. Matter23, 249802 (2011). doi:10.1088/0953-8984/23/24/249802ADS
35. Kapcia, K., Robaszkiewicz, S.: Stable and metastable phases in the atomic limit of the extended Hubbard model with intersite density-density interactions. Acta Phys. Pol. A 121, 1029 (2012)
36. Kapcia, K., Kłobus, W., Robaszkiewicz, S.: Charge orderings and phase separations in the atomic limit of the extended Hubbard model with intersite density-density interactions. Acta Phys. Pol. A 118, 350 (2010)
37. Robaszkiewicz, S., Micnas, R., Chao, K.A.: Thermodynamic properties of the extended Hubbard model with strong intra-atomic attraction and an arbitrary electron density. Phys. Rev. B 23, 1447 (1981). doi:10.1103/PhysRevB.23.1447
38. Micnas, R., Robaszkiewicz, S.: Thermodynamics of local-pair superconductors with anisotropic lattice structures. Phys. Rev. B 45, 9900 (1992). doi:10.1103/PhysRevB.45.9900
39. Micnas, R., Robaszkiewicz, S., Kostyrko, T.: Thermodynamic and electromagnetic properties of hard-core charged bosons on a lattice. Phys. Rev. B 52, 6863 (1995). doi:10.1103/PhysRevB.52.6863
40. Kłobus, W., Kapcia, K., Robaszkiewicz, S.: Magnetic orderings and phase separations in the zero-bandwidth limit of the extended Hubbard model with intersite magnetic interactions. Acta Phys. Pol. A 118, 353 (2010)
41. Murawski, S., Kapcia, K., Pawłowski, G., Robaszkiewicz, S.: On the phase diagram of the zero-bandwidth extended Hubbard model with intersite magnetic interactions for strong on-site repulsion limit. Acta Phys. Pol. A 121, 1035 (2012)
42. Murawski, S., Kapcia, K.J., Pawłowski, G., Robaszkiewicz, S.: Some properties of two dimensional extended repulsive Hubbard model with intersite magnetic interactions—a Monte Carlo study. Acta Phys. Pol. A 126, A-110 (2014). doi:10.12693/APhysPolA.126.A-110
43. Pawłowski, G., Robaszkiewicz, S.: Superconducting properties and phase separation effects in systems with local pairing. Acta Phys. Pol. A 91, 409 (1997)

44. Gerisch, T., Rieckers, A.: Limiting Gibbs states and phase transitions of a bipartite mean-field Hubbard model. J. Stat. Phys. 91, 759 (1998). doi: 10.1023/A:1023089930061MATHMathSciNet

45. Gerisch, T., Münzner, R., Rieckers, A.: Canonical versus grand-canonical free energies and phase diagrams of a bipolaronic superconductor model. J. Stat. Phys. 91, 1021 (1998). doi:10.1023/B:JOSS.0000033152.73539.a2

46. Bursill, R.J., Thompson, C.J.: Variational bounds for lattice fermion models II. Extended Hubbard model in the atomic limit. J. Phys. A: Math. Gen. 26, 4497 (1993). doi:10.1088/0305-4470/26/18/017MATHMathSciNet

47. Robaszkiewicz, S.: Magnetism and charge ordering in the Mott insulators. Acta Phys. Pol. A55, 453 (1979)

48. Kapcia, K., Kłobus, W., Robaszkiewicz, S.: Interplay between charge and magnetic orderings in the zero-bandwidth limit of the extended Hubbard model for strong on-site repulsion. Acta Phys. Pol. A 121, 1032 (2012)

49. Mancini, F., Plekhanov, E., Sica, G.: Spin and charge orderings in the atomic limit of the U-V-J model. J. Phys. Conf. Ser. 391, 012148 (2012). doi: 10.1088/1742-6596/391/1/012148

CHAPTER 5

UNCONVENTIONAL HIGH TEMPERATURE SUPERCONDUCTORS

Jiangping Hu

Keywords: CupratesIron-based superconductors, Unconventional high Tc superconducotors, Superexchange

INTRODUCTION

Almost three decades ago, the first family of unconventional high T_c superconductors, cuprates [1], was discovered. The discovery triggered intensive research and has fundamentally altered the course of modern condensed matter physics in many different ways. However, even today, after tens of thousands of papers devoted to the materials have been published, understanding their superconducting mechanism remains a major open challenge. Researchers in this field are sharply divided and disagree with each other on many issues arranging from minimum starting models to basic physical properties that are relevant to the cause of superconductivity. There is even a growing skepticism whether there are right questions that can be asked to settle the debate on the superconducting mechanism.

Many reasons can be attributed to the failure of answering the question of how superconductivity arises in cuprates. For example, material complexity makes theoretical modeling difficult, rich physical phenomena blind us from distinguishing main causes from side ones, and insufficient theoretical methods leave theoretical calculation doubtable. However, beyond all these difficulties and the absence of consensus, the lack of successfully realistic guiding principles to search for new high T_c superconductors from theoretical studies is the major reason. The failure was witnessed in the surprising discovery of the second family of high T_c superconductors, iron-based superconductors [2], in 2008. Today, those who are theory builders and those who are material synthesizers still remain disentangled.

Can valuable leads be provided from the theoretical side ahead of the potential discovery of the third family of high T_c superconductors? It is conceivable that the hope to settle high T_c mechanism relies on a positive answer to this question. Here, we believe that it is the time to seek a positive answer based on the following two reasons. First, in the past 7 years, the intensive research on iron-based superconductors has brought much new information. For those who believe that cuprates and iron-based superconductors should share a common high T_c mechanism, an opportunity to settle the debate arises as it is the first time that the traditional inductive reasoning becomes available in research. On one side, iron-based superconductors and cuprates share many common features, but on the other side they are not clones of each other. The similarities and differences can thus speak promising clues. Second, from the past massive searching efforts, it has become increasingly clear that unconventional high T_c superconductors are rare materials. Moreover, for the two known families, their superconductivities are carried robustly on CuO_2 layers in cupates and on FeAs/Se layers in iron-based superconductors respectively. The simultaneous existence of the rareness and robustness suggests that the unconventional high T_c superconductivity is tied to special ingredients in the electronic world, which define the gene of unconventional high T_c superconductivity. Thus, using inductive reasoning to identify the gene can open a new window to search for high T_c superconductors.

In this article, by taking the assumption that a common superconducting mechanism is shared by both known high T_c superconductors, we elucidate a recently emergent path to end the deadlock in solving high T_c mechanism by implementing inductive reasoning to reexamine the high T_c problem [3, 4]. This path stems from a simple framework that unifies cuprates and iron-based superconductors based on previous understandings in repulsive interaction or magnetically driven high T_c mechanisms. It suggests that iron-based superconductors are simply the counterpart of cuprates to realize robust extended s-wave pairing symmetries in a square lattice. Both materials share a key ingredient, the gene of unconventional high T_c superconductivity: a quasi two dimensional electronic environment in which the *d*-orbitals of cation atoms that participate in strong in-plane couplings to

Citation: Jiangping Hu, Unconventional high temperature superconductors, DOI 10.1007/s11434-016-1037-7.

the *p*-orbitals of anion atoms are isolated near Fermi energy. This environment allows the antiferromagnetic (AFM) superexchange couplings mediated through anions, the source of superconducting pairing, to maximize their contributions to superconductivity. Creating such a gene is tied to special arrangements between local electronic structures and crystal lattice structures, which explains why cuprates and iron-based superconductors are special and high T_c superconductors are so rare. The framework can be explicitly tested in future experiments as it leads to an explicit prediction to realize high T_c superconductivity in the Co/Ni-based materials with a quasi two dimensional hexagonal lattice structure formed by trigonal bipyramidal (TBP) complexes [3]. The new materials are predicted to be high T_c superconductors with a d±id pairing symmetry. If verified, the prediction will establish powerful guiding principles to search for high T_c superconductor candidates, as well as to settle the debate on unconventional high $_{Tc}$ superconducting mechanism.

QUESTIONS FOR UNCONVENTIONAL HIGH T_C SUPERCONDUCTIVITY

Implementing inductive reasoning to understand both cuprates and iron-based superconductors, we lay out the high T_c problem with the following three subsequent questions:

- What is the common interaction responsible for high T_c superconductivity in both families?
- What are the key ingredients to make both families special to host high T_c superconductivity?
- Where and how can we search for new high T_c superconductors?

The three questions are highly correlated. They form a self-contained unit to reveal high T_c superconducting mechanism.

In the past, the first question was the central question. Its answer was debated wildly. The second question was largely ignored. However, after the discovery of iron-based superconductors, it becomes clearer that the second question should be the central piece. While most researches have concentrated on these two families of high T_c superconductors, it is equally important to answer why numerous materials, which are similar to cuprates or iron-based superconductors in many different ways, do not exhibit high T_c superconductivity. Therefore, the essential logic here is that whatever our answer to the first question is, the answer must provide an answer to the second question. The answer to the second question can provide promising leads to answer the third question. An explicit theoretical prediction of new high T_c superconductors and its experimental verification can finally justify the answer of the first question to end the debate on high T_c mechanism.

THE ANSATZ TO THE FIRST QUESTION

We start with the first question. Our proposed answer to the first question is that only the superexchange AFM interactions mediated through anions are responsible for generating superconductivity in both families of high T_c superconductors. We call this ansatz as the selective magnetic pairing rule [4] in the repulsive interaction or magnetically driven superconducting mechanisms. One may argue that this answer is somewhat trivial as it has been accepted in a variety of models for cuprates [5, 6]. However, as we will discuss below, the answer is highly non-trivial in iron-based superconductors because their magnetisms are involved with different microscopic origins. Three main reasons to support this rule are:

- It naturally explains the robust d-wave pairing symmetry in cuprates and the robust s-wave pairing symmetry in iron-based superconductors;
- It is supported by a general argument that without the existence of mediated anions in the middle, the short-range Coulomb repulsive interactions between two cation atoms can not be sufficiently screened to allow superconducting pairing between them;
- It places strict regulations on electronic environments that can host high T_c superconductivity and thus results in a straightforward answer to the second question.

The case of Cuprates

As we have pointed out above, the rule is a familiar ansatz in cuprates. It has provided a natural explanation to the d-wave pairing symmetry [7], arguably the most successful theoretical achievement in the studies of cuprates. In fact, historically, in determining the pairing symmetry of cuprates, the d-wave pairing symmetry was theoretically predicted before the emergence of major experimental evidence [8–10].

Here we briefly review the main theoretical approaches in obtaining the d-wave pairing symmetry in curpates. There are two types of approaches to obtain the d-wave pairing symmetry based on effective models built in a two-dimensional Cu square lattice (Fig. 1a). One is the traditional weak coupling approach. This approach starts with a closely nested Fermi surfaces in which the spin-density wave (SDW) instability can take place by onsite electron–electron repulsive interaction (the Hubbard interaction) [7, 8]. The other is the strong interaction approach. It starts directly with short-range magnetic exchange interactions. In cuprates, the magnetic exchange interactions are the nearest neighbor (NN) AFM superexchange interactions mediated through oxygen atoms [5, 9, 10]. Both approaches consistently predict d-wave superconducting states.

The consistency can be attributed to the following simple pairing symmetry selection rule: the pairing symmetry is selected by the weight of its momentum space form factor on Fermi surfaces [11]. This rule is based on the following observation in repulsive interaction or magnetically driven high T_c superconducting mechanism: the superconducting pairings are dominated on bonds with the strongest effective AFM exchange couplings. This rule has been emphasized in the second type of models with local AFM superexchange interactions [10, 12]. In the case of cuprates, the decoupling of the NN AFM superexchange interaction in the pairing channel results in two possible pairing symmetries: an extended s-wave with a superconducting order in the reciprocal space $\Delta s_{(k)} \propto \cos k_X + \cos k_Y$ and a d-wave with $\Delta_d(k) \propto \cos k_X - \cos k_Y$. With the Fermi surface shown in Fig. 2c, the d-wave form factor has a much larger amplitude on the Fermi surfaces than the extended s-wave. Thus, the d-wave is favored by opening much larger superconducting gaps to save more AFM exchange energy in the superconducting state. This rule is also behind the weak coupling approach based on the Hubbard model in cuprates [7]. As the Hubbard model only includes the onsite repulsive interactions and its kinetic part is dominated by the NN hopping, the leading effective AFM exchange couplings are also generated on the NN bonds. In fact, considering the AFM fluctuations near half-filling in the Hubbard model, the effective electron–electron interaction mediated by the AFM fluctuations in the pairing channel has the following property [7]: it starts with a large repulsive onsite interaction followed by an attractive interaction between two NN sites, and then oscillates between repulsive and attractive with a rapid decay as increasing the space distance. This property essentially tells us that the pairing is also dominated on the NN bonds.

The Case of iron-Based Superconductors

Comparing an FeAs/Se layer with a CuO_2 layer, as shown in Fig. 1a, b, we notice several important differences: (1) the As/Se atoms in the former are located exactly below or above the middle points of the four-Fe squares; (2) the distance between two NN Fe atoms is very short, which is only about 2.8 Å. This value is close to the lattice constant of the body-centered cubic Fe metal; (3) the distance between two next NN (NNN) Fe atoms is about 3.8 Å, which is close to the distance of two NN Cu atoms in the CuO_2 layer. These differences suggest that the magnetic exchange couplings between two NNN Fe atoms, just like those between two NN Cu atoms, are mediated by the *p*-orbitals of As/Se atoms. Thus the magnetic couplings between two NNN sites are dominated by superexchange mechanism. However, two *d*-orbitals between two NN Fe atoms have large overlap which causes direct hoppings and direct magnetic exchange couplings. Therefore, the NN exchange magnetic couplings have a different microscopic mechanism from the NNN ones. These differences explain why the effective magnetic models in iron-based superconductors are much complex and exhibit both itinerant and local types of magnetic characters [14].

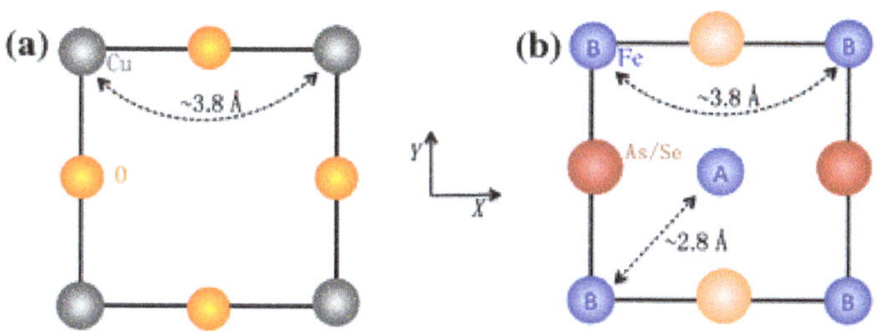

Figure 1: (Color online) The comparison of the lattice unit cells between cuprates and iron-based superconductors. **a** The unit cell and lattice constant of the CuO_2 layer in cuprates; **b** The unit cell and lattice constant of the FeAs/Se layer which includes two irons marked as A and B

The short NN distance and the existence of direct magnetic exchange mechanism also have a profound effect on the superconducting pairings. In cuprates, one can argue that the repulsive interaction between two NN Cu atoms can be ignored because of the existence of oxygen atoms in the middle, which create a large local electric polarization to screen the effective Coulomb interaction. This allows pairing to take place on the NN bonds. However, if there is a direct hopping between two atoms, there is no local electronic polarization to screen the Coulomb interactions between them. Thus, in iron-based superconductors, the repulsive interaction between two NN Fe sites must be large so that the pairing between NN bonds is essentially forbidden. But the physics between two NNN Fe sites are the same as those between two NN sites of cuprates. The effective Coulomb interaction between two NNN sites is screened by the strong electronic polarization created by As/Se atoms.

We can picture the above discussion in a simple manner. Considering the original two-iron unit cell as shown Fig. 1b, we label the two Fe sites in the unit cell as A and B respectively so that the Fe square lattice composes of two square sublattices, A and B. Each sublattice can be considered as an analogy of the Cu square lattice of cuprates, The pairings between the two lattices are forbidden due to the existence of strong repulsive interactions. The pairing exists only within each sublattice. Namely, as illustrated in Fig. 2b, the pairings are only allowed between different 2-Fe unit cells and are forbidden within the unit cells. Such an analogy allows us to apply the same pairing symmetry selection rules to predict the pairing symmetry of iron-based superconductors. If we draw the Fermi surfaces, as shown in Fig. 2d, in the Brillouin zone of the two Fe unit cell, which is also the Brillouin zone with respect to each sublattice, the Fermi surfaces are located either at the corner (*M*) or at the center (Γ). As shown in Fig. 2d, the form factor of the extended s-wave Δs(k) has a large weight on Fermi surfaces. Thus, the extended s-wave is clearly favored. The picture does not depend on the presence or absence of hole pockets at Γ points.

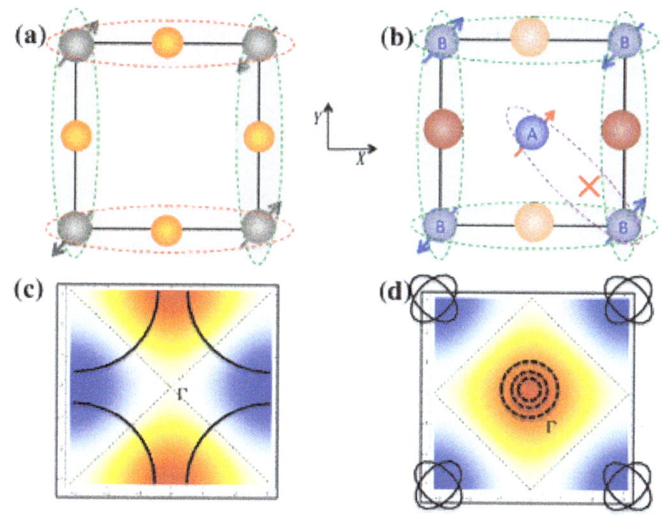

Figure 2: The comparison of superconducting pairings between cuprates and iron-based superconductors in both real and momentum spaces. **a** The real space pairing configuration in the d-wave superconducting state of cuprates; **b** The real space pairing configuration in the extended s-wave superconducting state in iron-based superconductors (the red multiplication sign indicates the forbidden pairing between A and B sublattices); **c** The Fermi surfaces of cuprates and the weight distribution of the d-wave order parameter in the momentum space (red and blue colors represent regions with large positive and negative values respectively); **d** The typical Fermi surfaces of iron-based superconductors and the weight distribution of the extended s-wave order parameter in the momentum space. The Fermi surfaces at Γ with dashed lines are hole pockets which can be absent in iron-chalcogendies [13].

The above discussion suggests that iron-based superconductors are simply a counterpart of cuprates to realize the extended s-wave pairing symmetry in a square lattice. The extended s-wave in iron-based superconductors endures the same robustness as the d-wave in cuprates. The robust s-wave symmetry in iron-based superconductors has been supported by overwhelming experimental evidence accumulated in the past several years [16, 16, 17]. This understanding explains the missing part in the previous theoretical studies which failed to obtain the robust s-wave pairing. In the previous studies based on weak coupling approaches [17], the repulsive interaction between the A and B sublattices is not seriously considered and only onsite repulsive interactions are considered in calculating pairing symmetries. With only onsite repulsive interaction, the effective attractive interactions are generated in both NN and NNN bonds. In general, the NN bonds favor the d-wave pairing symmetry [18] and the NNN bonds favor the extended s-wave symmetry. Thus, pairing symmetries from these models become very sensitive to the detailed parameters and Fermi surface properties [17, 18]. The same sensitivity also exists in the models based on local AFM J_1–J_2 exchange couplings [12]. With the existence of both J1J1, the NN AFM exchange couplings, and J_2, the NNN AFM exchange couplings, the phase diagram is very rich [12, 19]. The robust s-wave is only obtained when J_1 is argued to be inactive in providing pairing [20].

Summarizing above discussions, iron-based superconductors and cuprates can be unified in one superconducting mechanism. The former provides extreme valuable information to distinguish the roles of different magnetic interactions in providing superconducting pairing. The robust s-wave pairing symmetry in iron-based superconductors, just like the d-wave in cuprates, is a strong indiction to support the AFM superexchange couplings as the dominant sources for pairing.

THE ANSWER TO THE SECOND QUESTION

As we have mentioned earlier, the challenge is that the answer to the first question has to result in a natural answer to the second question. To show that this is the case for the above ansatz, we first discuss explicit conditions posed by the answer to the first question. Then, we discuss how both cuprates and iron-based superconductors fulfill these conditions. Finally, we address why it is difficult to satisfy these conditions and explain why unconventional high T_c superconductors are rare.

Conditions and Rules for Unconventional High T_c Superconductivity

In order to generate the strong AFM superexchange couplings and maximize their contributions to high T_c superconductivity, we can argue the following specific requirements for potential high T_c candidates:

- The necessity of cation–anion complexes: as the AFM superexchange couplings are mediated through anions, the potential candidates must include structural units constructed by cation–anion complexes. Within the units, there must be shared anions between two neighboring complexes. Moreover, strong chemical bondings between two anions should be forbidden as they generally destroy the AFM exchange processes.
- The orbital selection rule: the orbitals of cation atoms that participate in strong chemical bondings with anion atoms to generate strong AFM superexchange couplings must play a dominant role near Fermi energy. The best electronic environment for high T_c superconductivity is achieved when these orbitals are isolated near Fermi energy. Namely, the band structures near Fermi energy should be dominated by the orbitals of cation atoms whose kinematics are generated through the couplings to anions. We will show that this requirement essentially answers why cuprates and iron-based superconductors are special to host high T_c superconductivity. It is the most powerful rule to narrow our search for potential high T_c candidates. Following this rule, we can combine symmetry analysis and density functional theory (DFT) to search for new high T_c electronic environments. This rule has been implicated in cuprates as an orbital distillation effect based on the observation that the higher T_c is achieved when $d_{x^2-y^2}$-orbitals are dressed less by d_{z^2} orbitals in cuprates [21].
- The pairing symmetry selection rule: we have explicitly discussed this rule above. This rule allows us to link pairing configurations in real and momentum spaces directly. Following this rule, we may be able to design structures to realize superconducting states with specific pairing symmetries.
- Electron–electron correlation and half-filling: the atomic orbitals in cation atoms that can produce strong AFM superexchange couplings require to balance their spatial localization and extension. Moreover, in general, the strong AFM superexchange couplings are achieved when the orbitals are close to be half-filling. Thus, the half-filled *3d* orbitals in transition metal elements are clearly the best choices.
- Dimensionality: for *d*-orbitals, due to their two-dimensional nature in the spatial configuration, the orbital selection rule naturally demands a quasi two dimensional electronic environment. In an electronic band structure with strong three-dimensional band dispersions, it is difficult to maintain a purified orbital character. While one may argue that it is possible to satisfy these requirements in quasi one dimensional electronic environments, finding such an example is extremely difficult.

Summarizing these conditions and rules for transition metal based compounds, we can specifically define the gene of high T_c superconductors as a quasi two dimensional electronic structure in which the *d*-orbitals of cation atoms that participate strong in-plane chemical bonding with the *p*-orbitals of anion atoms are isolated near Fermi energy. In the following two subsections, we show that both cuprates and iron-based superconductors are special materials to carry such a gene.

The Case of Curpates

Cuprates belong to perovskite-related structural materials. The perovskite-related structures are the most popular and stable structures in nature. In a perovskite-related structure, the basic building block is the cation–anion octahedral complex shown in Fig. 3a. In cuprates, the CuO_6 octahedral complexes form two dimensional CuO_2 layers to provide a quasi two dimensional electronic structure. In a pure CuO_6 octahedral complex, the five *d*-orbitals of the Cu atom are split into two groups by crystal fields, t_{2g} and eg, as shown in Fig. 3c. The energies of the two $_{eg}$ orbitals, d_{z^2} and $d_{x^2-y^2}$, due to their strong couplings to the surrounding oxygen atoms, are higher. Moreover, in the CuO_2 layer, the energy of d_{z^2} orbital is lowered either by the Jahn–Teller effect or by the absence of apical oxygen atoms. Thus, the local energy configuration at cation sites is described according to Fig. 3d in which the $d_{X^2-Y^2}$ orbital sits alone at the top.

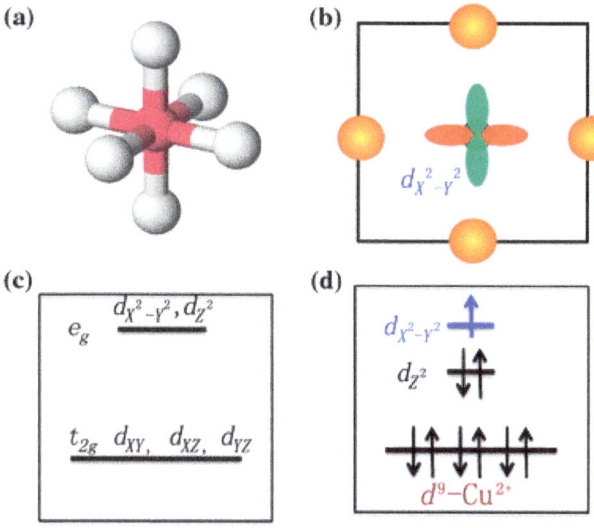

Figure 3: Local electronic environment and selected orbitals in cuprates. a The sketch of an octahedral complex; b The coupling configuration of the selected $d_{X^2-Y^2}$ orbital in CuO_2 layers; c The crystal field splitting of cation d-orbitals in an octahedral complex; d The true local energy configurations at Cu sites in curpates to indicate that the blue orbital, $d_{X^2-Y^2}$, is selected in the d^9 filling configuration.

It is easy to notice that only the $d_{X^2-Y^2}$ orbital has strong in-plane couplings to the *p*-orbitals of oxygens to mediate strong AFM superexchange couplings. Namely, only the electronic band attributed to the $d_{X^2-Y^2}$ orbital can support high T_c superconductivity. To isolate the $d_{X^2-Y^2}$ orbital near Fermi energy, nine electrons on the d shell are required. Thus, the gene of high T_c superconductivity can only be satisfied in a d^9 filling configuration at cation sites, which explains why $Cu^{2+}$is a natural choice. As a matter of fact, in the past several decades, numerous transition metal compounds with perovskite-related structures were discovered. Except curpates, none of them exhibits high T_c superconductivity.

The Case of Iron-Based Superconductors

The electronic physics of iron-based superconductors locates on the two dimensional FeAs/Se layers. The layers are constructed by edge-shared tetrahedral $FeAs_4(Se_4)$ complexes shown in Fig. 4a. The four coordination tetrahedral complex, just slightly less popular than the octahedral complex, is another important structure unit to form crystal lattices.

In a tetrahedral complex, as shown in Fig. 4c, the t_{2g} orbitals have higher energy than the e_g orbitals because of their strong couplings to anions. Under such a configuration, one may jump to argue that a d^7 filling configuration can make all t2g orbitals near half-filling to satisfy the gene requirements. However, the argument is misleading because of the following two major reasons. First, the crystal field energy splitting in a tetrahedral complex between the t_{2g} and the e_g orbitals is much smaller than the one in the octahedral complex. Second, the $d_{x^2-y^2}e_g$ orbital has very large dispersion due to the short NN Fe–Fe distance in FeAs/Se layers. Therefore, the simple argument can not exclude $d_{x^2-y^2}e_g$ orbitals near the Fermi energy.

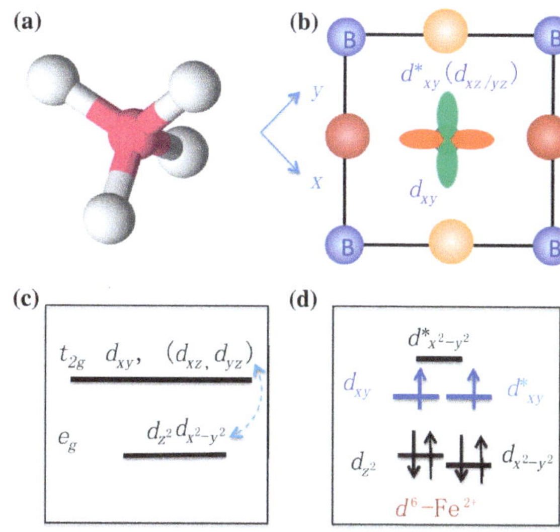

Figure 4: Local electronic environment and selected orbitals in iron-based superconductors. **a** The sketch of an tetrahedral complex; **b** The coupling configurations of the selected d_{xy}-type of orbitals to anion atoms in FeAs/Se layers; **c** The crystal field splitting of cation d-orbitals in an tetrahedral complex; **d** The local energy configurations at Fe sites in iron-based superconductors (the blue orbitals are isolated in d^6 filling configuration to dominate electronic physics near Fermi energy).

However, if we carefully examine the 2-Fe unit cell, because of the short distance between two NN Fe atoms, the local electronic environment of an Fe atom is not only affected by the four surrounding As/Se atoms in the tetrahedral complex but also the four neighboring Fe atoms. In fact, the d_{xz} and d_{yz} orbitals are strongly coupled to the $d_{x^2-y^2}e_g$ -orbitals of the neighboring Fe atoms. Thus, a more complete picture is that the d_{xz} and d_{yz} orbitals form two molecular orbitals. One of them, which has $d_{x^2-y^2}$ symmetry character, strongly couples to the $d_{x^2-y^2}e_g$ -orbitals of the neighboring Fe atoms. The coupling pushes this orbital to higher energy. The other, which has d_{xy} symmetry character, remains to a pure orbital with strong couplings to the surrounding As/Se atoms. Therefore, the more accurate local energy configuration is given by Fig. 4d, in which there are two d_{xy} type of orbitals in the middle in which one of them is formed by $d_{xz/yz}$ orbitals. These two orbitals can host possible high T_c superconductivity. With this configuration, we immediately determine that the $3d^6$ configuration of Fe^{2+} is special to satisfy the gene requirements.

The above energy configuration has been hidden behind the simplified effective two-orbital models constructed for iron-based superconductors [22]. Near Fermi energy, the two-orbital effective model was shown to capture the band dispersions of the five-orbital models that was derived by fitting DFT calculations [23, 24]. If we check the symmetry characters of the two orbitals in the two-orbital model, both of them have d_{xy} symmetry characters rather than $d_{xz/yz}$ interpreted in the original paper [22].

The above analysis suggests that the electronic structure in iron-based superconductors realizes the high T_c gene. As a matter of fact, we also notice that there are a variety of materials based on other transition metal elements with identical structures to iron-based superconductors. However, none of them exhibits high T_c superconductivity.

THE ANSWER TO THE THIRD QUESTION

A clear message from above discussion is that the genes of high $_{Tc}$ superconductivity stem from very special collaborations between the local electronic physics of cation–anion complexes and crystal structures. We can argue that symmetry play the key role behind the collaboration. In fact, we can argue that it is the symmetry collaboration between local complex and global crystal structures to make it possible to realize high T_c genes.

Octahedral/Tetrahedral Complexes and Square Lattice Symmetry

Both octahedral and tetrahedral complexes have a fourfold rotation principal axis. Their *d*-orbitals are classified locally by C_4 and S_4 rotation symmetries respectively. If a d orbital can be isolated in a band structure, it should have a similar classification in constructed crystal structures. This argument suggests that a square lattice symmetry is required to fulfill the gene conditions for materials constructed by octahedral and tetrahedral complexes. Both cuprates and iron-based superconductors indeed have square lattice symmetry. The selected orbitals that produce high T_c genes are classified identically in the symmetry groups of the crystal lattices and their local complexes. This correspondence allows them to be isolated in the electronic structures near Fermi energy without messing up with other orbitals.

The octahedral or the tetrahedral complexes are the most common structures in nature. They can form many different two dimensional crystal lattices. If we consider crystal structures formed by these complexes beyond the square lattice symmetry, such a correspondence is absent and different orbital characters generally get mixed. Thus, it is difficult to make the targeted orbitals to be isolated in band structures of non-square lattices formed by these two complexes, such as trigonal or hexagonal lattice structures, to fulfill the gene conditions. This explains why cuprates and iron-based superconductors are close to be unique systems to host the high T_c genes in materials constructed by octahedral and tetrahedral complexes.

Prediction of Trigonal/Hexagonal High T_c Electronic Environments

The symmetry collaboration argument suggests that if we want to create a high T_c gene in trigonal/hexagonal lattice structures, we may have to search lattices built by cation–anion complexes with threefold or sixfold principal rotation axis. Thus, we examine trigonal bipyramidal complexes (TBP) shown in Fig. 5a, which is a five coordination complex and carries a threefold principal rotation axis. The two dimensional hexagonal structure formed through conner-shared TBP, shown in Fig. 5b, has appeared in Mn-based $YMnO_3$ [25, 26] and Fe-based $Lu_{1-x}Sc_xFeO_3$ [27].

The explicit prediction is that a d^7 filling configuration, which can be realized by Co^{2+} or Ni^{3+} cations, fulfills the gene conditions of high T_c superconductivity in a material that carries above two dimensional hexagonal layers. Moreover, the pairing symmetry selection rule predicts that the superconducting states in these materials close to the d^7 filling configuration have a d±id dpairing symmetry.

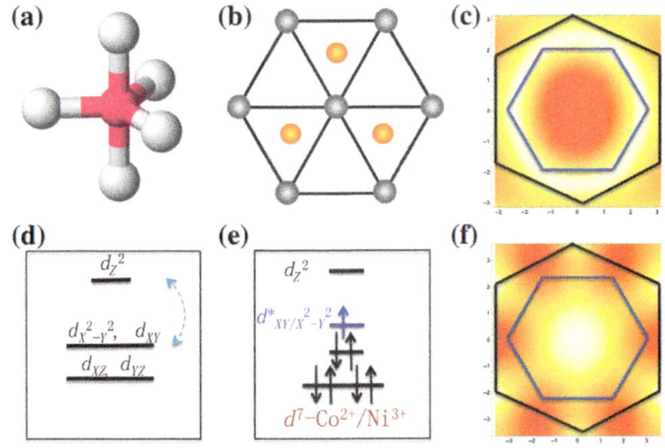

Figure 5: Predicted Co/Ni-based hexagonal lattices constructed by trigonal bipyramidal (TBP) complexes: **a** The sketch of a TBP complex; **b** The two-dimensional hexagonal layer formed by TBP complexes; **c** The weight distribution of an extended s-wave and Fermi surfaces (red color indicates large absolute values); **d** The crystal field splitting of cation d-orbitals in a TBP complex; **e** The local energy configurations at cation Co/Ni sites in an hexagonal layer; **f** similar to (**c**), the weight distribution of d±id-wave and Fermi surfaces.

The crystal field energy splitting of the *d*-orbitals in TBP is shown in Fig. 5. The d_{z^2} orbital has the highest energy due to its strong couplings to the two apical anions. The double degenerate $d_{x^2-y^2}$ and d_{xy} orbitals are strongly coupled to the in-plane anions. The double degenerate d_{xz} and d_{yz} orbitals have the lowest energy and are only weakly coupled to anions. As the hexagonal lattice is formed by three corner-shared TBPs, the $d_{x^2-y^2}$ and d_{xy} orbitals form two molecular orbitals. One of them can strongly couple to the d_{z^2} orbital so that the degeneracy is lifted. As the d_{z^2} orbital has higher energy, the coupling lowers the energy level of the coupled molecular orbital. The other is completely isolated from other orbitals and can be selected to provide the desired high T_c electronic environment. A local energy configuration is described by Fig. 5e. Thed7d7 filling configuration can fulfill the gene conditions. The DFT calculation on such a structure confirms this picture [3].

Around the d^7 filling configuration, a quasi two dimensional band structure is formed and electronic physics near Fermi energy is dominated by a single band attributed to the selected orbital. The band has a Fermi surface shown in Fig. 5c, f. If we apply the pairing symmetry selection rule, as the pairing should be dominated on the NN bonds in the cation trigonal lattice, for the extended *s*-wave pairing, the form factor of the gap function in the momentum space is given by $\Delta_s \propto \cos k_y + 2\cos\frac{\sqrt{3}}{2}k_x\cos\frac{1}{2}k_y$, , and for the d±id-wave pairing, the factor is given by $\Delta_d \propto \cos k_y - \cos\frac{\sqrt{3}}{2}k_x\cos\frac{1}{2}k_y \pm i\sqrt{3}\sin\frac{\sqrt{3}}{2}k_x\sin\frac{1}{2}k_y$. Figure 5c, f illustrates the overlap between the amplitude of the two form factors with Fermi surfaces. The degenerate d±id-waves collaborate well with Fermi surfaces near half filling. Therefore, the system supports a robust d±id d-wave pairing superconducting state.

The superconducting transition temperature can be estimated by comparing the energy scales of the couplings between cations and anions in complexes. The Cu–O couplings in the octahedral complex of cuprates are more than twice stronger than the Fe–As/Se couplings in the tetrahedral complex of iron-based superconductors. The ratio of the maximum T_{cs} observed in these two families is in the similar order. In the TBP complex, the coupling strength sits between them and is about 2/3 of those in cuprates. Thus, the maximum T_c that can be realized in the TBP structure is expected to be around 100 K as the maximum T_c in cuprates can reach 160 K.

Materials constructed by the TBP complexes are very limited. The Co/Ni based materials described above have not been synthesized. Thus, it is an explicit prediction to be tested in future experiments.

DISCUSSION

The above prediction, if verified, justifies our answer to the first question. But most importantly, the verification can open the door to theoretically design and search for new unconventional high T_c superconductors. A general search procedure can be: (1) design a possible lattice structure that can be constructed by certain cation–anion complexes; (2) use symmetry tools to understand local electronic physics; (3) perform standard DFT calculations to obtain band structures and its orbital characters; (4) apply the gene requirements to determine conditions and likelihood on the existence of high T_c superconducting environment; (5) design realistic materials whose lattice structures can be stabilized.

In designing electronic environments for high T_c superconductivity, there are helpful clues and possible directions. For example, we can ask whether we can design crystal structures for all 3*d*transition elements to realize high T_c superconductivity. As the *d*-orbitals which are responsible for high T_c superconductivity must make strong couplings to anion atoms, they typically gain energy in the crystal field environment, which explains why all high T_c superconductors, including predicted Co/Ni-based materials, are involved with the second half part of the 3*d*transition elements in the periodic table. Whether we can overcome this limitation to make specific designs for the first half 3*d* transition elements, in particular, Mn and Cr, is a very intriguing question. Another example is to ask whether we can design superconducting states with particular pairing symmetries as we have explicit rules to determine pairing symmetries. We have noted that cuprates and iron-based superconductors are examples of the d-wave and the extended s-wave pairing symmetry in a square lattice. Our predicted material is a realization of the d-wave pairing symmetry in trigonal/hexagonal lattice structures. Thus, for example, we can ask a specific question about how to realize an extended s-wave in the trigonal/hexagonal lattice structures.

We have mainly focused on the 3*d* orbitals which are known to produce the strongest correlation effect. However, even if carrying less electron–electron correlation effect, we can also consider other type of orbitals at cation sites, including 4*d*, 5*d*, 4*f*, 5*f* and even higher level*s*-orbitals. We can search for materials as potential unconventional superconductors in which the kinematics of these orbitals near Fermi energy can be isolated and is generated through strong couplings to the *p*-orbitals of anions. In general, as long as there is a charge transfer energy gap between orbitals of cations and anions, the AFM superexchange coupling should be generated. Thus, moderate high T_c may be achieved. For 5*d* and 5*f* orbitals, because of large spin orbital couplings, the orbitals can be reconfigured to have drastically different real space configurations. This may result in more possible designs on crystal lattice structures to generate superconducting states. One example is Sr_2IrO_4 [28], which can be considered as a lower-energy scale clone of cuprates [29, 30]. For the *s*-orbitals, as they are symmetric in space, we may design a cubic-type three dimensional lattice structure to achieve the conditions.

In summary, cuprates and iron-based superconductors can be unified in a framework based on repulsive interaction or magnetically driven high T_c mechanisms. This unification leads to important rules to regulate electronic environments required for unconventional high T_c superconductivity. The rules can guide us to search for new high T_c superconductors. Following these rules, we made an explicit prediction about the existence of high T_c superconductivity in the Co/Ni-based two dimensional hexagonal lattice structure constructed by trigonal bipyramidal complexes. Verifying this prediction can pave a way to establish unconventional high T_c mechanism.

Acknowledgments

The author acknowledges helpful discussions with D.L. Feng, X. H. Chen, H. Ding and D. Scalapino. The author thanks X.X. Wu, C.C. Le and J. Yuan for carrying out numerical calculations. The work was supported by the National Basic Research Program of China, National Natural Science Foundation of China and the Strategic Priority Research Program of Chinese Academy of Sciences.

Conflict of Interest

The authors declare that they has no conflict of interest.

REFERENCES

1. Bednorz JG, Muller KA (1986) Possible high Tc superconductivity in the Ba—La—Cu—O system. Z Phys B 64:189–193
2. Kamihara Y, Watanabe T, Hirano M et al (2008) Iron-based layered superconductor LaO1x Fx FeAsδx ¼ 0:05 0:12Þ with Tc ¼ 26 k. J Am Chem Soc 130:3296–3296
3. Hu JP, Le CC, Wu XX (2015) Predicting unconventional high temperature superconductors in trigonal bipyramidal coordinations. Phys Rev X 5:041012
4. Hu JP, Yuan J (2015) Robustness of s-wave pairing symmetry in iron-based superconductors and its implications to fundamentals on magnetically-driven high temperature superconductivity. arXiv:1506.05791
5. Anderson PW, Lee PA, Randeria M et al (2004) The physics behind high-temperature superconducting cuprates: the ‘plain vanilla’ version of RVB. J Phys Condens Matt 16:R755–R769
6. Scalapino DJ (1999) Condensed matter physics—the cuprate pairing mechanism. Science 284:1282–1283
7. Scalapino D (1995) The case for dx2-y2 pairing in the cuprate superconductors. Phys Rep 250:329–365
8. Bickers NE, Scalapino DJ, Scalettar RT (1987) CDW and SDW mediated pairing interactions. Int J Mod Phys B 1:687–695

9. Gros C, Poilblanc D, Rice TM et al (1988) Superconductivity in correlated wavefunctions. Physica C 153:543–548
10. Kotliar G, Liu J (1988) Superexchange mechanism and d-wave superconductivity. Phys Rev B 38:5142–5145
11. Hu JP, Ding H (2012) Local antiferromagnetic exchange and collaborative fermi surface as key ingredients of high temperature superconductors. Sci Rep 2:381
12. Seo KJ, Bernevig BA, Hu JP (2008) Pairing symmetry in a twoorbital exchange coupling model of oxypnictides. Phys Rev Lett 101:206404
13. Dagotto E (2013) Colloquium: the unexpected properties of alkali metal iron selenide superconductors. Rev Mod Phys 85:849–867
14. Dai P, Hu JP, Dagotto E (2012) Magnetism and its microscopic origin in iron-based high-temperature superconductors. Nat Phys
15. 8:709–718
16. Richard P, Qian T, Ding H (2015) Arpes measurements of the superconducting gap of Fe-based superconductors and their implications to the pairing mechanism. ArXiv:1503.07269
17. Fan Q, Zhang WH, Liu X et al (2015) Plain s-wave superconductivity in single-layer FeSe on SrTiO3 probed by scanning tunnelling microscopy. Nat Phys 11:946–952
18. Hirschfeld PJ, Korshunov MM, Mazin II (2011) Gap symmetry and structure of Fe-based superconductors. Rep Prog Phys 74:124508
19. Maier TA, Graser S, Hirschfeld PJ et al (2011) d-wave pairing from spin fluctuations in the Kx Fe2y Se2 superconductors. Phys Rev B 83:100515
20. Yu R, Zhu J, Si Q (2014) Orbital-selective superconductivity, gap anisotropy and spin resonance excitations in a multiorbital t j1 j2 model for iron pnictides. Phys Rev B 89:024509
21. Fang C, Wu YL, Thomale R et al (2011) Robustness of s-wave pairing in electron-overdoped Að1yÞ Feð2xÞ Se2 (a ¼ K, Cs). Phys Rev X 1:011009
22. Sakakibara H, Suzuki K, Usui H et al (2013) Three-orbital study on the orbital distillation effect in the high tc cuprates. Phys Proc 45:13–16
23. Hu J, Hao N (2012) S4 symmetric microscopic model for ironbased superconductors. Phys Rev X 2:021009
24. Kuroki K, Onari S, Arita R et al (2008) Unconventional pairing originating from the disconnected fermi surfaces of superconducting LaFeAsO1x Fx. Phys Rev Lett 101:087004
25. Graser S, Maier TA, Hirschfeld PJ et al (2009) Near-degeneracy of several pairing channels in multiorbital models for the Fe pnictides. New J Phys 11:5016–5016
26. Yakel HL, Koehler WC, Bertaut EF et al (1963) On the crystal structure of the manganese III trioxides of the heavy lanthanides and yttrium. Acta Cryst 16:957–962
27. Smolenskii GA, Bokov VA (1964) Coexistence of magnetic and electric ordering in crystals. J Appl Phys 35:915–918
28. Masuno A, Ishimoto A, Moriyoshi C et al (2015) Expansion of the hexagonal phase-forming region of LuScFeO by containerless processing. Inorg Chem 54:9432–9437
29. Crawford MK, Subramanian MA, Harlow RL et al (1994) Structural and magnetic studies of Sr2 IrO4. Phys Rev B 49:9198–9201
30. Wang F, Senthil T (2011) Twisted hubbard model for Sr2 IrO4: Magnetism and possible high temperature superconductivity. Phys Rev Lett 106:136402
31. Yan YJ, Ren MQ, Xu HC et al (2015) Electron-doped Sr2 IrO4: an analogue of hole-doped cuprate superconductors demonstrated by scanning tunneling microscopy. Phys Rev X 5:041018

CHAPTER 6 ELECTRONIC ENERGY LEVELS IN HIGH-TEMPERATURE SUPERCONDUCTORS

H. P. Roeser, D. T. Haslam, J. S. López, M. Stepper, M. F. von Schoenermark, F. M. Huber, A. S. Nikoghosyan

Keywords: High-temperature superconductor, Superconductor crystal structure, Superconducting current channel, Superconducting unit area

INTRODUCTION

High-temperature superconductors (HTSC) are known since the discovery by Bednorz and Müller in 1986 and the number of different HTSC materials is still increasing. Modern techniques allow very detailed investigations even down to the atomic level within the unit cell of a HTSC crystal. Nevertheless the physical phenomenon is still not fully understood. One important aspect is the role of optimum doping to achieve the maximum critical transition temperature T_c. For example, varying in cuprates the oxygen content or the heat treatment of the materials dramatically change their transition temperatures, critical current densities, magnetic fields and other properties. This paper is an attempt to establish a link between T_c and the doping and crystal structure in HTSC materials.

CRYSTAL STRUCTURE AND PARTICLE IN THE BOX CONCEPT

The energy level estimate for electronic excitation in an atom is given by

$$E = \frac{h^2}{8m_e(d_A)^2} \approx 6.7 \times 10^{-19}\ \text{J} \approx 4\ \text{eV} \qquad (1)$$

where m_e is the electron mass and d_A the atomic diameter leading to a typical value of 4 eV for $d_A = 0.3$ nm [1]. Equation (1) can be considered as the lowest energy level E_1 of a particle in a box (PiB) with a quantum well width equal to the diameter of the atom.

The parent materials of high-temperature superconductors, e.g. cuprates, are usually electrical insulators. They resist the flow of electric current because their valence electrons are tightly bonded to their atoms. Therefore, these materials should show no electronic excitation for conducting electrons or holes, but only the electronic excitation in an atom. A very well studied example is the anti-ferromagnetic insulator La_2CuO_4, which is the parent compound of many HTSCs. This ternary compound becomes superconducting by doping with an alkaline earth metal (Ba, Sr, Ca, Na, K, ...) and/or introduction of oxygen deficiency or interstitial oxygen excess within the crystal. It is well known that these measures create Cu3+-ions in the CuO2 plane of the doped unit cell of cuprates being partly responsible for superconducting carriers with mass $M_{eff} = 2m_e$. This is true for p-type (e.g. hole-doped $La_{2-\Delta}Ca_{1+\Delta}Cu_2O_{6+\delta}$) as well as n-type (e.g. electron-doped $Pr_{1-\Delta}LaCe_{\Delta}CuO_{4-\delta}$) HTSCs. De

Detailed measurements have been done for most of the HTSCs to optimize the doping for maximum T_c. It has been found that for each HTSC a homogeneous doping distribution is necessary and that there exist an optimum doping level with a parabolic decrease in T_c for lower and higher doping levels forming a bell-shaped curve. These doped materials are now conductors and therefore should show an electronic excitation for conducting electrons or holes. In this case the energy level estimate for a HTSC should read in analogy to (1)

$$E_1 \approx \frac{h^2}{8M_{eff}x^2} = 3.3 \times 10^{-21}\ \text{J} \approx 20\ \text{meV} \qquad (2)$$

where x is the distance between doped unit cells projected into the superconducting CuO_2 plane for cuprates. This leads to a typical value of 20 meV for x = 3 nm and $M_{eff} = 2m_e$ (see Sect. 4). Crystals built by tetragonal unit cells prefer to form a square homogeneous doping distribution whereas crystals built by orthorhombic unit cells could also form a hexagonal homogeneous doping distribution [2, 3]. In both cases the doping pattern in the CuO_2 plane offers straight lines or arrays of equidistant doped unit cells throughout the whole crystal.

Citation: H. P. Roeser, D. T. Haslam, J. S. López, M. Stepper, M. F. von Schoenermark, F. M. Huber, A. S. Nikoghosyan, Electronic Energy Levels in High-Temperature Superconductors, DOI: 10.1007/s10948-010-0850-5.

Therefore the electronic excitation could be related to the ground state E1 of a 1-dimensional particle in the box (PiB), which scales with the size of the doping separation x.

Equation (2) contains only one particle in the box, which ignores the fact that multi-layered CuO_2 planes (n > 1) per chemical formula offer even higher critical transition temperatures than single CuO_2-layer cuprates. A simple approach to account for this is the analogy between (2) and the Fermi energy E_F and its relation to the carrier density Nc given by

$$E_F = \frac{h^2}{8M_{\mathrm{eff}}} N_c^{2/3} \left[\frac{3}{\pi}\right]^{2/3} \tag{3}$$

If the carrier density increases by a factor n, the Fermi energy increases by $E_F \sim (nN_c)^{2/3} = n^{2/3}(N_c)^{2/3}$. This transforms (2) into

$$E_1 \approx \frac{h^2 n^{2/3}}{8M_{\mathrm{eff}}x^2} \tag{4}$$

On the other hand, a superconductor is characterized by its transition temperature T_c and the thermal energy of the crystal E_{th} related to it. Therefore, E_{th} is a function of kT_c expressed by $E_{th} = f(kT_c)$. We are interested in a correlation between Eth and the energy E_1, which is given by the doping separation x in a straight line of equidistant doped unit cells through the whole crystal. Thus, E_1 is replaced by $f(kT_c)$ and (4) can be written as

$$(2x)^2 n^{-2/3} = \frac{h^2}{2M_{\mathrm{eff}} f(kT_c)} \tag{5}$$

In the following, more than 20 HTSC materials will be analyzed by plotting $(2x)^2n^{-2/3}$ versus $1/T_c$. The doping separation x will be derived from the experimentally determined optimum doping for maximum T_c. The procedure for the analysis will be described in detail for the HTSC family $HgBa_2Ca_{n-1}Cu_nO_{2n+2+\delta}$.

HBCCO FAMILY

Since the discovery of $HgBa_2CuO_{4+\delta}$ in 1993 [4], which is the first member of the Mercury cuprate homologous series $HgBa_2Ca_{n-1}Cu_nO_{2n+2+\delta}$, a lot of effort has been spent to prepare bulk samples of HBCCO containing up to n = 6 layers of CuO_2. At normal pressure the mercury based component with n = 3 exhibits the highest $T_c \approx 135$ K. Therefore this HTSC family has been studied in great detail by many research groups to get insight into understanding HTSCs towards the achievement of hopefully higher transition temperatures [5]. The structural arrangements of the first three members of this family are illustrated in Figs. 1, 4 and 6. For all members the crystal structure is tetragonal and the CuO_2 planes are super flat [6]. The cell parameters and the doping levels at T_c or below have been used because in particular the doping values could change significantly from T = 300 K down to T < T_c. The synthesis and the quality of preparation for the different members of the Hg-family are very important to avoid transition temperatures which are caused

(a) by a mixture or the intergrowths e.g. of Hg-1212 and Hg-1223 and

(b) by multi doped members like $Hg_{0.98}Ba_{2.01}Ca_{1.01}Cu_{2.00}O_{6+\delta}$ [7],

instead of being the result of a pure single phase. The oxygen above O_{2n+2} stoichiometry is located in the Hg layers as shown in Figs. 1, 4 and 6.

Hg-1201

Figure 1 shows the tetragonal crystal structure of a singlephase $HgBa_2CuO_{4+\delta 1}$. The interstitial oxygen excess atom is placed in the Hg plane that forms the top and bottom of the unit cell (HgO_δ). Synthesizing single-phase Hg-1201 compounds with various oxygen contents over a range of 4.03– 4.14 showed a maximum $T_c = 95 \pm 1$ K at an optimized doping level of $\delta_1 \approx 0.08$ [4, 8]. The excess oxygen was homogeneously distributed.

The undoped Hg-1201 unit cell contains four oxygen atoms (4O2−), which require eight electrons resulting in the electronic configuration $1Hg^{2+} + 2Ba^{2+} + 1Cu^{2+} + 4O^{2-}$. This material has no Cu^{3+}-element and is not superconducting unless it is doped with an additional oxygen atom.

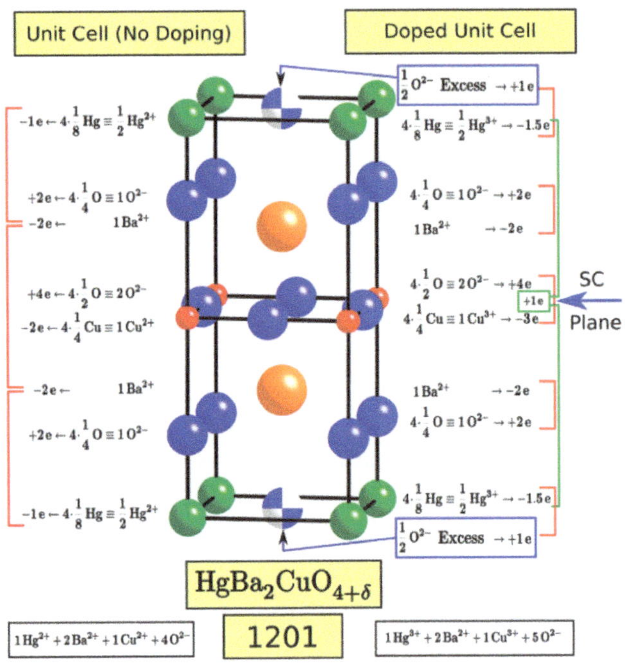

Figure 1: Unit cell of Hg-1201 illustrating the pure and oxygen doped electronic configuration.

A doped unit cell with $HgBa_2CuO_5$ needs two more electrons, which are provided by transforming Hg^{2+} into Hg^{3+} and Cu^{2+} into Cu^{3+} because Hg^{3+} and Cu^{3+} are the next elements in line with the lowest ionization energy (Fig. 2). This effect is described in literature as charge transfer from HgO_δ to CuO_2 planes. The HgO_δ plane is considered to act as "charge reservoir" and the oxygen excess is "doping" Cu^{3+}-holes into the CuO_2 plane.

A uniform distribution of doped unit cells with oxygen excess atoms arrange in a square distribution of Cu^{3+}- ions in the CuO_2 plane. The distance between two Cu3+- ion positions is always given by $x_1^2 = (z_1^2 + z_2^2)a^2$ with $z_1, z_2 = 0, 1, 2, \ldots$. and the lattice constant a forming an HTSC unit area of $(z_1^2 + z_2^2) = \Sigma$ of CuO_2 unit elements [3 and references therein]. The HTSC unit area has four doped unit cells with Cu^{3+}-ions at its corners counting as one doped element, and the doping density is thus given by $(\Sigma)^{-1}$. The optimum oxygen excess value is derived from the bell-shaped Tc versus δ1 curve, which is centered at $\delta_1 = 0.0825 \pm 0.005$ [8]. This results in a doping density of $(\Sigma)^{-1} = (4 + \delta_1)/4 - 1 \equiv 2.06\%$. According to the square pattern distribution for tetragonal unit cells Σ should by formed by $\Sigma = 48.5 \approx (z_1^2 + z_2^2) = (7^2 + 0^2) = 49$ CuO_2 unit elements. Therefore the doping distance results in $x_1 = \sqrt{49}a = 2.71 \times 10^{-9}$ m.

The length x_1 can also be calculated via the diagonal of the CuO_2 unit area by $x_1^2 = (z_3 a\sqrt{/2})^2 + (z_4 a\sqrt{/2})^2 = (2 \times 3.5^2 + 2 \times 3.5^2)a^2$ with $z_3 = z_4 = 3.5$. The analysis, mentioned in Sect. 2 for n = 1, results in $(2x)^2 n^{-2/3} T_c = 2.79 \times 10^{-15}$ m² K. Figure 3 shows the superconducting CuO_2 plane and the distribution of the unit cells with Cu^{3+}-ions. The distribution of the oxygen excess atoms has the same pattern in the Hg plane exactly above and below.

Figure 3: Homogeneous doping distribution of Hg-1201 with δ = 0.08 in the CuO_2 plane. The linear array of doped unit cells forms superconducting unit areas with four doped unit cells at each corner and superconducting current channels.

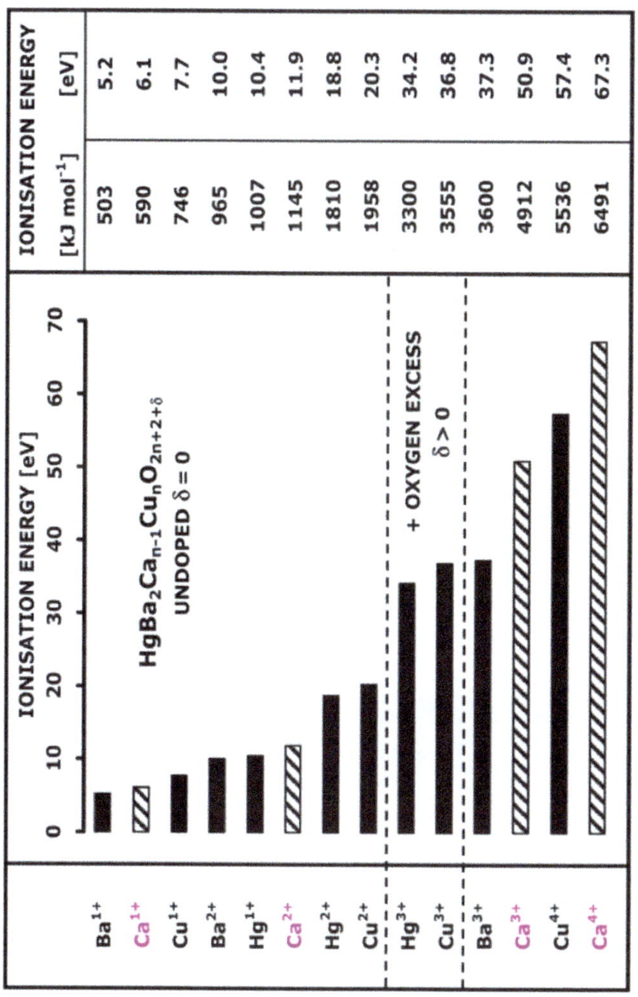

Figure 2: Ionization energy levels of the different atoms in $HgBa_2Ca_{n-1}Cu_nO_{2n+2+\delta}$.

Hg-1212

Figure 4 shows the tetragonal crystal structure of a singlephase $HgBa_2CaCu_2O_{6+\delta2}$ (Hg-1212). Again, the interstitial oxygen excess atom is placed in the Hg plane. The undoped Hg-1212 unit cell contains six oxygen atoms ($6O^{2-}$), which require 12 electrons resulting in the electronic con- figuration $1Hg^{2+} + 2Ba^{2+} + 1Ca^{2+} + 2Cu^{2+} + 6O^{2-}$ with two CuO_2 planes. This material has no Cu3+-element and is not superconducting unless it is doped with an additional oxygen atom. Doping a unit cell with one oxygen atom transforms the electronic configuration into $1Hg^{3+} + 2Ba^{2+}$

$+ 1Ca^{2+} + 1Cu^{2+} + 1Cu^{3+} + 7O^{2-}$ and only one CuO_2 plane contains a Cu^{3+}-ion. Because there is no preference for one of the two CuO2 planes the Cu3+-ion will alternate between the two CuO_2 planes. This means that half of the oxygen doping will serve the upper CuO_2 plane and the other half the lower plane. Therefore the Cu3+-ion doping density for one CuO_2 plane will be determined by $0.5\,\Sigma^{-1} = 0.5[(6 + \delta_2)/6^{-1}]$. Assuming a homogeneously oxygen excess doping distribution in the Hg plane given by Σ^{-1} results in a square oxygen excess doping distance d_2 with $(d_2)^2 = \Sigma a^2 = (z_1^2 + z_2^2)a^2$. The square doping distribution of Cu^{3+}-ions in each CuO_2 plane will be given by $(x_2)^2 = (z_3^2 + z_4^2)a^2$ with the relation $x_2 = \sqrt{2}d_2$ as illustrated in Fig. 5. Therefore the two doping distributions result in the relation $(z_3^2 + z_4^2) = 2(z_1^2 + z_2^2)$ with a strong restriction by choosing appropriate z1 to z4 numbers.

The transition temperatures for single-phase Hg-1212 reported in literature have a relatively sharp value of 126 ± 1.5K[6, 9–12], whereas the oxygen doping varies by $\delta_2 = 0.21 \pm 0.01$ probably caused by the bell-shaped T_c

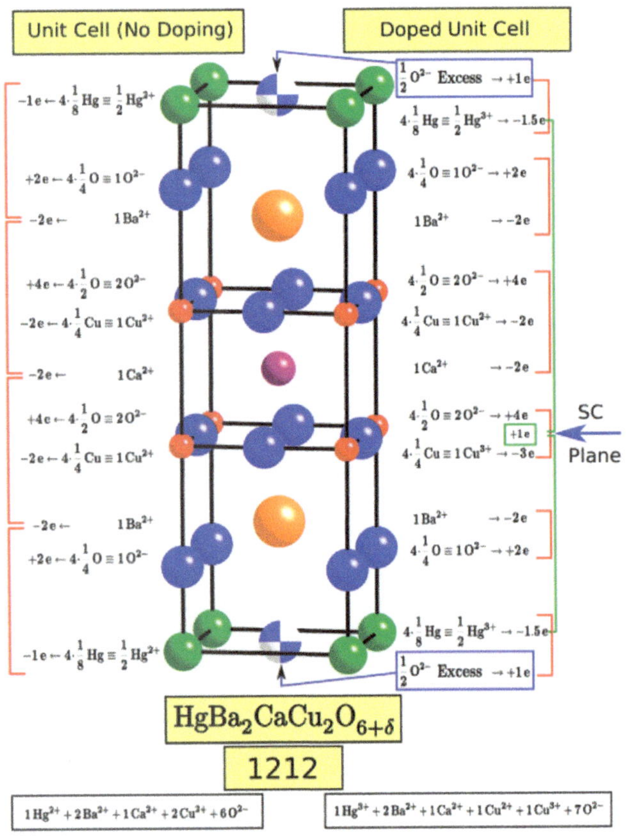

Figure 4: Unit cell of Hg-1212 illustrating the pure and oxygen doped electronic configuration. The Cu^{3+}-ion will alternate between the two CuO_2 planes resulting in two superconducting CuO_2 planes.

versus $_{x_2}$ curve. This results in a density for oxygen excess of $\Sigma^{-1} = 3.50 \times 10^{-2}$ or $\Sigma = 28.6 \approx 29 = (5^2 + 2^2)$

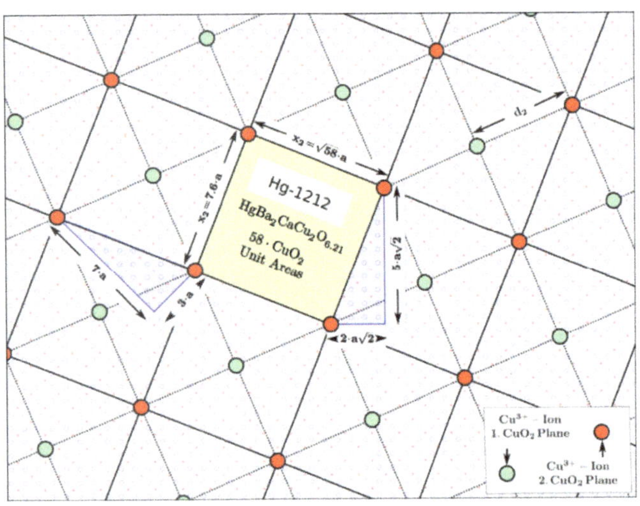

Figure 5: Red and filled circles represent the Cu^{3+}-ion distribution in the upper CuO_2 plane, green and open circles represent the Cu^{3+}-ion distribution in the lower CuO_2 plane of Hg-1212. Both together represent the oxygen excess distribution in the Hg plane separated by d_2. The Cu^{3+}-ion separation is given by $x_2 = \sqrt{2}d_2$

and $d_2 = \sqrt{29}a = 2.08 \times 10^{-9}$ m. The Cu^{3+}-ion distribution per CuO_2 plane is given by $x_2 = \sqrt{2}d_2 = \sqrt{58}a = \sqrt{(7^2 + 3_2)}a = 2.94 \times 10^{-9}$ m. The length x_2 can also be calculated via the diagonal of the CuO2 unit area by $x_2^2 = (z_5 a\sqrt{2})^2 + (z_6 a\sqrt{2})^2 = (8 + 50)a^2$ with $z_5 = 2$ and z6 = 5. Both CuO2 planes have the same Cu^{3+}-ion distribution but they are shifted to match the homogeneous oxygen distribution and current channels in both CuO_2 planes as illustrated in Fig. 5. The final calculation with n = 2 results in $(2x_2)^2 n^{-2/3} T_c = 2.75 \times 10^{-15}\ \mathrm{m}^2$ K.

Hg-1223

Figure 6 shows the crystal structure of the unit cell for a single-phase $HgBa_2Ca_2Cu_3O_{8+\delta3}$ (Hg-1223) including the electronic configuration for the undoped and doped unit cells. As for Hg-1201 and Hg-1212 the oxygen excess atom can only serve one CuO_2 plane to create a Cu^{3+}-ion. Therefore only 1/3 of the homogeneously oxygen excess doping distribution with density (Σ^{-1}) will be responsible for the Cu_{3+}-ion distribution in each CuO_2 plane. This requires a nearly perfect regular triangle doping structure for the oxygen excess atoms on top/bottom of the unit cell with side length d_3 as well as for the Cu^{3+}-ion distribution in each CuO_2 plane with side length x_3. But the center points of tetragonal unit cells cannot form a perfect regular triangle structure. Therefore we are looking for the closest solution with the condition that the Cu^{3+}-ion distribution should be lined up in straight lines with equidistant Cu^{3+}-ion positions.

The fact that only one of the four Cu atoms at the corner of a unit cell can be transformed into a Cu^{3+}-ion the triangular pattern requires the consideration that for the distance

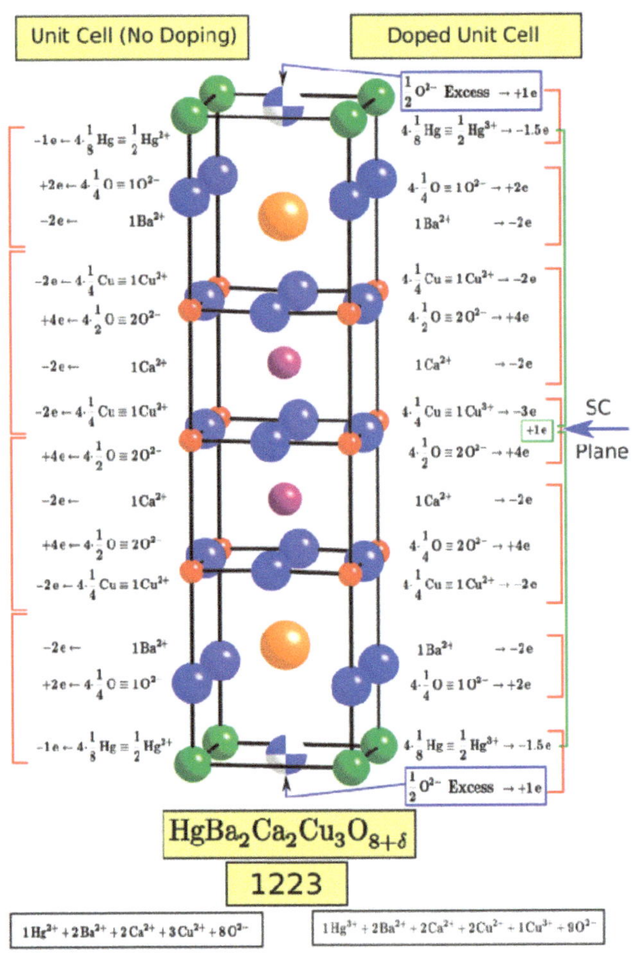

Figure 6: Unit cell of Hg-1223 illustrating the pure and oxygen doped electronic configuration. The Cu^{3+}-ion will alternate between three CuO2 planes resulting in three superconducting CuO_2 planes.

between two doped unit cells a position shift should be taken into account so that a position shift so that a Cu^{3+}-ion must

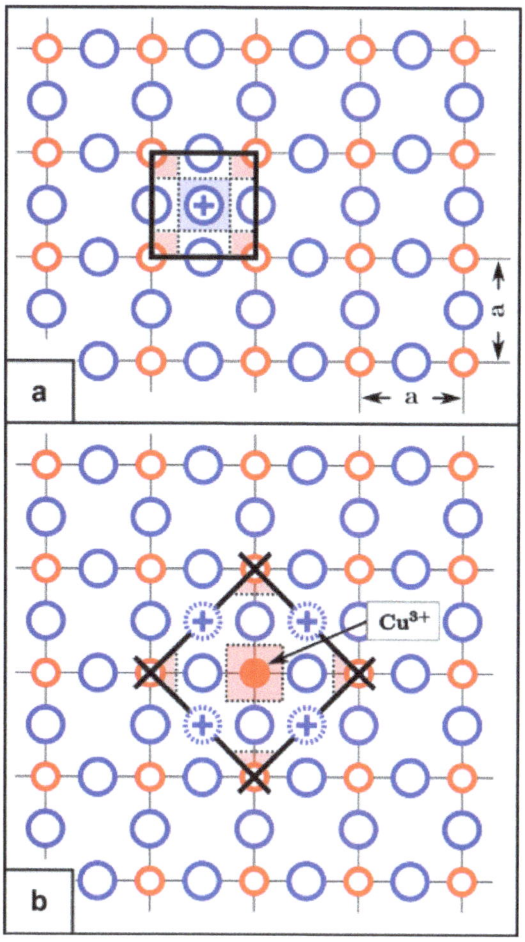

Figure 7: The positions of Cu^{3+}-ions in the CuO_2 plane and oxygen excess atoms at top/bottom of the unit cell relative to each other. (a) For a doped unit cell there exist four possible Cu^{3+}-ion positions at the corners. Each of them lies within an area of $(a/4)^2$ given by the different ionic radii of $R(O^{2-}) \approx 0.125$ nm and $R(Cu^{3+}) \approx 0.07$ nm. (b) For a given Cu^{3+}-ion position the oxygen excess atom can be located in one of four neighboring unit cells.

lay within the area $(a/4)^2$ (Fig. 7). In an ideal case the unit area for the oxygen excess distribution (A_{Oxy}) is given by two regular triangles with four oxygen doped unit cells at the corners and an area of $A_{Oxy} = \sqrt{(3/4)}(d_3)^2 = \sqrt{(3/4)}(z_1^2 + z_2^2)(a/4)^2$ or calculated via the diagonal of the CuO_2 plane $(d_3)^2 = (z_3^2 + z_4^2)(a\sqrt{2}/4)^2$ with $z_1, z_2 = 0, 1, 2, \ldots$. The oxygen excess density (Σ^{-1}) is defined as the ratio between a2 and A_{Oxy} with $\Sigma = \sqrt{(3/4)}(z_1^2 + z_2^2)$. The Cu^{3+}-ion distance x_3 is given by $x_3 \approx \sqrt{3}d_3$ with a superconducting unit area of $A_{SC} \approx \sqrt{(3/4)}(x_3)^2 \approx 3/2 d_3 x_3$.

An optimum doping level δ_3 between 0.33 and 0.41 associated with a transition temperature range of $T_c = 134 \pm 2$ K has been reported [6, 11–13]. The large range of $\delta_3 = 0.37 \pm 0.04$ is probably caused by a broad bell-shaped form. The oxygen excess doping density results in $\Sigma^{-1} = 8.37/8 - 1 = 4.6 \times 10^{-2} \equiv 4.6\%$ with a separation of $d_3 = 5a$ given by $d_3^2 = (z_1^2 + z_2^2)a^2 \approx 25a^2$. Figure 8 illustrates the numerical best solution for $d_3^2 = (3.5^2 + 3.5^2)a^2 = (3.5a\sqrt{2})^2 = 24.5a^2$ with $d_3 = 4.95a$ and $x_3 = \sqrt{3}d_3 = 8.57a \approx 8.5a = 3.27 \times 10^{-9}$ m. The final calculation with n = 3 results in (2x3)2 × n−2/3Tc = 2.76 × 10−15 m2K.

The final calculation with n = 3 results in $(2x_3)^2 \times n^{-2/3}T_c = 2.76 \times 10^{-15}$ m^2K.

Hg-1234, Hg-1245 and Hg-1256

The rise in T_c with the number of superconducting CuO_2 planes up to n = 3 stimulated research groups to synthesize Hg-1234, Hg-1245 and Hg-1256 with n = 4, 5 and 6 CuO_2 planes per unit cell. Experimental results demonstrated [6, 11] that

(a) it is very difficult to synthesize pure single phases for n > 3,

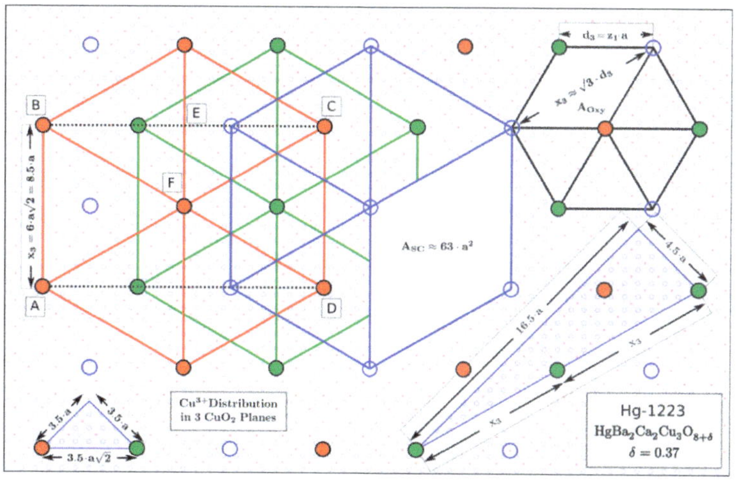

Figure 8: Red and filled circles represent the Cu^{3+}-ion distribution in the upper CuO_2 plane, green and slightly shaded circles represent the Cu^{3+}-ion distribution in the middle and blue and open circles in the lower CuO_2 plane of Hg-1223. All together represent the oxygen excess distribution in the Hg plane separated by d_3. The Cu^{3+}-ion separation is given by $x_3 \approx \sqrt{3}d_3$.

MATERIAL	T_c [K] (exp.)	Crystal Structure tet [nm]	Oxygen Excess δ	Oxygen Excess Σ^{-1} [%]	Oxygen Excess x [nm]	CuO_2 Planes n	$(2x)^2 \cdot n^{-2/3} \cdot T_c$ m_t [m^2K] $\cdot 10^{-15}$
$HgBa_2CuO_{4+\delta}$ Hg-1201	95 ± 1	a = 0.388 c = 0.950	0.0825 ± 0.005	2.0	2.71 ± 0.1	1	2.79
$HgBa_2CaCu_2O_{6+\delta}$ Hg-1212	126 ± 1.5	a = 0.386 c = 1.265	0.21 ± 0.01	3.5	2.94 ± 0.1	2	2.75
$HgBa_2Ca_2Cu_3O_{8+\delta}$ Hg-1223	134 ± 2	a = 0.385 c = 1.610	0.37 ± 0.04	4.6	3.27 ± 0.2	3	2.76

Figure 9: Experimental data of optimum doping level δ for maximum T_c for Hg-1201, Hg-1212 and Hg-1223.

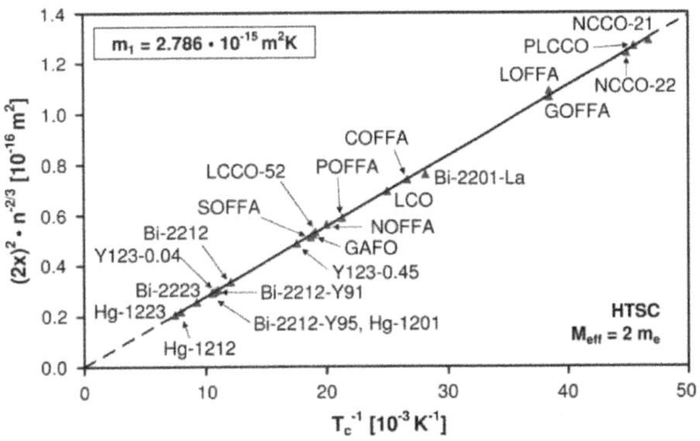

Figure 10: Correlation between doping distance x and the inverse of the critical transition temperature T_c in HTSCs. The value n is the number of superconducting CuO_2 planes per chemical formula within a unit cell. The data for Hg-HTSCs are analyzed in this work. All others are taken from [2, 3, 14 and references therein].

(b) they are superconducting but the transition temperature is decreasing to onset values of $T_c = 105$ K (n = 4), $T_c = 101$ K (n = 5) and $T_c = 95$ K (n = 6), and

(c) it has not been possible so far to increase the oxygen excess density compared to Hg-1223 with δ = 0.41. In fact the doping reduced to δ = 0.32 for Hg-1245 to achieve superconductivity at all.

Therefore there are no parabolic bell-shape curves available to determine the maximum T_c for optimum doping. According to Fig. 9 the optimum doping of HBCCO and n > 3 for reaching the maximum of a bell-shaped curve should be well above $\Sigma^{-1} \equiv 5\%(\delta >> 0.4)$, which seems to be impossible to reach so far [6].

P-type and N-type HTSCs

Additionally to the mercury cuprate homologous series the authors have investigated different types of cuprates (p-type, n-type, with tetragonal as well as orthorhombic unit cell crystal structures, oxygen excess, oxygen deficiency, single and double doped devices) and iron-based HTSCs [2, 3, 14 and references therein]. The analysis demonstrates that also in other HTSC families the doping distance x in the superconducting plane is a physical length of interest and shows a strong correlation to the inverse of the transition temperature T_c as illustrated in Fig. 10. The correlation can be written in the form

$$(2x)^2 n^{-2/3} = m_1 \frac{1}{T_c} \tag{6}$$

where n is the number of superconducting planes, e.g. CuO_2 planes in cuprates, per chemical formula within a unit cell. Using a weighted linear regression a straight line fits the data points with slope m1 = $(2.786 \pm 0.029) \times 10^{-15}$ m² K in the range 20 K ≤ T_c ≤ 134 K and a very small ordinate intercept value of $(-1.69 \pm 3.33) \times 10^{-19}$ m². Comparing (5) and (6) with $M_{eff} = 2m_e$ results in $f(kT_c) = (3.133 \pm 0.033)kT_c \approx \pi kT_c$ according to

$$\frac{m_1 k}{kT_c} = \frac{h^2}{2M_{\text{eff}} f(kT_c)} \tag{7}$$

With the correlation curve in Fig. 10 (4) and (5) transform into

$$(2x)^2 n^{-2/3} 2M_{\text{eff}} \pi kT_c \approx h^2$$

or

$$\frac{(2x)^2 n^{-2/3} M_{\text{eff}}\, kT_c}{h} \frac{}{2} \approx \frac{\hbar}{2} \tag{8}$$

Material	T_c [K] exp.	$(2x)^2 \cdot n^{-2/3}$ [10^{-18} m²]	$(2x)^2 \cdot n^{-2/3} \cdot T_c$ [10^{-15} m²K]
$Nd_{1.85}Ce_{0.15}CuO_{3.98}$ NCCO-21	21.4 ± 0.5	129.0	2.762
$Pr_{0.88}LaCe_{0.12}CuO_{3.96}$ PLCCO	22 ± 2	126.8	2.789
$Nd_{1.84}Ce_{0.16}CuO_{3.98}$ NCCO-22	22.3 ± 0.5	124.1	2.767
$LaO_{0.89}F_{0.11}FeAs$ LOFFA	26 ± 1	109.0	2.834
$GdO_{0.83}F_{0.17}FeAs$ GOFFA	26 ± 1	106.5	2.769
$Bi_2Sr_{1.6}La_{0.4}CuO_{6.1}$ Bi-2201-La	35.5 ± 2.5	76.0	2.699
$CeO_{0.84}F_{0.16}FeAs$ COFFA	37.5 ± 1	74.0	2.774
$La_2CuO_{4.08}$ LCO	40 ± 1.5	69.4	2.776
$PrO_{0.89}F_{0.11}FeAs$ POFFA	47 ± 1	59.0	2.772
$NdO_{0.89}F_{0.11}FeAs$ NOFFA	50 ± 1	56.3	2.813
$La_{1.82}Ca_{1.18}Cu_2O_{6.02}$ LCCO-52	52.4 ± 2	53.3	2.794
$GdFeAsO_{0.85}$ GAFO	52.9 ± 0.6	52.4	2.773
$SmO_{0.9}F_{0.1}FeAs$ SOFFA	53.5 ± 1	51.3	2.743
$YBa_2Cu_3O_{6.55}$ Y123-0.45	57 ± 2	48.7	2.777
$Bi_2Sr_2CaCu_2O_{8.18}$ Bi-2212	83 ± 1	33.6	2.786
$Bi_2Sr_2Ca_{0.92}Y_{0.08}Cu_2O_{8.16}$ Bi-2212-Y91	91 ± 1	30.5	2.773
$YBa_2Cu_3O_{6.96}$ Y123-0.04	93 ± 1	29.8	2.772
$HgBa_2CuO_{4+\delta}$ Hg-1201	95 ± 1	29.4	2.791
$Bi_2Sr_2Ca_{1-\Delta}Y_{\Delta}Cu_2O_{8+\delta}$ Bi-2212-Y95	95 ± 1	29.2	2.770
$Bi_2Sr_2Ca_2Cu_3O_{10.22}$ Bi-2223	108 ± 2	25.6	2.767
$HgBa_2CaCu_2O_{6+\delta}$ Hg-1212	126 ± 1.5	21.8	2.744
$HgBa_2Ca_2Cu_3O_{8+\delta}$ Hg-1223	134 ± 2	20.6	2.755

Figure 11: Experimental data of high-temperature superconductor materials with optimum doping for maximum transition temperature

T_c. The analysis of Hg-based HTSCs is described in this work. Other materials are analyzed in [2, 3, 14] and references therein using the same procedure.

In this respect the theoretical value for the slope m_1 could be given by

$$m_1 \approx \frac{h^2}{2\pi k M_{\text{eff}}} \approx 2.778 \times 10^{-15}\ \text{m}^2\ \text{K} \qquad (9)$$

DISCUSSION

Considering a particle with mass M, which is confined inside a box of width x, its wave function satisfies the freeparticle Schrödinger equation. Confinement of a particle may be interpreted as a standing wave in which an integer number (n = 1, 2,...) of half-wavelengths must fit into the box $n1/2\lambda_{dB} = x$, where $\lambda_{dB} = h/p$ represents de Broglie wavelength of the particle [1]. This results in $p = h/\lambda_{dB} = (nh)/(2x)$ and the kinetic energy E_{kin} for the standing wave, a solution of the time-independent Schrödinger equation, is given by

$$E_n(\text{PiB}) = \frac{p^2}{2M} = \frac{n^2h^2}{8Mx^2} \qquad (10)$$

Comparing (8) and (10) it looks like that the lowest energy level E_1(PiB) of the superconducting carrier with $M = M_{eff} = 2m_e$ is connected to the transition temperature T_c by $E_1 = \pi k T_c$.

To prove that (8) is universally applicable more HTSC families should be investigated e.g. the thallium homologous series with oxygen excess and deficiency and the strontium series with infinite layer structures. Up to now, more than 20 different HTSC materials match the linear curve quite well (Fig. 11). So far it is not possible to predict new HTSC candidates with much higher T_c because (8) contains two free parameters, x and Tc. And it is not known which one determines the exact location on the linear graph, because for all materials there exist only one optimum value for x or T_c respectively.

It is of interest to note that conventional superconductors N_b, V, Ta and Hg also show a strong correlation between the bond length x and the inverse of the critical superconducting transition temperature T_c in the form $(2x)^2 N_{eff} = m_2 1/T_c$. Here Neff is the number of electrons in the outermost s-shell. The slope of the fitted straight line has a quite different value of $m_2 \approx 3.0 \times 10^{-18}$ m^2 K [15].

CONCLUSION

The mercury cuprate homologous series $HgBa_2Ca_{n-1}CunO_{2n+2}+\delta$ belongs to the oxygen excess doped HTSCs. The oxygen excess atom transforms a Cu^{2+}-ion into a Cu^{3+}-ion within a unit cell, which is responsible for the superconducting state in the CuO_2 plane. Assuming a homogeneous doping distribution the doping density δ can be translated into a doping distance x between doped unit cells. The same procedure can be applied to different types of cuprates (p-type, n-type, with tetragonal as well as orthorhombic unit cell crystal structures, oxygen excess, oxygen deficiency, single and double doped devices) and iron-based HTSCs. For more than 20 HTSCs the correlation between the doping distance x and the critical transition temperature shows the very first link between T_c and the crystal structure and geometry in form of a straight line. The correlation could be interpreted as the lowest energy level of a particle in the box of width x with the value $E_1 = \pi k T_c$ being the kinetic energy of a standing wave with wavelength $\lambda_{dB} = 2x$.

Acknowledgements

We would like to thank J. Vernerey, H. Hall andA. Bohr for discussions and valuable comments.

REFERENCES

1. Rohlf, J.W.: Modern Physics from α to Z0. Wiley, New York (1994)
2. Roeser, H.P., Hetfleisch, F., Haslam, D.T., López, J.S., Stepper, M., Vernerey, J., Huber, F.M., von Schoenermark, M.F., Nikoghosyan, A.S.: Acta Astronaut. 66, 637–642 (2009)
3. Roeser, H.P., Haslam, D.T., Hetfleisch, F., López, J.S., von Schoenermark, M.F., Stepper, M., Huber, F.M., Nikoghosyan, A.S.: Acta Astronaut. 67, 546–552 (2010)
4. Putilin, S.N., Antipov, E.V., Chmaissem, O., Marezio, M.: Nature 362, 226–228 (1993)
5. Zhao, X., et al.: Adv. Mater. 18, 3243–3247 (2006)
6. Capponi, J.J., et al.: Physica C 235–240, 146–149 (1994)
7. Tsuchiya, T., Fueki, K.: Physica C 288, 47–56 (1997)
8. Fukuoka, A., et al.: Physica C 265, 13–18 (1996)
9. Radaelli, P.G., Wagner, J.L., Hunter, B.A., Beno, M.A., Knapp, G.S., Jorgensen, J.D., Hinks, D.D.: Physica C 216, 29–35 (1993)
10. Schilling, A., Cantoni, M., Guo, J.D., Ott, H.R.: Nature 363, 56– 58 (1993)
11. Chmaissem, O., et al.: Physica C 217, 265–272 (1993)
12. Kareiva, A., Barkauskas, J., Mathur, J.S.: J. Phys. Chem. Solids 61, 789–797 (2000)
13. Dai, P., Chaloumakos, B.C., Sun, G.F., Wong, K.W., Xin, Y., Lu, D.F.: Physics C 243, 201–206 (1995)
14. Huber, F.M., Roeser, H.P., von Schoenermark, M.F.: J. Phys. Soc. Jpn. 77(Suppl. C), 142–144 (2008)
15. Roeser, H.P., Haslam, D.T., López, J.S., Stepper, M., von Schoenermark, M.F., Huber, F.M., Nikoghosyan, A.S.: Acta Astronaut. 67, 1333–1336 (2010)

CHAPTER 7

EXPERIMENTAL DEMONSTRATIONS OF HIGH-Q SUPERCONDUCTING COPLANAR WAVEGUIDE RESONATORS

LI HaiJie, WANG YiWen, WEI LianFu, ZHOU PinJia, WEI Qiang, CAO ChunHai, FANG YuRong, YU Yang & WU PeiHeng

Keywords: Superconducting coplanar waveguide resonator, resonance frequency, quality factor

Superconducting coplanar waveguide (CPW) resonators have been extensively studied for years, specifically as radiation detectors [1–3], parameter amplifiers [4], and the data bus of superconducting qubits [5–10]. Compared with other low temperature detectors, e.g. transition edge sensors (TESs) [11,12], superconducting tunnel junctions (STJs) [13,14], superconducting nanowire single photon detectors (SNSPDs) [15], the primary advantage of the CPW devices is that they are relatively easy to be multiplexed into large arrays. In principle, a series of CPW resonators with different frequencies can be measured using a single feedline, which greatly simplifies the system structure and relevant cryogenic electronics.

Particularly, a superconducting CPW resonator with sufficiently-high quality factor can serve as a sensitive radiation detector. Physically, the incoming photons break Cooper pairs in the superconducting resonator and create quasiparticle excitations. This changes the kinetic inductance [16] and resonance frequency of the CPW resonator, which can be easily measured experimentally. In theory, higher quality factor allows more sensitive detection of photons. Also, CPW resonators with high quality factors (i.e. with narrower 3 dB-bandwidths) imply that more resonators (with different resonant frequencies) can be multiplexed on-chip for a fixed bandwidth. This is essentially important for the development of large sensor-arrays used in astronomy [1,17–19].

Here, we report an experimental demonstration of an absorption-type of superconducting CPW resonators, which can be used as the microwave kinetic inductance detectors (MKIDs) [1,19] with high quality factors. Our device is made from a Nb film (t~160 nm) deposited on a high resistance Si subtract by magnetron sputtering. The designed circuit is generated by the usual photolithography. The measured loaded quality factor of our resonator at 20 mK is $Q_l = 1.1654\times10^6$ at the resonant frequency $f_0 = 1.8575$ GHz. We also measured the temperature dependence of the resonance frequency in the range of base temperature (20 mK) to around 1180 mK.

DEVICE AND MEASUREMENT SYSTEM

The microscope photograph of our meandered CPW λ/4 transmission line resonator is shown in Figure 1. The resonator shorts to the ground plane at one end, and capacitively couples to a through line at another end. In our design, niobium is selected as the metal film material, due to its well superconductivity at low temperature. The characteristic impedance Z_0 of through line is designed to be 50 Ω, in order to avoid the reflections of the driving signals. In fact, for a single layer dielectric the impedance can be analytically calculated using [20]

$$Z_0 = \frac{30\pi}{\sqrt{\epsilon_{e,t}}}\frac{K(k')}{K(k)}. \quad (1)$$

Here, $\epsilon_{e,t}$ is the effective dielectric constant with a correction of the film thickness t, $K(k) = \int_0^{\pi/2} d\theta / \sqrt{1-k^2\sin^2\theta}$ is the complete elliptic integral with modulus k. The parameters k and k' are determined by the resonator's geometry parameters [20]: the center strip width w, slot width between the centerstrip and the ground plane s, metal film thickness t and the thickness h of the silicon substrate. Physically, the fundamental resonance frequency f_0 is determined by the overall length of the CPW resonator l and the effective substrate dielectric parameter ε:

$$f_0 = \sqrt{\frac{2}{\epsilon+1}}\frac{c}{4l}. \quad (2)$$

Citation: HaiJie Li, YiWen Wang, LianFu Wei, PinJia Zhou, Qiang Wei, ChunHai Cao, YuRong Fang, Yang Yu, PeiHeng Wu, Experimental demonstrations of high-Q superconducting coplanar waveguide resonators, DOI: 10.1007/s11434-013-5882-3.

Here, c is the speed of an electromagnetic wave in a vacuum. In our design, the length of the CPW resonator is set as 15.814 mm, corresponding to a fundamental resonance frequency f_0 = 1.8575 GHz. Also, the center strip width w_t is set as 20 μm and the slot width s_t between the center strip and the ground plane is 15 μm. Note that the CPW resonator is decoupled from the through line so its impedance is not necessary to be 50 Ω.

Several aspects, e.g. coupling strength between the through line and the resonator, film and dielectric losses, and also the radiation loss, may influence the quality factor Q of the CPW resonator. For the present device, i.e. niobium film deposited on the silicon substrate, the film and dielectric losses are very small at very low temperature ($T \ll T_c$, T_c = 9.2 K for niobium). The radiation loss is also negligible, as the line width of the transmission line is very narrow. As a consequence, at $T \ll T_c$ the quality factor Q_c, related mainly to the coupling capacitance, dominates the loaded (total) quality factor Q_l defined by

$$\frac{1}{Q_l} = \frac{1}{Q_i} + \frac{1}{Q_c}. \tag{3}$$

Here, Q_i is the internal (unloaded) quality factor, accounting for the energy leakage due to all other losses. The coupling region of our device is designed as a coupling capacitor: a part of the CPW resonator is parallel and close to the through line. Although Q_c cannot be calculated from coupling strength analytically, numerical simulations show that a weak coupling leads to a high quality factor. Therefore, the designed Q_c can be adjusted by changing either the length of the CPW resonator parallel to the through line or the distance between them. Certainly, the coupling strength should not be set extremely small; otherwise the CPW resonator will not be excited for observation. The numerical results, considering our current fabrication techniques, show that, if the geometry parameters for the coupling region are set as those in Figure 1(c), then the desired quality factor could be up to 10^6.

After the above design and relevant numerical simulations, we fabricated our CPW resonator devices by magnetron sputtering and photolithography technique. Experimentally, a 160 nm thick niobium film was deposited on a 500 μm thick silicon substrate, and then contact exposure as well as reflect ion etching are used to make the desired patterns on the film.

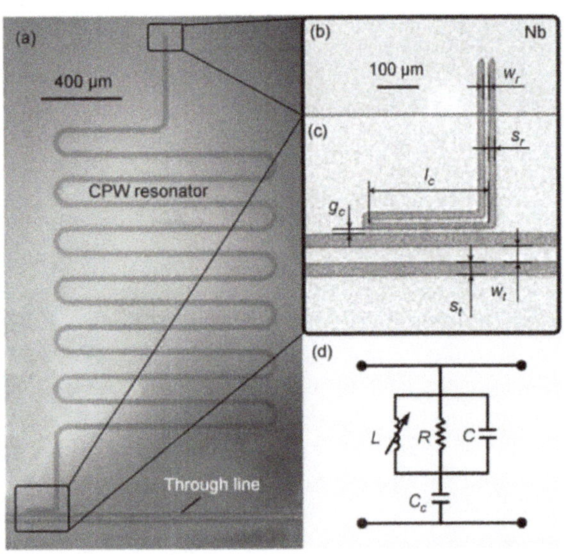

Figure 1: (a) Microscope photograph of the key structure of our device. Light and dark areas are niobium film and silicon substrate, respectively. The through transmission line is used for excitation and readout. The meandered CPW resonator has an overall length of 15.814 mm, corresponding to a fundamental frequency around 1.8575 GHz. (b) One end of the CPW resonator is shorted to the ground plane. (c) The magnified coupling region and main design parameters: The length of the capacitor elbow l_c is 160 μm and the distance between through line gap and the capacitor elbow gap gc is 7 μm. The width of center conductor w_r and the gap between center conductor and ground s_r are both 7 μm. (d) The equivalent circuit of the measured device.

The device can be treated equivalently as a measured parallel lumped RLC resonant circuit, shown in Figure 1(d), where R, L and C characterize the small resistance, inductance and capacitance of the measured CPW resonator, respectively. Cc is the coupling capacitance between through line and the CPW resonator. Note that the total inductance of the CPW resonator: $L = L_m + L_k$, consists of two parts: the magnetic inductance L_m and the kinetic inductance L_k [19]. In particular, the latter one depends strongly on the temperature of the superconducting device. With such a property, the CPW resonator could be used as an attractive candidate of sensitive radiation detector.

The CPW resonator chip is glued (with GE varnish) onto a gold-plated sample block made of oxygen-free high purity copper and wire bonded to microwave transmission lines on a printed circuit board which was carefully designed to ensure a good performance up to 6 GHz microwave signals.

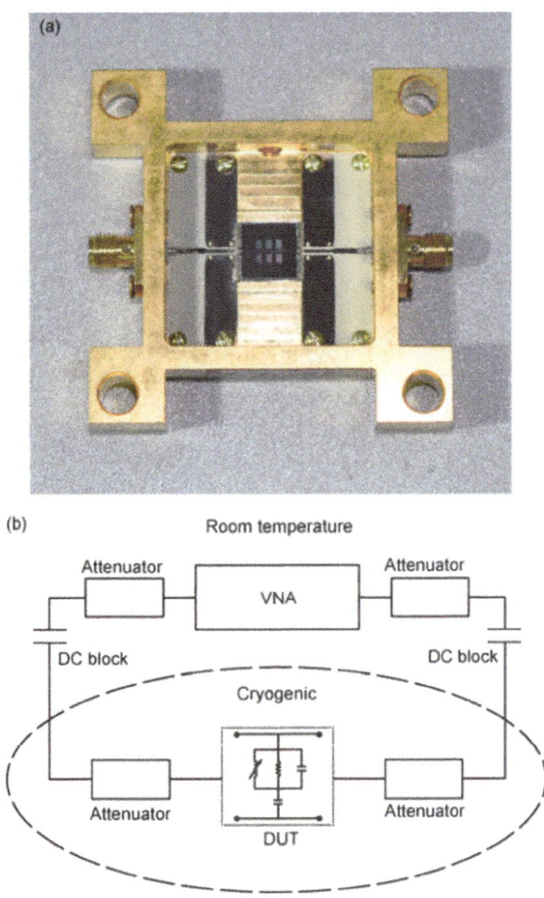

Figure 2: Inner view of the sample block (a) and schematics of the measurement system (b).

Additional bonds are placed at all chip sides for good ground-connection. During the measurements the sample block is placed at the mixing chamber plate in our dilution refrigerator. Figure 2(b) shows the schematics of our measurement system. Cryogenic coax cables connect the sample block to room temperature circuits. Attenuators and DC blocks are positioned appropriately to reduce unwanted circuit noises. The device is measured with the Agilent E5071C vector network analyzer (VNA). The raw data collected consists of complex transmission amplitude S_{21} for 1601 points in the desired frequency interval. Since the data is not calibrated, losses in the coax cables are included.

EXPERIMENTAL RESULTS

By sweeping the frequency and adjusting the power of VNA, the S21 data were obtained and a resonance dip near the designed position was identified. It is observed that the shape of the resonance dip strongly depends on the driving power. For higher power, various nonlinear effects reveal and the resonance features of the device are complicated. While within the lower power regime, a maximum resonance dip depth was observed for an optimal driving power [21]. Indeed, it is seen from Figure 3 that, at the driving power -46 dBm an optimal resonance dip with a sufficiently high signal to noise ratio was found around 1.8575 GHz. Note that the present data include all the losses in the coax cables, which are estimated to be about –14 dB (at 1.8575 GHz). The inset of Figure 3 shows a sharper dip, corresponding to the same measurement with a wider bandwidth view.

From the obtained S21 data, the loaded (total) quality factor of the CPW resonator can be calculated as

$$Q_l = \frac{f_0}{\Delta f}, \tag{4}$$

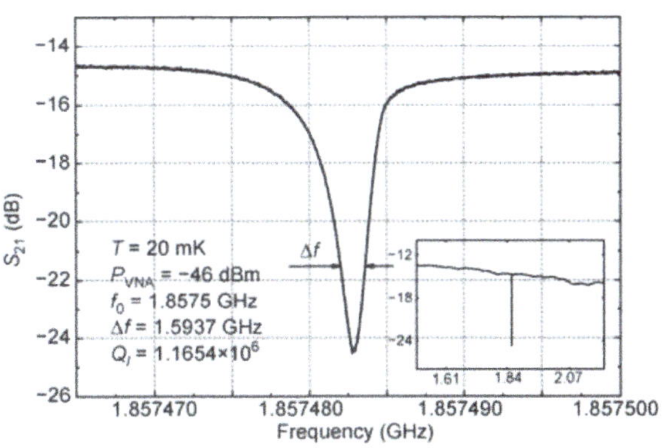

Figure 3: The observed transmission dip. From the obtained S_{21} data, one can calculate the loaded quality factor of the measured CPW resonator: $Q_l = f_0/\Delta f = 1.1654\times10^6$. The inset shows a sharper resonance dip, corresponding to the same measurements with a wider bandwidth view.

with f_0 being the resonance frequency and Δf the 3 dB bandwidth. Q_l includes both the source and load dissipation of the device. In our case, f_0 =1.8575 GHz and Δf = 1.5937 KHz were observed. Thus, the loaded quality factor is calculated as $Q_l = 1.1654\times10^6$. Also, Q_i and Q_c can be individually calculated from the minimal values of the measured S_{21} parameters at the resonance point, i.e. $Q_i = Q_l/|S'_{21}|$. And $Q_c = Q_l(1-|S'_{21}|)$, with = $\min\{S_{21}\}$. For our device, we found Qi = 3.2845×106 and $Q_c = 1.8063\times10^6$.

Furthermore, we measured the temperature dependence of the resonance frequency and quality factor by slowly enhancing the system temperature, e.g. from 30 to 1180 mK. We can see clearly from Figure 4 that, with increasing temperature, the center frequency shifts to higher value and the resonance dip becomes broader and shallower. Phenomenally, these behaviors can be explained in the framework of the two-level systems (TLSs) theory [22,23], in which the unsaturated TLSs residing on the metal and dielectric surfaces are believed to be a significant loss mechanism. The resonant interaction of TLSs with electric field (microwave) leads to a temperature dependent variation of the substrate dielectric constant. As a consequence, both dielectric constant ε and the surface inductance L_s determine the resonance frequency as

$$\frac{\delta f}{f} = \frac{1}{2}\frac{\delta L_s}{L_s} - \frac{F}{2}\frac{\delta \epsilon}{\epsilon}, \tag{5}$$

where F is a filling factor and $L_s = L_m + L_k(0)\times H(T)$. The magnetic inductance Lm remains essentially a constant in the concerned temperature range. While, H(T) represents the temperature dependence of the kinetic inductance L_k, which is strongly influenced by the temperature variation near the critical temperature T_c. However, at the present ultra-low temperature $T \ll T_c$ (T_c = 9.2 K for niobium), the kinetic inductance can still be treated as a constant. Thus, the observed resonance frequency shifts were mainly related to the temperature-dependent dielectric constant. Indeed, it was

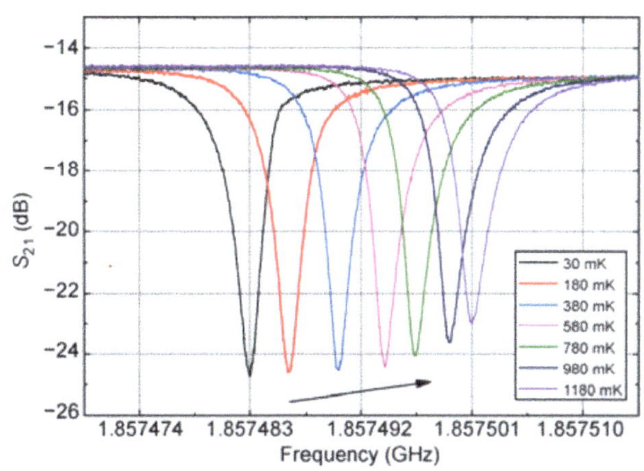

Figure 4: The increasing resonance frequency and decreasing quality factor of the CPW resonator versus the increasing system temperature for $T \ll T_c$.

shown that the dielectric constant decreases with temperature, and thus the resonance frequency of the CPW resonator increases. Certainly, if the device works at the temperature closing to T_c, then the effects due to TLSs saturate and the temperature dependence of kinetic inductance is dominant. Therefore, the increasing L_k decreases the resonance frequency. Basically, the surface resistance R_s increases with temperature monotonically, therefore the resonance dip is broader and shallower (and thus the quality factor decreases) for the increasing temperature.

DISCUSSION AND CONCLUSIONS

The main advantage of superconducting CPW resonator device is that it can be easily integrated. That is, multiple resonators operating at different frequencies can be read out through a common feedline. In fact, the number of the multiplexed resonators depends on several factors, e.g. dvice quality factors and manufacture accuracy. In principle, a device with higher-Q and better manufacture accuracy allows for a smaller frequency separation and thus more resonators multiplexed on one chip. In order to make a simple presentation of frequency domain multiplexing scheme, we also fabricated a device with six resonators coupled to a common through line. Each resonator is designed with a different length, corresponding to a different resonance frequency. Figure 5 shows the measured results of such a multiplexed device, where six resonance dips can be observed expectably. One can see from the Table 1 that, the measured resonance frequency f_m agrees well with the designed resonance frequency f_t. Specifically, the designed length of the resonator No. 5 is the same as that measured in Figure 3. The small difference (about 1.1 MHz) between the two measured resonance frequencies indicates that our devices are sufficiently accurate. In summary, we have designed and fabricated a high quality niobium quarter-wave CPW resonator. The transmission of microwave is measured with a vector network

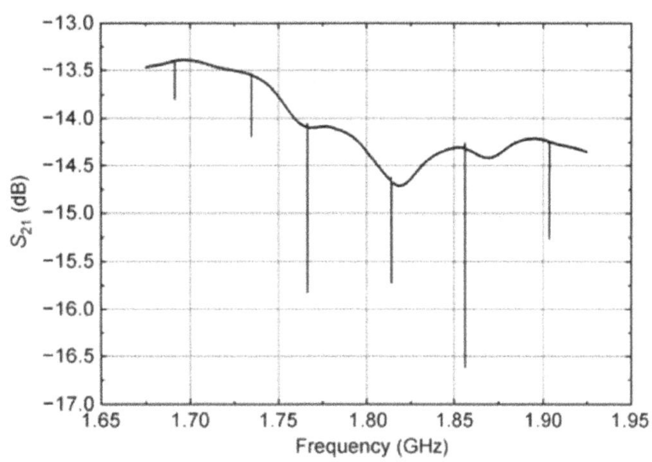

Figure 5: Resonance dips for six resonators multiplexed on a chip.

Table 1: Designed and measured resonance frequencies of six CPW resonators on a chip

No.	l_d (μm)	f_t (GHz)	f_m (GHz)
1	17414	1.6958	1.6914
2	17014	1.7357	1.7346
3	16614	1.7775	1.7664
4	16214	1.8213	1.8139
5	15814	1.8674	1.8563
6	15414	1.9159	1.9037

analyzer and a sharp resonance dip is observed at designed position. At 20 mK, the measured quality factor of the device is $Q_l = 1.1654\times10^6$, which is significantly high and could be applied to sensitive radiation detections and quantum information processing.

This work was supported by the National Natural Science Foundation of China (90921010, 11174373) and National Basic Research Program of China (2010CB923104).

REFERENCES

1. Day P K, LeDuc H G, Mazin B A, et al. A broadband supercon-ducting detector suitable for use in large arrays. Nature, 2003, 425: 817–821
2. Mazin B A, Bumble B, Day P K, et al. Position sensitive X-ray spectrophotometer using microwave kinetic inductance detectors. Appl Phys Lett, 2006, 89: 222507
3. Day P K, Leduc H G, Goldin A, et al. Antenna-coupled Microwave kinetic inductance detectors. Phys Rev A, 2006, 559: 561–563
4. Tholén E A, Ergül A, Doherty E M, et al. Nonlinearities and parametric amplification in superconducting coplanar waveguide resonators. Appl Phys Lett, 2007, 90: 253509
5. Wallraff A, Schuster D I, Blais A, et al. Strong coupling of a single photon to a superconducting qubit using circuit quantum electrodynamics. Nature, 2004, 431: 162–167
6. Yu L B, Tong N H, Xue Z Y, et al. Simulation of the spin-boson model with superconducting phase qubit coupled to a transmission line. Sci China Phys Mech Astron, 2012, 55: 1557–1561
7. Amsüss R, Koller C, Nöbauer T, et al. Cavity QED with magnetically coupled collective spin states. Phys Rev Lett, 2011, 107: 060502
8. Zhou Z W, Chen W, Sun F W, et al. A survey on quantum information technology. Chin Sci Bull, 2012, 57: 1498–1525
9. Qian, Y, Zhang Y Q, Xu J B. Amplifying stationary quantum discord and entanglement between a superconducting qubit and a data bus by time-dependent electromagnetic field. Chin Sci Bull, 2012, 57: 1637–1642
10. Kubo Y, Diniz I, Grezes C, et al. Electron spin resonance detected by a superconducting qubit. Phys Rev B, 2012, 86: 064514
11. Irwin K, Hilton G, Wollman D, et al. X-ray detection using a super-conducting transitionedge sensor microcalorimeter with electrothermal feedback. Appl Phys Lett, 1996, 69: 1945–1947
12. Chervenak J, Irwin K, Grossman E, et al. Superconducting multiplexer for arrays of transition edge sensors. Appl Phys Lett, 1999, 74: 4043–4045
13. Peacock A, Verhoeve P, Rando N, et al. Single optical photon detectionwith a superconducting tunnel junction. Nature, 1996, 381: 135–137
14. Shen D D, Zhu R, Xu W W, et al. Character and fabrication of Al/ Al2O3/Al tunnel junctions for qubit application. Chin Sci Bull, 2012, 57: 409–412
15. You L X, Shen X F, Yang X Y, et al. Single photon response of superconducting nanowire single photon detector. Chin Sci Bull, 2010, 55: 441–445
16. Tinkham M, Emery V. Introduction to Superconductivity.

2nd ed. New York: Dover Publications, 1996

17. Jiang Y, Liang M, Jin B B, et al. A simple Fourier transform spectrometer for terahertz applications. Chin Sci Bull, 2012, 57: 573–578
18. Gao L, Guo J, Wang Y H, et al. A 23 GHz high-temperature superconducting microstrip filter for radio astronomy. Chin Sci Bull, 2009, 54: 3485–3488
19. Mazin B A. Microwave kinetic inductance detectors. PhD Thesis. California: California Institute of Technology Pasadena, 2004
20. Wadell B C. Transmission Line Design Handbook. Norwood, MA: Artech House Publishers, 1991
21. De Visser P J, Withington S, Goldie D J. Readout-power heating and hysteretic switching between thermal quasiparticle states in kinetic inductance detectors. J Appl Phys, 2010, 108: 114504
22. Gao J, Daal M, Martinis J M, et al. A semiempirical model for two-level system noise in superconducting microresonators. Appl Phys Lett, 2008, 21: 212504
23. Wisbey D S, Gao J, Vissers M R, et al. Effect of metal/substrate interfaces on radio-frequency loss in superconducting coplanar waveguides. J Appl Phys, 2010, 108: 093918

CHAPTER 8

THE ELECTRONIC STRUCTURE OF NB_3AL/NB_3SN, A NEW TEST CASE FOR FLAT/STEEP BAND MODEL OF SUPERCONDUCTIVITY

Yanlong Ding, Shuiquan Deng, Yong Zhao

Keywords: Nb_3Al/Nb_3Sn, Electronic structure, Flat/steep band model, First-principle

INTRODUCTION

Superconductors have found widespread applications ranging from civil to military uses. The superconductor-based Magnetic Resonance Imaging (MRI) technique used in many hospitals, the superconducting super-collider project in Texas USA, the Large Hadron Collider (LHC) in CERN, the ITER project, the Experimental Advanced Superconducting Tokomak (EAST) in China and so on are all examples where superconductors are necessary materials. Another important application is the superconducting magnetic levitation system, which not only saves much of the electrical energy but also saves the space and weight of the vehicle and dramatically increases its speed [1]. Such important applications require deeper and exact understanding of the properties, e.g. mechanical and critical ones, of a superconductor.

The flat/steep band model [2, 3] for superconductivity has been proposed over many years in attempting to understand superconductivity from a chemistry point of view. The previous studies [4] indicate that the electronic structures of conventional elemental and complex superconductors show good agreement with this model. However, this model met a non-trivial challenge in the case of Calcium, although it finally proved to be the shortcoming of the first-principle methods instead of that of the flat/steep band model [5]. Nb_3Al/Nb_3Sn is a typical A15 superconductor, which though has been intensively studied, respectively, remains a challenge to the existing theories regarding their strain and scale-dependent critical properties such as the upper critical magnetic field and current density. On the other hand, Nb_3Al and Nb_3Sn have very important applications on the thermonuclear fusion experimental reactor as magnetic plasma-control materials for their overall merit factors. The results of earlier studies on the preparations, crystal structures, electronic structures, and various properties of A15 compounds have been systematically reviewed by many authors [6–9]. With respect to the electronic structures of A15 compounds, the earlier studies based on the first-principle methods [10–12] have revealed very flat bands around the Fermi level, which implies a van-Hove singularity in A15 system. However, the earlier experimental studies have never observed any Peierls-like structural instability in a wide temperature range. This fact itself indicates that either the flat band feature is over estimated as in the case of Calcium [5] or there are some other factors which counteract the effect of the "flat/bands" or both reasons may play a role. In this work, we study the electronic structure of Nb_3Al/Nb_3Sn using the state-of-the-art first-principle method.

COMPUTATIONAL DETAILS

The electronic structure of Nb_3Al/Nb_3Sn has been calculated using the DFT theory as implemented in the Vienna ab initio simulation package (VASP) [13, 14]. The exchange–correlation potential has been calculated at the generalized gradient approximation (gga) level using the Perdew, Burke and Ernzerhof (PBE) formalism [15]. The $10 \times 10 \times 10$ and $20 \times 20 \times 20$ Monkhorst grids have been used to sample the first BZ for structural relaxation and self-consistent calculations, respectively. The cutoff energy for the plane wave expansion is chosen as 320 eV.

Citation: Yanlong Ding, Shuiquan Deng, Yong Zhao, The electronic structure of Nb3Al/Nb3Sn, a new test case for flat/ steep band model of superconductivity, DOI 10.1007/s40534-014-0045-z.

RESULTS AND DISCUSSION

Crystal Structure

Nb_3Al/Nb_3Sn is a typical A15 compound with a space group Pm3¯nPm3¯n. The unit cell structure of the compound is shown in Fig. 1. The structure belongs to a simple cubic crystal system and contains six Nb atoms and 2 Al or Sn atoms in one conventional crystallographic unit cell. Each unit cell contains six A-type atoms which lie on the surface of the unit cell and form chains along the axis directions, respectively. On each chain, the A-type atoms distribute equidistantly. The two B-type atoms occupy the corners and the center of a unit cell. The A-type and B-type atoms occupy the 6c and 2a Wyckoff positions of the space group, respectively.

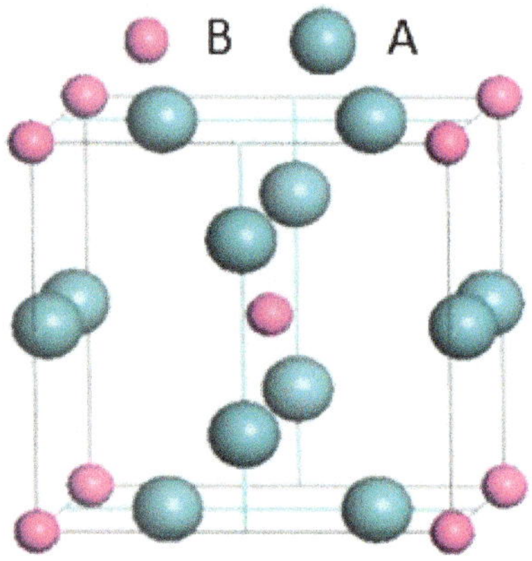

Figure 1: The crystal structure of an A15 compound

Electronic Structure

The electronic density of states (DOS) of Nb_3Al and Nb_3Sn is shown in Fig. 2. As can be seen from Fig. 2, except for tiny shifts and changes of the peak positions and heights, the overall agreement of the DOS curves of Nb_3Al and Nb_3Sn is very good. The DOS around the Fermi level has a sharp peak, which implies a dramatic change of physical properties if the Fermi level is changed by whatsoever means, e.g. doping or stress. By decomposing the electronic states into atomic centered states, we have found that the states around the Fermi level are essentially contributed by the Nb d-states. It is thus obvious that the superconducting properties are crucially decided by the Nb atoms. The energetically lowest peaks at −5.6 and −7.9 eV for Nb_3Al and Nb_3Sn, respectively, are due to the s-states of the B-type atoms. These results agree generally with those of the earlier studies [12, 16].

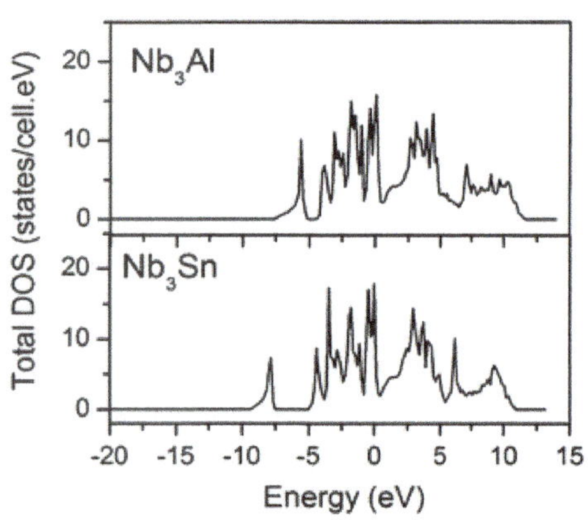

Figure 2: Density of states of Nb_3Al and Nb_3Sn

The first BZ for Nb_3Al and Nb_3Sn is shown in Fig. 3 where the symmetry lines and points are indicated. The coordinates of the symmetry points are (0, 0, 0), (1, 0, 0), (1, 1, 0), (1, 1, 1), for Γ, X, M, and R, respectively. The calculated band structures for Nb_3Al and Nb_3Sn are shown in Figs. 4 and 5, respectively. As for the DOS curves, the band structures of Nb_3Al and Nb_3Sn are quite similar. In both cases, the lowest bands are from the s, p states of the B-type atoms. The bands lying in the range of −3.2 to 5 eV for Nb_3Sn and those between −3 and 4.5 eV for Nb_3Al are contributed by the d-states of Nb. These bands split into the bonding and anti-bonding groups. As shown in Figs. 4 and 5, the bonding bands are flatter than the anti-bonding bands.

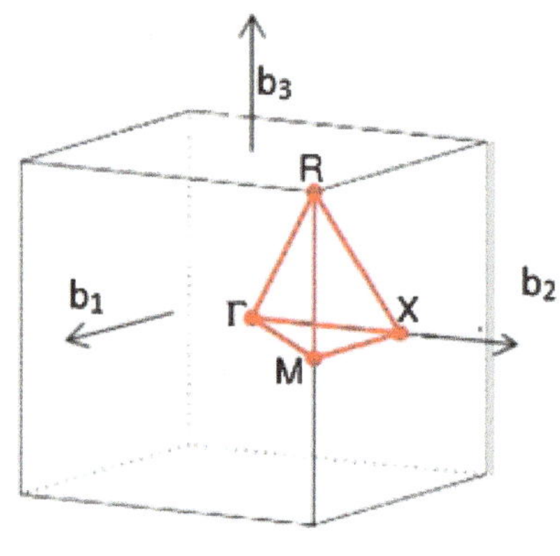

Figure 3: The first BZ for Nb_3Al and Nb_3Sn structures.

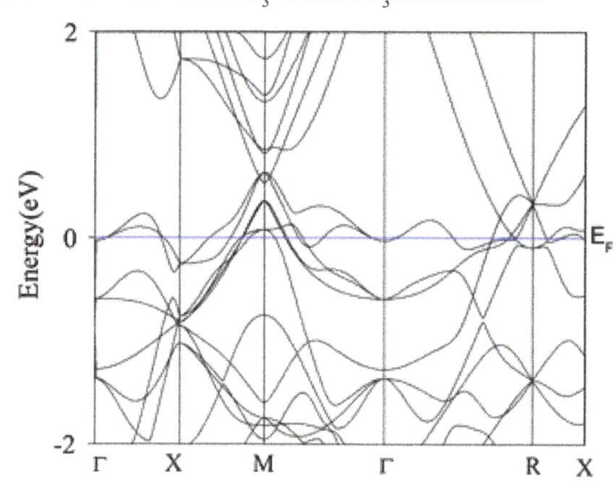

Figure 4: Electronic band structure of Nb_3Al along symmetry directions.

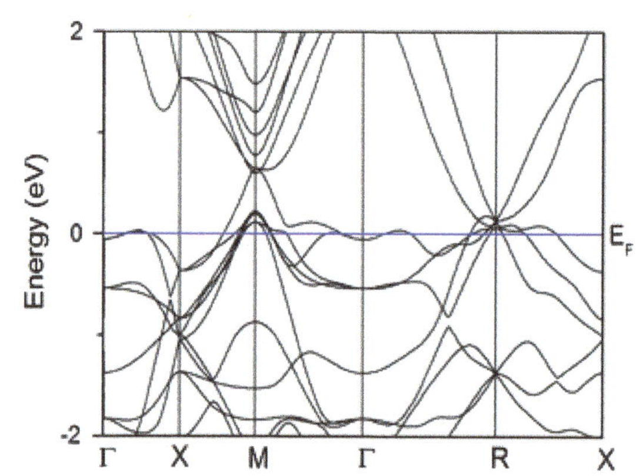

Figure 5: Electronic band structure of Nb_3Sn along symmetry directions.

As revealed in earlier studies [10–12], the bands around the Fermi level of Nb_3Al and Nb_3Sn appear rather flat which is, however, someway an artificial fact due to the chosen large energy scale in representing the band structure. Moreover, the earlier results are not consistent with each other with respect to the accurate position of the flat bands relative to the Fermi level [10–12]. Our investigations on the bands around the Fermi level indicate that the energy bands nearby Γ and R points are significantly flatter than some other bands, while those around the M point are obviously steeper. These steep bands play an important role in counteracting the possible structural instability caused by the high DOS (E_F). In this sense, Nb_3Al and Nb_3Sn fit more to the "flat/steep" band model than to the van-Hove singularity scenario. It is evident that more quantitative experimental data are needed to judge the theoretical results. In particular, the metric details of the Fermi surface are necessary to draw a solid conclusion.

ENDING REMARKS

Our studies indicate that the importance of the steep band was not realized in earlier work. The steep band character in the A15 system was neglected due to the use of large energy scale. Our studies indicate clearly that the steep bands locate around the M point in the first BZ (Fig. 4). The obtained results reveal that Nb_3Al/Nb_3Sn fits more to the "flat/steep" band model than to the van-Hove scenario.

Acknowledgements

This work is financially supported by the Science Foundation for International Cooperation of Sichuan Province (2014HH0016), the Fundamental Research Funds for the Central Universities (SWJTU2014: A0920502051113-10000), National Magnetic Confinement Fusion Science Program (2011GB112001).

REFERENCES

1. See for example: http://www.superconductors.org/uses.htm
2. Deng S, Simon A, Ko¨hier J (1998) Supercon-ductivity and chemical bonding in mercury. Angew Chem Int Ed 37(5):640–642
3. Deng S, Simon A, Ko¨hier J (2005) Pairing mechanisms viewed from physics and chemistry. Struct Bond 114:103–141
4. Deng S, Simon A, Ko¨hier J (2008) Calcium d states: chemical bonding of CaC6. Angew Chem Int Ed 47:6703–6706
5. Deng S, Simon A, Ko¨hier J (2002) Flat-band steep band scenario and superconductivity the case of calcium. Solid State Sci
6. 2(1):31–38
7. Weger M, Goldberg B (1973) Lattice and Electronic Properties of the b-Tungstens. Solid State Phys 28:1–177
8. Dew-Hughes D (1975) A-15 compounds: a review. Cryogenics 15(8):435–454
9. Godeke A (2006) A general scaling relation for the critical current density in Nb3Sn. Supercond Sci Technol 19(10):100–116
10. Glowacki BA (1999) Niobium aluminide as a source of highcurrent superconductors. Intermetallics 7:117–140
11. Mattheiss LF, Weber W (1982) Electronic structure of cubic V3Si and Nb3Sn. Phys Rev B 25(4):2248–2269
12. Klein BM, Boyer LL, Papacon-stantopoulos DA (1978) Selfconsistent augmented- plane-wave electronic-structure calculations for the A15 compounds V3X and Nb3X, X = Al, Ga, Si, Ge, and Sn. Phys Rev B 18(12):6411–64338
13. Pickett WE, Ho KM, Cohen LM (1979) Electronic properties of Nb3Ge and Nb3Al from self-consistent pseudotential. Phys Rev B 19(4):1734–1750
14. Kresse G, Furthmu¨ller J (1996) Efficient iterative schemes for ab initio total-energy calculations using a plane-wave basis set. Phys Rev B 54(16):11169–11186
15. Kresse G, Furthmu¨ller J (1996) Efficiency of abinitio total energy calculations for metals and semiconductors using a plane-wave basis set. Comput Mater Sci 6(1):15–50
16. Perdew JP, Burke K, Ernzerhof M (1996) Generalized gradient approximation made simple. Phys Rev Lett 77(3):865–3868
17. Paduani C (2007) Electronic Structure of the A3B Compounds: A = Nb: B = Al, Ga, Ge, Sn and In. Braz J Phys37(3B):1073–1079

CHAPTER 9

MAGNETORESISTANCE OSCILLATIONS IN LA-BASED METALLIC GLASS

WANG YongTian, LIU ZongDe and WANG WeiHua

Keywords: Metallic glass, superconducting, resistance oscillation

The magnetoresistance oscillations of low-dimensional superconductors, such as nanowire, nanobelt, or granular systems, have been extensively studied [1,2]. This is one of the most intriguing problems in condensed-matter physics, and this technology is widely used to develop new kinds of superconducting quantum interference devices (SQUIDs), such as SQUIDs based on carbon nanotubes [3] and SQUIDs made via the DNA templating of superconducting nanowires [4]. However, the low-dimensional samples in these studies are either single or quasi-1D. An oriented 2D nanoarray with a large aspect ratio and flexible substrate is highly desirable.

In this letter, we report the fabrication of a simple and flexible superconducting device based on a La-based metallic glass. Below the superconducting transition temperature, T_c, the ribbon, which has high intrinsic flexibility and strength, exhibited a clear periodic modulation in its differential resistance as a function of a perpendicular magnetic field. The oscillations resulted from a two-dimensional array composed of La nanocrystals with high aspect ratios and uniform orientations. These arrays spontaneously assembled on the free surface of the glass ribbon upon quenching, which is not the case for other nanoparticles and nanowires [5-7]. The work presents an advanced application of the device for an implementation using metallic glasses as the precursors for the nano-material [8–11].

EXPERIMENTAL

The $La_{60}Fe_{30}Al_{10}$ ribbons were prepared using a single roller melt-spinning apparatus at 5–25 m/s in an argon atmosphere. The structure was characterized via X-ray diffraction (XRD) in a MAC M03 XHF diffractometer with CuKa radiation. Thermal analysis was performed using a Perkin-Elmer DSC-7 differential scanning calorimeter (DSC). The topography of the ribbon was studied using an atomic force microscope (AFM, Digital Instruments NanoScope IIIa D-3000 AFM). The magnetism and standard four-point electrical resistance measurements were performed using a PPMS 6000.

RESULTS

Figure 1 shows the XRD patterns of the $La_{60}Fe_{30}Al_{10}$ ribbons for various wheel velocities. For the ribbon spun at 25 m/s, the XRD pattern of the bottom surface (in contact with the copper roller) shows a broad maximum peak indicating an almost amorphous phase. The sharp crystallization peak of the inset DSC curve confirms its glass-like nature. No amorphous phase can be detected from the ribbon free surface, but a strong and broadening crystalline peak is superimposed on the XRD curve. When the spin velocity decreases to 10 and 5 m/s, the intensity of the peak decreases. Through closer examination of the XRD pattern, one can find a much stronger La (002) peak, which suggests that the hexagonal La phase has a well-defined texture with the c-axis predominantly perpendicular to the ribbon plane [12]. According to the Scherrer formula [13,14], the peak is broadened because of the formation of ultrafine nanocrystals.

Figure 2(a)-(d) shows the AFM micrographs of the $La_{60}Fe_{30}Al_{10}$ ribbon spun at 5 m/s (left) and 25 m/s (right). The bottom shows the corresponding cross-section analysis (1 μm×1 μm). For the ribbon at 5 m/s, some nanocrystals with random distribution were observed on the free surface, and the size and distribution of the nanocrystals were far from uniform. However, when the wheel velocity was increased to 25 m/s, a large number of columelliform nanocrystals with a lateral width of ~100 nm were assembled into a compact 2D film over the whole ribbon free surface. The improved texture also resulted in the greatly enhanced intensity of the La (002) peak. The precipitation of Rare-Earth (RE) particles is a common

Citation: Wang Y T, Liu Z D, Wang W H. Magnetoresistance oscillations in La-based metallic glass. Chinese Sci Bull, 2011, 56: 3984–3987, doi: 10.1007/ s11434-011-4764-9.

phenomenon for the RE-Fe-Al glass-forming system [11,12]. The La nanocrystals on the surface have a stronger temperature gradient in the direction perpendicular to the ribbon plane [11]. Therefore, these La nuclei tend to grow normal to the ribbon plane, and finally assemble into the film.

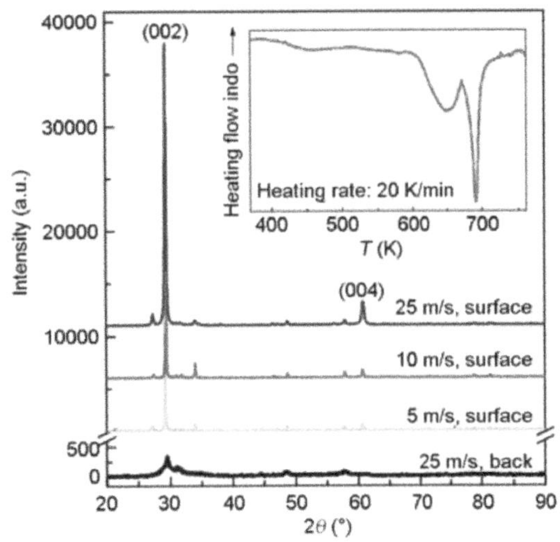

Figure 1: XRD patterns for the $La_{60}Fe_{30}Al_{10}$ ribbons at various wheel velocities. The inset shows the DSC curve of the ribbons at 25 m/s. The heating rate is 20 K/min.

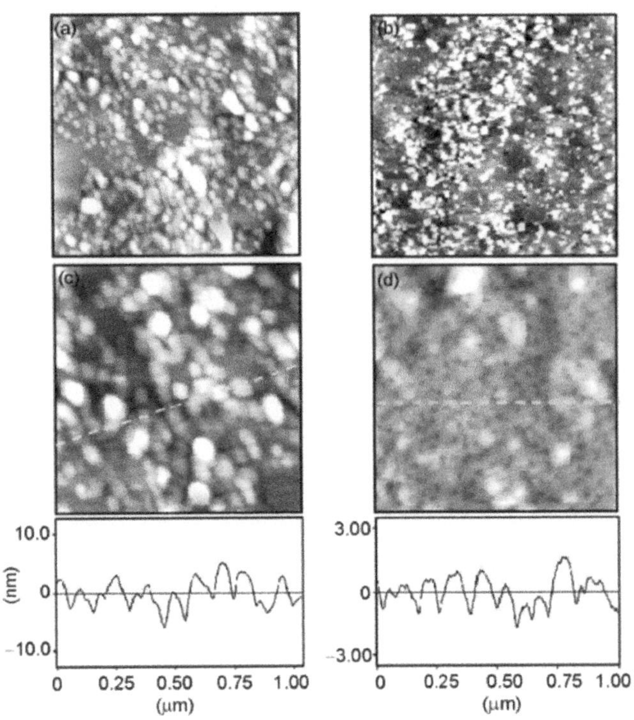

Figure 2: Topographic AFM images of the La60Fe30Al10 ribbon free surface and corresponding cross-section analysis. (a) 5 m/s, 2 μm × 2 μm; (b) 25 m/s, 10 μm×10μm; (c) 5 m/s, 1 μm × 1 μm; (d) 25 m/s, 1 μm × 1 μm.

Figure 3 shows the relative resistance, R/R_{10K}, versus the temperature, T, of the ribbon in different magnetic fields. The left inset shows the standard four-probe setup for ribbon in-plane and transverse resistance measurements (R_s and R_T). The R_T is 10 times less than the in-plane resistivity R_s. Both RT and R_s show metallic behaviors from room temperature to T_c, with residual resistance ratios, R(300 K)/R(10 K), of 3.2 and 3.5, respectively. This indicates that direct ohmic contact is established between the deposited Ag electrode and ribbon [15]. At H=0 T, the R shows a sharp drop near 5 K before decaying to zero, indicating the ribbon is entering a superconducting state. The initial temperature of superconductivity is near the reported hexagonal α-La polycrystalline with T_c=4.9 K [16]. The transition range is broad, ΔT≈2 K, which is significantly greater than that in similarly prepared bulk La-based alloys. This indicates we are approaching the one dimensional limit, which is typical for a 1D superconductor because of its susceptibility to thermal fluctuation [4,5]. The right inset of Figure 3 shows the H dependence of the resistance with H perpendicular to the ribbonfree surface at 2, 4, and 6 K, respectively. Above T_c≈5 K, both R_s and R_T remain constant when the magnetic field increases, whereas below T_c the destruction of the superconducting state by the magnetic field at 2 and 4 K results in different stages with slight slope

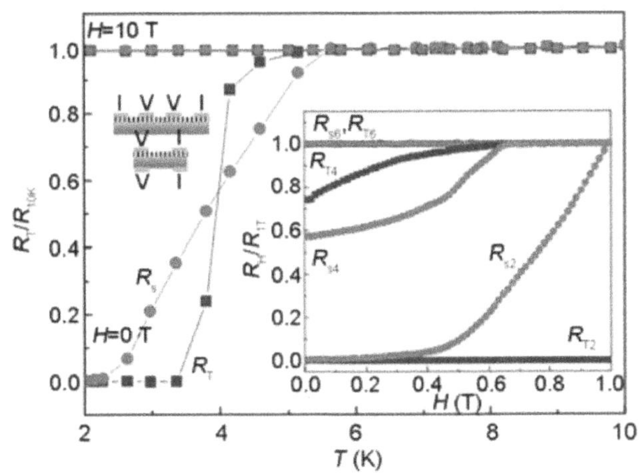

Figure 3: The relative resistance, R/R_{10K}, versus temperature measurements of the ribbon spun at 25 m/s. The left inset diagram shows the standard four-probe setup for surface and transverse resistance measurements (R_s and R_T), respectively. The four orange regions denote the silver plate electrodes. The right inset shows the resistance R_s and RT versus magnetic field measurements at 2, 4, and 6 K, respectively.

Changes Until the Critical Field

An analysis of the differential magnetoresistance dR/dH with respect to H is shown in Figure 4. Note that differentiating the curve to magnify the slight slope change is a powerful and widely-used data analysis method [5]. This figure shows that, for ribbons spun at 5 m/s, the dR/dH includes three distinct regimes, which are the same as those of the superconducting granular film with disordered grains [17] No oscillations are observed. However, for ribbons at 25 m/s, dR/dH has well-defined oscillations. The oscillation period DH = 0.36 T is reproducible and independent of temperature and the magnetic history of the ribbon, i.e., sweeping the magnetic field up or down. Although it fluctuates greatly, the amplitude of the oscillations does not show a systematic dependence on H below the critical temperature T_c, above which significant oscillations are no longer detected.

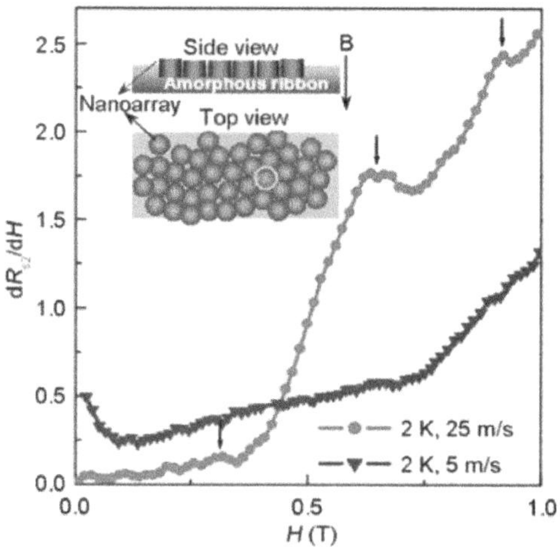

Figure 4: The derivative dR_s/dH vs. H at 2 K for the ribbons spun at 5 and 25 m/s, respectively. The inset shows the diagram of La-based glass ribbon. The circle denotes the superconducting screening current loops.

DISCUSSION

The resistance oscillations in conventional SQUIDs and Little-Parks (LP) resistance oscillations in thin-walled cylinders are the most well-known examples of quantum interference [4]. Taking into account the circular La nanorods and T-independent period, it seems that there are mesoscopic superconducting ring-like structures in the ribbon surface, which may induce the periodic magnetoresistance oscillation in the superconducting regime, such as Little-Parks-like oscillations. Our findings deviate significantly from the expected outcome because the standard LP oscillations are expected in a superconducting hollow cylinder near T_c. This deviation may occur because La nanorods with a high aspect ratio are not fully superconducting. Even for H = 0 T, the core of the nanorod shows a finite resistance below 2 K. The broadening and the finite resistance below the onset of superconducting transition are likely the consequence of the size confinement in 1D created by the thermally-activated phase slips near T_c [2] or the quantum phase slips far below T_c [2]. As the magnetic field increases, even at temperatures well below T_c, there is a superconducting cylindrical shell surrounding the La nanorod. In the superconducting-dominated states, the magnetic field creates Meissner screening currents and through them suppresses the energy barrier for the phase slips. The net current is a combination of normal and super currents. The periodic oscillations are produced by the latter, which likely reflected the oscillations of the screening currents around the La nanorod as a function of the magnetic field.

The observed magnetoresistance oscillation is currently not well understood. We suggest the following quantitative explanation for LP oscillations [18], in which the period of the oscillations, DH, is controlled by the superconducting flux quantum Φ_0=h/2e (here e is the electron charge and h is Plank's constant), divided by the cross-sectional area S enclosing the magnetic field, DH • S=Φ_0=2.07×10–15 Wb, for a superconducting circle loop or cylinder. The sum of the externally-applied magnetic flux Φ and the flux induced by the superconducting screening currents surrounding the La nanorods is quantized in units of the flux quantum Φ_0. The screening currents are absent when $\Phi=n\Phi_0$, but the screening currents reach their maximum amplitude when $\Phi=(n+1/2)\Phi_0$, resulting in a minimum value of T_c for the higher flux values. Experimentally, the LP oscillations of T_c can be observed as a periodic variation of the R with H at a fixed temperature. Therefore, the dR/dH is characterized by the regular appearance of minima or maxima in Hn. The measured period, DH = 0.36 T, theoretically corresponds to a superconducting current loop with R_1=43 nm, which is in good agreement with the La nanorod experimental radius (the green circle, 40–50 nm) measured via AFM (see the inset of Figure 4). Thus, the oscillation periods are attributed to the superconducting screening currents surrounding the La nanorods, i.e. the dominant cell area in the La-based array. Note that although the oscillation is reproducible for different samples, the period of oscillation is not well defined. To further test the prediction, we compare the La nanoarray with a superconducting granular Sn film and wire [17]. Because the 2D film is macroscopic, and there is a wide distribution of loop size, the oscillation of the superconducting screening currents will average out and no oscillations will be detected. This is similar to the case for ribbons with disordered La nanocrystals (spun at 5 m/s). However, the 1D granular Sn wire with reduced dimensionality begins to show a series of periods [17]. Analogously, the oriented La nanoarray originated from the ribbon spun at 25 m/s, its degree-of-freedom is even less than that in the ribbon at 5 m/s. Therefore, there is only one dominant superconducting loop around the La nanorod, which is responsible for the main oscillation periods.

CONCLUSION

In summary, a La nanoarray with a large aspect ratio and uniform orientation was assembled directly from a $La_{60}Fe_{30}Al_{10}$ glass alloy upon quenching. This nanoarray is similar in function to a simple and flexible SQUID-like device. The present single-step formation method eliminates the conventional cumbersome process and offers an easy way to fabricate advanced functional materials using metallic glass as precursors.

This work was supported by the National Natural Science Foundation of China (51101056, 50731008 and 50921091), the National Basic Research Program of China (2007CB613904), the Fundamental Research Funds for the Central Universities (09Q44), and the State Key Lab of Advanced Metals and Materials (2010Z-02).

REFERENCES

1. Wang J, Shi C T, Tian M L, et al. Proximity-induced superconductivity in nanowires: Minigap state and differential magnetoresistance oscillations. Phys Rev Lett, 2009, 102: 247003
2. Tian M L, Wang J, Zhang Q, et al. Superconductivity and quantum oscillations in crystalline Bi nanowire. Nano Lett, 2009, 9: 3196–3202
3. Cleuziou J P, Wernsdorfer W, Bouchiat V, et al. Carbon nanotube superconducting quantum interference device. Nat Nanotech, 2006, 1: 53–59
4. Hopkins D S, Pekker D, Golbart P M, et al. Quantum interference device made by DNA templating of superconducting nanowires. Science, 2005, 308: 1762–

1765

5. Johansson A, Sambandamurthy G, Shahar D. Nanowire acting as a superconducting quantum interference device. Phys Rev Lett, 2005, 95: 116805
6. Hasselbach K, Mailly D, Kirtley J R. Micro-superconducting quantum interference device characteristics. J Appl Phys, 2002, 91: 4432
7. Xu Q B, Castillejos R P, Li Z F, et al. Fabrication of high-aspect-ratio metallic nanostructures using nanoskiving. Nano Lett, 2006, 6: 2163–2165
8. Inoue A, Shimizu T, Yamaura S, et al. Development of glassy alloy separators for a proton exchange membrane fuel cell (PEMFC). Mater Trans, 2005, 46: 1706–1710
9. Schroers J, Lohwongwatana B, Johnson W L, et al. Gold based bulk metallic glass. Appl Phys Lett, 2005, 87: 061912
10. Kim J J, Choi Y, Suresh S, et al. Nanocrystallization during nanoindentation of a bulk amorphous metal alloy at room temperature. Science, 2002, 295: 651–657
11. Wang Y T, Bai H Y, Pan M X, et al. Giant enhancement of magnetocaloric effect in metallic glass matrix composite. Sci China Ser G-Phys Mech Astron, 2008, 51: 337–348
12. Sun Z G, Löser W, Eckert J, et al. Phase separation in Nd60−xYxFe30Al10 melt-spun ribbons. Appl Phys Lett, 2002, 80: 772
13. 13 Guo W H, Kui H W. Bulk nanostructured alloy formation with controllable grain size. Acta Mater, 2004, 8: 2117–2121
14. Wang W H, Dong C, Shek C H. Bulk metallic glasses. Mater Sci Eng R, 2004, 44: 45–89
15. Stankovich S, Dikin D A, Dommett G H B, et al. Graphene-based composite materials. Nature, 2006, 442: 282–286
16. Agyeman K, Muller R, Tsuer C C. Alloying effect on superconductivity in amorphous lanthanum-based alloys. Phys Rev B, 1979, 19: 193–198
17. Herzog A V, Xiong P, Dynes R C. Magnetoresistance oscillations in granular Sn wires near the superconductor-insulator transition. Phys Rev B, 1998, 58: 14199–14202
18. Vloeberghs H, Moshchalkov V V, Haesendonck C V, et al. Anomalous Little-Parks oscillations in mesoscopic loops. Phys Rev Lett, 1992, 69: 1268–1271

CHAPTER 10

A NOVEL SUPERCONDUCTING QUANTUM INTERFERENCE DEVICE FOR BIOMAGNETIC MEASUREMENTS

Zhang Shulin, Zhang Guofeng, Wang Yongliang, Zeng Jia, Qiu Yang, Liu Ming, Kong XiangYan & Xie XiaoMing

Keywords: SQUID Bootstrap circuit (SBC), gradiometer, magnetocardiography (MCG), fetal MCG, magnetoencephalography (MEG)

Superconducting quantum interference devices (SQUIDs) have been widely used to detect the very weak biomagnetic fields from picotesla (pT) down to femtotesla (fT), such as human magnetoencephalography (MEG), magnetocardiography (MCG) and fetal MCG [1]. Besides the high sensitive SQUID sensors which are needed for successfully implementation of biomagnetic measurements, the other key problem is to detect such weak signals under an environment with very strong background fields. At present, different SQUID systems are employed: SQUID magnetometers for the measurements performed in super magnetically shielding rooms (MSRs) or SQUID gradiometers with different antenna figurations in moderate MSRs or unshielded environments [2,3].

In this work, we report a helium-cooled gradiometer based on a novel SQUID named SQUID bootstrap circuit (SBC) [4–6]. By using the SBC gradiometer, biomagnetic measurements was performed in a moderate MSR. Besides MCG and MEG signals, fetal MCG signals were also detected. SBC is a voltage-biased SQUID direct readout scheme. By increasing the SQUID flux-to-current coefficient and the dynamic resistance with two branches, the flux-to-voltage coefficient $(\partial V/\partial \Phi)_{SBC}$ of SBC is enhanced to suppress the noise from the preamplifier.

Figure 1(a) shows a schematic diagram of our designed SBC. Using a voltage feedback coil of inductance L2 and a shunt resistor R_s in series, the $(\partial V/\partial \Phi)_{SBC}$ can be enhanced by increasing the dynamic resistance. In order to suppress the external magnetic noise in the SBC chip area, a dual-loop washer-SQUID of a gradiometer configuration was designed. The input coil couples external magnetic flux to the SQUID by connecting a first-order axial wire-wound antenna, as shown in Figure 1(b). The antenna has a configuration of 1:1 with a baseline of 50 mm and 18 mm in diameter. In order to enhance the noise suppression, a thin niobium shielding tube was used to cover the SQUID chip, as indicated in Figure 1(b). Figure 1(c) shows the measured field spectral density of the SBC gradiometer in our MSR. The noise level was measured to be 5 fT/$\sqrt{Hz}$ at 100 Hz.

Based on the SBC gradiometer, a biomagentic measurement system was set up in our MSR. The SBC gradiometer was mounted in a special nonmagnetic glass fiber cryostat. The distance between the lower pick-up coil of the antenna and the outside bottom of the cryostat is 16 mm. In order to reduce the external magnetic disturbances, the data acquisition and processing system was placed outside the MSR. By using the SBC system, MCG, fetal MCG and MEG signals were detected. The inset of Figure 2 shows the real-time MCG signals of an adult recorded with a sampling rate of 2560 Hz. The bandwidth is 0.1–150 Hz. By averaging data of 40 cycles, MCG signal with a signal to noise ratio of 40 dB was obtained. The R peak of the MCG signal was approximately 40 pT (Figure 2).

The inset of Figure 3 shows the recorded real-time fetal MCG signal with a sampling rate of 1 kHz. The gestational age of the fetus is 30 weeks. The high-frequency noise and baseline drifts were removed by combination of a 90 Hz low-pass FIR filter and the wavelet transformation. Figure 3 shows the averaged fetal MCG signal of 36 cycles. Comparing to the above measured adult MCG signal, the fetal MCG signal

Citation: Zhang S L, Zhang G F, Wang Y L, et al. A novel superconducting quantum interference device for biomagnetic measurements. Chin Sci Bull, 2013, 58: 2917–2919, doi: 10.1007/s11434-013-5968-y.

was weak and only about 0.8 pT. For the MEG measurement, auditory evoked magnetic field was detected. The stimulus was a pure sinusoidal tone with a frequency of 1 kHz, a duration of 100 ms and an interval of 1.1 s. The inset of Figure 4 shows the averaged MEG trace by 100 times with a sampling rate of 375 Hz. High frequency noise was removed by a low-pass filter of 35 Hz. From the trace, the prominent N100m peak can be clearly recognized [7]. The amplitude was about 0.14 pT.

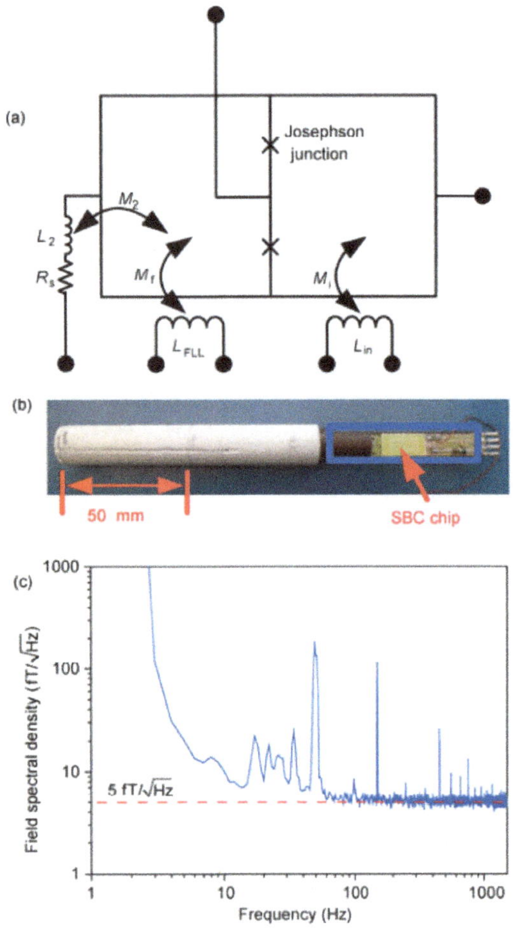

Figure 1: SBC gradiometer. (a) Schematic diagram of the designed SBC; (b) photograph of the gradiometer module; (c) noise spectrum of the SBC gradiometer.

From the above results, amplitudes of all the measured signals were smaller than those reported before [8–10]. For the MCG and fetal MCG signals, the smaller results attribute to the shorter gradiometer baseline of 50 mm [11], which is also the reason why the maternal MCG signal was not observed during the fetal MCG measurements. For the MEG signal, it is because of the non-specially designed MEG cryostat [12]. In conclusion, a SBC gradiometer with a noise level of 5 fT/$\sqrt{\text{Hz}}$ at 100 Hz was developed and applied for the biomagnetic measurements. Besides the MCG and MEG signals, a fetal MCG signal was also measured for the first time in China. Further optimization of the SBC gradiometer and development of a multichannel biomagnetic system will be studied in the near future.

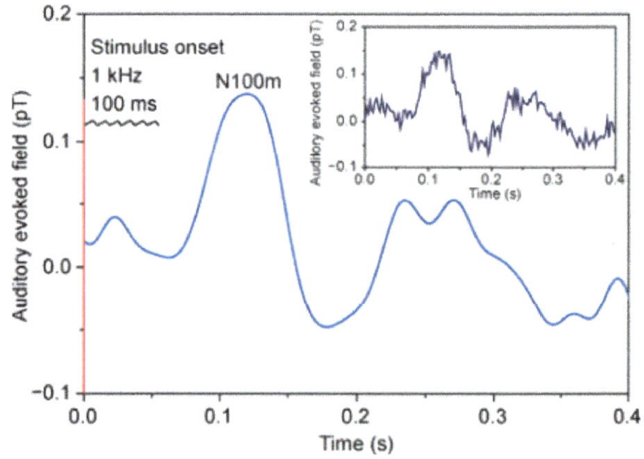

Figure 4: Auditory evoked magnetic field averaged by 100 times (inset: signal without filter).

The authors thank Professor Yi Zhang from the Research Center Jülich, Germany, for his enthusiastic encouragement and fruitful discussion. This work was supported by the Knowledge Innovation Program of the Chinese Academy of Sciences (KGCX2-YW-906, KGCX2-EW-105), "Hundred Talents Program" of the Chinese Academy of Sciences and Strategic Priority Research Program (B) of the Chinese Academy of Sciences (XRD04020300).

REFERENCES

1. Sternickel K, Braginski A I. Supercond Sci Tech, 2006, 19: S160– S171
2. Pizzella V, Penna S D, Gratta C D, et al. Supercond Sci Tech, 2001, 14: R79–R114
3. Vrba J, Haid G, Lee S, et al. Clin Phys Physiol Meas, 1991, 12: 81–86
4. Xie X M, Zhang Y, Wang H W, et al. Supercond Sci Tech, 2010, 23: 065016
5. Zhang Y, Zhang G F, Wang H W, et al. IEEE Trans Appl Supercon, 2011, 21: 501–504
6. Zhang G F, Zhang Y, Zhang S L, et al. Physica C, 2012, 480: 10–13
7. Hämäläinen M, Hari R, Ilmoniemi R J, et al. Rev Mod Phys, 1999, 30: 413–497
8. Schneider S, Abraham-Fuchs K, Reichenberger H, et al. Physiol Meas, 1993, 14: A55–A60
9. Bick M, Stemickel K, Panaitov G, et al. IEEE Trans Appl Supercon, 2001, 11: 673–676
10. Comani S, Srinivasan V, Alleva G, et al. Phys Med Biol, 2007, 53: N87–N97
11. Zhang Y, Wolters N, Lomparski D, et al. IEEE Trans Appl Supercon, 2005, 15: 631–634
12. Knuutila J E T, Ahonen A I, Hämäläinen M S, et al. IEEE Trans Magn, 1993, 29: 3315–3321

CHAPTER 11 | NB_3AL SUPERCONDUCTOR

Zhao Liu, Yongliang Chen, Lupeng Du, Pingyuan Li, Yajing Cui, Xifeng Pan and Guo Yan

Keywords: Nb3Al, Superconductivity, Powder metallurgy, Mechanical alloying

INTRODUCTION

With development of modern transportation and pursuit of high speed in various vehicles, superconducting maglev is promising in taking an important role in the future transportation system. Selection of a proper superconducting wire is vital in fabrication of superconducting magnet. Nb_3Al, due to its excellent superconducting properties, has drawn comprehensive attention since 1980s [1, 2]. A15 structure materials are famous for their superconductivities and Nb_3Al has the higher critical temperature among all the A15 structure materials [3, 4]. Furthermore, high value of critical current density (J_c) under high applied field makes it a promising substitution material for the fabrication of high-field magnet. The current existing fabrication method, powder-in-tube (PIT), makes it possible for a large-scale application of Nb_3Al prepared by powder metallurgy.

In International Thermonuclear Experimental Reactor (ITER) [5], compared with the well-developed Nb_3Sn, Nb_3Al possesses better stress/strain tolerance. This makes it an ideal choice for the fabrication of the toroidal field (TF) coil in terms of the large size of the D-shape TF [6,7]. At the moment, the relatively successful methods to prepare Nb_3Al wire are restricted to high-temperature methods, because of the difficulty of the phase formation. Therefore, in order to acquire Nb_3Al in the right stoichiometry, high-temperature generation specialized instruments are needed, such as RHQT [8]. In this work, powder metallurgy was used to synthesize Nb_3Al and the effect of mechanical alloying on the phase formation was investigated.

METHODS

The experimental process includes into two parts. In the first part, normal hand grinding method was used to mill the Nb, Al powder. The mole ratio of Nb/Al = 3/1 was weighed, placed in an agate mortar and pestle, and mixed by gentle hand grinding for 0.5 h. The mixtures were pressed into round pellet with a diameter of 11.5 mm and a thickness of 1.5 mm using hardened steel die with uniaxial 12 MPa pressure. Then, the pellet was sealed into a vacuum-fused quartz tube with a vacuum degree of 5.5×10^{-3} Pa. Finally, the samples were heat treated in a tube furnace and argon was vented to avoid permeability or leakage.

In the second part, mechanical alloying was adopted and a SPEX [9] mixer 8000 high-energy mill was used. All other experimental procedures were the same.

The crystal structure was studied by powder X-ray diffraction (XRD) using an X'Pert MRD diffractometer with Cu Kα radiation. All observed reflections were indexed. Lattice constants were determined from LeBail refinements. Microstructure and composition of the sample were analyzed using a field emission scanning electron microscope (FESEM) equipped with an energy-dispersive X-ray analysis (EDX). DC magnetization was measured with a SQUID magnetometer (MPMS, Quantum Design) and resistivity measurements were performed with a physical property measurement system (PPMS, Quantum Design).

RESULTS AND DISCUSSION

Figure 1 shows the XRD patterns of Nb/Al reaction samples sintered at 800 °C for different times. It can be observed that the main phases can be indexed as $NbAl_3$ and Nb. According to the phase diagram, Nb is inferred as the α-Nb which is the Nb-rich solid solution of aluminum in niobium. Figure 2 shows the XRD patterns of the samples sintered for 10 h at different sintering temperatures. Compared with the data of Fig. 1, it is obvious that both the changes of temperature and extension of heating duration make no difference to the resultant phase formation. Figure 2 shows the XRD patterns of the samples sintered for 10 h at different sintering temperatures. By comparison, it is apparent that either the change of temperature or extension of heating duration makes no difference to the resultant phase formation.

Citation: Zhao Liu, Yongliang Chen, Lupeng Du, Pingyuan Li, Yajing Cui, Xifeng Pan, Guo Yan, Nb3Al superconductor, DOI 10.1007/s40534-014-0036-0.

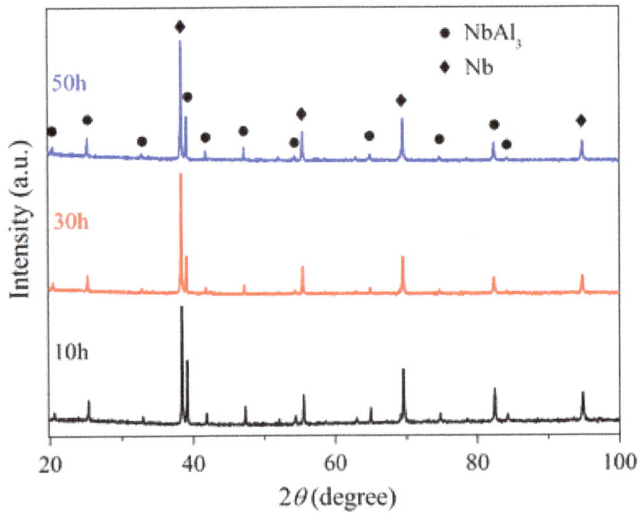

Figure 1: XRD patterns of samples sintered at 800 C

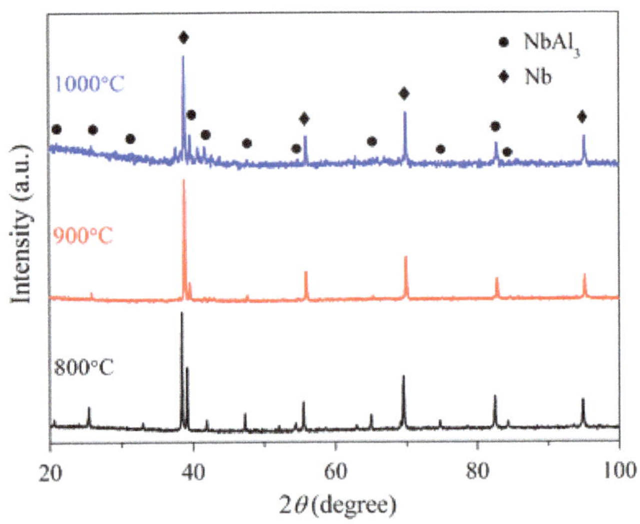

Figure 2: XRD pattern of samples sintered at 10 h for different temperatures.

Thus, it is concluded that the reaction between Nb and Al results in the formation of $NbAl_3$under the selected heat-treatment conditions. Furthermore, $NbAl_3$ is the most thermodynamically stable of the Nb–Al intermediate phases and so the kinetics of conversion to Nb_3Al by reaction with Nb can be expected to be slog. Consequently, it is concluded that, in the range of 800–1,000 °C, the temperature and duration of heat treatment are not the controlling factors of the formation of Nb_3Al.

The SEM image of sample sintered at 800 °C for 50 h is displayed in Fig. 3. It can be seen from the picture that the sample is very porous and the size of the biggest grain in this field of view is more than 10 μm. The very coarse morphology of the sample would give the diffusion reaction a very long diffusion distance. Considering the very high formation enthalpy [9] of Nb_3Al, the long diffusion distance between the two grains makes it impossible for the formation of Nb_3Al using hand milling. The EDX data on grains (see Table 1) support the conclusion that the big grains are $NbAl_3$.

Figure 3: SEM image for the sample sintered at 800 °C for 50 h.

Table 1: EDX results of a random point on the grain

Element	Weight (%)	Atom (%)
Al	44.0	73.0
Nb	56.0	27.0

Figure 4 shows the XRD patterns of Nb_3Al samples prepared for different spex milling and sintering times at 900 °C. The main phase can be indexed as the A15 structure Nb_3Al. The high-purity Nb_3Al sample was obtained when the spex milling time reached to 5 h, and spex milling for 10 h and sintering for 10 h gave the optimal sample. The sample that was spex milled for only 3 h contained a great quantity of Nb. This can be explained by short spex milling time; the Nb/Al powders lacked enough extra driving forces for diffusion reaction. Furthermore, it can be seen from all the four patterns that no $NbAl_3$ phase was found, thus indicating that spex milling is effective in increasing the activity of the two powders by improving the mixing and reducing the diffusion distance.

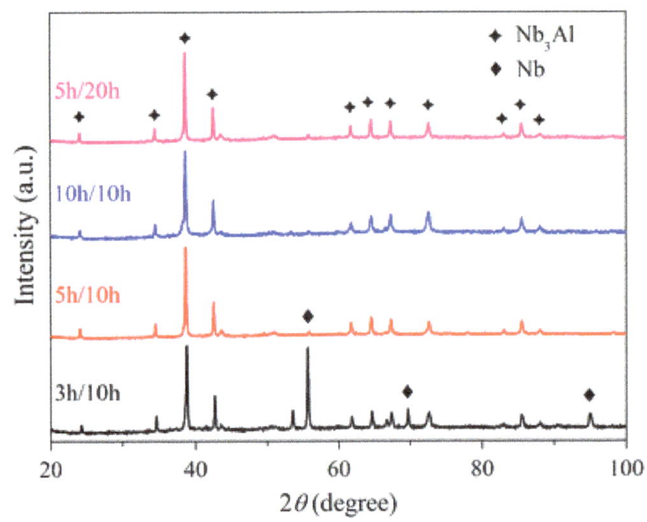

Figure 4: XRD patterns of the samples prepared for different milling and sintering times at 900 °C

Figure 5 shows the temperature dependence of magnetization for the sample spex milled for 5 h and sintered for 20 h, and the critical temperature (T_{conset}) is 14.2 K which is 4 K lower than the optimal value. The impurity phases and the porous microstructure may give rise to the poor superconducting property, although little impurity phase was indexed from the XRD patterns. It should be noted that no superconducting transition of Nb was detected, which further confirms that the Nb has sufficiently reacted. However, it is important to realize that the Nb_3Al single phase range is relatively wide according to the phase diagram. In addition, the heat-treatment temperature selected in this work can only give an Nb_3Al sample with the stoichiometry of only 20 at.% Al. Therefore, the deviation from the right stoichiometric concentration of Aluminum may be another reason that accounts for the low T_c of the sample, though the sample is an almost-pure single phase as the XRD reflects.

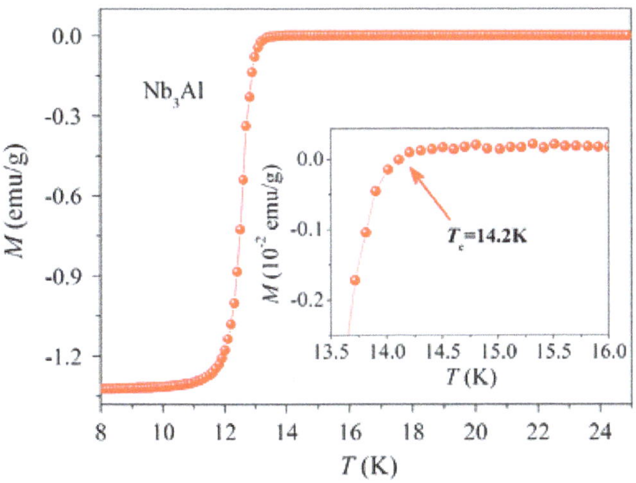

Figure 5: The temperature dependence of magnetization for the Nb_3Al sample, and inset shows the enlarged view of the superconducting transition.

CONCLUSIONS

- The composition of Nb/Al powder used in the experiment is not in the Nb_3Al solid solution homogeneity region in phase diagram of Nb–Al binary alloy, but the as-grown sample appears to be single phase.
- 800 °C is not adequate in terms of kinetics for the equilibration of Nb_3Al phase formation.
- Spex milling is effective in increasing activity of the powders by improving the mixing of the powders, simultaneously reducing the diffusion distance. By adoption of spex milling and the conventional PIT method, fabrication of Nb_3Al magnet under low temperature can be realized.
- The impurities and the deviation of the right stoichiometric concentration of aluminum in Nb_3Al are probably responsible for the poor quality of superconducting transition.
- Combination of spex milling and the PIT method may be one of the solutions to prepare Nb_3Al wire under low temperature in the future.

ACKNOWLEDGMENTS

This work was supported by the National Magnetic Confinement Fusion Science Program (Grant No. 2011GB112001), the Program of International S&T Cooperation (Grant No. 2013DFA51050), the National Natural Science Foundation of China (Grant Nos. 11104224, 11004162, 51377138, and 51302224), the Research Fund for the Doctoral Program of Higher Education of China (Grant No. 20110184120029), the Fundamental Research Funds for the Central Universities (Grant Nos. 2682013ZT16, SWJTU11ZT31, 2682013CX004, and SWJTU11BR063), and the Science Foundation of Sichuan Province (Grant Nos. 2011JY0031 and 2011JY0130).

REFERENCES

1. Yamada Y, Ayai N (1999) Development of Nb3Al superconductors for International Thermonuclear Experimental Reactor (ITER). Cryogenics 39:115–122
2. Koizumi N, Takeuchi T, Okuno K (2005) Development of advanced Nb3Al superconductors for a fusion demo plant. Nucl Fusion 45:431–438
3. Takeuchi T (2000) Nb3Al conductors for high-field applications. Supercond Sci Technol 13:R101–R119
4. Okuno K, Shikov A, Koizumi N (2004) Superconducting magnet system in a fusion reactor. J Nucl Mater 329–333:141–147
5. Seeber B (2007) Critical current of a Nb3Sn bronze route conductor under uniaxial tensile and transverse compressive stress. Supercond Sci Technol 20:S184–S188
6. Ando T (1996) Design consideration of the ITER-TF coil with a react-and-wind technique using Nb3Al conductor. IEEE Trans Magn 32:2324–2327
7. Fukuzaki T, Takeuchi T, Banno N (2002) Effect of rapid heating, quenching and transformation conditions on the superconducting properties and microstructure of Jelly-Roll processed Nb3Al superconductors. Supercond Sci Technol 15:1404–1409
8. Koizumi N, Ando T, Sugimoto M et al (1998) Development of a 13-T and 40-kA Nb3Al conductor for toroidal coils of fusion reactors. Fusion Eng Des 41:277–281
9. Calka A, Wexler D, Zhou J et al (1998) Mechano-synthesis of niobium–aluminum superconductors for the high field applications. Mater Sci Forum 269–272:1043–1048

CHAPTER 12

EXPLORATION OF NEW SUPERCONDUCTORS AND FUNCTIONAL MATERIALS

Hideo Hosono, Keiichi Tanabe, Eiji Takayama-Muromachi, Hiroshi Kageyama, Shoji Yamanaka, Hiroaki Kumakura, Minoru Nohara, Hidenori Hiramatsu and Satoru Fujitsu

Keywords: Superconductivity, iron pnictide, new superconductors, superconducting wire, superconducting tape, functional material, powder in tube

INTRODUCTION

Since the discovery of superconductivity by Heike Kamerling Onnes in 1911 [1], it has been the dream of researchers to realize a room temperature superconductor. Although the fundamental theoretical framework for superconductivity was established in 1957 by the Bardeen–Cooper–Schrieffer (BCS) theory, there exists no theory which can quantitatively predict the critical temperature (T_c) even now [2]. Thus, looking for high T_c superconductors is like a voyage in a big ocean without a precise compass, i.e., researchers have to move ahead believing their sense and/or intuition referring to what theorists say. In this sense, looking for high T_c superconductors is a truly challenging subject and typical 'all or nothing' research.

It is a historical fact that materials leading to breakthroughs have been discovered in most cases by chance amidst concentrated research efforts undertaken with a unique but flexible view. This is particularly true for the exploration of new superconductors. The group of one of the present authors (HH) discovered superconductivity in LaFePO in 2006 [3] and $LaFeAsO_{1-x}F_x$ in 2008 [4] through LaNiPO ($T_c = 3$ K) [5] in 2007 in the course of exploring magnetic semiconductors, which started from his extensive research of transparent p-type semiconductors LaCuO*Ch*(where *Ch* = S and Se) with the same crystal structure as the so-called 1111-type layered compounds. P-type conduction in LaCuO*Ch* originates from the mobile holes at the top of the valence band which is composed of *Ch* p orbitals and Cu 3*d* orbitals [6]. It was his idea to novel magnetic semiconductors to utilize strong *d–p* interactions in LaCuO*Ch* by replacing the nonmagnetic Cu^+ ion with a magnetic 3*d* transition metal cation with a + 2 charge state. In order to keep electro-neutrality upon this substitution, Ch^{-2} is required to be replaced by Pn^{3-}[7]. This is the reason why his group started to examine the electronic and magnetic properties of La*TM*O*Pn* (where *TM* = 3*d* transition metal, *Pn* = P and As). This effort resulted in the discovery of iron-based superconductors (IBSCs) through the concentrated effort to find high performance p-type transparent semiconductors, which is a branch of his research home ground, transparent oxide semiconductors [8].

The discovery of IBSCs was accepted with surprise by the condensed matter community because iron, with a large magnetic moment, was widely believed to be most harmful to the emergence of superconductivity. Extensive research into these materials started globally, particularly in China [7]. As a result, the discovery of IBSCs was chosen as the breakthrough of the year in 2008 by *Science Magazine* and the paper [4] reporting $T_c = 26$ K in $LaFeAsO_1{-x}F_x$became the most cited report among all the original research papers published in 2008.

In early 2009, the Japanese Government announced the launch of a new large funding program—FIRST (**F**unding Program for World-Leading **I**nnovative **R**&D on **S**cience and **T**echnology). The aim of the FIRST Program is to advance the kind of leading-edge research and development that will strengthen Japan's international competitiveness while contributing to society and people's welfare through the application of its results. Hideo Hosono's proposal 'exploration for novel superconductors and relevant functional materials, and development of superconducting wires for industrial applications' was selected as one of 30 projects out of ~800 applications covering a very broad area of science and technology.

It was his expectation to find novel functionalities and materials with high potential through this tough and really challenging work just as IBSCs were found though the

Citation: Hideo Hosono, Keiichi Tanabe, Eiji Takayama-Muromachi, Hiroshi Kageyama, Shoji Yamanaka, Hiroaki Kumakura, Minoru Nohara, Hidenori Hiramatsu and Satoru Fujitsu, Exploration of new superconductors and functional materials, doi:10.1088/1468-6996/16/3/033503.

exploration of magnetic semiconductors. Hideo Hosono organized the research team to be mainly composed of solid state chemists who have experience and achievements not only in superconductors, but also in the relevant functional materials. Since research in finding new superconductors typically belongs to the domain of condensed matter physics, this team organization is a unique feature of this project. It was his belief that excellent solid state chemists will find new properties by serendipity, even if they fail to succeed in the hunt for new high T_c materials. This philosophy was set at the beginning of the project, i.e., 'All or something!'.

This article reviews the major research achievements obtained in our FIRST Project performed over the 4 years from March 2010 through March 2014, along with some background for the research. We have examined more than 1000 materials to seek new superconductors. The fraction of success was relatively small (~3%), but just as we expected at the outset. So far, unsuccessful results in this field have not been presented in an academic journal. In this review, we have listed the records of materials that we examined in this project, including the unsuccessful materials, based on a consensus among the members of the research team that the consideration of unsuccessful trials will be good fertilizer for future research.

OVERVIEW

At the beginning of this project, Hosono laid down five research targets. These were: (1) the discovery of a new superconductor with $T_c > 77$ K, (2) the development of new superconductors with high performance, (3) the development of related materials with outstanding functions, (4) the development of meter-class superconducting wires with a critical current density $J_c > 10^5$ A cm^{-2} based on IBSCs or other novel materials, and (5) the production of prototype Josephson junction and SQUID devices using IBSC thin films. This project consists of six research groups as listed in table 1. Four groups (HH, EM, HKa and SY) have concentrated on the exploration of new superconductors, and the other two groups (KT and HKu) have concentrated on the development of superconducting wires and tapes. The HH group collaborated with five other groups in the field to look for superconductors (MN) and the discovery of catalysis using electride (MH, TS, AS and SK). The HKu group collaborated with YK in the field of superconducting wires. Each group has made an effort to achieve its purpose using its special skills.

Table 1. Organization of FIRST Hosono Project.

	Team Leader	Mission
Exploration of Superconductors	Hideo Hosono (Core-Researcher) (Tokyo Institute of Technology)	Exploration of superconductors of transition metal compounds with layered structure collaborated with Minoru Nohara (Okayama University)
		Development of superconducting device by thin film technique
		Development of novel functional materials (especially C12A7 electride) collaborated with Michikazu Hara (Tokyo Institute of Technology), Tohru Setoyama (Mitsubishi Chemical Corporation), Alex Shluger (University College London) and Sung Wng Kim (Sungkyunkwan University)
	Eiji Takayama-Muromachi (National Institute of Materials Science)	Exploration of superconductors using high pressure synthesis technique and their characterization
	Hiroshi Kageyama (Kyoto University)	Exploration of superconductors using low temperature synthesis technique and their characterization
	Shoji Yamanaka (Hiroshima University)	Exploration of superconductors with layered or clathrate structure and their characterization
Development of Superconducting Wire	Keiichi Tanabe (Sub-Core-Researcher) (International Superconductivity Technology Center	Development and evaluation of superconducting wire and superconducting device prepared by thin film technique
	Hiroaki Kumakura (National Institute of Materials Science)	Development and evaluation of superconducting wire prepared by PIT method collaborated with Yoichi Kamihara (Keio Universty)

Though the FIRST Project did not discover a new superconductor with $T_c > 77$ K (56–58 K maximum), over 100 new superconductors have been developed and characterized as new types of IBSC (112) have been found, a new dopant into IBSC (H^-) has been employed to induce superconductivity, intercalation type compounds have been found, cobalt-based and titanium-based superconductors have been found, etc. The search for new superconductors is the most important target of the FIRST Project, and over 40 researchers in four groups have worked on this mission. They have examined more than 1000 materials to seek new superconductors. The number of new superconductors found is a relatively small percentage, as envisioned at the onset, which is part of the motivation for the extensive search. We believe that listing all materials examined, including both successes and failures, is meaningful for the people who work in this field or will join this field in the future, and thus show these in table 2. The details of some representative results will be described in section 3.

Table 2. Exploration of new superconductors in this project. (a) Materials which exhibited superconductivity. (b) Materials which exhibited no superconductivity.

(a) Materials which exhibited superconductivity

[shaded box]: noted in this review, (HP): synthesized using high pressure technique

IBSC

	Material	T_c (K)	Comment
1	$SmFeAsO_{1-x}H_x$ [9]	56	large solubility limit of dopant H^- (HP)
2	$NdFeAsO_{1-x}H_y$ [9]	54	large solubility limit of dopant H^- (HP)
3	$CeFeAsO_{1-x}H_x$ [10]	48	large solubility limit of dopant H^- (HP)
4	$LaFeAsO_{1-x}H_x$ [11]	36	two T_c domes (HP)
5	$SmFeAs_{1-y}P_yO_{1-x}H_x$ [12]	<56	T_c decreased by replacing with P (HP)
6	$LaFe_{1-y}Zn_yAsO_{1-x}H_x$ [13]	<36	T_c decreased by replacing with Zn (HP)
7	$LaFeAsO_{0.85}H_x$ (x =0-0.85) [14]	35	indirect electron doping by V_O and H^- (HP)
8	$LaFeAsO_{1-x}$ [15]	22	indirect electron doping by V_O (HP)
9	$CaFe_{1-x}Co_xAsH$ [16]	23	direct electron doping into CaFeAsH (HP)
10	$Ca_{1-x}La_xFeAsH$ [17]	47	indirect electron doping into CaFeAsH (HP)
11	$Ca_{1-x}Sm_xFeAsH$ [17]	46	indirect electron doping into CaFeAsH (HP)
12	$CaFeAsF_{1-x}$ [18]	27	indirect electron doping (HP)
13	$Sr_{1-x}La_xFe_2As_2$ [19]	23	indirect electron doping (HP)
14	$(Ba,La)Fe_2As_2$ [20]	22.4	indirect electron doping (thin film)
15	$(Ba,Ce)Fe_2As_2$ [21]	13.5	indirect electron doping (thin film)
16	$(Ba,Pr)Fe_2As_2$ [21]	6.1	indirect electron doping (thin film)
17	$(Ba,Nd)Fe_2As_2$ [21]	5.8	indirect electron doping (thin film)
18	$(Sr,La)Fe_2As_2$ [22]	20	indirect electron doping (thin film)
19	$(Ca,La)Fe_2(As,P)_2$ [23]	45	bulk SC achieved by co-doping of La and P
20	$Ba(Fe,Pt)_2As_2$ [24]	24	dome-like relation between T_c and dopant Pt
21	$(Ca,La)FeAs_2$ [25]	34	new type IBSC: 112
22	$(Ca,La)Fe(As,Sb)_2$ [26]	43	increase in T_c by co-doping
23	$(Ca,RE)Fe(As,Sb)_2$ [27]	47	highest T_c in 112 type (RE=La, Ce, Pr, Nd)
24	$Ca_{10}(Ir_4As_8)(Fe_2As_2)_5$ [28]	16	analogous to $Ca_{10}(Pt_4As_8)(Fe_{2-x}Pt_xAs_2)_5$
25	$Na_{0.65}Fe_{1.93}Se_2$ [29]	37	intercalating Na into FeSe
26	$(Na,NH_3)Fe_2Se_2$ [29]	42	co-intercalating Na and NH_3 into FeSe
27	$LaFeAs(O,C_2)$	27	electron doping by C_2 (HP)

Analogous structure to IBSC

	Material	T_c (K)	Comment
28	$SrAl_2Si_2$	4.6	IBSC 122 type structure (HP)
29	NbSiAs [30]	8.2	square net of Si
30	$BaTi_2Sb_2O$ [31]	3	square net of Ti (measured under HP)
31	$BaTi_2(Sb_{1-x}Bi_x)_2O$ [32]	1.2	square net of Ti, SC by doping of isovalent ion.
32	$BaTi_2(Sb_{1-x}Sn_x)_2O$ [33]	2.5	square net of Ti, SC by doping of aliovalent ion
33	$La(Co_{1-x}Fe_x)_2B_2$ [34]	4	square net of Co
34	$(La_{1-x}Y_x)Co_2B_2$ [34]	4	square net of Co
35	$BaNi_2(As,P)_2$ [35]	3.3	analogous to 122 type of IBSC
36	$Ba(Ni,Cu)_2As_2$	3.3	analogous to 122 type of IBSC
37	LnNiAs$O_{1-x}H_x$ [36]	3.7	H doping to LnNiAsO (Ln=La-Nd) (HP)
38	$NdNi_{0.64}Bi_2$	4	square net of Ni and Bi
39	YNi_xBi_2 [37]	4	square net of Ni and Bi
40	$LaNi_xBi_2$ [37]	4	square net of Ni and Bi
41	$CeNi_xBi_2$ [37]	4	square net of Ni and Bi
42	LaNiBN [38]	4.1	square net of Ni (HP)
43	CaNiBN [38]	2.2	square net of Ni (HP)
44	LaPtBN [38]	6.7	square net of Pt (HP)
45	$La_3Ni_2B_2N_3$ [38]	15	square net of Ni (HP)
46	$LaPd_2As_2$ [39]	1	collapsed 122 structure
47	$LaPd_2Sb_2$ [40]	1.4	$CaBe_2Ge_2$ structure
48	$SrPt_2Sb_2$ [41]	2.1	$CaBe_2Ge_2$ structure
49	$BaPt_2Sb_2$ [42]	1.9	$CaBe_2Ge_2$ structure
50	$SrPd_2Bi_2$	2.2	$CaBe_2Ge_2$ structure
51	$CaPd_2Bi_2$	2.6	$CaBe_2Ge_2$ structure
52	$SrPt_2Bi_2$	2.6	$CaBe_2Ge_2$ structure
53	La_2Sb [43]	5	square net of La

Other layered structure

	Material	T_c (K)	Comment
54	$La_3Ru_8B_6$ [44]	3.2	homologous structure
55	CrNbN	11	Cr layer (multiphase)

56	$Ca_2Al_3Si_4$ [45]	6.4	high pres. phase of Ca-Al-Si (HP)
57	Mg_4AlSi_3 [46]	5.2	new layered structure (HP)
58	$Sr(Si_{1-x}Ni_x)_2$	3	Si layer
59	$Ba(Si_{1-x}Cu_x)_2$	3	Si layer
60	$LaSi_2H_{0.03}$	3	H doping to $ThSi_2$ structure (T_c increased)
61	$BaGe_2H_{0.27}$	4	H doping to $ThSi_2$ structure (T_c decreased)
62	$LaGe_{1.7}$	2	$ThSi_2$ structure with cage formed by sp^2 orbital
63	$Zr_2Ru_3Si_4$ [47]	5.7	analogous structure to ZrRuP superconductor
64	MgPtSi [48]	2.5	honeycomb lattice
65	$SrAuSi_3$ [49]	1.54	noncentrosymmetric structure
66	Li_2IrSi_3 [50]	3.8	noncentrosymmetric structure
67	LaIr*Pn* [51]	5.3	noncentrosymmetric structure (*Pn* = P and As)
68	LaRhP [51]	2.5	noncentrosymmetric structure
69	$Bi_4O_4S_3$ [52]	4.5	double BiS plane
70	$LnO_{1-x}F_xBiS_2$ (*Ln*=La,Ce,Nd) [53]	3 (10)	triple BiS plane. (measured under HP)
71	KMo_6Se_8	9	new Chevrel phase compound
72	Nb_5Ir_3O	10.5	O doped Mn_5Si_3 structure
73	$Ba_2(Bi_{1-x}Sb_x)_3$ [54]	4.3	square/honeycomb lattice
74	Y_2CrC_3	1.5	Cr/C layer
75	$SrNiSn_3$	5	noncentrosymmetric structure
76	β-ZrNF	15	ion exchange of β-ZrNCl by F
77	$Zr_5(N_4O)F_6$	2-5	low SVF
Intercalation to layered structure			
	Material	T_c (K)	Comment
78	TiNCl← amines [55]	17	intercalation of long alkylene diamine →high T_c
79	TiNCl← alkali metal (Li-Rb) [56]	18.0	highest T_c by Na intercalation
80	TiNCl← alkali metal + solvent [56]	10.5	highest T_c by Na+THF co-intercalation
81	TiNBr← alkali metal [57]	17.2	highest T_c by K intercalation
82	ZrNCl← K + THF, PC [58]	16	highest T_c by K+THF co-intercalation
83	α-ZrNCl← alkali metal	10	superconductivity of α-type structure
84	HfNCl← *AE*(Ca-Ba), solvent [59]	26	highest T_c by Ca+THF co-intercalation
85	*M*NCl←Yb, Eu (*M*: Zr, Hf) [60]	24.3	highest T_c by Eu+NH_3 co-intercalation into HfNCl
86	Hf_2N_2O←Li	4.6	La_2O_3 type structure
Chalcogenides			
	Material	T_c (K)	Comment
87	1T-TaS_2	3	suppressing CDW (HP)
88	$AuTe_2$ [61]	2.3	SC by dissociation of Te_2 dimer (CdI_2 structure)
89	$(Au,Pt)Te_2$ [62]	4	SC by dissociation of Te_2 dimer (CdI_2 structure)
90	$(Au,Pd)Te_2$	3	SC by dissociation of Te_2 dimer (CdI_2 structure)
91	$(Ir,Pt)Te_2$ [63]	3.1	SC by dissolving of orbital ordering (CdI_2 structure)
92	Cu_xIrTe_2	3	SC by dissolving of orbital ordering (CdI_2 structure)
93	$(Ir,Rh)Te_2$ [64]	2.6	SC by dissolving of orbital ordering (CdI_2 structure)
94	Ir_xSe_2 [65]	8	pyrite structure (HP)
95	$Ir_{1-x}Rh_xSe_2$ [66]	10	pyrite structure (Rh doping) (HP)
96	$Ir_{0.95-x}Rh_xTe_2$ [67]	4.6	pyrite structure (small effect of Rh doping) (HP)
Others			
	Material	T_c (K)	Comment
97	$NbBe_2$	2.6	Laves phase
98	MoC_{1-x} [68]	12	formed $MoC_{0.681}$ under 6 GPa
99	$MoC_{0.75}$ [69]	13	formed $MoC_{0.75}$ under 17 GPa
100	YFe_2SiC	3.5	having YFe_2Si framework
101	ScC_xO_y	8	having C-C bonding (low SVF)
102	$Ti_{4-x}N_3$	<5	decomposition of TiNCl in NH_3
103	$Zr_{4-x}N_3$	<10	decomposition of ZrNCl in NH_3
104	$Hf_{4-x}N_3$	<6	decomposition of HfNCl in NH_3
105	$Mg(Mg_{1-x}Al_x)Si$ [70]	6.2	anticotunnite or TiNiSi structure (HP)
106	$Ca(Al, Si)_2$ [71]	2.6	Laves phase (HP)
107	Nb_4MSi (M=Ni, Co, Fe) [72]	7.7	2 dimensional network of Nb
108	$Ba_3Ir_4Ge_{16}$ [73]	6.1	aving cage structure
109	$Ba_4Ir_8Ge_{28}$ [73]	3.2	having cage structure
110	Ca_2InN [74]	0.6	doping of In^{-1} into Ca_2N
111	SnAs	2	hole in As4*p* orbital
112	$CuZr_2$	1	
113	$HfZr_2$	1	

(b) Materials which exhibited no superconductivity

[shaded box]: possible to obtain the target phase, [empty box]: impossible to obtain the target phase

IBSC (Square net of Fe)				
MgFeAsH	SrFeAsH	BaFeAsH	EuFeAsH	MgFePH
CaFePH	$CaFeAsF_{1-x}$	$CaFeAsF_{1-x}O_x$	$CaFeAsF_{1-x}H_x$ [9]	$Ca_{1-x}La_xFeAsF$
$Sr_{1-x}La_xFeAsF$	$Ca_{1-x}Na_xFeAsH$	LaFeAsO:N	$CaFeAsH_{1-x}O_x$	LaFeOSb
CaFeOSe	SrFeOSe	KFeSeF	LaFeSiF	$Ba(Fe,Re)_2As_2$
$Ca(Fe,Au)_2As_2$ (thin film)	$Ca(Fe,Mg)_2As_2$	$Ca(Fe,Al)_2As_2$	$Ca(Fe,Pt)_2As_2$ [75]	$Ca(Fe,Ru)_2As_2$
$(Ca,La)Fe_2As_2$	$YFe_{2-x}Co_xGe_2$	$BaFe_{2-x}Zn_xAs_2$	$BaFe_2Sb_2$	$BaFe_2Se_2$
$(Fe,Mn)_2AlB_2$	$(Fe,Co)_2AlB_2$	$(Fe,Ni)_2AlB_2$	$(Fe,Cr)_2AlB_2$	$ZrFe_2Si_2$
$ZrFe_2B_2$	$LuFe_{2-x}Si_2$	$YFe_{2-x}Si_2$	$TlFe_{2-x}Co_xSe_2$	$TlFe_{2-x}Ni_xSe_2$
$Tl_{1-x}Fe_2Se_2$	$MgFe_2As_2$	$CdFe_2As_2$	YFe_2B_2	$Fe_{2-x}Cu_xAs$
(Fe,Pt)Se	Fe(Te,As)	Fe(Se,As)	$FeP_{0.5}X_{0.5}0.5$ (X = Cl, Br)	$FeSe_{1-x}X_x$ (X = P, Cl)
$FeSb_{1-x}$	La-$Ca_{10}(Ir_4As_8)(Fe_2As_2)_5$	Ge- $Ca_{10}(Ir_4As_8)(Fe_2As_2)_5$	P- $Ca_{10}(Ir_4As_8)(Fe_2As_2)_5$	Re- $Ca_{10}(Ir_4As_8)(Fe_2As_2)_5$
K- $Ca_{10}(Ir_4As_8)(Fe_2As_2)_5$	Sr- $Ca_{10}(Ir_4As_8)(Fe_2As_2)_5$	Ba- $Ca_{10}(Ir_4As_8)(Fe_2As_2)_5$	$Ca_{10}(M_4As_8)(Fe_2As_2)_5$ (M=Pd,Rh)	$Ba_{10}(Pt_4As_8)(Fe_2As_2)_5$
$Sr_{10}(Pt_4As_8)(Fe_2As_2)_5$	$Ca_{10}Pt_{5.8}Fe_{8.2}Sb_{18}$	Ba_2InO_3FeAs	Sr_2InO_3FeAs	Sr_2FeO_3FeAs
Ba_2FeO_3FeAs	$Sr_4V_2O_5Fe_2As_2$	$Bi_2SrFe_2O_4Se_2$	$La_3O_2Fe_4As_4$	$Ca_3F_2Fe_4As_4$
$La_2O_3Fe_2(S_{1-x}F_x)_2$	$La_2O_3Fe_2(Se_{1-x}F_x)_2$	$La_2O_3Fe_2(Se_{1-x}As_x)_2$	$La_2O_3Fe_2(Se_{1-x}Ge_x)_2$	$La_2O_3(Fe_{1-x}Co_x)_2Se$
$La_2O_3(Fe_{1-x}Mn_x)_2Se$	$NdFe_xSb_2$	$Fe_{1-x}Cu_xAs_{0.5}$	$EuFeAsO_{1-x}$	$Fe_{1-y}Si_2$
$(Fe_{1-x}Mn_x)_{1-y}Si_2$	$(Fe_{1-x}Co_x)_{1-y}Si_2$	$(Fe_{1-x}Mo_x)_{1-y}Si_2$	AgFeAs	$La(Fe_{1-x}Co_x)_2Si_2$
$Fe_{0.82}(Si_{1-x}Ge_x)_2$	$RFeC_2$	$Sc(Fe,Co)C_2$	$Sc(Fe,Ni)C_2$	$La(Fe,Co)Si_2$
$La(Fe,Mn)Si_2$	$LaFe_{0.5}Sb_2$	$CaFe_4As_3$	$CaFe_3AgAs_3$	$CaFe_3CuAs_3$
$CaFe_3LiAs_3$	$La_3O_4Fe_4As_3$	$Fe(Se,As)_2$	$(Fe,Mn)As_2$	$(Fe,Ru)As_2$
$(Fe,Rh)As_2$	$(Fe,Co)As_2$	$(Fe,Ni)As_2$	$(Fe,Pd)As_2$	$La_{3-x}RE_xCu_2O_{6-y}Fe_2As_2$
$Sr_3Fe_2Cu_2Se_2O_5$	$Sr_{2.6}K_{0.4}Fe_2Cu_2Se_2O_5$	$La_3O_4Fe_4As_4$	$Ca_3Fe_4As_4F_3$	Ca_2FeOsO_6
$Ba_{1-x}K_xFe_2X$ (X = S, Se)	$NaFe_2O_4$	$BaFe_{2-x}Co_xSe_3$	$Tl_{1-x}Ba_xFeSe_2$	$La_2Fe_2Se_2O_{3-x}H_x$
$La_2Fe_2S_2O_{3-x}H_x$	CaBaFe2Pn2O (Pn = As Sb, Bi),	$SrFe_2Pn_2O$ (Pn = As, Sb, Bi)	$BaFe_2Pn_2O$ (Pn = As, Sb, Bi)	$BaFe_2Se_2O$
t-FeS				

Analogous structure to IBSC material				
Square net of transition metal elements		$BaSc_2Pn_2O$ (Pn=As, Sb, Bi)	$SrTi_2As_2O$	$BaTi_2(As,Sb)_2O$ [31]
$EuTi_2As_2O$	$BaTi_2Sb_2O$	$(SrF)_2Ti_2Bi_2O$ [76]	$(SrF)_2Ti_2(Sb_{1-x}Bi_x)_2O$	$Zr_2Ti_2As_2H$
$Zr_2Ti_2Pn_2H$ (Pn = Sb, Bi)	$ZrTiPn$ (Pn = Sb, Bi)	ZrTiAs	BaV_2Ge_2O	BaV_2Sn_2O
BaV_2Pn_2O (Pn = As, Sb, Bi)	$LnCrAsO$ (Ln=La-Nd, Sm) [77]	SrCrAsF	EuCrAsF	$La_{1-x}CaMnAsO_{1-y}F_y$
$Ce_{1-x}CaMnAsO_{1-y}F_y$	$Sm_{1-x}CaMnAsO_{1-y}F_y$	SrMnAsF	BaMnAsF	EuMnAsF
$LaMnAsO_{1-x}H_x$	$CeMnAsO_{1-x}H_x$	$DyMnAsO_{1-x}H_x$	$LaZn_{1-x}Mn_xAsO_{0.75}$	$CaMn_2As_2$
$CaMn_2Sb_2$	$CaMnBi_2$	$(Ca,La)Mn_2Sb_2$	$BaMn_2As_2$	$(Ba,K)Mn_2As_2$
$BaMnRuAs_2$	$BaMnFeAs_2$	$BaMnCoAs_2$	$BaMnRhAs_2$	$BaMnIrAs_2$
$BaMnBi_2$	$LaMn_2Si_2$	$Sr(Mn,Fe)_2P_2$	$Sr(Mn,Cr)_2P_2$	$Sr(Mn,Al)_2P_2$
$Sr(Mn,Zn)_2P_2$	$(Ca,Eu)Mn_2P_2$	$La_2O_2Mn_2Te_2O$	$CaMn_2Pn_2O$ (Pn = As, Sb, Bi)	SrMn2Pn2O (Pn = As, Sb, Bi)
$BaMn_2Pn_2O$ (Pn = As, Sb, Bi)	$BaMn_2Se_2O$	$Ca_{10}(Pt_4As_8)(Mn_{2-x}Pt_xAs_2)_5$	$LaCoAsO_{1-x}H_x$	$(Ca,La)CoAs_2$
$(Sr_{1-x}Ca_x)Co_2Ge_2$	$BaCo_2Ge_2$	YCo_2B_2	YCo_2Ge_2	$LaCo_{0.6}Sb_2$
$LaCo_2Si_2$	$La(Co_{1-x}Ni_x)_2Si_2$	$LaCo_2(B_{1-x}Si_x)_2$ [37]	$LaCo_2Ge_2$	$NdCo_xSb_2$
$GdCo_2B_2$	$LaCo_2B_2C_x$	$ZrCo_2Si_2$	$HfCo_2Si_2$	$HfCo_2Ge_2$
$CaCo_2Pn_2O$ (Pn = As, Sb, Bi)	$SrCo_2Pn_2O$ (Pn = As, Sb, Bi)	$BaCo_2Pn_2O$ (Pn = As, Sb, Bi)	$Ca_{10}(Pt_4As_8)(Co_{2-x}Pt_xAs_2)_5$	$(Ca,La)NiAs_2$
CaNiAsH	$Ba_{1-x}La_xNi_2As_2$	MgNiGe [78]	$CaNi_{1-x}Mn_xGe$ [79]	$CaNi_{1-x}Mn_xGeH$ [80]
Y_3NiSi_3	LaNiAs	La_3NiSi_3	$CeNiBiO_{1-x}$	$ZrNi_{0.75}P$
$ZrNi_{0.75}As$	KNi_2Se_2	$ZrNi_2Si_2$	RhNiP	$GdNi_xBi_2$
$Ni_3(Te_{1-x}Se_x)_2$	$CaNi_2Pn_2O$ (Pn = As, Sb, Bi)	$SrNi_2Pn_2O$ (Pn = As, Sb, Bi)	$BaNi_2Pn_2O$ (Pn = As, Sb, Bi)	$Ca_{10}(Pt_4As_8)(Ni_{2-x}Pt_xAs_2)_5$
Cu_2As	Cu_2Sb	MgCuAs	$Ca_2Cu_6P_5$	$BaCu_{1.8}As_2$
$BaCu_2As_2$	$Ba_2Cu_3P_4$	$BaCu_6As_2$	$YCuSb_2$	$YCuAs_2$
CeCuPO	$GdCu_{1.25}P_{1.75}$	$ZrCu_2P_2$	$ZrCu_2As_2$	ZrCuSiAs
$HfCu_2P_2$	$Hf_{0.5}Cu_{1-x}Mn_xP$	$HfCuSi_2$	$HfCuGe_2$	BiOCuS
$Ca_2Cu_2Pn_2O$ (Pn = As, Sb, Bi)	$Sr_2Cu_2Pn_2O$ (Pn = As, Sb, Bi)	$Ba_2Cu_2Pn_2O$ (Pn = As, Sb, Bi)	$La_3(O,F)_2Cu_4As_4$	$Bi_2LnO_4Cu_{2-x}Fe_xSe_2$ (Ln=lamtanide elements)
$BaZnBi_2$	$K_2NbO_{3-x}F_{1+x}$	$(Ca,La)RuAs_2$	$La_3Ru_2B_2N_3$ [38]	LaRuBN [38]

LaRhBN [38]	(La, Sr)Pd_2As_2	LaIrBN [38]	(Sr,K)Pt_2As_2	(Ca,La)$PtAs_2$
$[Sr_2MO_2][Ti_2As_2O]$, $[Sr_3M_2O_5][Ti_2As_2O]$ (M = Cr, Mn, Fe, Co, Ni, Cu)				
Square net of traditional elements		TiSiAs	$SrGa_2$	$BaGa_2$
NbGeAs	TaGeAs	$BaNiSn_3$	$LaMn_xSb_2$	La_2O_2Sb
Ce_2O_2Sb	$CeAgSb_2$	$Nd(Fe_{1-x}Co_x)_ySb_2$	La_2O_2Te	Y_2O_2Bi [81]
La_2O_2Bi [81]	$LaAgBi_2$	Ce_2O_2Bi [81]	$CeMn_yBi_2$	$CeZn_yBi_2$
$CeCu_yBi_2$	$CeAgBi_2$	Pr_2O_2Bi [81]	Nd_2O_2Bi [81]	Sm_2O_2Bi [81]
Eu_2O_2Bi	Gd_2O_2Bi	Er_2O_2Bi [81]	Yb_2O_2Bi	Y_5Si_3
La_5Si_3				
Other layered structure				
Honeycomb lattice		Mg_2PtSi	SrPdAs	(Sr,Y)PtAs
SrPtSb	BaPtSb	SrPdSb	YPtAs	LaPtSb
MgAgAs	SrCoP	SrNiP	SrCuP	SrPdP
SrAgP	SrIrP	SrAuSn	SrPtSn	SrAuSb
CaAuP	SrAuP	SrAuSi		
Mn_5Si_3 structure		V_5P_3	$V_5P_3N_x$	V_5As_3N
Nb_5Si_3	Ta_5Si_3	V_5Ge_3	$Nb_5Ge_3N_x$	$Ta_5Ge_3N_x$
$Nb_5Ge_3C_x$	Zr_5Pb_3	Ta_5Ir_3O		
Mg_4AlSi_3 structure		$LiMg_4Si_3$	$NaMg_4Si_3$	$SrMg_4Si_3$
$BaMg_4Si_3$	YMg_4Si_3	$BaGe_3Mg_4$		
B, Si layer		CrB_2	MnB_2	FeB_2
ZnB_2	$SrNiSi_3$	$Ba(Si_{1-x}Zn_x)_2$	$Ba(Si_{1-x}Ni_x)_2$	$SiTe_2$
$SmGa_{2-x}Si_x$	CaCuSi	$YbGa_{0.9}Si_{1.1}$	$LaGa_{2-x}Si_x$	Lu_2AlSi_2
$CaSrSi_4$	$Ba_3Ga_{0.7}Si_{4.3}$	$Sr_{1-x}Na_xAl_2Si_2$	$Ba_{1-x}K_xAl_2Si_2$	$Ca_{1-x}Na_xAl_2Si_2$ [45]
$SrAl_{2-x}Si_{2+x}$	$BaSi_2$ (HP phase)			
Skutterdite structure		$Sr_xRhGe_{1.5}Se_{1.5}$	$Ba_xIrGe_{1.5}S_{1.5}$	$Ba_xIrGe_{1.5}Se_{1.5}$
$BaRh_2Ge_4S_6$ [82]	$BaRh_2Ge_4Se_6$ [82]	$BaIr_2Ge_4S_6$ [82]	$BaIr_2Ge_4Se_6$ [82]	$La_xIrGe_{1.5}S_{1.5}$
$La_xRhGe_{1.5}S_{1.5}$	$LaFe_4Bi_{12}$	$LaCo_4Bi_{12}$	$LaNi_4Bi_{12}$	$LaRu_4Bi_{12}$
$LaRh_4Bi_{12}$	$LaPd_4Bi_{12}$	$LaIr_4Bi_{12}$	$LaPt_4Bi_{12}$	
Others		$Y(Co,Fe)_3B_2$	$Y(Co,Ni)_3B_2$	Ca_2N [83]

Ca_xBN (film) [84]	$Ba(Ge_{1-x}Cu_x)_2$	Cr_2N	CuNCN	NbCrN
TiNF	Ti_2SbP	Zr_2SbP	Cr_2Ti	Cr_2Zr
Cr_2Hf	CrTaN	$NiTe_2$	$Ni(Te_{1-x}Se_x)_2$	$NaNiF_3$ [85]
SmNiAs	ZrAs	Ni_xBi_3	$CaRuO_3$ [86]	$CaRhO_3$ [87]
$CaOsO_3$	$SrOsO_3$	$BaOsO_3$	$CaRhO_3$ (HP phase) [88]	*MA*N phase (*M*=Ti,V,Zr,Nb,Hf,Ta) (*A*=Si,P,S,Ga,In,Sn,Pb)
$La_3Ni_2O_7$	Sr_2SnO_{4-x}	Ba_2SnO_{4-x}	SnO: F	$Cu(Al,Mg)O_2$
$SrCr_2O_4$	$LaSrMnO_3H$	$(CuCl)LaNb_2O_{7-x}F_x$	Bi_2WO_6	$Ho_{0.25}Sr_{0.75}FeO_{2-d}$
$Na_2La_2Ti_3O_{10-x}$	$Sr_2TiO_3F_2$	$CaCu_3Cr_4O_{12}$	$SrCuO_2$: F	
Intercalation to layered structure				
FeOCl	VOCl	TiOCl	CrOCl	TiNCl ← mono-amine [55]
Zr_2N_2S [89]	Hf_2N_2S [89]	$Ca_xHf_2N_2S$	α-HfNBr	Ti_2PTe_2 ← Ag, Cu, Zn [90]
$Bi_2PdO_4F_x$	$Bi_{1.67}Pb_{0.33}PtO_4F_x$	Pb_3O_4 ← F	$Zr_5(N_4O)F_{6-x}$	Ti_2N_2S
Ti_2N_2O	$Hf_5(N,O)_5F_{6-x}$	Zr_2N_2O	Graphite ← KH	graphite ← B
$CsLaTe_2O_5Cl$				
perovskite and its related structure				
antiperovskite structure		Mn_3AlC	Mn_3SnC	Mn_3AlN
Mn_2CrN	Mn_3ZnN [91]	$AlNi_3C$	$Mn_3ZnN_{1-x}C_x$	$Mn_3InN_{1-x}C_x$
$Mn_3Sb_{1-x}Sn_xN$	$Mn_3Ag_{1-x}Sn_xN$	Ca_3SiO	Ca_3SnO	Ca_3PbO
$SrAg_3P$	$SrIr_3P$	$SrAu_3P$	Mg_2CNi_4	Ca_4OSb_2
$RhNCr_3$				
double perovskite structure		Ca_3OsO_6	Ca_2MnOsO_6	Ca_2CoOsO_6
Ca_2ZnOsO_6	Sr_3OsO_6	Sr_2LiOsO_6 [92]	Sr_2MnOsO_6	Sr_2FeOsO_6 [93]
Sr_2CoOsO_6 [93]	Sr_2NiOsO_6	Sr_2YOsO_6	Sr_2PbOsO_6	Sr_2DyOsO_6
$Ba_2(Sr_2,Ca_2)FeOsO_6$	$Ba_2YScOsO_6$	Sr_4FeMoO_8		
normal perovskite		$SrGeO_3$ [94]	$(Ca_{1-x}Y_x)GeO_3$	$(Sr_{1-x}La_x)GeO_3$ [94]
$CaTiO_{3-x}H_x$	$SrTi(O,H)_3$ (thin film)	$EuTiO_{3-x}H_x$	$CaVO_3$:H	$SrVO_3$:H

	[95]			
$SrCrO_2H$ [96]	$Sr_{1-x}Ba_xCrO_2H$ (x = 0.1-0.4)	$Ln$$CrO_3$ (Ln = La, Sm, Nd)	$BaTiO_{3-x}$	$BaZrO_3$:H
$SrSnO_{3-x}$	$SrNbO_2N$	$BaSnO_{3-x}$	$NaIrO_3$	$Na_{1-x}Ca_xIrO_3$
$SrMoO_{2.5}N_{0.5}$	$BaBiS_3$			
other perovskite		$A_2Ru_2O_5$	$LaBaMn_2O_6$	$YBaMn_2O_6$
$A_2Cr_2O_5$				
cuprate superconductor and its related structure				
(Sr,Eu)CuO_2	(Sr,Pb)CuO_2	Bi_2CuO_4_CaH_4	Bi_2CuO_4-CuF2	BiOCuS
YBCO:H	$(Cu_2S_2)Sr_3Fe_2O_5$	$Sr_3FeNiCu_2Se_2O_5$	$Sr_3Ni_2Cu_2Se_2O_5$	$Sr_3Cu_4Se_2O_5$
$Bi_2YCu_2O_4Se_2$	$Bi_2SrCu_2O_4Se_2$	$SrLa_2Cu_4Se_2O_5$	$A_2CuO_x(OH)_y$	$LaSrTiO_4$
Sr_2CrO_4	Ca_2CrO_4	$Sr_3Cr_2O_7$	$Ca_3Cr_2O_7$	$A_4Cr_3O_{10}$ (A = Sr, Ca)
$Sr_{2-x}Na_xCrO_4$	$YCaCrO_4$	$Sm_{1-x}K_xMn_nO_{3n+1}$ (n = 1, ∞)		
clathrate structure				
$Ca_3Ni_{12}B_6$	$Sr_3Ni_{12}B_6$	KC_{24}	Li_xC_{60}	K_xC_{60}
Ca_xC_{60}	Ba_xC_{60}	Mg_2C_{60}	Ba_3C_{60} [97]	Ca_xAlN
Li_xSi_{136}	$Na_{32}Si_{136}$ [98]	K_xSi_{136}	Mg_xSi_{136}	Ca_xSi_{136}
Ba_xSi_{136}				
Pyrite structure				
FeS_{2-x}	$PtSb_2$	(Pt,Pd)Sb_2	(Pt,Ir)Sb_2 [99]	Pt(Sb,As)$_2$
Pt(Sb,Te)$_2$	Pt(Sb,Sn)$_2$	(Pt,Au)Sb_2	$IrSb_2$	
other materials				
carbide		CaC_2	BaC_2	YC_2
nitride		Ti_3N_4	Zr_3N_4	Hf_3N_4
VCrN	Mg_xAlN	MnN (thin film)	CoN (thin film)	
oxide		TiO_x (anatase) [100]	Cr_3O	Cr_nO_{2n-1} (n = 3, 4)
MnO_2 (nanosheet)	Co_3O_4: H	NiO: H	GeO_2 (HP phase)	h-WO_3: H
$BaScO_2H$	$Ti_3O_{4.9}N_{0.1}$	$Ti_4O_{6.99}N_{0.01}$	(Ir,Ru)O_2	CuAl(O,N)$_2$
$LaSiO_3H$	$NaTi_2O_4$	$Bi_4Ti_3O_{12}$: H	$Sc_4Ti_3O_{12}$	$Cs_{0.7}Ti_{1.825}O_4$: F
$La_2Ti_2O_{7-x}$	Ba-Ti-O glass + e	$Y_2Ti_2O_{2-x}$	$Sr_2TiO_3F_2$	$Cr_{1-x}Ti_xO_2$
$Cr_{1-x}V_xO_2$	nSrO • V_2O_3 (n= 4, 2, 1)	$Bi_xV_8O_{16}$	CaV_2O_5	LiV_2O_4
$LaVO_3$: H	$LaVO_4$: H	$NaCr_2O_4$ [101]	$Ca_{1-x}Na_xCr_2O_4$	$CaCr_2O_4$
$BaCr_2O_4$	$SrCu_3Cr_4O_{12}$	$BaCu_3Cr_4O_{12}$	$YCu_3Cr_4O_{12}$	$LaCu_3Cr_4O_{12}$
$CeCu_3Cr_4O_{12}$	$LuCu_3Cr_4O_{12}$	$EuCu_3Cr_4O_{12}$	$BiCu_3Cr_4O_{12}$	$Cr_{1-x}Mn_xO_2$
Ag_2MnO_2	Pb_2MnO_4: F	$LiMn_2O_4$	$Sr_3Co_2O_6$ [102]	$Ba_3Co_2O_6(CO_3)_x$
$Sr_{3-x}Y_xCo_2O_6$ [103]	$Sr_{3-x}Ca_xCo_2O_6$	$Sr_3Co_{2-x}Zn_xO_6$	$Sr_3Co_{2-x}Fe_xO_6$	Y-ZrO_{2-x}
$SrNb_2O_6$: F	$Ba_5Nb_4O_{15-x}$	$Ba_2Nb_5O_9$: F	$PbMoO_4$: N	$Ca_2Nb_2O_{6-x}F_{1+x}$
Sr_2NbO_3N	$Li_2Ir_{1-x}Ru_xO_3$ [104]	$La_4Ru_2O_{10}$	$Bi_3Ru_3O_{11}$	$Sr_3Ru_2O_7C_{12}$
$Sr_3Ru_2O_7F_2$	$LaCu_3Ru_4O_{12}$	$CdRh_2O_4$ [105]	$Cd_{1-x}Na_xRh_2O_4$	$La_2Pd_2O_5$
$BaSbO_3$	CsW_2O_6	$SrWO_2N$	$LiOsO_3$ [106]	$Na_{1-x}Ca_xOsO_3$
$KOsO_3$	Ca_3LiOsO_6 [107]	$Ca_2Os_2O_7$	$Ca_5Os_2O_{12}$	Ba_2OsO_5
Bi_2OsO_6	$Pb_2Os_2O_7$	$Pb_2Ir_2O_7$	Ag_2BiO_3	$Ba_{1-x}K_xBiO_3$ (junction)
chalcogenide		Cu_xS	Cu(S,As)	KMo_3S_3
$SrPt_3S$	$BaPt_3S$	InV_6Se_8	$Pt_xBi_2Se_3$	KMo_3Se_3
$Cu_xBi_2Se_3$	$Cu_xBi_2Te_3$	$Pd_xBi_2Te_3$	Li_xZrTe_3	SnTe
pnictide		CoP	NiP	(Co,Ni)P
$La_xIr_4P_{12}$	Mg_xAlP	$BaNi_2P_4$	$BaPd_2P_4$	$(Y_{1-x}Eu_x)$P
$SrPt_3Sb$	$BaPt_3Sb$	La_3TiSb_5	La_3VSb_5	La_3CrSb_5
La_3MnSb_5	La_3TiBi_5	La_3VBi_5	La_3CrBi_5	La_3ZtBi_5
La_3HfBi_5				
silicide, germanide, stannide, plumbide		(Ca, Mg)Si_2	$Fe_3Al_2Si_4$	$MgAl_2Si_2$
Mg_4AlSi_6	$Mg_{1.3}Ga_{0.7}Si$	Li_xPtSi_3	$BaMg_2Si_2$	$Ti_2Ru_3Si_4$
BaPdSi	SrPdSi	$BaMg_2Ge_2$	Rh_3Ge_7	$BaMg_2Sn_2$
$BaMg_2Pb_2$				
halide		$K_{1+x}TiF_4$	$K_{1-x}TiF_4$	$Zr_5(N_4Na)F_6$
$Zr_5(N_4H$-$)F_6$	$CsSnI_3$	La_2TeI_2		
metal		$ZrMn_2$	Zr_5Pb_4	

This project produced superconducting wires and tapes with $J_c > 10^5$ A cm^{-2} by the powder-in-tube (PIT) method and has developed efficient magnetic pinning centers for thin film type wires and tapes. Furthermore, the FIRST Project has succeeded in preparing Josephson junction and SQUID devices by using epitaxial thin films of IBSCs, clarifying the physical properties of IBSCs including small anisotropy and high durability in magnetic fields. The induction of a metallic state from an insulating parent material of IBSCs by the electrostatic method was also a result of the research on IBSC thin film devices. We will describe these details in section 4.

The new functional materials developed in the project are rather diverse. The discovery of a highly efficient catalyst for ammonia synthesis, the $12CaO{\cdot}7Al_2O_3$ (C12A7) electride, is the most remarkable result [108], which has an impact not only in the academic community, but also in industry. The discovery of the spontaneous decomposition of carbon dioxide gas on the C12A7 electride surface [109], the preparation of stable perovskite titanium oxy-hydride [110], the development of a bipolar oxide semiconductor and its complementary circuit device [111], the development of a new class transparent oxide conductor, $SrGeO_3$ [94], the discovery of the first Slater insulator, $LiOsO_3$, which is a ferroelectric metal [106], the development of a material showing a new type of giant magnetoresistance, $NaCr_2O_4$ [94], and the discovery of the two-dimensional electride properties in Ca_2N [83], are also representative results. It is interesting that some of these harvests have resulted in the exploration of new superconductors. These results are to be introduced briefly in section 5.

This project has reported these results in more than 330 original papers and numerous oral and poster presentations including over 170 invited and plenary talks at international meetings. Researchers from the FIRST Project have also applied for over 30 patents.

NEW SUPERCONDUCTORS

In the FIRST Project, four research groups have concentrated on exploring and evaluating novel superconductors, and contributed to the progress of the research of superconductivity through discovery of many novel superconductors and phenomena. We will review these results.

Iron-based Superconductors

The history of IBSCs started in 2006 when LaFePO with $T_c = 5$ K was found by Hosono's group [3]. Only a few researchers took notice of this new type of superconductor based on iron with a large magnetic moment. In 2008, the discovery of superconductivity in $LaFeAsO_1{-}xFx$ with$T_c = 26$ K by the same group [4] rekindled global interest in this area and opened a new frontier of superconductivity. At the early stage, this superconductor family was called 'pnictide superconductors'. However, researchers now call them 'iron-based superconductors' because several measurements and evaluations have clarified that they all have a similar electronic structure where the *3d* electrons derived from Fe ion dominate the Fermi level to play a primary role in superconductivity.

Magnetism had long been believed to be incompatible with superconductivity. Thus, the use of elements with a large magnetic moment, typically Fe, Ni and Co, was intentionally avoided in the field of superconductivity. Hence, the discovery of a high-T_c superconductor based on iron impacted research in this field. Condensed matter scientists encountered a new frontier of superconducting materials.

Immediately after Hosono's group reported an increase in T_c (to 43 K) for La FeAs $O_{0.89}F_{0.11}$under high pressure [112], two groups in China [113–115] reported a higher T_c (=55 K) for $SmFeAsO_1{-}xFx$ under ambient pressure. By now, it is reported that the highest T_c values (55–58 K) in non-cuprate bulk superconductors are observed for some IBSCs as $SmFeAsO_{0.74}F_{0.26}$[116], $SmFeAsO_{0.85}$ [117], $SmFeAsO_{0.8}H_{0.2}$ [9], and $Gd_{0.8}Th_{0.2}SmFeAsO$ [118].

In 2008, Johrendt's group in Germany reported (Ba,K) Fe_2As_2 ($T_c = 38$ K) [119]. Thanks to the ease of growing its single crystals, which have a lateral size of several millimeters using Sn or FeAs as a flux, its physical properties have been well elucidated. In addition, researchers discovered several types of IBSCs and characterized their superconducting properties [120–127].

In this section, we review the recent progress of IBSCs focusing on the results of this project.

Features of IBSCs

Crystal structure

Although approximately 100 IBSCs have been reported, their parent materials may be classified into seven types in terms of crystal structures (figure 1). These materials contain a common structural unit of the Fe*Pn* (or Fe*Ch*) layer formed by the square net of Fe^{2+} (as the formal charge), which is tetrahedrally coordinated by four pnictogen (*Pn*) and/or chalcogen (*Ch*) atoms (see figure 1(a)). Unlike cuprate superconductors, where the parent materials are Mott insulators, this layer shows metallic conductivity without doping. An insulating blocking layer is composed of *M*, *M*O or *M*F etc, where *M* indicates a metallic element such as an alkali, alkaline earth, or rare earth metal that lies between Fe*Pn* (or Fe*Ch*) layers. Similar to cuprates, this layered structure provides quasi-two-dimensional carrier transport properties, although the magnitude of anisotropy rather differs depending on the blocking layer. The local structure of the Fe*Pn* layer is affected directly by the atomic (or ionic) size of *M* because *M* elements in the blocking layer bond to Fe elements. The crystal structures and the brief introductions of the seven different parent materials for IBSCs are described below.

(i) 1111-type materials (LnFePnO, Ln: lanthanide, AEFeAsF, AE: alkaline earth, Pn: P, As).

1111-type compounds have the same structure as LaFeAsO, and are the prototype version of IBSCs. Due to their atom composition ratios, they are called ‹1111-type›. Figure 1(b) shows their crystal structure, which is a ZrCuSiAs-type structure [128, 129] with a tetragonal *P4/nmm* space group. Although LaFePO and LaFeAsO, along with their crystal structures, were identified by Zimmer *et al* 20 years ago [130], they were discovered to be superconductors in 2006 [3] and 2008 [4], respectively. Moreover, their two-dimensionality is relatively high among the seven types, and only this group has T_c values above 50 K as a bulk form.

The 1111-type compound is composed of an alternating stack of positively charged *Ln*O layers and negatively charged FeAs (or FeP) layers along the *c*-axis. As mentioned above, the local structure of the FeAs layer is the same in all types of IBSCs. The distance between the FeAs layers corresponds to the length of the *c*-axis (~0.8–0.9 nm). The formal valence state of each atom is Ln^{3+}, Fe^{2+}, As^{3-}, and O^{2-}. Fe^{2+} contains six electrons in its *3d* orbital, and these electrons play an essential role in driving the superconductivity and magnetism. The electronic and magnetic properties of La*TMPn*O (*TM*:

3*d* transition metal (Cr–Zn)) are summarized in table 3 [120]. The 1111-type compounds composed of Fe and Ni reveal superconductivity.

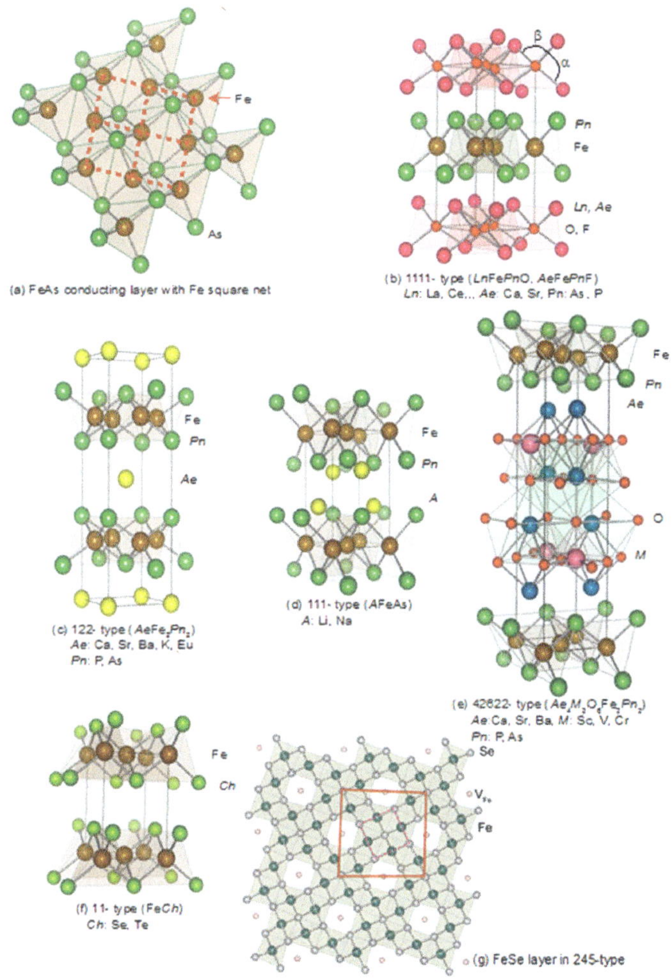

Figure 1. Crystal structures of iron based superconductors. (a) Structure of the FeAs conducting layer, which is common to all IBSCs. The dotted line indicates the Fe square net. (b) 1111-type, (c) 122-type, (d) 111 type (e) 42622-type and (f) 11-type structures. (g) Structure of $K_{0.775}Fe_{1.613}Se_2$ from the [001] direction in the $5^{1/2} \times 5^{1/2} \times 1$) cell showing fully occupied Fe sites decorated with ordered vacancy sites. The dotted square indicates the basal plane of a 122-type unit cell. Reprinted with permission from [125].

Table 3. Summary of electromagnetic properties of La*TMPn*O. (*TM*: 3*d* transition metal, and *Pn*=P or As.)

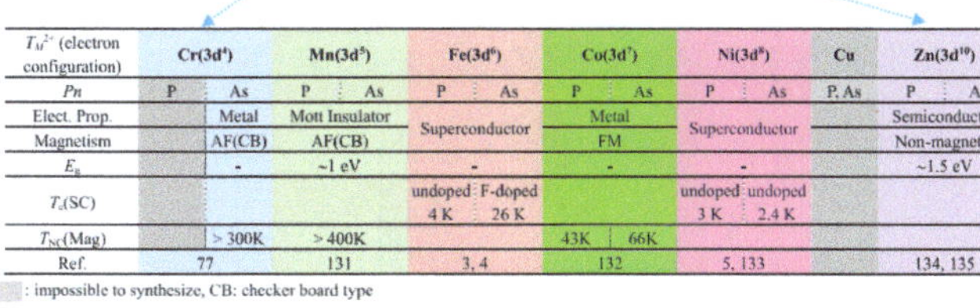

T_M^{2+} (electron configuration)	Cr(3d⁴)		Mn(3d⁵)		Fe(3d⁶)		Co(3d⁷)		Ni(3d⁸)		Cu	Zn(3d¹⁰)	
Pn	P	As	P	As	P	As	P	As	P	As	P, As	P	As
Elect. Prop.		Metal	Mott Insulator		Superconductor		Metal		Superconductor			Semiconductor	
Magnetism		AF(CB)	AF(CB)				FM					Non-magnetic	
E_g		-	~1 eV		-		-		-			~1.5 eV	
T_c(SC)					undoped 4 K	F-doped 26 K			undoped 3 K	undoped 2.4 K			
T_N(Mag)		> 300K	> 400K				43K	66K					
Ref.	77		131		3, 4		132		5, 133			134, 135	

(grey): impossible to synthesize, CB: checker board type

The lanthanide elements from La to Gd can occupy the *Ln* site for the 1111-type of material with *Pn* = P [130]. In the case of *Pn* = As, La to Ho and Y can also occupy the *Ln* site [136]. Additionally, $Ca(Fe_1{-}xCox)AsF$ (T_c = 22 K) is a fluoride-containing superconductor of this type [138].

In this project, the effect of a hydride ion as a new electron donor to this type was studied earnestly and the results are described in section 3.1.2.

(ii) *122-type materials* (*MFe_2Pn_2, M: alkali, alkaline earth or Eu*)).

122-type materials have a '$ThCr_2Si_2$' type crystal structure with a tetragonal *I4/mmm* space group [137]. This group contains the largest number of compounds among the five parent families.

In the case of $AEFe_2P_2$, not only alkaline earth elements but also lanthanides (La–Pr, Eu) can occupy the *AE* site. In $AEFe_2As_2$, the *AE* site can be occupied by alkaline earth, alkali metal, or Eu^{2+}. Figure 1(c) shows the crystal structure of the 122-type. The layer composed of *AE* ions, which is thinner than the *Ln*—O layer of 1111-type, is sandwiched between the FeAs conducting layers. The distance between the FeAs layers of 122-type (0.5–0.6 nm) is shorter than that of 1111-type (0.8–0.9 nm). Because the nearest FeAs layers face each other with a mirror plane, the lattice parameter *c* is twice the FeAs–FeAs distance. The lattice parameter *a*(~0.4 nm) is almost the same as that of 1111-type. Consequently, both 1111 and 122-type materials have similar Fe–Fe distance in the FeAs layer. Since single crystals of several millimeters can be obtained using Sn or FeAs as a flux, the physical properties of 122-type are well evaluated compared to other types of IBSCs. Johrendt's group of Germany was the first to report superconductivity for 122-type materials [119].

In this project, lanthanide element doped 122 superconductors were prepared and evaluated in their bulk and thin-film forms, and these results are described in section 3.1.3.

(iii) 111-type materials (AFePn, A: alkali metal).

While an *AE* ion (alkaline earth ion with formal charge of 2+) is alternately sandwiched between Fe*Pn* layers in 122-type, 111-type compounds contain two *A* ions (*A*: Li^+, Na^+) between Fe*Pn* layers in a unit cell. The crystal structure of this type is known as 'CeFeSi' type, with a tetragonal *P4/nmm* space group (figure 1(d)). This type is compatible with the structure of 1111-type where all the oxygen atoms are removed, and the *Ln* site is occupied by Li^+ or Na^+. Wang *et al* [139] (T_c = 18 K: LiFeAs) and Parker *et al* [140] (T_c = 10 K: NaFeAs) first reported superconductivity for 111-type materials.

iv) *Materials with thick blocking layer* (32522-type ($AE_3M_2O_5Fe_2Pn_2$, *M*: Al, Sc)), (42622-type ($AE_4M_2O_6Fe_2Pn_2$, *M*: Sc, V, Cr)), (homologous type ($Ca_{n+1}Sc_nO_yFe_2As_2$: n = 3, 4, 5)).

The distance between the Fe*Pn*/*Ch* layers is in the order of the 1111, 122, 111 and 11-types. In contrast, these three types of iron oxy-pnictide have a thick blocking layer composed of a quasi-perovskite structure assembled by MO_5 pyramids and *AE* (see figure 1(e) for the 42622-type ($Sr_4Sc_2O_6Fe_2As_2$)). The FeAs–FeAs distance is 1.55 nm and 2.45 nm for $Sr_4Sc_2O_6Fe_2As_2$ and $Ca_6(Sc_{0.4}Ti_{0.6})_5OyFe_2As_2$, respectively. The groups of Kishio and Shimoyama have studied these types of materials systematically [141–143]. The highest T_c reported so far is 43 K [143]. Considering the thick blocking layer, this type should have the highest two-dimensionality, but the concrete value of anisotropic properties has not yet been reported because of difficulty of single crystal growth. The 32522-type has been proposed as a promised parent material [144,145], and the emergence of superconductivity in the 32522-type was reported by Shirage *et al* in 2011 for $(Ca_3Al_2O_{5-}y)(Fe_2Pn_2)$ (Pn = As (T_c = 30.2 K) and P (T_c = 16.8 K)) [146].

(v) *Materials containing additional arsenic* (Ca_1−xLaxFeAs_2), $(Ca_{10}(M_4As_8)(Fe_2As_2)_5$, ($M$: Pr, Ir)).

These new types of iron pnictide superconductor were found by Nohara's group of this project. The details for (Ca_1−xLaxFeAs_2) (T_c = 43 K) and $(Ca_{10}(M_4As_8)(Fe_2As_2)_5$, ($M$: Pr, Ir)) ($T_c$ = 38 K) are described in sections 3.1.4 and 3.1.5, respectively.

(vi) *11-type materials* (Fe_1+xCh, Ch: *Se, Te*).

The 11-type crystal has the simplest structure among the parent compounds and is essentially the alkali metal-free 111-type. This crystal structure is known as 'α-PbO› type with a tetragonal$P4/nmm$ space group (figure 1(f)). A typical 11-type superconductor is β-FeSe (T_c = 8 K) [147]. Medvedev *et al* reported the 11-type may exhibit a high T_c (=37 K) under 8.9 GPa [148].

Furthermore, FeSe attracts attention as one of the candidates showing higher T_c than boiling temperature of liquid N_2. Several groups in China reported that the monolayer of FeSe deposited on a $SrTiO_3$ substrate showed high T_c (65 K) in 2012 and they raised T_c to 100 K [149–153]. Though this superconductivity emerges so far only for monolayers of FeSe deposited on a $SrTiO_3$ substrate, a new route to high T_c materials is expected to be found.

(vii) *245-type materials* (A_1−$x$$Fe_2$−$y$Se2: A = *K, Cs, Rb, Tl*).

In 2010, Guo *et al* reported a potassium-intercalated iron selenide superconductor with relatively high T_c value (30 K) [154]. The crystal structure changed from 11 to quasi-122-type upon intercalation, of which the space group is assigned to $I4/m$ due to vacancy ordering as shown in figure 1(g). Though Guo *et al* noted its chemical notation as K$x$$Fe_2Se_2$, the detailed chemical and structural analyses for its optimal material showed the composition to be$A_{0.8}Fe_{1.6}Se_2$ (=$A_2Fe_4Se_5$) and ordering of Fe vacancies with a $\sqrt{5}\times\sqrt{5}\times1$ supercell in the 122-type crystal structure [155]. This type of material shows a wide range of non-stoichiometry, and with low Fe concentration is an antiferromagnetic insulator. The superconductivity in AxFe$y$$Se_2$emerges in the proximity of an antiferromagnetic (AFM) Mott insulating state, similar to the cuprate high temperature superconductors [156]. Many unique properties let us classify this as an independent type apart from the 122-type. Ivanovskii reviewed this material [157].

In this project, we intercalated Na to FeSe employing the ammonothermal method, which cannot be prepared using a conventional thermal treatment at high temperatures. The result is described in section 3.1.6. Using this type of material with low Fe content as a Mott insulator, we examined the effect of the electric field on its electrical transport properties. The results are described in section 4.2.5.

Electronic structure

Figure 2 shows the photoemission spectra of LaFeAsO and $LaFeAsO_{0.94}F_{0.06}$, and calculated partial density of states (PDOS) for Fe 3d and As 4p [158]. The Fermi level (E_F) controlling the transport property is primarily formed by a complex tangle of five Fe 3d orbitals, due to the small contribution of As, which is unlike cuprate superconductors where only Cud_{x2-y2}contributes to the E_F. With five bands comprising E_F, multi-pockets, i.e., a disconnected Fermi surface (FS), appear on the FS. The energy levels of d_{x2-y2}, d_{xy}, and d_{yz} are sensitive to both changes in the symmetry of the $FeAs_4$ tetrahedron and the carrier density. Such an electronic structure dominates high T_c with the unique pairing mechanism.

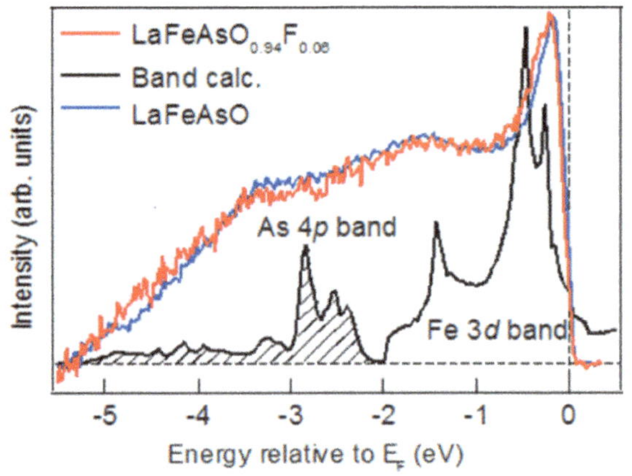

Figure 2. Photoemission spectra of $LaFeAsO_{1-}xFx$ and calculated PDOS. The near-$E_F$peak and the weak peak at about −1.5 eV corresponding to Fe 3d bands survive, and a broad peak corresponding to the As 3p band appears in the range −(3 ~ 4) eV. Reprinted with permission from [158]. Copyright 2008 by the Physical Society of Japan.

At the early stage of the theoretical approach for the pairing mechanism, several physicists [159–166] suggested the possibility of spin fluctuation mediated pairing, where the spin fluctuation arises around the nesting vector (π, 0) (see figure 3 [160]). The spin fluctuation mediates $s_{\pm}$-wave pairing, where the gap function has s-wave symmetry, but its sign is reversed between the electron and hole Fermi surfaces.

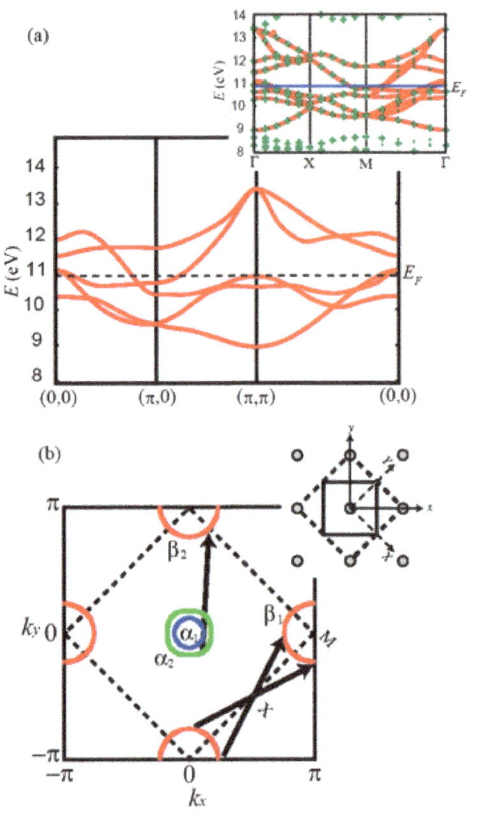

Figure 3. (a) The band structure of the five-band model in the unfolded BZ, where the interlayer hoppings are included. To compare

with the ten-band model (thick red lines in the inset; the symbols are the present local-density approximation results), note the original (dashed lines) and the unfolded (solid lines) BZ shown in (b). (b) Fermi surface for $n = 6.1$ (with the interlayer hoppings ignored), with the arrows indicating the nesting vectors. The inset depicts the original (dashed lines) and reduced (solid lines) unit cell in real space. Reprinted with permission from [160]. Copyright 2008 by the American Physical Society.

In contrast, recent experimental results showed that high T_c is revealed when the nesting is degraded, or even in the absence of the nesting by heavy doping of impurities [11, 149, 151, 154,167]. To explain the robust superconducting state against impurities, Kontani and Onari [168] proposed a mechanism of the S_{++}-wave superconducting state induced by orbital fluctuations, due to the phonon-mediated electron–electron interaction. On the other hand, Suzuki *et al*[169] succeeded in reproducing the general trend of composition dependence of T_c in*Ln*FeAsO$_1$–*x*H*x* (*Ln*: La, Ce, Sm and Gd) by the diagonal (next nearest neighbor) electron hopping model, where the next nearest neighbor (diagonal) hoppings between iron sites dominate over the nearest neighbor ones, plays an important role in the enhancement of the spin fluctuation and thus superconductivity. The theoretical and experimental evaluation for the superconducting mechanism will continue from now on.

Electronic Phase Diagram

Unlike cuprate superconductors whose parent materials are Mott insulators, the parent materials of IBSCs are antiferromagnetic metals with sufficient conduction carriers. Hence, it is considered that carrier doping into IBSCs mainly alter the FS, which in turn leads to suppression of antiferromagnetism.

Here we mainly describe the electronic phase diagram for the 1111-type (*Ln*FeAsO, Ln: rare earth element). The parent materials for the 1111-type have tetragonal crystal structure at room temperature, but transform into orthorhombic structure at lower temperatures. In LaFeAsO, Pauli paramagnetism (PM) is shown around room temperature, and changes into AFM at a slightly lower temperatures ($T_N \sim 140$ K) than that of the structural transitions ($T_s \sim 160$ K) [170, 171].

Generally, superconductivity occurs in the tetragonal phase and not in the orthorhombic phase due to antiferromagnetic ordering in the orthorhombic phase. With doping (e.g., substituting O for F in *Ln*FeAsO), the tetragonal–orthorhombic transition temperature decreases and is accompanied by suppression of the AFM state and superconductivity emerges in succession. Electrons are doped into the bulk, when an element with more valence electrons is substituted. In contrast, holes are doped by substituting an element with fewer valence electrons. In many cases of both 1111 and 122-type, it is possible to substitute Fe or As in the conducting layer and *Ln*, *AE*, O or F in the blocking layer for other elements. The former and the latter are called 'direct doping' and 'indirect doping', respectively.

The critical temperature (T_c) increases, reaches a maximum, and then decreases as the dopant level increases. Since the decrease in T_c in the over doping level is due to the precipitation of the secondary phase as SmOF in SmFeAsO$_1$–*x*F*x*, the proposed phase diagram for 1111-type doped with F does not show the correct T_c behavior in the over doping region [172–176]. In contrast, Hanna *et al* [9] prepared SmFeAsO$_1$–*x*H*x* and showed its optimal T_c (=55 K) at $x = 0.20$ and decrease in T_c by additional doping (over doping) without precipitation of the secondary phase, indicating a wide superconducting dome in 1111-type. Figure 4(a) shows the schematic phase diagram for the 1111-type.

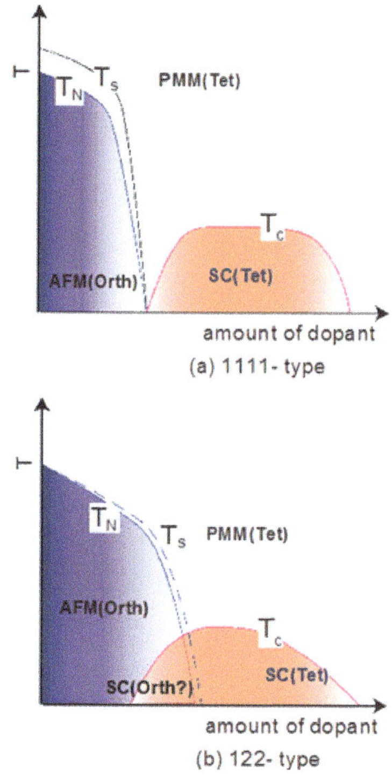

Figure 4. Schematic temperature versus composition phase diagram for (a) 1111-type and (b) 122-type (T_s: structural transition temperature; T_N: magnetic transition temperature; AFM: antiferromagnetic phase; PMM: paramagnetic metal phase; SC: superconducting phase). Reprinted with permission from [125]. Copyright 2012 by Maney Publishing.

The first IBSC reported was formed by electron doping as LaFeAsO$_1$–*x*F*x* where F substituted the O site as ($F_O^{\bullet} + e'$) [4]. In addition to the substitution of oxygen sites by F ($T_c = 55$ K for SmFeAsO$_{0.9}$F$_{0.1}$) [115], various routes for electron doping have been reported, including the formation of an oxygen vacancy ($V_O^{\bullet} + e'$: $T_c = 55$ K for SmFeAsO$_{0.85}$) using a high-pressure synthesis [117, 177, 178] substitution of H$^-$ for an O^{2-} site ($H_O^{\bullet} + e'$: $T_c = 55$ K for SmFeAsO$_{0.8}$H$_{0.2}$) [9], substitution of Th for *Ln* ($Th_{Ln}^{\bullet} + e'$: $T_c = 56$ K for Gd$_{0.8}$Th$_{0.2}$FeAsO) [118,179], and substitution of Co, Ni or Ir for Fe ($Co_{Fe}^{\bullet} + e'$: $T_c = 14$ K for LaFe$_1$–*x*Co*x*AsO [180, 181],$T_c = 22$ K for I$_1$–*x*Co*x*AsF [137], $Ni_{Fe}^{\bullet\bullet} + 2e'$: $T_c = 6$ K for LaFe$_1$–*x*Ni*x*AsO [182], $Ir_{Fe}^{\bullet\bullet} + e'$: T_c $T_c = 18$ K for SmFe$_1$–*x* Ir*x*AsO [183]). The optimal amount of doping is 0.1–0.2/Fe atoms for each route, and indirect doping appears to be more effective than direct doping in achieving a high T_c, which should be due to less structural perturbation to the conducting layer.

For the 122-type, the shape of the electronic phase diagram is similar to the 1111-type as a general trend. The remarkable difference between the 1111- and 122-types is whether the AFM and superconducting phases are distinctly overlapped or not as shown in figure 4(b). In the 1111-type, the regions showing AFM and superconductivity are separated or barely overlap, whereas the 122-type materials have AFM regions with a high T_c. The optimal T_c is apparently located around the temperature corresponding to the extrapolation of the SDW curve to zero temperature, i.e., a superconducting dome appears around the quantum critical temperature of SDW [184]. The emergence of SC by doping of isoelectronic dopant, such as P for As, is also a unique property of the 122-type.

The comparison in doping between the 122- and the 1111-type is shown in table 4.

Table 4. Typical example of emergence of superconductivity by doping (T_c and composition).

	1111		122	
Doping type	indirect	direct	indirect	direct
	55 K ($SmFeAsO_{0.9}F_{0.1}$) [115]	14 K ($LaFe_{0.89}Co_{0.11}AsO$) [180]	45 K ($Ca_{0.83}La_{0.17}Fe_2(As,P)_2$ [23]	22 K ($BaFe_{1.8}Co_{0.2}As_2$) [186]
Electron	55 K ($SmFeAsO_{0.85}$) [117]	22 K($CaFe_{0.9}Co_{0.1}AsF$) [137]		19 K ($BaFe_{1.908}Ni_{0.092}As_2$) [187]
	55 K ($SmFeAsO_{0.8}H_{0.2}$) [9]	6 K ($LaFe_{0.96}Ni_{0.04}AsO$) [182]	22 K ($Sr_{0.6}La_{0.4}Fe_2As_2$) [19]	24 K ($BaFe_{1.886}Rh_{0.114}As_2$) [188]
	56 K ($Gd_{0.8}Th_{0.2}FeAsO$) [118]	18 K ($SmFe_{0.89}Ir_{0.11}AsO$) [183]	22 K ($Ba_{0.93}La_{0.07}Fe_2As_2$) [20]	18 K ($BaFe_{1.914}Pd_{0.086}As_2$) [188]
Hole	13 K(?) ($Nd_{0.8}Sr_{0.2}FeAsO$) [185]	no	38 K ($Ba_{0.6}K_{0.4}Fe_2As_2$) [119]	no
Isoelectronic	no	no	no	30 K ($BaFe_2(As_{0.68}P_{0.32})_2$) [189]
22 K ($Ba(Fe_{0.56}Ru_{0.44})_2As_2$) [190]				

Properties

Compared to MgB_2 and cuprates, IBSCs have several distinct characteristics. It is included in the unique characteristic of IBSCs that the Fe 3*d* multi-orbital form Fermi surface described in (b) and the parent material is the antiferromagnetic metal described in (c).

Generally, T_c decreases upon doping with magnetic impurities such as Fe, Ni, and Co. In the case of cuprate superconductors, T_c of $YBa_2Cu_3O_{7-y}$ decreases from 90 K to 50 K by substituting Ni (17%) for Cu, and that of $La_{1.85}Sr_{0.15}CuO_{4-y}$ also decreases from 40 K to 4.2 K by substituting Ni (5%) for Cu [191]. The substitution of such elements for Fe on FeSCs with an optimal state shows a similar trend. The superconductivity of $NdFeAsO_{0.89}F_{0.11}$ (T_c = 48 K) disappears by substituting Co (>11%) or Mn (>4%) [192]. In contrast, the emergence of superconductivity by substitution of Co^{2+} ($3d^7$), Ni^{2+} ($3d^8$) or other transition metals for Fe^{2+}($3d^6$) in the non-superconducting parent material described in (c) is also a unique nature for IBSC.

The high T_c, large upper critical field (H_{c2}) and small anisotropy are important merits in applying IBSCs practically. Table 5 summarizes these values of IBSCs along with those of $MgB_2$and cuprates. Only cuprates achieve higher T_c than the boiling point of liquid N_2 (77 K). It has been reported that the anisotropic ratio of the resistivity ($\gamma\rho$) of the 122-type IBSCs is compatible with that of MgB_2 and smaller than that of cuprates. The $H_{c2}(0)$ of IBSCs is higher than that of MgB_2, but is smaller than that of a typical cuprate. The anisotropic ratio of the H_{c2},γ_H, of IBSCs is smaller than those of MgB_2 and cuprates. The $H_{c2}(0)$ is defined as the upper critical field at 0 K. The $\gamma\rho$ means the ratio of the resistivity along the crystal axes directions, $a(\rho(a))$ and c ($\rho(c)$) measured just above T_c. The application of IBSCs to superconducting wires and devices will be described in section 4.

Table 5. Properties of high T_c superconductors.

	IBSCs	MgB_2	Cuprates
Parent material	antiferromagnetic metal ($T_N \sim 150$ K) (excepting for the 245-type)	non-magnetic metal	antiferromagnetic insulator ($T_N \sim 400$ K)
Orbitals composing Fermi level	five Fe 3*d* orbitals	two B2*p* orbitals	single Cu3*d* orbital
T_c	56 K ($Gd_{0.8}Th_{0.2}SmFeAsO$) [118] 38 K ($Ba_{0.6}K_{0.4}Fe_2As_2$) [119]	39 K (pure MgB_2) [196]	92 K ($YBa_2Cu_3O_{7-\delta}$) [199] 105 K ($Bi_2Sr_2Ca_2Cu_3O_{10}$) [200] 134 K ($HgBa_2Ca_2Cu_3O_8$) [201]
$H_{c2}(0)^{//ba}$	~90 T ($(Ba_{0.55}K))Fe_2As_2$) [193]	~40 T ($MgB_{1.87}C_{0.13}$) [197]	210 T ($YBa_2Cu_3O_{7-\delta}$) [202]
	~150 T ($SmFeAsO_{0.7}F_{0.25}$) [193]	~18 T (pure MgB_2) [197]	400 T ($Bi_2Sr_2CaCu_2O_8$) [202]
$\gamma_\rho = \rho(c)/\rho(ab)$	3–5 ($Ba(Fe_{0.936}Co_{0.074})_2As_2$) [194] 8-10 ($SmFeAsO_{0.7}F_{0.25}$) [94]	~3.5 (pure MgB_2) [198]	30-250 ($YBa_2Cu_3O_{7-\delta}$) [203–205] >1000 ($Bi_2Sr_2CaCu_2O_8$) [206]
$\gamma_H = H_{c2}(0)^{//b}/H_{c2}(0)^{//c}$	1.1-1.2 ($(Ba_{0.55}K))Fe_2As_2$) [193] ~1.5 ($SmFeAsO_{0.7}F_{0.25}$) [195]	~4 ($MgB_{1.87}C_{0.13}$) [197] ~6 (pure MgB_2) [197]	6 ($YBa_2Cu_3O_{7-\delta}$) [202] 21 ($Bi_2Sr_2CaCu_2O_8$) [202]

[a] $H_{c2}(0)^{//ab}$: estimated by extrapolating the H_{c2}-T curve, where H_{c2} is measured by applying a magnetic field along the *ab* plane.

Hydrogen as an Electron Donor

The most effective route to achieve high T_c is 'indirect' and 'electron' doping into the 1111-type (see table 4). As mentioned in the previous section, the solubility limit of F, which is used for the 'indirect' and 'electron' dopant, is 20% at most. Thus, one could not know the effect on the emergence of SC by the impurity doping for its whole SC dome. So we have explored an appropriate dopant with high solubility limit.

Out of some candidates of monovalent anions ($X_O^{\bullet}$ + e) we choose the hydride ion (H^-) as an electron dopant which could substitute for O^{2-} [9, 12]. Though it is generally considered that the ionic state of hydrogen is a proton (H^+) in condensed matter, hydrogen is also present as the hydride ion (H^-) in stable materials such as LaH_2, CaH_2 and NaH. Its ionic radius depends on the environment, i.e., 208 pm in Pauling's estimation, 129 pm in NaH and 106 pm in CaH_2. We considered that H^- should be able to dissolve into 1111-type superconductors with replacement of O^{2-} because the blocking layer is composed of a lanthanide cation which can form stable hydrides. To prevent the evaporation of hydrogen during the preparation process, we employed a high pressure synthesis technique, i.e., under 2 GPa at 1200 °C. The experimental results clearly showed a high solubility limit of H^- doping. While the impurity phases as SmAs and/or SmOF precipitate in $SmFeAsO_{1-x}F_x$ ($x > 0.15$) [207, 208], such phases could not be observed in $SmFeAsO_{1-x}H_x$ ($x \leq 0.4$). Figure 5 compares the hydrogen content (x) and the deficient amount of oxygen (y) in the prepared samples per the chemical formula ($SmFeAsO_{1-y}Hx$) as a function of nominal x in the starting mixture [9]. The former value was determined by TG-MS, and the latter was measured using EPMA. For nominal $x \leq 0.4$, the hydrogen content agrees with y and the nominal x, indicating the oxygen site (O^{2-}) was successfully substituted with hydrogen (H^-). Figure 6 shows the electronic phase diagram of $SmFeAsO_1-xH_x$ superimposed with that of $SmFeAsO_{1-x}F_x$, with the fluorine content x measured by EPMA as reported by Köhler and Behr [207], in which the T_c versus x plots of $SmFeAsO_{1-x}H_x$ and $SmFeAsO_{1-x}F_x$ overlap at $x < 0.15$, indicating that hydrogen gives indirect electron doping to the FeAs layer just like fluorine. While the solubility limit of fluorine in the oxygen site is restricted to less than 20% ($x = 0.2$) [172, 207, 208], that of hydrogen can reach 40% for not only Sm-1111 but also all Ln-1111 compounds. The wider substitution range is useful for the optimization of the electron-doping level to induce superconductivity and to complete the electronic phase diagram, including the overdoped region.

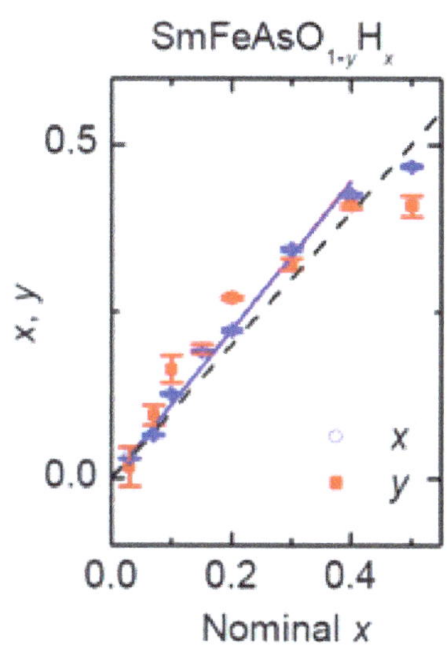

Figure 5. Oxygen deficiency content determined by EPMA (y) and hydrogen content estimated by TG-MS (x) in $SmFeAsO_1yHy$ as a function of nominal x in the starting mixture. The measured x is almost equal to y and nominal x, indicating that the deficiency of the oxygen site is wholly compensated for by the occupation of hydrogen. Reprinted with permission from [9].

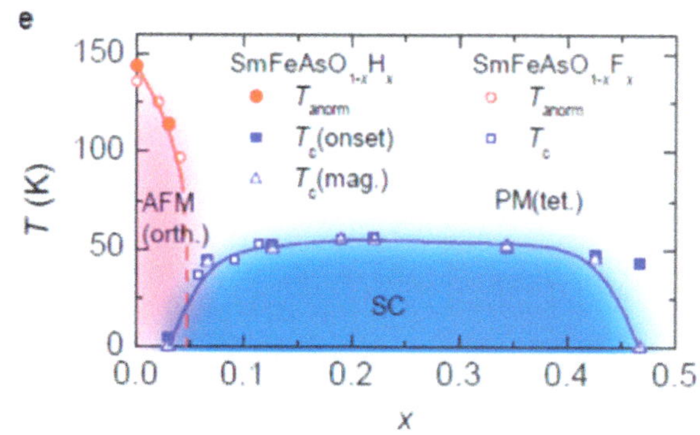

Figure 6. x–T diagram of $SmFeAsO_{1-x}H_x$ superimposed by that of $SmFeAsO_{1-x}F_x$. Reprinted with permission from [9].

The unique feature of revealed SC is observed in $LaFeAsO_{1-x}H_x$ [11]. While T_{c-x} plots for $LnFeAsO_{1-x}H_x$ (Ln = Ce, Sm, Gd) exhibit a single T_c dome as shown in figure 6 (for Sm), $LaFeAsO_1-xHx$ has a two T_c dome structure (see figure 7). When the amount of dopant (x) is increased, the SC emerged from $x > 0.04$, T_c increased and reached a maximum at $x = 0.1$ (T_c = 26 K) and then decreased for a while. Surprisingly after reaching the minimum at $x = 0.2$, T_c increased again and showed a broad peak (maximum T_c = 36 K at $x = 0.35$), which could not be attained by F-doping. The two domes merged into a wider single dome with the optimal T_c(=45 K at $x = 0.3$) by applying a high pressure of 3 GPa. The characteristics of these two T_c domes in $LaFeAsO_1-xHx$ are listed in table 6.

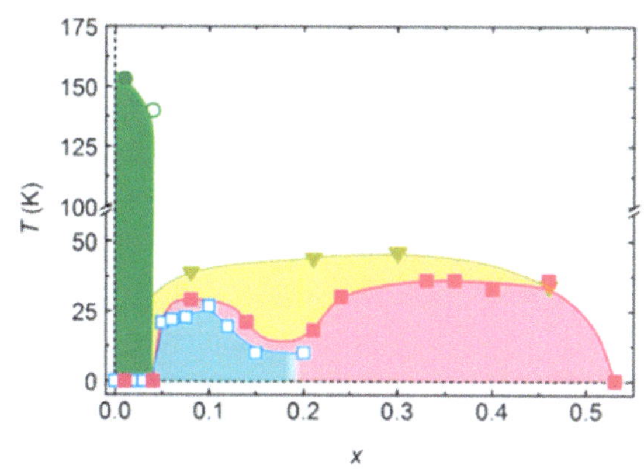

Figure 7. Electronic phase diagram for $LaFeAsO_{1-x}H_x$ (filled symbols) and $LaFeAsO_{1-x}F_x$(open symbols). The Tc under ambient pressure (squares) and 3 GPa (inverted triangles) was determined from the intersection of the two extrapolated lines around the superconducting transition and T_s (circles) was taken as the anomaly kink in the resistivity-temperature (ρ–T) curve. Reprinted with permission from Macmillan Publishers Ltd: [11],

Table 6. Characteristics of two domes in $LaFeAsO_{1-}xHx$.

Dome	First	Second
x	$0.05 \leqslant x \leqslant 0.2$	$0.2 \leqslant x \leqslant 0.5$
Exponent, n	$0.0 \leqslant n \leqslant 2.3$	$0.7 \leqslant n \leqslant 2.0$
T_c^{max}	29 K	36 K
T_c sensitivity to x	High	Low
Under high pressure	Unified	Unified
FS nesting between hole and electron pockets	Strong	Weak
DOS(E_F)	No shoulder	Shoulder

The exponent (n) was estimated from the curve fitting of ρ-T plots to $\rho = AT^n + \rho_0$ near T_c.

Iimura *et al*, who found this two dome structure, considered initially that the superconductivity in the low x region (first dome) was due to the spin fluctuation mechanism and that in the high x region (second dome) was due to the orbital fluctuation mechanism. Their density functional theory (DFT) calculations showed the strong FS nesting between the hole (at Γ point) and the electron (at M point) pockets at the low x region, which was the most important glue in the spin fluctuation model (see figures 8(a)–(h)). On the other hand, the FS nesting weakened with increasing x, while the difference between the energy levels of Fe 3d bands (3d_{xy}, 3d_{yz}, and 3d_{xz}) decreased and became almost zero (degeneracy) at x = 0.30–0.35 where the maximum T_c of the second dome occurs (see figure 8(i)).

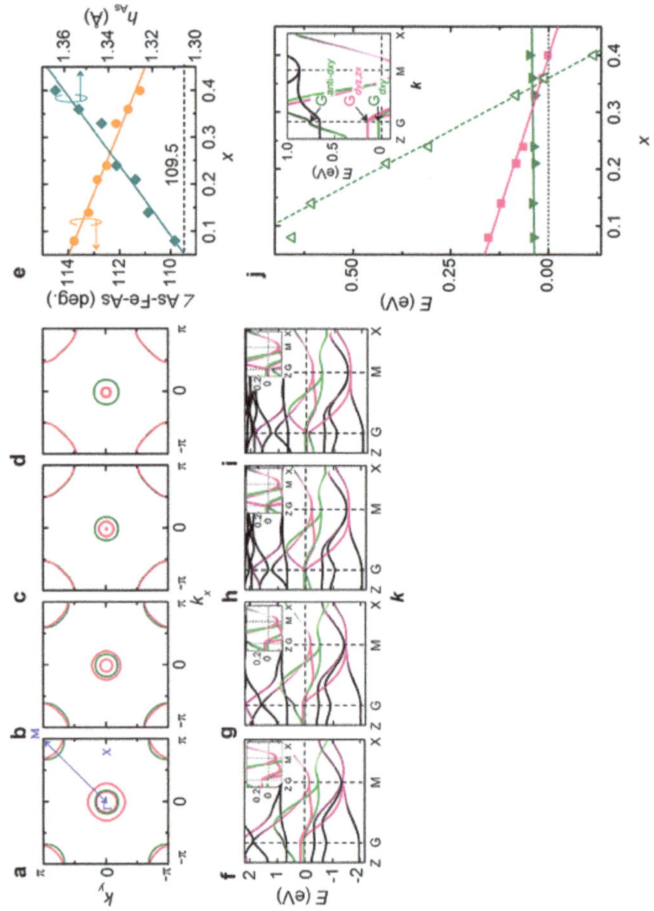

Figure 8. Electronic structure of LaFeAsO$_{1-}x$Hx. (a)–(d) Two-dimensional Fermi surface of LaFeAsO$_{1-}x$Hx with x = 0.08 (a), 0.21 (b), 0.36 (c) and 0.40 (d). The blue arrow represents the nesting vector in the (π–π) direction. The contribution of Fe-d_{xy} and $d_{yz,zx}$ orbitals are colored green and pink, respectively. (f)–(i) Band structures of LaFeAsO$_{1-}x$H$x$with x = 0.08 (f), 0.21 (g), 0.36 (h) and 0.40 (i). Insets show close-up views of the low energy region. The contribution of Fe-d_{xy} and $d_{yz,zx}$ orbitals are colored green and pink, respectively. (j) Variation in energy level of relevant Fe 3d bands at G point with x. The inset is the band structure of LaFeAsO$_{0.92}$H$_{0.08}$. The G$_{d_{xy}}$ (filled green inverted triangles) and G$_{anti-}d_{xy}$ (open green triangles) signify the bonding and anti-bonding states, respectively, for a bond primary composed of two Fe-d_{xy} orbitals in a unit cell. Also shown is the energy level of degenerate $d_{yx,zx}$ band (G$d_{yz,zx}$ indicated by filled pink squares). The solid and dashed lines are as a visual guide. Reprinted with permission from Macmillan Publishers Ltd: [11], Copyright 2012.

The importance of degeneracy in IBSCs is deduced from the empirical plots reported by Lee *et al*, where the more the bonding angle of As–Fe–As (α: see figure 1(b)) is near 109.5° (regular FeAs$_4$ tetrahedron), the higher the T_c revealed [209]. The regular FeAs$_4$ tetrahedron is achieved by the degeneracy of Fe 3d bands and hence it should be reasonable to consider that the orbital fluctuation mechanism to mediate superconductivity emerges effectively in such a condition. The effect of the degeneracy of Fe 3d bands for the emergence of superconductivity is clear from the comparison of a lanthanide cation substituted with La. The α of the parent phases of LnFeAsO$_1$−xHx are 114° (La), 112° (Ce), 111° (Sm) and 110° (Gd) and these values increase with doping of F or H. As shown in figure 9, the material of which α is close to 109.5° achieves the degeneracy of Fe 3d bands and regular tetrahedral structure by a small amount of doping and shows T_c maximum simultaneously. From this viewpoint, Iimura *et al* proposed that the high T_c of 1111-type superconductors was mainly due to the orbital fluctuation mechanism.

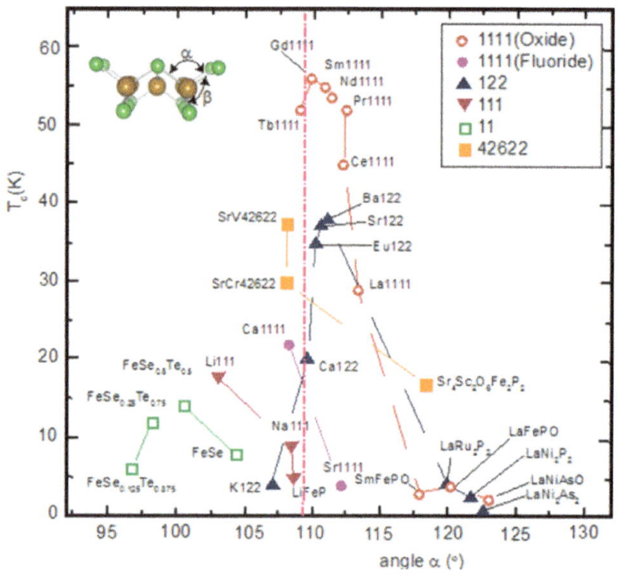

Figure 9. Correlation between T_c and bonding angle α of Pn(Ch)–Fe–Pn(Ch). α is adopted mainly from the data of the parent materials measured at room temperature. *Tc*shows the highest reported value. Reprinted with permission from [125].

On the other hand, the experimental result using inelastic neutron scattering for LaFeAsO$_{1-x}$D$_x$ suggested the presence of spin fluctuation in both low and high x superconducting regions with different wave number [210]. Such spin fluctuation disappeared at the boundary of x = 0.2. The relationships between T_c and E_R of several cuprate materials and IBSCs containing these two regions are plotted in figure 10, where E_R is the measured value indicating the strength of spin fluctuation. For superconductors deriving from the spin fluctuation mechanism, it is known that the value of E_R/k_BT_c is 4–6. As shown in figure 10, all plots line up roughly on the straight line with the gradient of 5.7, which indicates that the spin fluctuation mechanism acts primarily to reveal superconductivity in not only the low x but also the high x region. The theoretical calculations based on the random-phase approximation indicate that the spin fluctuations at$x$ = 0.1 are due to intra-orbital nesting within Fe 3$d_{yz,zx}$, whereas the spin fluctuations at x = 0.4 originate from intra-orbital nesting within Fe 3$dx_{2-}y_2$. These results suggest that the orbital multiplicity plays an important role in the doping and/or material dependence of T_c of the IBSCs.

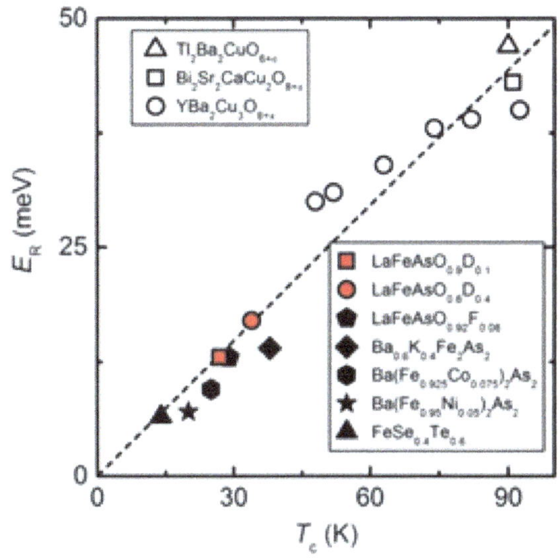

Figure 10. The T_c dependence of E_R in iron pnictides (filled symbols) and cuprates (open symbols). The red square and circles are the present data for the samples with $x = 0.1$ and 0.4, respectively. The dashed line is the averaged slope of $5.7k_BT_c$. Reprinted with permission from [210]. Copyright 2013 by the American Physical Society.

The discovery of a two-SC-dome structure implies the existence of responsible two parent materials. Hiraishi *et al* found a parent phase in the high x region with different magnetic structure from that on $x = 0$ using the multi-probe method composed of neutron, muon and synchrotron x-ray beams complementarily [211]. It was clarified by the muon spin relaxation (μSR) that new magnetic ordering phase developed with increasing dopant over $x \sim 0.4$. From the experimental results using neutron and synchrotron x-ray diffraction, it was identified that this magnetic phase has a different antiferromagnetic ordering from that on $x = 0$ and the structural transition occurs at $x \sim 0.5$ where the magnetic transition temperature shows maximum. Such features are shown in figure 11. This new magnetic phase is tentatively assigned to the parent phase in the high x region.

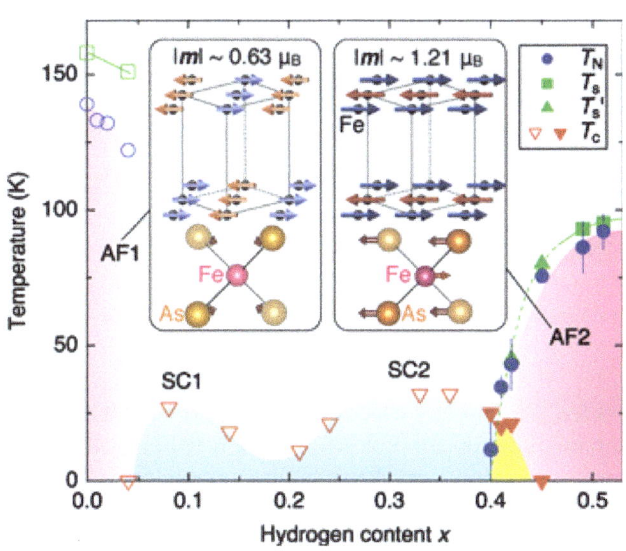

Figure 11. Magnetic, structural and superconducting phase diagram of LaFeAsO$_1$_xHx. The original parent compound with $x = 0$ exhibits a structural transition at $T_s = 155$ K, followed by an antiferromagnetic state (AF1) at $T_N = 137$ K. With increasing x, two superconductivity domes appear: $0.05 \leq x \leq 0.20$ (SC1) with $T_{c,max} = 26$ K, and $0.20 \leq x \leq 0.42$ (SC2) with $T_{c,max} = 36$ K. Eventually, another antiferromagnetic phase (AF2) appears in the range $0.40 \leq x \leq 0.51$. In the advanced parent compound at $x = 0.51$, structural and magnetic transitions occur at $T_s \sim 95$ K and $T_N = 89$ K, respectively. T_s' indicates the c axis upturn temperature observed in x-ray measurements. The filled and open marks are obtained from the present and previous results, respectively. The magnetic structures of AF1 (left) and AF2 (right) are shown with their magnetic moments m, where the solid lines represent the tetragonal cell. The displacements of the Fe and As atoms across the structural transitions are schematically described by the arrows on the $FeAs_4$ tetrahedra from the view of the orthorhombic long axis, in which the Fe and As atoms move by 0.07 Å (0 Å) and 0.06 Å (0.01 Å) in $x = 0.51$ ($x = 0$), respectively. The error bars represent the uncertainty in the least-squares fitting routines.

In theoretical approach, Suzuki *et al* successfully explained the composition dependence forLnFeAsO$_{1-x}$H$_x$ (Ln: La, Ce, Sm, Gd) [169]. They showed that, besides the Fermi surface nesting, a peculiar motion of electrons, where the next nearest neighbor (diagonal) hopping between iron sites dominates over nearest neighbor ones, plays an important role in the enhancement of the spin fluctuation and thus superconductivity. According to their explanation, the crossover between the Fermi surface and this 'prioritized diagonal motion' regime occurs smoothly with doping in the Ce, Sm and Gd case, while the two regimes are separated to give a double dome T_c phase diagram in the La case. The feature of this change is schematically drawn in figure 12.

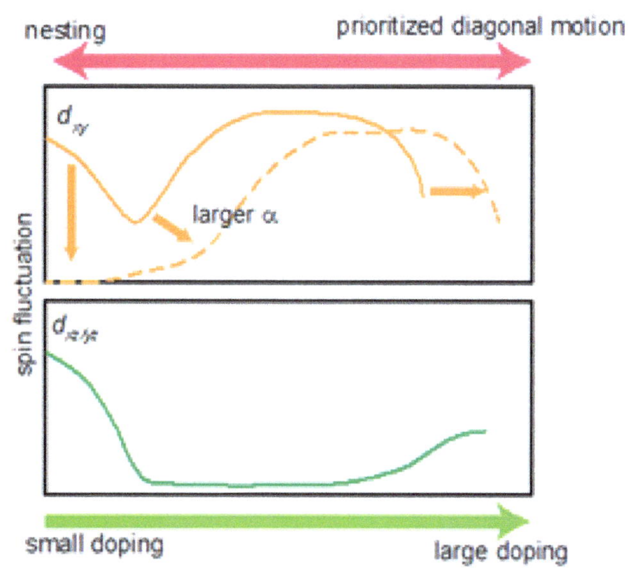

Figure 12. Schematic figure of the spin fluctuation contribution to superconductivity. Reprinted with permission from [169]. Copyright 2014 by the American Physical Society.

The hydrogen doping technique was applied to the alkaline earth metal 1111-type materials, such as CaFeAsF. It reveals superconductivity by replacing Fe with Co ($T_c = 26$ K) [137]. Hanna*et al* realized the electron doping by partially replacing F with H and heating them in He to selectively eliminate H as CaFeAsF$_{1-x}$H$_x$ → CaFeAsF$_{1-x}$ + x/2H$_2$ [18]. The T_c of 29 K was attained by forming an F vacancy (20% of the F site). In the case of direct electron doping into CaFeAsH, the substitution of Fe with Co led this material to become a superconductor (maxT_c = 29 K) [16]. Indirect electron doping (substitution of Ca with La) showed higher T_c than that of Co doping as shown in figure 13 [17]. The material obtained by substitution of 20% of Ca with La showed the highest T_c (47 K) in IBSCs without a rare-earth element as the main component.

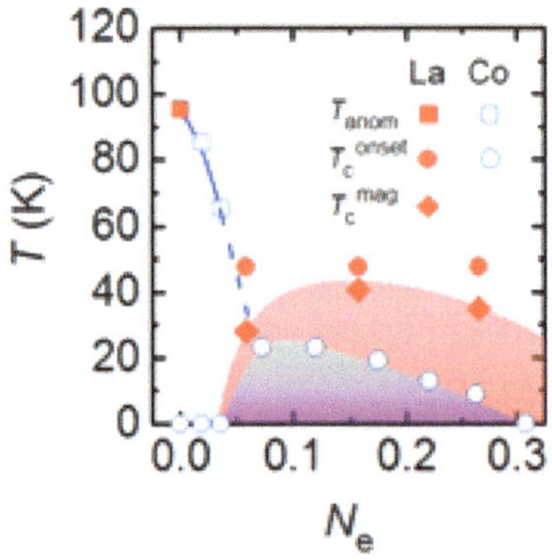

Figure 13. (a) Electronic phase diagram of $Ca_{1-x}La_xFeAsH_{1-y}O_y$ as a function of the total number of doped electrons per iron ($N_e = x-y$), superimposed on that of $CaFe_{1-x}Co_xAsH$. Reprinted with permission from [17]. Copyright 2014 by the Physical Society of Japan.

Rare-earth doped 122 iron Arsenides

Rare-earth (*RE*) doping of 122-type iron arsenides has been intensively studied as part of the FIRST Project. The major achievements in this regard include the high-pressure synthesis of $(Sr_{1-x}La_x)Fe_2As_2$ with a maximum T_c of 22 K [19] and the growth of thin films of $(Sr_{1-x}La_x)Fe_2As_2$ (maximum T_c = 20.8 K) [22] and $(Ba_{1-x}La_x)Fe_2As_2$ (maximum T_c = 22.4 K) by pulsed laser deposition (PLD) [20, 21]. Further, $(Ba_{1-x}RE_x)Fe_2As_2$ with *RE* = Ce, Pr and Nd has also been obtained by means of a nonequilibrium film growth process [21]. The success of *RE*doping enabled us to compare the phase diagrams of electron (La) and hole (K) doped $BaFe_2As_2$as well as those of indirectly (La or K) and directly (Co) doped $BaFe_2As_2$. Another achievement is the simultaneous La and P doping of $CaFe_2As_2$ [23]: melt-grown crystals of $(Ca_{1-x}La_x)Fe_2(As_{1-y}P_y)_2$ exhibited bulk superconductivity with a maximum T_c of 45 K for $0.12 \leq x \leq 0.18$ with $y = 0.06$, while $(Ca_{1-x}La_x)Fe_2As_2$ without phosphorus exhibited filamentary superconductivity [23]. Characteristic two-dimensional Fermi surfaces were observed by means of angle-resolved photoemission spectroscopy (ARPES) in the 45 K phase of La- and P-doped $CaFe_2As_2$ [212].

To induce superconductivity in 122-type $AEFe_2As_2$ (where *AE* = alkaline-earth elements), both types of doping carrier, i.e., holes and electrons, are typically used by substituting appropriate aliovalent elements, which include alkali metals (*A*) such as K, which substitute for *AE* sites as in hole-doped $(Ba_{1-x}K_x)Fe_2As_2$ [119], and transition metals (*TM*) such as Co, which substitute for Fe sites as in electron-doped $Ba(Fe_{1-x}Co_x)_2As_2$ [186, 193]. The doping can be classified into two types for the 122-type $AEFe_2As_2$, namely, 'indirect doping' for doping at sites other than the Fe sites and 'direct doping' for doping at the Fe sites [21]. Because the superconducting FeAs and intermediary *AE* layers are spatially separated, direct doping has a major influence on carrier transport and thus superconductivity. Tables 7 and 8 summarize the maximum T_c of directly doped $AE(Fe_{1-x}TM_x)_2As_2$ and indirectly doped $(AE_{1-x}A_x)Fe_2As_2$, respectively. The indirectly hole-doped $(Ba_{1-x}K_x)Fe_2As_2$ exhibits a maximum T_c of 38 K, which is considerably higher than those for directly electron-doped $AE(Fe_{1-x}TM_x)_2As_2$. Therefore, we expected that a new indirect 'electron' doping at the *AE* sites for $AEFe_2As_2$ would lead to high-$T_c$superconductivity, as expected from the markedly higher T_c observed for indirectly electron-doped $SmFeAs(O_{1-x}F_x)$ (55 K) [115] than that for directly electron-doped $Sm(Fe_{1-x}Co_x)AsO$ (17 K) [232].

Table 7. The maximum superconducting transition temperature T_c (K) of transition metal (TM)-doped $AE(Fe_{1-x}TM_x)_2As_2$, where AE=Ca, Sr and Ba; and TM=Co, Ni, Ru, Rh, Pd, Ir and Pt. For instance, T_c = 20 K for $Ca(Fe_{1-x}Co_x)_2As_2$. Pt-doped $Ca(Fe_{1-x}Pt_x)_2As_2$ does not exhibit superconductivity [75].

	Fe	Co	Ni
Ca		20 [213]	15 [214]
Sr		19.2 [215]	9.8 [216]
Ba		23 [186, 193]	20.5 [217]
	Ru	**Rh**	**Pd**
Ca		14 [218]	10 [219]
Sr	19.3 [220]	21.9 [221]	8.7 [221]
Ba	22 [190]	23.2 [188]	19 [188]
		Ir	**Pt**
Ca		22 [222]	no SC [75]
Sr		24.2 [221]	16.5 [223]
Ba		28 [224]	23 [225]

Table 8. The maximum superconducting transition temperature T_c (K) of the alkali metal (A)-doped $(AE_{1-x}A_x)Fe_2As_2$, where *AE*=Ca, Sr and Ba; and *A*=Na, K, Rb and Cs.

	Na	K	Rb	Cs
Ca	26 [226]			
Sr	26 [227]	36.5 [228, 229]		37.2 [228]
Ba	34 [230]	38 [119]	23 [231]	

However, indirect electron doping of $Sr(Ba)Fe_2As_2$ by substituting the divalent Sr(Ba) sites with trivalent *RE* ions was difficult to perform by means of conventional solid-state reactions. While Muraba *et al* [19] and Wu *et al* [233] examined the La substitution for $SrFe_2As_2$ and $BaFe_2As_2$, respectively, solid-state reactions of the ingredient mixture for $(Sr_{1-x}La_x)Fe_2As_2$ or $(Ba_{1-x}La_x)Fe_2As_2$ did not yield the La-substituted 122 phase upon using the conventional glass-tube technique.

In contrast, the indirect RE doping of $CaFe_2As_2$ was possible by a conventional melt-growth technique for *RE* = La, Ce, Pr and Nd [234–237]. These materials exhibited superconductivity at T_c = 40–49 K: Saha *et al* [234] reported T_c = 47 K in $(Ca_{1-x}Pr_x)Fe_2As_2$; Gao *et al* [235] reported T_c = 42.7 K in $(Ca_{1-x}La_x)Fe_2As_2$; and Lv *et al* [236] reported T_c = 49 K in $(Ca_{1-x}Pr_x)Fe_2As_2$. These values of T_c are considerably higher than those reported for directly electron-doped $Ca(Fe_{1-x}TM_x)_2As_2$ and indirectly hole-doped $(Ca_{1-x}A_x)Fe_2As_2$ listed in tables 7 and 8, respectively. However, the shielding volume fractions of *RE*-doped $CaFe_2As_2$ were as low as <1% at 40 K [234, 235] or the shielding signal around 40 K was completely suppressed by the application of a weak magnetic field of 20 Oe [236].

Under such circumstances, Muraba *et al* have succeeded in the indirect La doping of $SrFe_2As_2$by applying a high-pressure synthesis process to obtain $(Sr_{1-x}La_x)Fe_2As_2$ polycrystals [19]. The ionic radius of La^{3+} (116 pm) is smaller than that of Sr^{2+} (126 pm) and therefore, it is natural that high-pressure synthesis is effective for obtaining La-substituted $SrFe_2As_2$. Muraba *et al* used a belt-type anvil cell to generate a pressure of 2 or 3 GPa at 1000 °C for 2 h and obtained solid solutions of $(Sr_{1-x}La_x)Fe_2As_2$ for $0 \leq x \leq 0.5$. Further, $(Sr_{1-x}La_x)Fe_2As_2$ exhibited bulk superconductivity over a narrow range around $x = 0.4$. The almost x-independent T_c (shown in figure 14) suggested the inhomogeneous replacement of La dopants at the Sr sites. Figure 14compares the electronic phase diagram of $(Sr_{1-x}La_x)Fe_2As_2$ [19] with those of $Sr(Fe_{1-x}Co_x)_2As_2$[215] and $(Sr_{1-x}K_x)Fe_2As_2$ [229]. Here, the doped carrier number per Fe, i.e., Δn/Fe = $x/2$ for $(Sr_{1-x}La_x)Fe_2As_2$ and $(Sr_{1-x}La_x)Fe_2As_2$ and Δn/Fe = x for $Sr(Fe_{1-x}Co_x)_2As_2$, is plotted in place of the doping composition x. The directly electron-doped $Sr(Fe_{1-x}Co_x)_2As_2$ exhibits superconductivity for $0.1 \leq \Delta n/\mathrm{Fe} \leq 0.2$ with a maximum T_c of 19 K at $\Delta n/\mathrm{Fe} \sim 0.1$, at which point the antiferromagnetic ordering vanishes [215]. The maximum T_c of the indirectly electron-doped $(Sr_{1-x}La_x)Fe_2As_2$ is slightly higher than but close to that of directly electron-doped $Sr(Fe_{1-x}Co_x)_2As_2$. These results contrast markedly with the observed higher T_c and wider superconducting range of the direct hole-doped system $(Sr_{1-x}K_x)Fe_2As_2$ (37 K) [228, 229], shown in figure 14, and $(Sr_{1-x}Cs_x)Fe_2As_2$ (37 K) [228]. The superconducting dome continues to the end member of this family, i.e., KFe_2As_2 [228, 229] and $CsFe_2As_2$ [228] at Δn/Fe = 0.5.

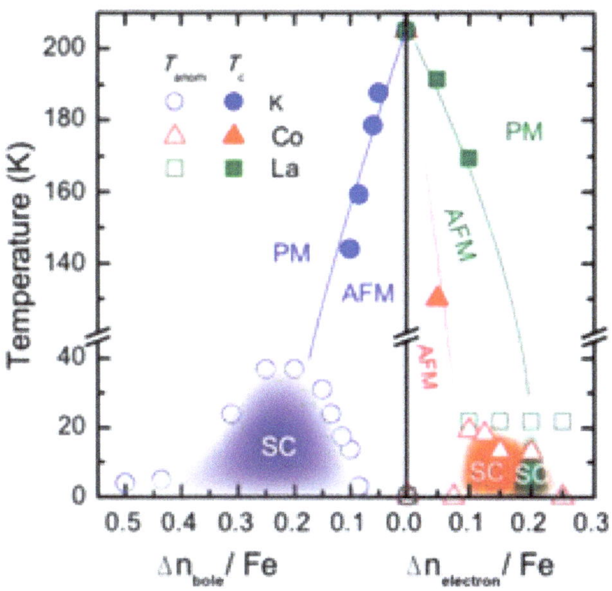

Figure 14. Electronic phase diagrams for electron-doped $(Sr_{1-x}La_x)Fe_2As_2$ [19] and $Sr(Fe_{1-x}Co_x)_2As_2$ [215] and for hole-doped $(Sr_{1-x}K_x)Fe_2As_2$ [229]. The ratio $\Delta n_{electron}$/Fe and Δn_{hole}/Fe denote the injected number of electrons and holes per Fe atom, respectively. Reprinted with permission from [19]. Copyright 2010 by the American Physical Society.

Another promising technique that we have developed for the indirect *RE* doping of $SrFe_2As_2$and $BaFe_2As_2$ is the non-equilibrium pulsed laser deposition (PLD) method [20–22]. Hiramatsu*et al* [22] and Katase *et al* [20, 21] have succeeded in the homogeneous doping of *RE* dopants in the films fabricated by PLD, and they have reported observing a superconducting dome from underdoped to overdoped regions irrespective of the largely different ionic radii of Ba^{2+} (142 pm), Sr^{2+} (126 pm) and La^{3+} (116 pm). Here, we mention that $CaFe_2As_2$ epitaxial films could not be obtained by PLD [238]. Thin films of $(Sr_{1-x}La_x)Fe_2As_2$ (thickness = 200 nm) were grown on $(La,Sr)(Al,Ta)O_3$ (LSAT) (001) single crystals at a film-growth temperature of 750 °C [22], while thin films of $(Ba_{1-x}RE_x)Fe_2As_2$ (thickness = 150–250 nm) were grown on MgO (001) single crystals at an optimized film-growth temperature of 850 °C [20, 21]. Further, $(Sr_{1-x}La_x)Fe_2As_2$ thin films were successfully obtained for $0.0 \leq x \leq 0.48$ and $(Ba_{1-x}La_x)Fe_2As_2$for $0.0 \leq x \leq 0.44$. The lattice parameters showed monotonic decrease with increasing x (La content), thereby indicating that the substitution of the La^{3+} ion for the Sr^{2+} and Ba^{2+} ions was achieved in the epitaxial films [20, 22]. Figure 15 shows the electronic phase diagrams of $(Sr_{1-x}La_x)Fe_2As_2$ [22] and $(Ba_{1-x}La_x)Fe_2As_2$ [20] epitaxial films. Here, the doped carriers per Fe (=x/2) for these two compounds are plotted for comparison with the phase diagram of the directly electron-doped $Sr(Fe_{1-x}Co_x)_2As_2$ [215]. The antiferromagnetic ordering at T_{anom}, which is determined by the anomaly in resistivity, is suppressed as the La content (x) increases. The maximum onset superconducting transition temperatures T_c^{onset} are 20.8 and 22.4 K at$x$/2 = 0.16 and 0.07 for $(Sr_{1-x}La_x)Fe_2As_2$ [22] and $(Ba_{1-x}La_x)Fe_2As_2$ [20], respectively. The maximum values of T_c^{onset} and the corresponding x values of $(Sr_{1-x}La_x)Fe_2As_2$ and $(Ba_{1-x}La_x)Fe_2As_2$ are very close to those reported for directly doped $Sr(Fe_{1-x}Co_x)_2As_2$ [215] (shown in figure 15) and $Ba(Fe_{1-x}Co_x)_2As_2$ [193] (not shown), respectively. In addition, the suppression rate of T_{anom} is similar for both compounds. These results sharply contrast with those of the 1111-type system, in which the maximum T_c of indirectly electron-doped $SmFeAs(O_{1-x}F_x)$ (55 K) [115] is considerably higher than that for directly electron-doped $Sm(Fe_{1-x}Co_x)AsO$ (17 K) [232]. These results can yield a clue to elucidate the mechanism of superconductivity in iron-based materials.

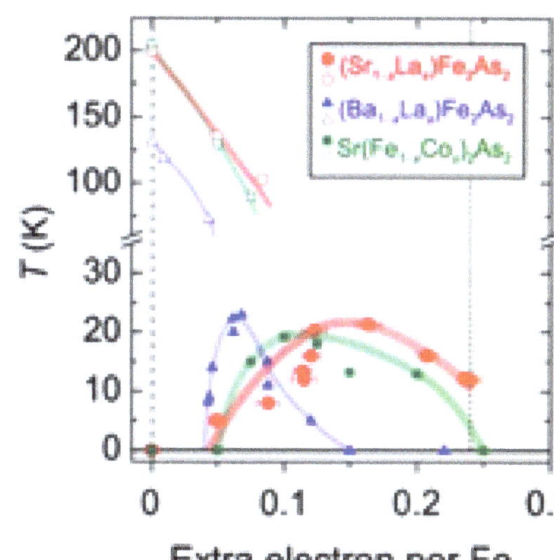

Figure 15. Electronic phase diagrams of $(Sr_{1-x}La_x)Fe_2As_2$ [22] and $(Ba_{1-x}La_x)Fe_2As_2$ [20] epitaxial films. The T_{anom} and T_c^{onset} values are indicated by open and closed symbols, respectively. Those of $Sr(Fe_{1-x}Co_x)_2As_2$ [215] are shown for comparison. Reprinted with permission from [22]. Copyright 2013 by IEEE.

Katase *et al* [21] have also succeeded in obtaining $(Ba_{1-x}RE_x)Fe_2As_2$ thin films with *RE* = Ce, Pr and Nd with reduced solubility limits of x = 0.29, 0.18 and 0.13, respectively. Thin films of $(Ba_{1-x}RE_x)Fe_2As_2$ exhibited superconductivity at 13.4, 6.2 and 5.8 K for Ce, Pr and Nd dopants, respectively [21]. The T_c values of *RE*-doped $SrFe_2As_2$ and $BaFe_2As_2$ are listed in table 9.

Table 9. The maximum superconducting transition temperature T_c (K) of rare-earth (*RE*) doped $(AE_{1-x}RE_x)Fe_2As_2$, where *AE*=Ca, Sr and Ba; and *RE*=La, Ce, Pr and Nd.

	La	Ce	Pr	Nd
Ca	45 [23]	37 [234]	49 [236]	
Sr	22 [19]			
Ba	22.4 [20]	13.4 [21]	6.2 [21]	5.8 [21]

A conventional melt-growth method can be used to achieve indirect *RE* doping of $CaFe_2As_2$, although the obtained samples do not exhibit bulk superconductivity. In this project, two kinds of indirect doping, i.e., aliovalent La doping at the Ca sites and isovalent P doping at the As sites, were combined to induce bulk superconductivity in $CaFe_2As_2$ [23]. The substitution of aliovalent La for Ca resulted in electron doping without leading to a change in the lattice parameters because the ionic radius of La^{3+} (116 pm) and Ca^{2+} (112 pm) are similar, while the substitution of isovalent P for As resulted in a decrease in the lattice parameters without leading to a change in the number of carriers. This enabled us to tune the number of charge carriers and lattice parameters independently, and thus to optimize superconductivity in $(Ca_{1-x}La_x)Fe_2(As_{1-y}P_y)_2$, which resulted in bulk superconductivity at $x = 0.17$ and $y = 0.06$. The resistivity shows a sharp drop at 48 K and becomes zero at 45 K [23], as shown in figure 16. A clear diamagnetic signal, together with robust diamagnetism against thes increase in magnetic field, is the reason for the bulk superconductivity at 45 K in this material [23]. Figure 17 shows the *T–x–y* phase diagram of $(Ca_{1-x}La_x)Fe_2(As_{1-y}P_y)_2$ [23]. Bulk superconductivity emerges in the range of $0.12 \leq x \leq 0.18$ and $y = 0.06$.

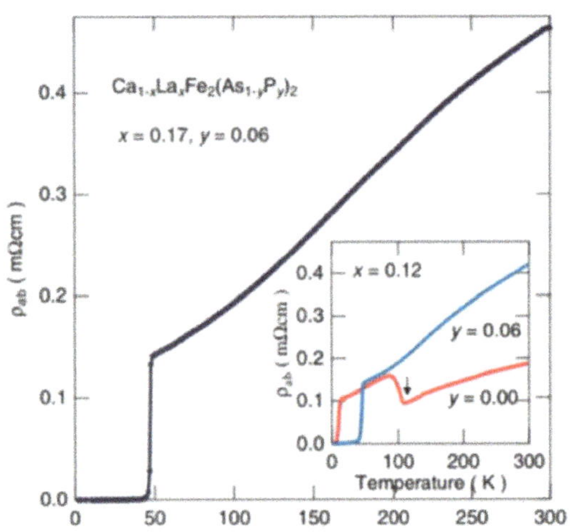

Figure 16. Temperature dependence of electrical resistivity ρ_{ab} for $(Ca_{1-x}La_x)Fe_2(As_{1-y}P_y)_2$ ($x = 0.17$ and $y = 0.06$) [23]. The resistivity starts to decrease at a transition temperature T_c^{onset} of 48 K and becomes zero below 45 K. The inset shows ρ_{ab} for $x = 0.12$ and $y = 0.00$ and 0.06. The arrow indicates the antiferromagnetic/tetragonal–orthorhombic structural transition.

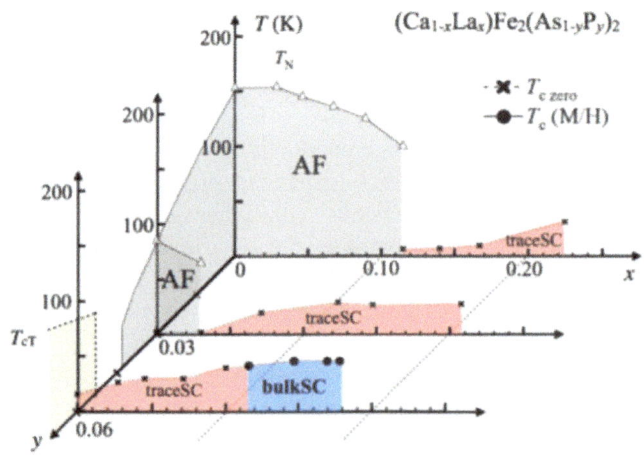

Figure 17. *T–x–y* electronic phase diagrams for $(Ca_{1-x}La_x)Fe_2(As_{1-y}P_y)_2$. *Tc(M/H)* denotes the bulk superconducting transition temperature determined from the magnetization. T_c^{onset} denotes the temperature below which the electrical resistivity becomes negligibly small. T_N represents the antiferromagnetic and structural transition temperature. T_{cT} denotes the transition temperature at which the high-temperature uncollapsed tetragonal (ucT) phase transforms into the low-temperature collapsed tetragonal (cT) phase. AF and SC indicate the antiferromagnetic and superconducting phases, respectively.

Angle-resolved photoemission spectroscopy (ARPES) [212] revealed that $(Ca_{0.82}La_{0.18})Fe_2(As_{0.94}P_{0.06})_2$ with $T_c = 45$ K possesses only cylindrical hole- and electron-like Fermi surfaces (FS). The size of the β hole-like FS is nearly the same as that of the ε electron-like FS, and both FS have a weak *kz* dispersion, thus giving rise to a quasi-nesting. This feature is similar to that for directly electron-doped $Ba(Fe_{1-x}Co_x)_2As_2$ ($T_c = 23$ K). Sunagawa *et al* [212] pointed out that a noticeable difference between $(Ca_{0.82}La_{0.18})Fe_2(As_{0.94}P_{0.06})_2$ and $Ba(Fe_{1-x}Co_x)_2As_2$ is the dimensionality of the inner hole-like FS; the inner hole-like FS (α_2) of the former shows a cylindrical shape, while that of the latter shows a strong *kz* dispersion and is closed near the Γ point. It has been suggested that the tendency toward quasi-nesting betweenaα_2 and β, together with β and ε, can induce high T_c in $(Ca_{0.82}La_{0.18})Fe_2(As_{0.94}P_{0.06})_2$ [212].

New type of 112 IBSC

The prominent achievements of the FIRST Project include the discovery of the 112-type iron arsenide superconductor $(Ca_{1-x}La_x)FeAs_2$ [25], whose structure is shown in figure 18, and the enhancement of superconducting transition temperature T_c up to 47 K by the simultaneous La and Sb doping of the 112 phase [26, 27]. In this subsection, we overview the crystal structure and superconducting properties of the newly discovered 112 phase.

The superconductivity of the 112 phase was first reported in $(Ca_{1-x}La_x)FeAs_2$ and $(Ca_{1-x}Pr_x)FeAs_2$ by Katayama *et al* [25] and Yakita *et al* [239], respectively. The substitution of a rare-earth element is essential to obtaining the 112 phase [25, 239]. A conventional melt-growth technique was used to obtain tiny single crystals [25, 239]. $(Ca_{1-x}La_x)FeAs_2$ exhibited superconductivity at T_c = 34 K [25], while $(Ca_{1-x}Pr_x)FeAs_2$ exhibited T_c of ~20 K with a broad resistive transition [239]. In a subsequent study, Sala *et al* [240] performed a high-pressure synthesis and obtained $(Ca_{1-x}RE_x)FeAs_2$ for *RE* = La–Gd. Moreover, $(Ca_{1-x}RE_x)FeAs_2$ for*RE* = Pr, Nd, Sm, Eu and Gd showed superconductivity at 10–15 K with a small shielding fraction of 5–20% [27, 240], while $(Ca_{1-x}Ce_x)FeAs_2$ did not exhibit superconductivity [27, 240]. The T_c of $(Ca_{1-x}RE_x)FeAs_2$ is summarized in table 10.

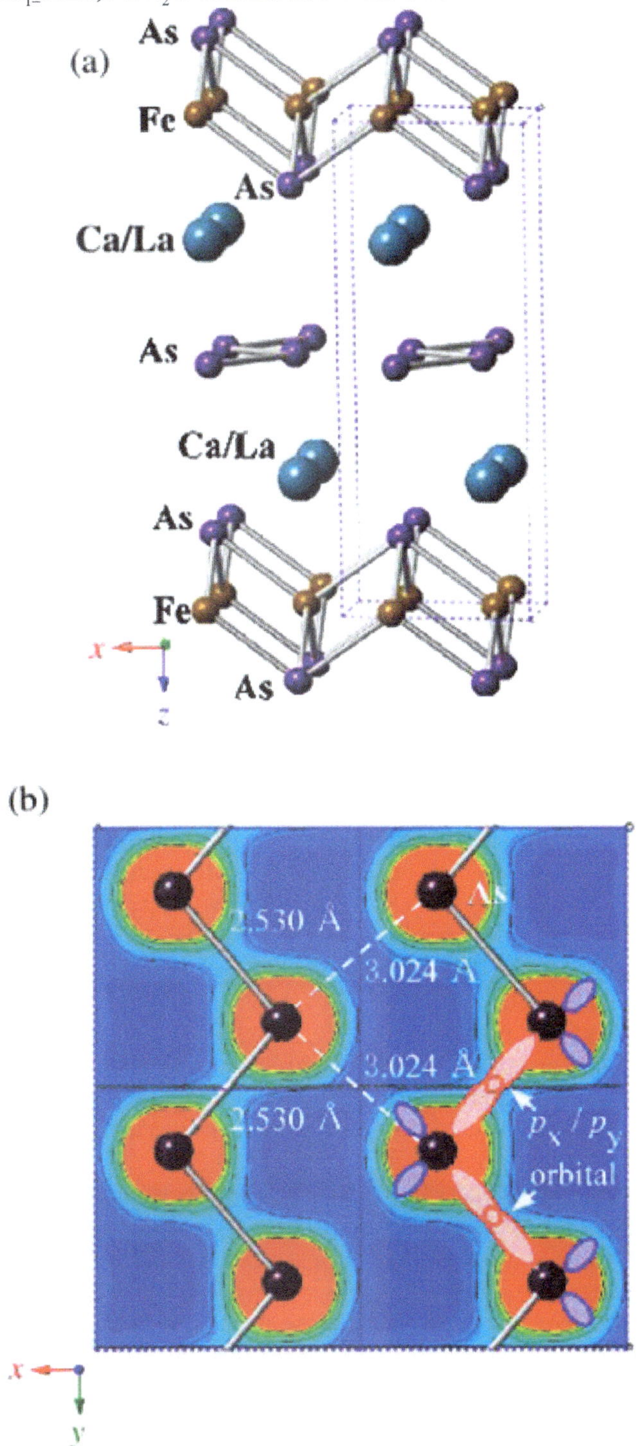

Figure 18. (a) The crystal structure of $(Ca_{1-x}La_x)FeAs_2$ (monoclinic, space group $P2_1$). (b) Top view of the arsenic zigzag chains. Arsenic $4p_x$ and $4p_y$ orbitals are schematically shown. The background color contour map shows the charge distributions obtained by synchrotron x-ray diffraction analysis. The charge accumulation between the adjacent As atoms in the zigzag chains suggests the formation of covalent bonds. Reprinted with permission from [25]. Copyright 2013 by the Physical Society of Japan.

Table 10. The superconducting transition temperature T_c (K) of RE-doped $(Ca_{1-x}RE_x)FeAs_2$ and *RE*- and Sb-doped $(Ca_{1-x}RE_x)Fe(As_{1-y}Sb_y)_2$ as determined by magnetic measurements.

RE	$(Ca_{1-x}RE_x)FeAs_2$	$(Ca_{1-x}RE_x)Fe(As_{1-y}Sb_y)_2$
La	34 [25]	47 [27]
Ce		43 [27]
Pr	20 [239]	43 [27]
Nd	11.9 [240]	43 [27]
Sm	11.6 [240]	
Eu	9.3 [240]	
Gd	12.6 [240]	

The 112-type $(Ca_{1-x}RE_x)FeAs_2$ compound crystallizes in a monoclinic structure with the space group of $P2_1$ (no. 4) [25] or $P2_1/m$ (No. 11) [239, 240]. The structure consists of alternately stacked FeAs and zigzag As bond layers with a Ca/La layer between them, as shown in figure18(a). The most prominent feature of this structure is the presence of the one-dimensional zigzag As chains along the *b*-axis, as shown in figure 18(b). The short As–As bond length of approximately 2.53 Å indicates the formation of arsenic single bonds where arsenic has a formal valence of As^- ($4p^4$ configuration). The presence of two unpaired electrons in As^- underlies the formation of two chemical bonds per As atom that yield a zigzag chain. In contrast, the arsenic at FeAs layers forms the As^{3-} valence state with the filled $4p^6$ configuration. Thus, the chemical formula of the 112 phase can be written as $(Ca^{2+}_{1-x} RE^{3+}_x)(Fe^{2+}As^{3-})As^- \cdot xe^-$ with excess charge of xe^-/Fe, which is injected into the superconducting FeAs layers. This formula can be compared with that of the 1111-type fluoride, $(Ca^{2+}_{1-x} RE^{3+}_x)(Fe^{2+}As^{3-})F^- \cdot xe^-$ [137, 241], where F^- with filled $2p^6$ orbitals forms an undistorted square network. In this manner, the 112-type structure can be related to the 1111-type structure of CaFeAsF. However, the chemical bonding of the intermediary layer is completely different between CaFeAsF and $CaFeAs_2$; the intermediary CaF layers consist of strong ionic bonds, while the CaAs layers consist of zigzag As chains with covalent bonds, which are weakly coupled to the adjacent Ca layers. Thus, the interlayer distance between the adjacent FeAs layers of $(Ca_{1-x}La_x)FeAs_2$ (~10.35 Å) [27] is considerably larger than that of CaFeAsF (~8.6 Å) [137], but is comparable to that of $Ca_{10}(Pt_4As_8)(Fe_{2-x}Pt_xAs_2)_5$ (~10 Å) with the Pt_4As_8 intermediary layers exhibiting a covalent nature [242, 243].

Kudo *et al* [26, 27] examined simultaneous doping, i.e., aliovalent La doping at the Ca sites and isovalent P or Sb doping at the As sites, and they observed that T_c increased up to 47 K for La- and Sb-doped $(Ca_{1-x}La_x)Fe(As_{1-y}Sb_y)_2$ [27]. Figure 19(a) shows the temperature dependence of the magnetization of $(Ca_{1-x}La_x)Fe(As_{1-y}Sb_y)_2$. The La-doped sample without Sb (y = 0) shows superconductivity at T_c = 34 K. The Sb doping results in an increase in T_c to 43 K for y = 0.01 and 47 K for y = 0.10. The enhancement of T_c is also evident in the electrical resistivity ρab of the La-doped y = 0.10 sample. The resistivity ρab exhibits a sharp drop at 49 K, and zero resistivity is observed at 47 K, as shown in figure 19(b). T_c was also enhanced by Sb doping of $(Ca_{1-x}RE_x)Fe(As_{1-y}Sb_y)_2$ for RE = Ce, Pr and Nd [27], as summarized in table 10. Figure 20shows the T–x phase diagram of $(Ca_{1\_}xREx)Fe(As_{1\_}ySby)_2$ [27]. The 112 phase can be obtained at $x \geq 0.15$ for $y = 0.0$. The superconducting transition temperature T_c is highest (35 K) at the lowest boundary of $x = 0.15$, and it decreases monotonically with the La content x. Superconductivity disappears at $x \geq 0.25$. The lower limit of x is extended down to $x = 0.12$ for$y = 0.10$, at which the highest T_c of 47 K is observed [27].

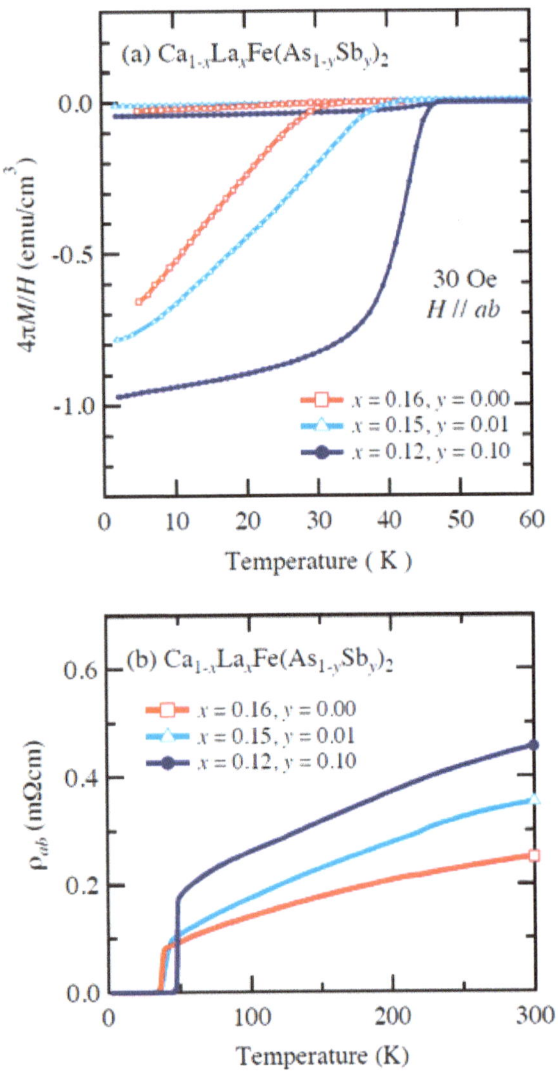

Figure 19. (a) Temperature dependence of the magnetization M of $(Ca_{1\_}xLax)Fe(As_{1\_}ySb_y)_2$ measured at a magnetic field H of 30 Oe parallel to the ab plane under zero-field-cooling and field-cooling conditions. (b) Temperature dependence of the electrical resistivity ρab of $(Ca_{1\_}xLax)Fe(As_{1\_}ySby)_2$ parallel to the ab-plane. Reprinted with permission from [27]. Copyright 2014 by the Physical Society of Japan.

Recently, Zhou *et al* [244] have successfully grown large single crystals of $(Ca_{1\_}xLax)FeAs_2$ with$T_c = 42.6$ K. Using these crystals, they estimated a critical current density J_c of 3.5×10^5 A cm^{-2}from the magnetic hysteresis loops at 5 K, which indicates a strong bulk pinning. The anisotropic upper critical field $Hc_2(0)$ was estimated to be 39 and 166 T for the out-of-plane ($H//c$) and in-plane ($H//ab$) directions, respectively. The H_{c2} anisotropy parameter γ was ~2 near T_c. The moderate anisotropy and high T_c indicates the potential of $(Ca_{1\_}xLax)$ $FeAs_2$ for practical applications.

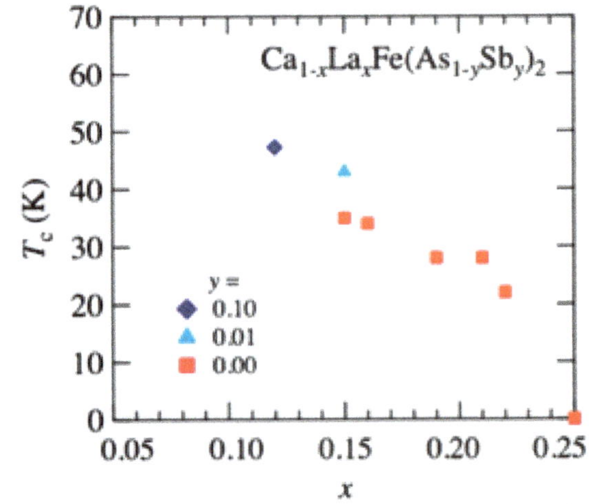

Figure 20. Dependence of the superconducting transition temperature Tc of $(Ca_{1\_}xLax)Fe(As_{1\_}ySby)_2$ on the La content x. Reprinted with permission from [27]. Copyright 2014 by the Physical Society of Japan.

10-3-8 and 10-4-8 Iron Arsenide Superconductors

The rich chemistry of arsenic allowed us to develop various iron arsenide superconductors. Arsenic exhibits a wide variety of chemical networks, which are known as catenation, depending on the number of valence electrons. For instance, neutral arsenic has three unpaired electrons ($4p^3$), and thus, it forms three single bonds per As atom, thereby resulting in a buckled honeycomb network of α-As (trigonal, space group $R\bar{3}m$). Monovalent arsenic (As^-) has two unpaired electrons ($4p^4$), and thus forms two single bonds per As atom, which results in either one-dimensional zigzag chains or cis–trans chains, or molecular-like As_4^- tetramers. The 112-type iron arsenide superconductor $(Ca_{1\_}xLax)FeAs_2$ consists of As zigzag chains [25], as described in the previous subsection. Cis–trans chains can be observed in $LaAgAs_2$ [245] and As_4 tetramers in skutterudite $CoAs_3$, for instance. Divalent arsenic (As^{2-}) has one unpaired electron ($4p^5$), and thus forms a single bond per As atom, which results in molecular-like As_2^{4-}dimers, as seen in pyrite-type $PtAs_2$. The 122-type iron arsenide superconductor $CaFe_2As_2$exhibits the formation of molecular As_2 between adjacent FeAs layers along the c-axis at the collapsed tetragonal phase transition [218, 234, 246]. The As_2 molecular bonds are broken in the uncollapsed tetragonal phase. Here, the formal valence of arsenic is As^{3-}, and the $4p$ orbitals are completely occupied, and thus, no direct chemical bonds are formed between As. The 10-3-8- and 10-4-8-type iron arsenide superconductors $Ca_{10}Pt_3As_8(Fe_{2\_}xPtxAs_2)_5$ and $Ca_{10}Pt_4As_8(Fe_{2\_}xPtxAs_2)_5$ consist of As_2 dimers in the Pt_3As_8 and Pt_4As_8 intermediary layers, respectively [242, 247, 248].

Superconductivity in the 10-3-8 and 10-4-8 compounds was first reported by Kakiya *et al* [242], Löhnert *et al* [247] and Ni *et al* [248] for Pt-based $Ca_{10}Pt_3As_8(Fe_{2\_}xPtxAs_2)_5$ and $Ca_{10}Pt_4As_8(Fe_{2\_}xPtxAs_2)_5$. The 10-4-8 compound exhibited a maximum T_c of 38 K [242, 249]. The 10-3-8 compound showed a lower T_c of 13 K [242] by Pt doping of $Ca_{10}Pt_3As_8(Fe_{2\_}xPtxAs_2)_5$. The T_c value was enhanced up to 30 K by La doping of $(Ca_{0.8}La_{0.2})_{10}(Pt_3As_8)(Fe_{2\_}xPtxAs_2)_5$ [249]. Three polymorphs have been identified in the 10-4-8 phase, i.e. tetragonal ($P4/n$), triclinic ($P\bar{1}$) and monoclinic ($P2_1/n$) structures, while the 10-3-8 compound crystallizes in a triclinic ($P\bar{1}$) structure. Hieke *et al* [250] reported the observation of the 10-3-8 phase for Pd-based $Ca_{10}Pd_3As_8(Fe_2As_2)_5$, which exhibited superconductivity at 17 K by La substitution for Ca.

Under the aegis of the FIRST Project, Kudo *et al* [28] discovered a new member of the 10-4-8 family with Ir_4As_8 intermediary layers, i.e. $Ca_{10}Ir_4As_8(Fe_{2-x}Ir_xAs_2)_5$. The compound crystallizes in the tetragonal structure with the space group *P*4/*n*, which is isotypic to one of the polymorphs of Pt-based 10-4-8. The crystal structure, shown in figure 21, consists of characteristic $IrAs_4$ squares, which are rotated alternately to form As_2 dimers. Figure 22 shows the temperature dependence of the in-plane electrical resistivity ρab of $Ca_{10}Ir_4As_8(Fe_{2-x}Ir_xAs_2)_5$ with x = 0.07–0.08 [28, 251]. The resistivity ρab reached zero at 17 K. Magnetic measurements demonstrated large shielding signals below T_c = 16 K. The small and almost *T*-independent Hall coefficient R_H, shown in the inset of figure 22, suggests overdoping, which most probably resulted in the low T_c. Another feature of interest is the kink in ρab at approximately 100 K. This kink is not due to antiferromagnetic ordering, since the singlet-peak structure of the ^{57}Fe-Mössbauer spectra remained unchanged down to 50 K (as shown in the inset of figure 22), but due to a structural phase transition [251].

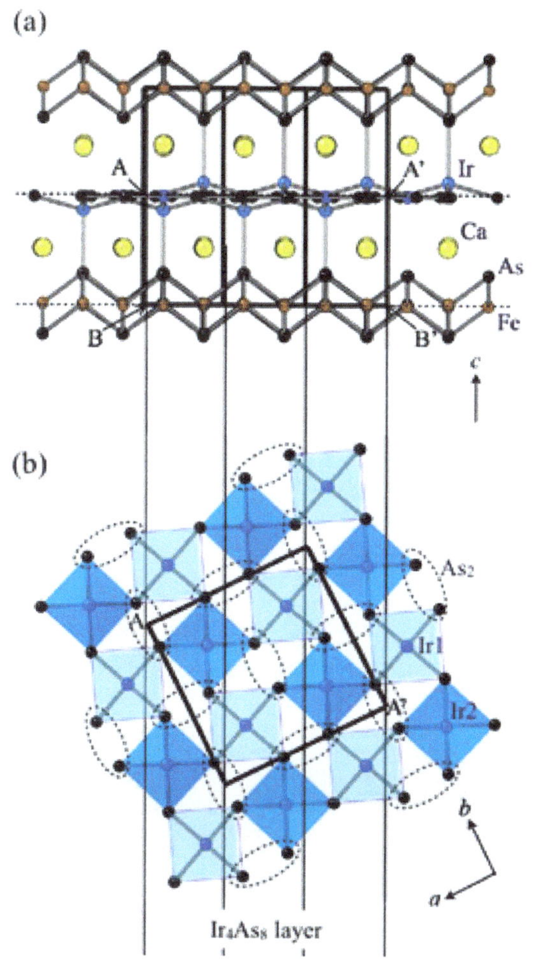

Figure 21. Crystal structure of 10-4-8 type $Ca_{10}Ir_4As_8(Fe_{2-x}Ir_xAs_2)_5$ with tetragonal structure (space group *P*4/*n*). (a), (b) Schematic overviews and the Ir_4As_4 layer, respectively. The blue and dark-blue hatches in (b) indicate $IrAs_4$ squares with coplanar Ir1 and non-coplanar Ir2, respectively. The dashed ellipsoids in (b) represent As_4 dimers. Reprinted with permission from Macmillan Publishers Ltd: [28], Copyright 2013.

Katayama *et al* [251] performed single-crystal x-ray diffraction and identified that the transition is characterized by the displacement of Ir2 at the non-coplanar sites along the *c*-axis, thereby resulting in doubled periodicity along the *c*-axis without breaking the *P*4/*n* symmetry. The structural phase transition suggests that either Ir charge or orbital degrees of freedom are active in the intermediary Ir_4As_8 layers. Sawada *et al*[252] performed ARPES measurements of $Ca_{10}Ir_4As_8(Fe_{2-x}Ir_xAs_2)_5$, thereby demonstrating that the Fe 3*d* electrons in the FeAs layers form hole-like and electron-like Fermi surfaces at the zone center and corners, respectively, as commonly observed in iron arsenide superconductors; Ir 5*d*electrons are metallic and glassy most probably due to atomic disorder related to the Ir 5*d*orbital instability.

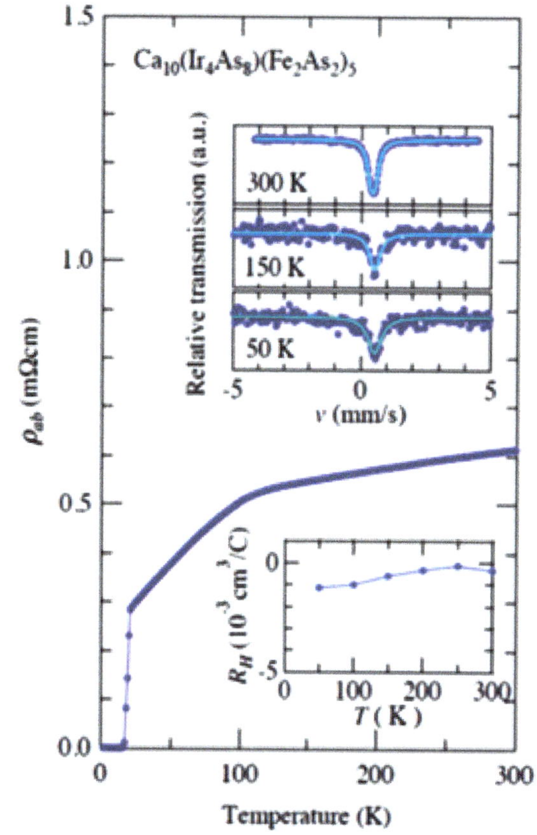

Figure 22. Temperature dependence of the electrical resistivity ρ_{ab} for $Ca_{10}Ir_4As_8(Fe_{2-x}Ir_xAs_2)_5$. The upper inset shows ^{57}Fe-Mössbauer spectra together with fitting curves. The lower inset shows the temperature dependence of the Hall coefficientR_H. Reprinted with permission from Macmillan Publishers Ltd: [28], Copyright 2013.

Another achievement of the FIRST Project is the growth of superconducting nanowhiskers of $Ca_{10}Pt_4As_8(Fe_{1.8}Pt_{0.2}As_2)_5$ by Li *et al* [253]. The typical whiskers obtained had a length of 0.1–2.0 mm, width of 0.4–5.0 μm and thickness of 0.2–1.0 μm. High-resolution transmission electron microscopy (TEM) images showed that the whiskers exhibited excellent crystallinity and that whisker growth occurred along the *a*-axis of the tetragonal (*P*4/*n*) structure. The whiskers exhibited superconductivity with T_c of 33 K, Hc_2 (*H*//*c*) of 52.8 T and J_c of 6.0 × 10^6 A cm^{-2} (at 26 K). Since cuprate high-T_c whiskers are fragile ceramics, the present intermetallic superconducting whiskers with high T_c are more suitable for device applications.

Intercalated FeSe Superconductors

The structurally simplest FeSe with space group *P4/nmm* has become a very attractive material. Though pure FeSe shows low T_c (8 K), it can be drastically enhanced by a factor of 5 under external high pressure [147, 148]. Moreover, the single-layer FeSe film epitaxially grown on $SrTiO_3$ substrates has a wide superconducting gap (~20 meV), which implies that T_c could be as high as 65 K in thin films despite the bulk T_c being ~40 K [149]. More recently, high T_c over 100 K has been reported in this type of single-layer FeSe [153]. Another high-T_c bulk SC derived from FeSe is intercalated $A_xFe_{2-y}Se_2$ with $ThCr_2Si_2$ structure (but classified to 245-type due to the exact composition) that is synthesized at a high temperature (~1300 K) but is only available for the large-sized monovalent metals such as *A* (K, Rb, Cs and Tl) [154–157, 254–256]. The bulk T_c of $AxFe_{2-y}Se_2$ is ~30 K and the average crystal structure is body-centered tetragonal (space group *I4/mmm*) [154]. However, the synthesis of FeSe intercalated by smaller alkali metals such as Li and Na has not succeeded using conventional high temperature processes.

The low-temperature method is another approach to intercalation, which is widely used and is suitable for intercalating alkali and alkaline earth metals, even those with small ionic radii. Many superconductors such as AxC_{60} and *AxMNX* (*A*: Li–K, Ca–Ba, Yb and Eu; *M*: Ti, Zr and Hf; and *X*: Cl, Br and I) were obtained through this method [55, 257]. Among them, the ammonothermal method, which uses liquid ammonia as a solvent and makes the starting materials react under high pressure in an autoclave, is a useful way to prepare the meta-stable and/or non-equilibrium materials. The relatively mild reaction keeps the host structure intact; therefore, the pure charge transfer without destroying the conductive layer is expected to favor the higher T_c.

Application of this method to intercalate *A* into FeSe was first carried out by a group from the Institute of Physics, China [258]. They obtained several superconductors with a higher T_c of 30–46 K (FeSe intercalated by Li, Na, K, Ca, Sr, Ba, Eu, and Yb) compared with the samples prepared by the high-temperature method. In this report, they demonstrated that the ammonothermal method was useful to synthesize the intercalated FeSe superconductors with relatively high T_c, while they did not mention remaining ammonia molecules or ions in their early stage of this research. Subsequently, they studied $KxFe_2Se_2$+NH_3 systematically and found two superconducting phases, $K_{0.3}Fe_2Se_2(NH_3)_{0.47}$ (T_c = 44 K) and $K_{0.6}Fe_2Se_2(NH_3)_{0.37}$ (T_c = 30 K), where they noted that the most important factor to control T_c was the potassium content [259]. Then, they prepared ammonia-free $K_{0.3}Fe_2Se_2$ and $K_{0.6}Fe_2Se_2$ by removing NH_3 at 200°C completely and showed the same T_c. The theoretical and/or empirical reason for higher T_c of $K_{0.3}Fe_2Se_2(NH_3)_{0.47}$ has not yet been reported.

In contrast, a group from the University of Oxford, UK, concentrated on the FeSe co-intercalated by Li and ammonia, and analyzed its precise crystal structure using the powder neutron diffraction [260, 261]. They synthesized $Li_{0.6}(ND_3)_{0.8}(ND_2)_{0.2}Fe_2Se_2$ (T_c = 43 K) by the ammonothermal method and showed its crystal structure as in figure 23. This crystal structure indicates that the resulting compound intercalates as not only the ammonia molecule but also the amide anion. Furthermore they found reversible adsorption and desorption of ammonia by controlling pressure of ammonia at <−10 °C and obtained $Li_{0.6}(ND_{2.7})_{1.7}Fe_2Se_2$ (ammonia-rich phase) with T_c = 39 K by exposing $Li_{0.6}(ND_3)_{0.8}(ND_2)_{0.2}Fe_2Se_2$ (ammonia-poor phase) to 1 atm of ammonia (ND_3) at −10 °C (see figure 24). By intercalating more ammonia, the spacing between Fe layers (*d*) increased from 8.26 Å of the poor phase to 10.59 Å of the rich phase. Though the T_c of the ammonia-poor phase is higher than that of the ammonia-rich phase, the lithium content is the same in both phases. This result looks to be inconsistent with Ying's result [259] who noted that the T_c depends on the intercalated alkali metal content. While the difference of T_c between the poor and the rich phase is not so large (4 K) compared with the case of potassium (14 K), the effect of intercalated amide anion is also unclear.

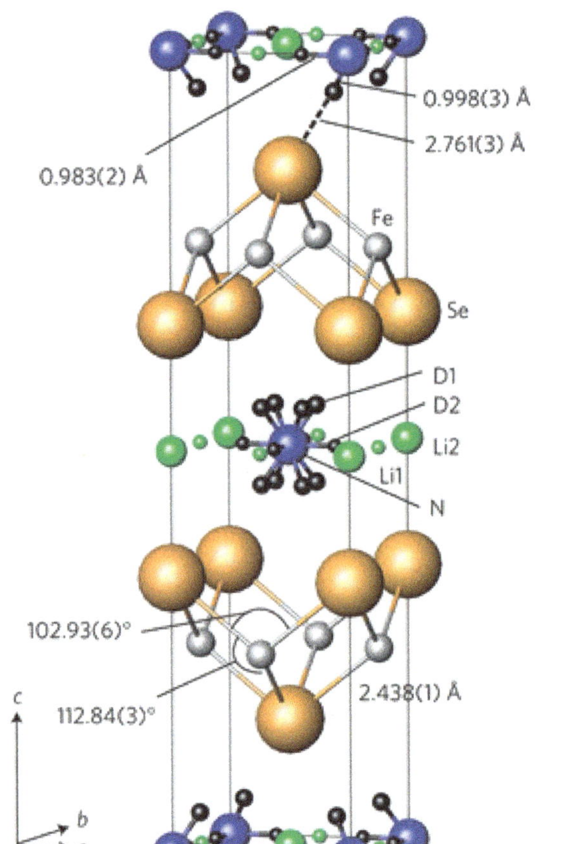

Figure 23. The 298 K crystal structure of $Li_{0.6(1)}(ND_2)_{0.2(1)}(ND_3)_{0.8(1)}Fe_2Se_2$. Refinement was against neutron powder diffraction data (GEM instrument). In the model each square prism of Se atoms contains either an $[ND_2]$ anion or an ND_3 molecule and these are both modeled as disordered over four orientations. The sizes of the spheres representing the Li atoms are in proportion to their site occupancies. Reprinted with permission from Macmillan Publishers Ltd: [260], Copyright 2013.

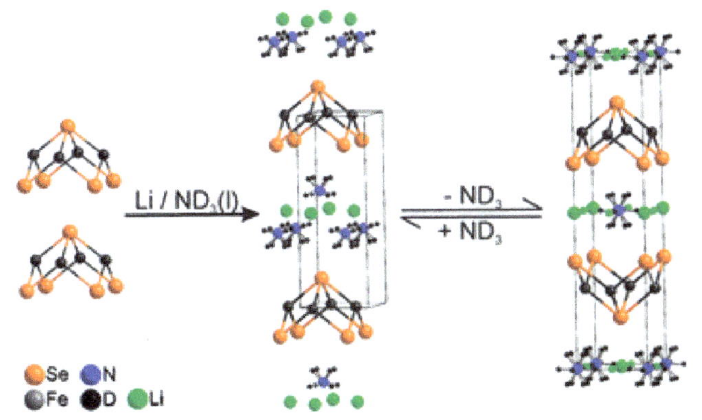

Figure 24. Schematic of the intercalation of lithium and ammonia into FeSe. Reprinted with permission from [261]. Copyright 2014 by the American Chemical Society.

In our research project, the group from Tokyo Institute Technology, Japan, synthesized three samples intercalated by sodium and ammonia as ammonia-free (phase I), ammonia-poor (phase II) and ammonia-rich phases (phase III) using the ammonothermal method and showed superconductivity with T_c = 37, 45 and 42 K, respectively (see figure 25) [29]. They prepared these phases by changing the Na/NH_3 ratio (0.03 to 0.3 mol^{-1}). For phase I, the reaction vessel was evacuated to ~10^{-2} Pa after immersion for 3 h at 223–243 K. The chemical compositions measured by EPMA were $Na_{0.65}Fe_{1.93}Se_2$ for phase I and $Na_{0.80}(NH_3)_{0.60}Fe_{1.86}Se_2$ for phase II. Phase III was so unstable (decomposed easily even at 250 K) that the composition could not be defined. From powder x-ray diffraction measurements, the spacing between Fe layers (*d*) was 6.8, 8.7 and 11.1 Å for phases I, II and III, respectively. The lower T_c and the wider *d* of the ammonia-rich phase (phase III) than those of ammonia-poor phase (phase II) are similar to the case of Li.

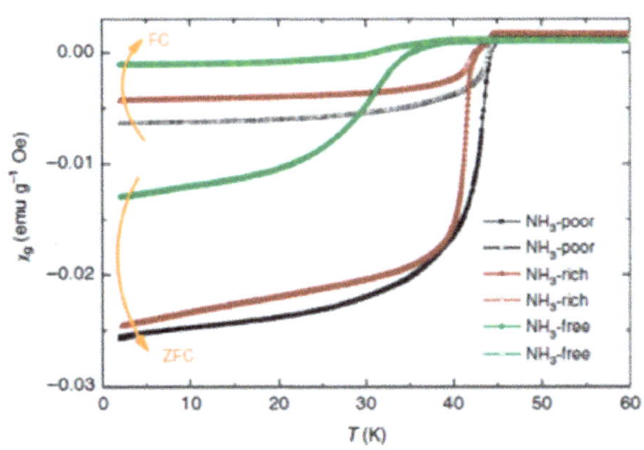

Figure 25. The magnetization curves of three Na/NH_3 intercalated FeSe measured with the zero-field-cooling (ZFC) and field-cooling (FC) modes at H = 10 Oe. Reprinted with permission from Macmillan Publishers Ltd: [29], Copyright 2014.

These results are summarized in table 11. It has been considered that the *d* value could be a possible guideline to explore the high T_c superconductor. In IBSCs, the *d* of FeSe (T_c = 8 K), LiFeAs (T_c = 18 K), $BaFe_2As_2$ (T_c = 38 K), and SmFeAsO (T_c = 56 K) are 5.5, 6.4, 6.5 and 8.7 Å, respectively, which looks similar to the general rule for cuprate superconductors. Expecting that the material with wider *d* could reveal higher T_c, Ogino *et al* [141–143] prepared IBSCs with a thick blocking layer with *d* = 15.5 to 24.5 Å. Unfortunately, the maximum T_c in their attempt was 43 K for $Ca_6(Sc,Ti)_5O_yFe_2As_2$ (*d* = 24.5 Å) (see section 3.1.1). In this case of these intercalates, the optimal *d* is 7.5–8.8 Å of the ammonia-poor phase. It looks as though *d* is not a critical factor to control T_c widely.

Table 11. Intercalated FeSe superconducting phases.

	NH_3-rich		NH_3-poor				NH_3-free	
			Metal-poor		Metal-rich			
	d	T_c	*d*	T_c	*d*	T_c	*d*	T_c
Li^+	~9.0	39	—	—	~8.3	44	—	—
Na^+	~11.1	42	—	—	~8.7	45	~6.8	37
K^+	~10.2	?	~7.8	44	~7.4	30	7.14	~44
AE^{2+}	—	—	8.0–8.4	35–40	~10.3	38–39	—	—
RE^{2+}	—	—	~8.1	42	~10.2	40–42	—	—

’–’: none, ‘?’: unknown.The separation of nearest Fe layers d (Å) and T_c (K) of intercalated (A/AE/RE)x(NH3)y(NH2) zFe2Se2 (A: alkali metals, AE: alkaline earth metals, RE: rare earth metals) superconductors synthesized by the ammonothermal method shown in the literature [29, 258–261].

An important feature of these intercalates is none or very small deficiency of Fe compared to those synthesized by the conventional high temperature method. In the case of conventionally prepared (Tl, K)Fe_xSe_2, superconductivity emerges over x = 1.7 and the compound with smaller Fe content is an antiferromagnetic insulator [156]. The maximum Fe content of the compound prepared by the high temperature method is reported to be x ~ 1.9 and the maximum T_c is 31 K. Zhang *et al* reported that the compound with excess Fe ($K_{0.87}Fe_{2.19}Se_2$) which was prepared by the Bridgman method using $Fe_{1+x}Se$ as a starting material revealed a sharp drop of resistivity at 44 K, but zero resistivity was observed at a lower temperature (25 K) and the shielding volume fraction at >25 K was very small (<1%) [262]. These results suggest that a small deficiency of Fe is favorable to achieve high T_c, which is due to the large amount of indirect electron doping efficiently. The ammonothermal method is profitable to prepare such a condition.

Superconductivity in Layered Titanium Compounds

Since the discovery of high-T_c superconductivity in a layered perovskite oxide $(La,Ba)_2CuO_4$ with a CuO_2 square lattice [263], numerous efforts have been dedicated to obtaining new superconducting families. Although this led to a series of discoveries of other high-T_c families such as MgB_2 [196], fullerenes [264] and iron pnictides [4], the transition temperatures never exceed liquid nitrogen temperature. In addition, the mechanism of high T_c superconductivity in copper oxides remains unsolved and is still under debate despite intensive investigations.

Searching parent structures having a d^1 square lattice may be a plausible strategy to access novel superconductivity, which is hole–electron symmetric with the d^9 square lattice. Sr_2VO_4 with V^{4+} (d^1), which is isostructural with La_2CuO_4 (figure 26(a)), is a promising candidate material though all attempts to inject carriers were so far unsuccessful [265, 266]. When the ligand field splitting energy (LFSE) for an octahedral environment is taken into account, however, one notices that the electronic structures of Sr_2VO_4 and La_2CuO_4 are quite different from each other. In the d^9 case, one unpaired electron occupies one of the anti-bonding e_g orbitals, $x^2 - y^2$, which is widely separated in energy from the z^2 orbital due to the Jahn–Teller effect and gives a half-filled electronic configuration. In the d^1 case, one electron occupies the non-bonding $t_{2g}(xy, yz, zx)$ orbitals. A certain octahedral distortion may be present, but is not enough to lift the orbital degeneracy significantly, so that the system should be better approximated by a 1/6-filled configuration.

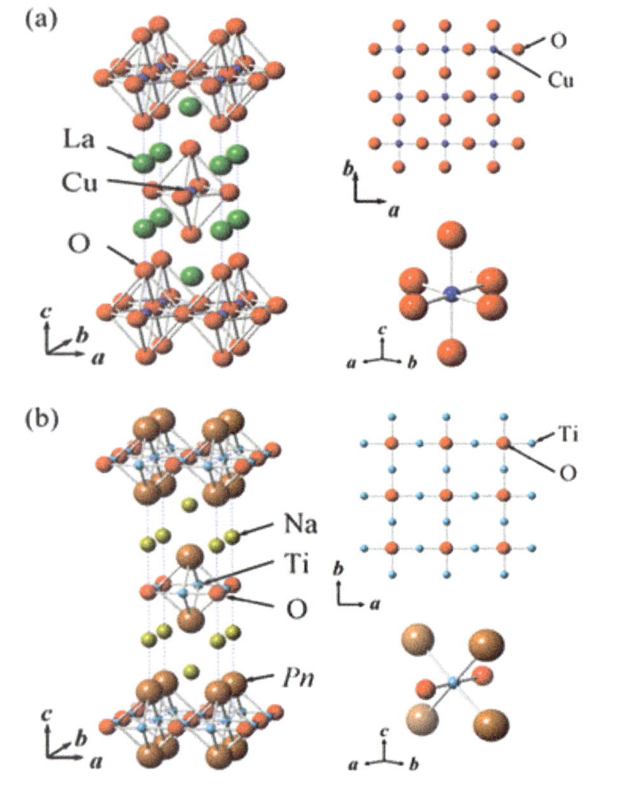

Figure 26. (a) Crystal structure of La_2CuO_4, CuO_2 sheets (top) and CuO_6 octahedron (bottom). (b) Crystal structure of $Na_2Ti_2Pn_2O$, Ti_2O sheets (top) and TiO_2Pn_4octahedron (bottom).

In order to tune the t_{2g} orbital levels more drastically (than possible in oxides), we looked for mixed anionic compounds, where a d^1 metal is coordinated octahedrally by two kinds of anion species. Due to the large difference between anions in terms of valence, electronegativity and ionic radius, a mixed-anion coordination geometry might provide a unique opportunity to splitt_{2g} orbitals to a greater extent. Such a situation is realized in layered titanium oxypnictides $Na_2Ti_2Pn_2O$ (Pn = As, Sb), the structure of which is illustrated in figure 26(b) [267]. The $Na_2Ti_2Pn_2O$ structure appears to be similar to that of La_2CuO_4, but contains an inverse Ti_2O square lattice. The trivalent titanium ion (d^1) has an octahedral TiO_2Sb_4 coordination, and the TiO_2Sb_4 octahedra share edges to form a two-dimensional network. Like cuprate superconductors, $[TiO_2Sb_2]^{2-}$ layers can be sandwiched by various block layers. For example, $BaTi_2As_2O$ with a Ba^{2+} layer, $(SrF)_2Ti_2As_2O$ with a $[(SrF)_2]^{2+}$ layer, and $(Ba_2Fe_2As_2)Ti_2As_2O$ with a $[Ba_2Fe_2As_2]^{2+}$ layer are reported (figure 27) [268–271].

Unlike cuprate superconductors, known 'parent' compounds in the layered titanium oxypnictides are already metallic without carrier doping [272, 273]. Instead of superconductivity, they show anomalies in magnetic susceptibility and electric resistivity likely ascribed to charge density wave (CDW) and/or spin density wave (SDW) transitions. The density wave transition temperatures T_{DW} are 330 K for $Na_2Ti_2As_2O$, 120 K for $Na_2Ti_2Sb_2O$ and 200 K for $BaTi_2As_2O$.

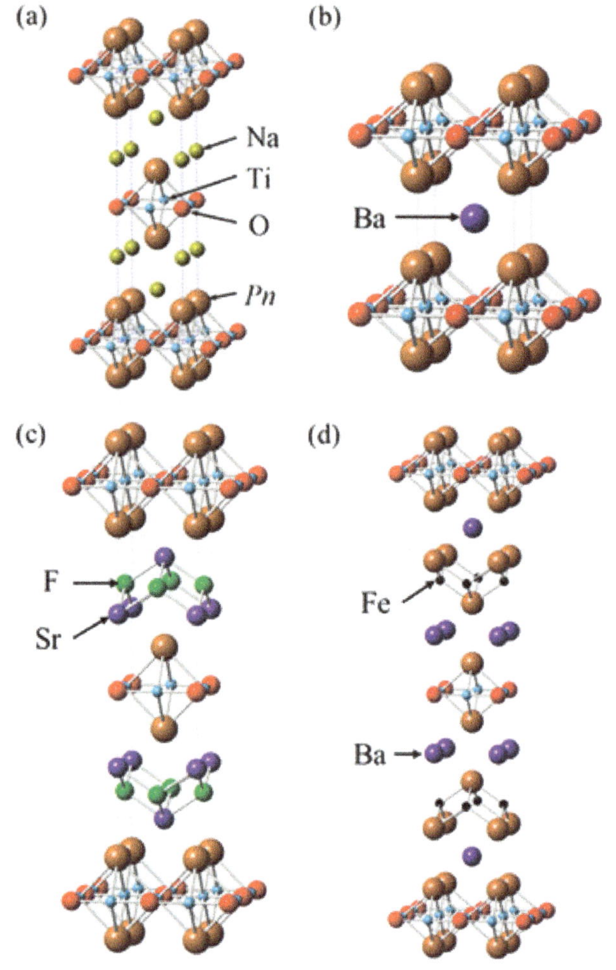

Figure 27. Crystal structures of (a) $Na_2Ti_2Pn_2O$, (b) $BaTi_2Pn_2O$, (c) $(SrF)_2Ti_2Pn_2O$, and (d) $(Ba_2Fe_2As_2)Ti_2As_2O$.

It is hence expected that the suppression of density wave phase is a key to induce superconductivity. In 2012, we prepared a new compound $BaTi_2Sb_2O$ and observed a bulk superconductivity transition at 1 K [31]. Doan *et al* independently showed the enhancedT_c of 5.5 K in Na-doped $BaTi_2Sb_2O$ [274]. These reports have sparked a lot of investigation into superconductivity in layered titanium oxypnictides. In this section, we demonstrate the present status of our understanding of this new superconducting family from both experimental [31–33, 275–284] and theoretical [285–289] viewpoints.

Superconductivity in $BaTi_2Sb_2O$

A pure phase of $BaTi_2Sb_2O$ was synthesized by the conventional solid-state reaction method using BaO (99.99%), Ti (99.9%) and Sb (99.9%) in stoichiometric quantity [31]. A pellet specimen was wrapped in tantalum foil, sealed in a quartz tube, and typically heated at 1000 °C for 40 h, followed by controlled cooling at a rate of 50 °C h^{-1} to room temperature. The product is air and moisture sensitive. $BaTi_2Sb_2O$ is tetragonal with lattice constants of a = 4.11039(2) Å,c = 8.08640(4) Å at room temperature. $BaTi_2As_2O$ has larger cell parameters, a = 4.046 Å,c = 7.272 Å. Figure 28(a) shows the result of the synchrotron x-ray diffraction refinement for $BaTi_2Sb_2O$ with the space group of *P4/mmm*. Magnetic susceptibility and resistivity for $BaTi_2Sb_2O$ (figures 29(a) and 4(b)) show a distinct anomaly at around T_{DW} = 50 K, which should be related to a density wave transition. Upon further cooling, the ρ–T curve showed zero resistivity, indicating a superconducting transition. A large diamagnetic signal associated with the shielding effect was observed at T_c = 1 K. The shielding volume fraction is as large as 58%, providing firm evidence for bulk superconductivity in $BaTi_2Sb_2O$. Heat capacity measurements revealed a distinct peak at around T_c, further supporting the bulk superconductivity.

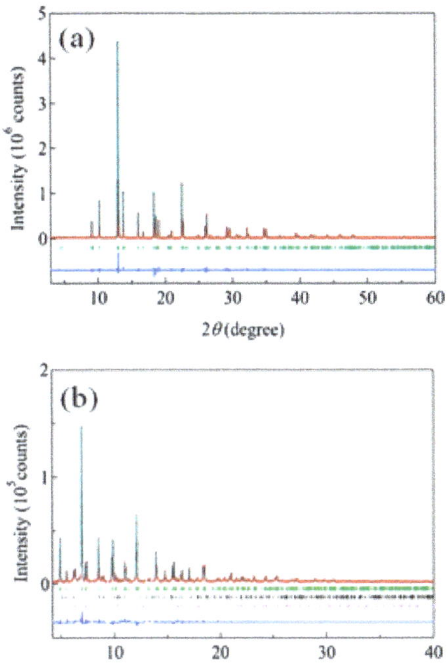

Figure 28. Refined synchrotron x-ray patterns of (a) $BaTi_2Sb_2O$ [31], and (b) $BaTi_2Bi_2O$ [33]. Reprinted with permission. Copyright 2012 for [31] and 2013 for [33] by the Physical Society of Japan.

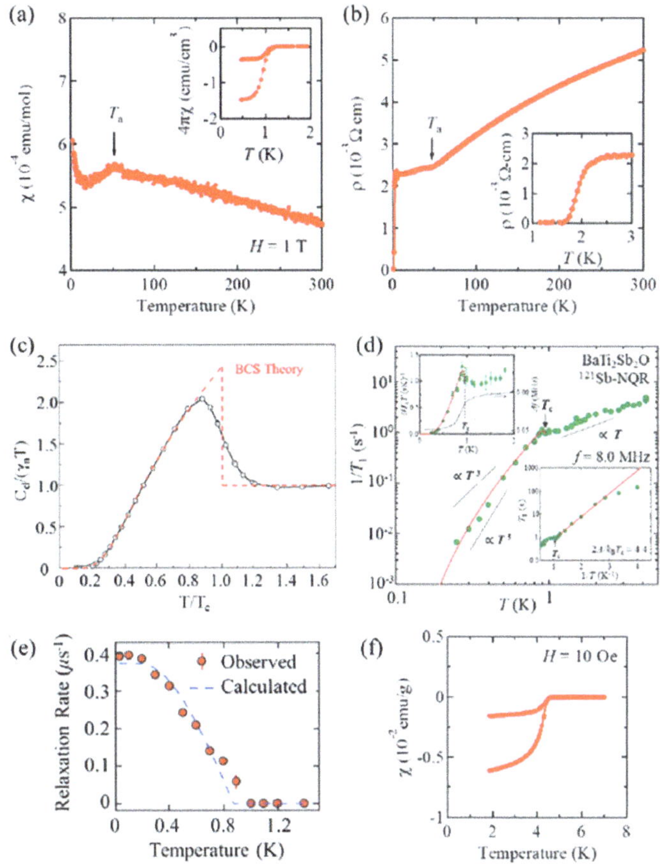

Figure 29. (a) Magnetic susceptibility and (b) electric resistivity of $BaTi_2Sb_2O$ (taken from [31]). (c) Specific heat capacity of $Ba_{0.85}Na_{0.15}Ti_2Sb_2O$ (taken from [278]). (d) ^{121}Sb-NQR result: spin-lattice relaxation rate of $BaTi_2Sb_2O$, taken from [275]. (e) μSR result: superconducting relaxation rate of $BaTi_2Sb_2O$. (f) Magnetic susceptibility of $BaTi_2Bi_2O$ (taken from [33]). Reprinted with permission. Copyright 2012 for [31] and 2013 for [33] by the Physical Society of Japan. Copyright 2013 for [275] and [278] by the American Physical Society.

Nature of Superconductivity in $BaTi_2Sb_2O$

It has been revealed from various experiments on $BaTi_2Sb_2O$ that the superconducting state is classified to a fully gapped *s*-wave state. The analysis of the specific heat at T_c gives $\Delta C(T_c)/\gamma T_c \sim 1.36$, which is consistent with the BCS (Bardeen–Cooper–Schrieffer) weak coupling limit value of 1.43 [31]. Gooch *et al* demonstrated that the temperature variation of the electronic heat capacity on $Ba_{1-x}Na_xTi_2Sb_2O$ (x = 0 and 0.15) is well described by a weak coupling BCS function with $2\Delta/k_BT_c = 2.9$, as shown in figure 29(c) [278]. Kitagawa *et al* reported that the temperature dependence of inverse spin–lattice relaxation time ($1/T_1$) in ^{121}Sb nuclear quadrupole resonance (NQR) shows a coherence peak just below T_c and decreases exponentially at low temperatures. From the slope of the plot, the magnitude of the superconducting gap is estimated to be $2\Delta/k_BT_c = 4.4$ (figure 29(d)) [275]. A superconductivity-induced muon relaxation rate σ_{sc}, which is proportional to the penetration depth as λ^{-2}, shows robust bulk superconductivity below $T_c \sim 1$ K (figure 29(e)) [277]. A fit to a BCS *s*-wave model in the weak coupling limit gave $T_c = 0.87 \pm 0.03$ K and $\sigma_{sc}(T = 0) = 0.37 \pm 0.01\ \mu s^{-1}$. Strong electron correlation in the present material is suggested from the Wilson ratio of $R_W = 2.21$, which is much larger than $R_W = 1$ for the free-electron approximation [31].

Density Wave State

While the superconducting state is described well by the weak coupling BCS scheme, the nature of the pairing mechanism, namely, whether the superconductivity in $BaTi_2Sb_2O$ is driven by the electron–phonon coupling or spin-fluctuation, remained unclear until recently. Theoretically, for $BaTi_2Sb_2O$, a magnetic instability associated with Fermi surface nesting, leading to a SDW state with a propagation vector of (π, 0) (figure 30(a)), was theoretically proposed by D Singh [290]. A sign-changing *s*-wave state within a scenario of spin-fluctuation mediated superconductivity was also suggested. The same type of bi-collinear antiferromagnetic state was shown for $Na_2Ti_2Sb_2O$ by X W Yan *et al*, while $Na_2Ti_2As_2O$ was suggested to have a blocked checkerboard antiferromagnetic state with a 2 × 2 magnetic unit cell (figure 30(b)) [291]. On the other hand, first-principles calculations of the phonon dispersions and electron–phonon coupling for $BaTi_2Sb_2O$ by Subedi [292] revealed lattice instability near the zone corners, which leads to a charge-density wave phase with a $\sqrt{2} \times \sqrt{2} \times 1$ superstructure arising from a coherent distortion corresponding to elongation or compression of the Ti squares without an enclosed O, such that the Ti squares with O rotate either clockwise or counterclockwise as shown in figure30(c).

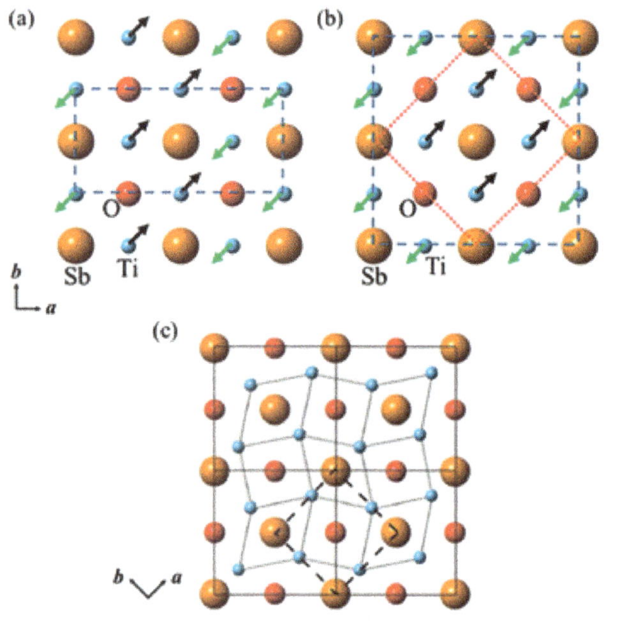

Figure 30. Theoretically proposed spin/charge density wave models for $BaTi_2Sb_2O$ or $BaTi_2As_2O$. (a) A bicolinear type and (b) blocked checkerboard type SDW model [290,291] (taken from [291]). (c)√2 × √2 × 1 superstructure as a result of CDW transition (taken from [292]). Reprinted with permission. Copyright 2013 for [291] by IOP Publishing and 2013 for [292] by the American Physical Society.

As shown in figure 31(b), zero-field muon spin relaxation measurements showed no significant increase in relaxation rate at the density wave ordering temperature, indicating that the density wave is of the charge rather than spin type [277]. $^{121/123}$Sb-NQR measurements revealed that the in-plane four-fold symmetry is broken at the Sb-site below $T_{DW} \sim 40$ K, without an internal field appearing at the Sb site, indicating a commensurate CDW ordering [275]. However, the absence of any superstructure peaks in high-resolution electron and neutron diffraction belowT_{DW} (figure 31(a)) signifies that the charge density wave does not involve modulation of atomic arrangement, implying a nontrivial nature of the CDW state. Recent in-depth structural studies have further suggested that $BaTi_2As_2O$ forms a symmetry-breaking nematic ground state that can be naturally explained as an intra-unit-cell nematic charge order with *d*-wave symmetry, pointing to the ubiquity of the phenomenon [293]. These findings, together with the key structural features in these materials being intermediate between the cuprate and iron-pnictide high-temperature superconducting materials, render the titanium oxypnictides an important new material system to understand the nature of nematic order and its relationship to superconductivity.

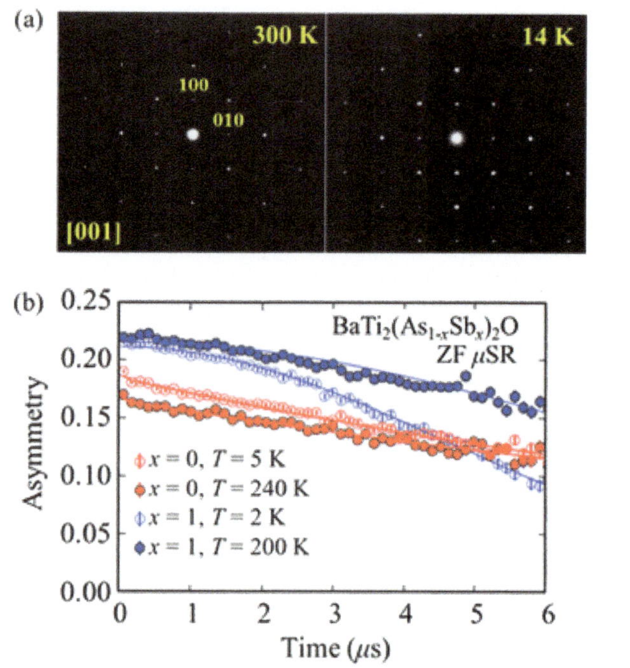

Figure 31. (a) Selected area electron diffraction patterns taken along the [001] axis for $BaTi_2Sb_2O$ at 300 K and 14 K. (b) Asymmetry versus time spectra from zero-field muon spin relaxation measurements on $BaTi_2As_2O$ and $BaTi_2Sb_2O$. Reprinted with permission from [277]. Copyright 2013 by the American Chemical Society.

Aliovalent cation/Anion Substitution

Since the 'parent' phase $BaTi_2Sb_2O$ is a 1 K superconductor, coexisting with the CDW state with$T_{DW} \sim 50$ K, it can be expected that the superconducting transition temperature can be enhanced by destabilizing the CDW phase. It was shown by Doan *et al* that the divalent Ba site in $BaTi_2Sb_2O$ is substitutable by monovalent Na ions up to 33% [274]. The T_c in $Ba_{1-}x$Na$x$$Ti_2Sb_2O$ gradually increases with increasing x (hole concentration) and attained a maximum T_c of 5.5 K at $x = 0.33$. Although the CDW state is gradually destabilized by the Na substitution, it still persists (e.g., $T_{DW} = 30$ K at $x = 0.25$): T_{DW} forms a downward concave curve in the region of a higher x, showing a saturation tendency. The alkaline-earth metal substitution by K [279] and Rb [280] is also effective in raising T_c. The K substitution ($Ba_{0.88}K_{0.12}Ti_2Sb_2O$) provides the highest T_c of 6.1 K, which is due to smaller chemical disorder because of the similarity in ionic radius between K^+ and Ba^{2+}.

It is also possible to control physical properties by aliovalent anion Sb substitution. We have prepared $BaTi_2(Sb_{1-}xSnx)_2O$ for $x \leq 2.5$ and obtained the electronic phase diagram as a function of x as shown in figure 32(b) [33]. A qualitative resemblance is seen between phase diagrams of hole doped systems, $BaTi_2(Sb_{1-}xSnx)_2O$ and $(Ba_{1-}xAx)Ti_2Sb_2O$ (A = alkali metals). The saturated behavior of T_c in the highly doped regime may be related to the robustness of the CDW phase. From quantitative point of view, however, $BaT_2(Sb_{1-}xSnx)_2O$ differs from $(Ba_{1-}xNax)Ti_2Sb_2O$. The maximum T_c in $BaTi_2(Sb_{1-}xSnx)_2O$ is 2.5 K at $x = 0.3$, which is remarkably smaller than $T_c = 5.5$ K in $(Ba_{1-}xNax)Ti_2Sb_2O$ at $x \sim 0.33$. The reduced T_c in the former system is attributed to the greater chemical disorder induced by the *Pn*-site substitution and also to the less pnictogen in the Sb/Sn atoms in comparison with Ba/Na atoms.

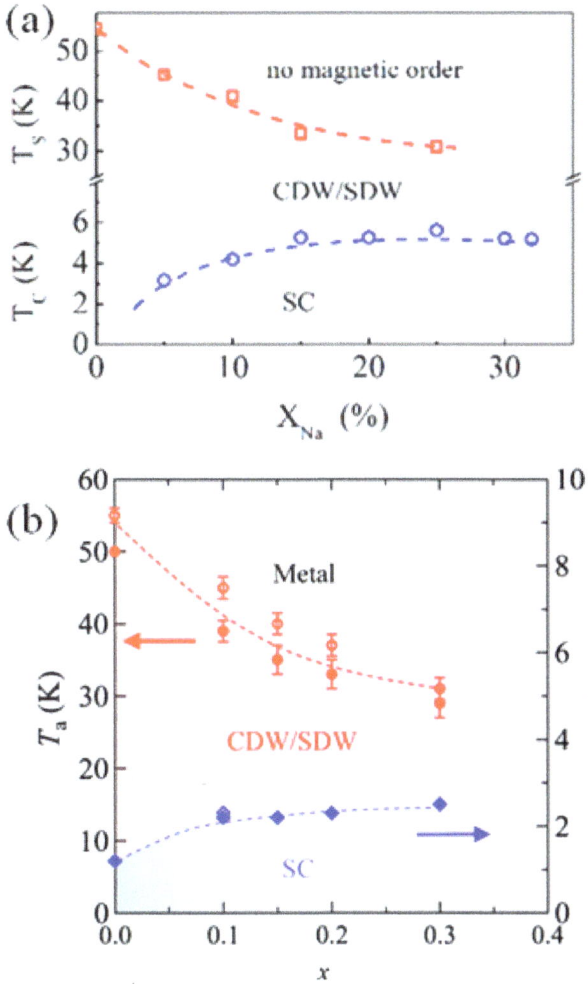

Figure 32. Phase diagram of (a) $Ba_{1-x}Na_xTi_2Sb_2O$ (taken from [274]) and (b) $BaTi_2(Sb_{1-x}Sn_x)_2O$ (taken from [33]). Reprinted with permission. Copyright 2014 for [33] by the Physical Society of Japan and 2012 for [274] by the American Chemical Society.

Isovalent Anion Substitution

The comparison between $BaTi_2As_2O$ (T_{DW} = 200 K, T_c = 0 K) and $BaTi_2Sb_2O$ (T_{DW} = 50 K, T_c = 1 K) led us to prepare the isovalent anion substitution system $BaTi_2Bi_2O$ so as to destabilize the CDW state and stabilize the superconducting state. It is shown from the structural refinement (figure 28(b)), $BaTi_2Bi_2O$, the first oxybismuthide in this family, is isostructural with $BaTi_2Sb_2O$ [31]. The magnetic susceptibility and electric resistivity indeed showed no anomaly indicative of CDW transition, meaning that hole doping is less effective in terms of destabilizing the CDW state than the isovalent anionic substitution. As a result, the enhanced T_c of 4.6 K is successfully observed in $BaTi_2Bi_2O$. The lattice parameters of $BaTi_2Bi_2O$ (a = 4.12316(4) Å, c = 8.3447(1) Å) are not so different from those of $BaTi_2Sb_2O$, indicating that the lattice expansion is not the primary factor to destabilize the CDW state. Compared with Sb, Bi is less electronegative and provides more covalent Ti–*Pn* bonding. Furthermore, the Bi 6*p*orbital is more diffuse than the Sb 5*p* orbital. Therefore, in $BaTi_2Bi_2O$ the Bi 6*p* orbital possibly contributes more to the density of states at the Fermi surface than the Sb 5*p* orbital in $BaTi_2Sb_2O$. It results in weaker nesting of the Fermi surface for $BaTi_2Bi_2O$ and suppressed CDW instability. The possession of three compounds $BaTi_2Pn_2O$ (*Pn*=As, Sb, Bi) allows us to prepare isovalent anionic solid solutions, $BaTi_2(As_{1-x}Sb_x)_2O$ ($0 \leq x \leq 1$) and $BaTi_2(Sb_{1-y}Bi_y)_2O$ ($0 \leq y \leq 1$) [32]. Despite the Vegard law behavior of the unit cell parameters in both solid solutions, it was unexpectedly observed a novel electronic phase diagram as summarized in figure 33(b), in marked contrast to that of the aliovalent substitution system (figure 32(a)). The gradual destabilization of the CDW state is seen going from x = 0 to x = 1 in $BaTi_2(As_{1-x}Sb_x)_2$, whereas superconductivity appears at T_c = 0.5 K for x = 0.9. The T_c increases with increasing y (Bi), takes a maximum value of 3.5 K for y = 0.2 and further Bi substitution lead to a decrease in T_c. For$y$ = 0.4 and 0.5, no superconductivity is seen down to the lowest temperature measured (1.85 K).

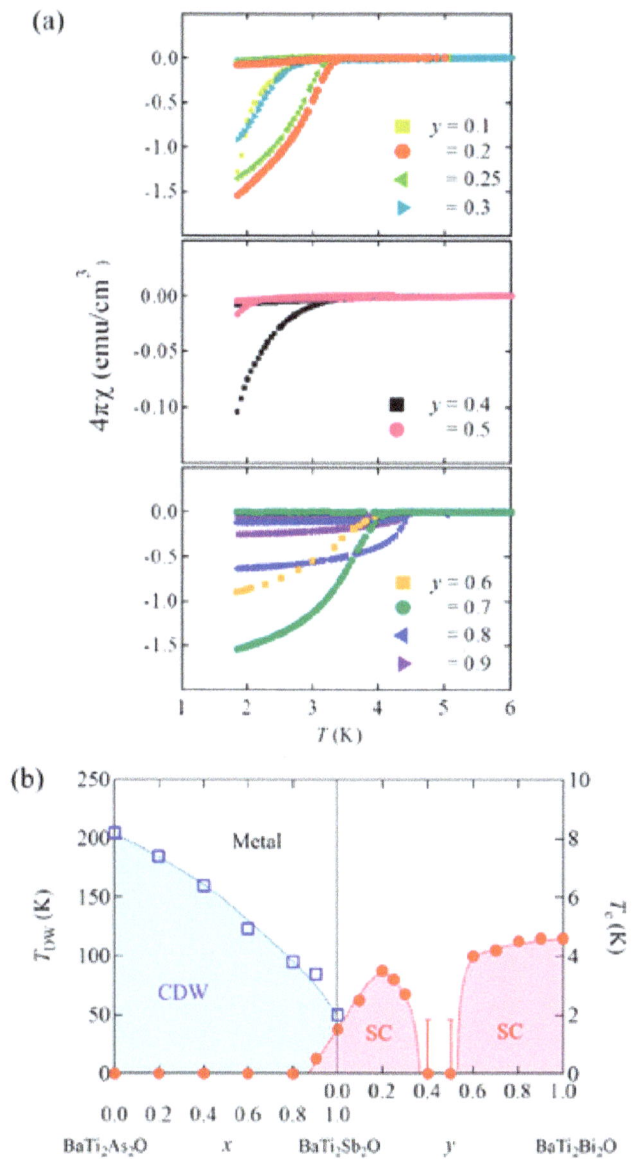

Figure 33. (a) Low-temperature magnetic susceptibility for $BaTi_2(Sb_{1-y}Bi_y)_2O$. (b) The electronic phase diagram of $BaTi_2(As_{1-x}Sb_x)_2O$ and $BaTi_2(Sb_{1-y}Bi_y)_2O$, demonstrating the presence of two superconducting phases. Reprinted with permission from [32]. Copyright 2013 by the Physical Society of Japan.

Most remarkably, superconductivity revives at y = 0.6, with an increased T_c toward y = 1 (4.6 K). The presence of the second superconducting state for $0.6 \leq y \leq 1$ is quite unusual and implies a novel mechanism behind the superconductivity. It is theoretically shown that a multiband structure involving three Ti 3*d* orbitals (*dxy*, dz_2, and dx_2y_2) is present at the Fermi surface

[290,294]. It is thus likely that the second superconducting phase is different in nature from the first superconducting phase. Recently, a two-dome structure in T_c has also been reported in the iron pnictide superconductor $LaFeAs(O_{1-}xHx)$ [11], where the multiband structure comprising of Fe $3d$ orbitals is proposed. It should be noted that the two-dome structure in $LaFeAs(O_{1-}xHx)$ is induced by the aliovalent substitution and the two superconducting regions are not separated. Hence, $BaTi_2Pn_2O$ might give a better opportunity to understand the nature of multiband Fermi surface.

Interlayer Interactions

In $BaTi_2Pn_2O$, the CDW state becomes destabilized as Pn = As (T_{DW} = 200 K) → Sb (T_{DW} = 50 K) → Bi (T_{DW} = 0 K). The magnetic susceptibility and electric resistivity of $(SrF)_2Ti_2Bi_2O$ (figure 29(f)) exhibit no signature of CDW transition [76]. However, zero-resistivity and diamagnetic signal are absent in $(SrF)_2Ti_2Bi_2O$ in spite of the fact that the in-plane lattice parameters are similar to each other. Since both $(SrF)_2Ti_2Bi_2O$ and $BaTi_2Bi_2O$ contain the $[Ti_2Bi_2O]^{2-}$ unit, it is possibly essential to consider the role of the A^{2+} unit, or the interlayer coupling in order to explain the difference in their superconducting properties. The $(SrF)_2^{2+}$unit is bulky with metal halide double layers, which provides an elongated interlayer distance of 10.685 Å, as compared with 8.345 Å in $BaTi_2Bi_2O$. Hence, the interlayer interaction, at least to a certain extent, is a key component for the appearance of superconductivity. Indeed, the application of external pressure on $Ba_{1-}xNaxTi_2Sb_2O$ results in the increase in T_c [286].

Mixed Anion Compounds

Most functional materials explored by us are single-anion compounds such as oxides, sulfides, chlorides, and bromides. Hence, exploring materials with mixed anion configurations around a transition metal would be promising toward realizing new or improved functional properties including high T_c superconductivity. For example, oxyhydride cubic perovskite $BaTiO_{3-}xHx(x \sim 0.6)$ with $TiO_{6-}nHn$ (n = 0, 1, 2) octahedral coordination (figure 34(a)), obtained by a topochemical reducing reaction using CaH_2 [268], shows a novel hydride exchangeability with hydrogen gas at moderate temperature [110]. Electron doping to Ti $3d$ t_{2g} band by hydride reduction makes $AETiO_{3-}xHx$ (AE = alkaline earth metal) metallic, with high conductivity of 10^2–10^4 S cm^{-1}, although superconductivity is absent [95]. $SrCrO_2H$ with CrO_4H_2 octahedra, directly prepared from high temperature and high pressure reaction, also adopts the cubic perovskite structure and exhibits the highest magnetic transition temperature among chromium oxides [96]. An oxychloride layered perovskite $(CuCl)LaNb_2O_7$ with Cu^{2+} (S = 1/2) in CuO_2Cl_4octahedral coordination (figure 34(b)), prepared by a topochemical ion-exchange reaction of $RbLaNb_2O_7$ [295], shows spin–liquid behavior with a finite gap in the excitation spectrum, due to quantum fluctuations enhanced by two-dimensional structure [296]. Ba_2BiSb_2, a hypervalent compound with a unique 'square-honeycomb' lattice (figure 34(c)) exhibits a CDW transition [54]. Isovalent anion substitution by Bi (i.e., $Ba_2Bi(Sb_{1-x}Bix)_2$ leads to the destabilization of the CDW phase and superconductivity with the maximum T_c of 4.4 K for Ba_2Bi_3 (x = 1) as shown in figure 34(d).

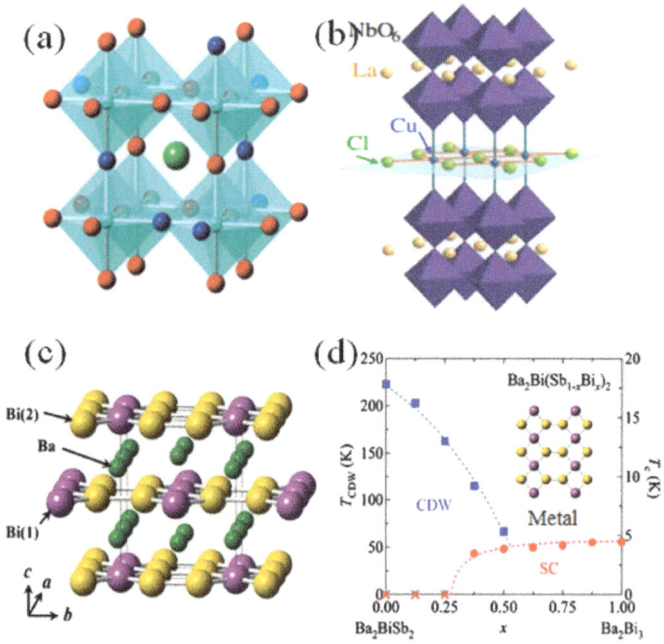

Figure 34. Crystal structure of (a) $BaTi(O,H)_3$, (b) $(CuCl)LaNbO_7$ and (c) $BaBiSb_2$. (d) The phase diagram of $Ba_2Bi(Sb_{1-}xBix)_2$. Reprinted with permission from [54]. Copyright 2014 by the Physical Society of Japan.

Conclusions

We have demonstrated recent studies on layered titanium compounds ATi_2Pn_2O (Pn = As, Sb, Bi) with Ti^{3+} in a square lattice, where Ti is coordinately octahedrally with four Pn and two O atoms. A newly synthesized $BaTi_2Sb_2O$ shows superconductivity at T_c = 1.2 K. The superconducting phase is of fully gapped BCS-type and is competing/coexisting with a CDW phase with T_{DW} of 1.2 K. The aliovalent cation/anion substitution leads to the stabilization of the superconducting state while the CDW state is destabilized only partially. The isovalent anion solid solution, $BaTi_2(As_{1-}xSbx)_2O$, shows a conventional phase diagram with the superconducting phase competing with the CDW phase. However, another isovalent anion solid solution, $BaTi_2(Sb_{1-}yBiy)_2O$, reveals the appearance of a second superconducting state for $0.6 < y$. The presence of the two superconducting phases strongly indicates the multi-orbital contribution to superconductivity, as also seen in iron arsenic superconductors. The nature of the second superconducting phase is not clear. This phase is possibly competing with other (presently unseen) phases. Further experiments as well as theories are necessary. It is important to note that superconductivity is only observed for A = Ba. Further exploratory studies may find superconducting materials with different blocking layers. Finally, we believe that mixed anionic materials are a fruitful playground for novel functional properties including high $T_c$superconductivity.

Intercalation Compounds with Layered and Cage-Like Structures

High-T_c superconductors recently developed have layered or cage-like structures, and include cuprates [263], iron pnictides [4], MgB_2 [196], alkali metal doped fullerides [297], and so on. They have intercalated structures composed of charged (doped) layers or frameworks coupled with charge reservoirs in the interlayer or cage-like space. In this study, we will also explore new superconductors, focusing on layered and cage-like structures with covalent networks, and doped electrons via intercalation. Layer structured metal nitride halides, and alloys with clathrate related structures have been developed; electrons are doped by means of intercalation using the interstices between the layers and the cages.

Intercalation Compounds of metal Nitride Halides

There are two kinds of layered polymorphs in metal nitride halides *MNX* (*M* = Ti, Zr, Hf;*X* = Cl, Br, I), α- and β-forms with the FOCl and the SmSI structures, respectively [298, 299]. The α-form layered polymorph has an orthogonal *M*N layer network separated by halogen layers as shown in figure 35(a). The β-form consists of double honeycomb-like *M*N layers sandwiched between close-packed halogen layers as shown in figure 35(b). Both polymorphs are band semiconductors with gaps larger than 2.4–4 eV. We have already reported that high-$T_c$superconductivity is obtained in β-HfNCl upon electron doping by intercalation of alkali metals. The highest T_c was observed in the lithium and tetrahydrofuran (THF) cointercalated compound $Li_{0.48}(THF)_y$HfNCl at 25.5 K [300]. The Zr homologue Li_xZrNCl also exhibits superconductivity at *Tc* ~ 14 K [301].

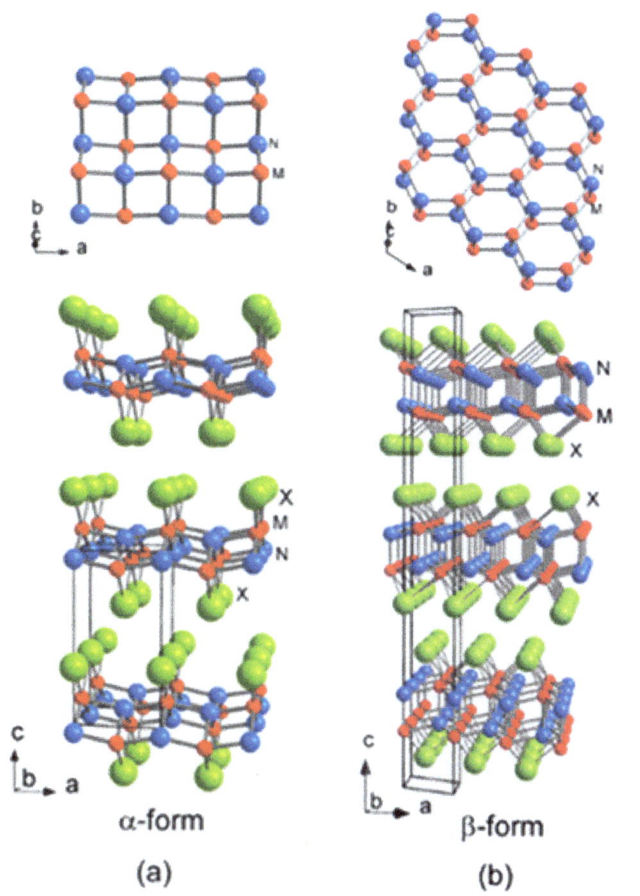

Figure 35. Schematic crystal structures of (a) the α- and (b) the β-forms of *MNX* (*M* = Ti, Zr, Hf; *X* = Cl, Br, I): small red balls, *M*; blue balls, N; and large green balls, *X*. The lower part of the illustration shows the views along the *b*-axes, and the upper part shows the two-dimensional nitride layers of each form. Reproduced from [299] with permission of The Royal Society of Chemistry.

The electron doped *M*NCl (*M* = Zr, Hf) shows unconventional superconductivity [302, 303]; the electron carrier concentration is very low, but the electron–phonon coupling constant observed ($\lambda_{e\text{-}ph} << 1$) is too small to explain the relatively high T_c values [304]. Unexpectedly large gap ratios have been observed by specific heat ($2\Delta/k_BT_c = 6.5$) [305] and tunneling spectroscopy ($2\Delta/k_BT_c = 7$–10) [306] measurements. The small isotope effect is also unconventional [307]. Since these superconductors do not contain any magnetic ions, a magnetic pairing mechanism is excluded. Charge and spin fluctuations have been proposed as the possible candidates for the pairing mechanism [308–310]. The β-form layered compounds have been highly electron-doped by using liquid ammonia solutions of alkaline earth as well as rare-earth metals. TiNCl with the α-form layered structure has been changed into superconductors by electron doping with alkali metals and organic bases such as pyridine and alkylene diamines.

Intercalation of Alkaline-earth and Rare-earth Metals in -HfNCl and -ZrNCl

In alkali metal (*A*) intercalated compounds, *A* atoms occupy the octahedral or trigonal prism sites between chlorine layers of β-*M*NCl [58, 311, 312]. The largest metal concentration attained is expected to be $x = 0.5$ for A_xMNCl (*M* = Zr, Hf). We are interested in the electron doping to a much higher concentration using multivalent metals. Alkaline earth (*AE* = Ca, Sr, Ba) [59] and rare earth metals (*RE* = Eu, Yb) [60] have been successfully introduced into the interlayer space of parent materials by using liquid ammonia solutions, which make it possible to study the effect of high doping concentration and additional magnetic spin, respectively. The dependence of the metal doping concentration *x* of $AE_x(Solv)_y$HfNCl (*AE* = Ca, Sr, Ba; *Solv* = NH_3, THF) onto the T_c is shown in figure 36 [59]. Note that the T_c is hardly influenced by the doping concentration *x* up to *x* ~ 0.4, corresponding to *x* ~ 0.8 for monovalent alkali metal intercalated compounds *Ax*(*Solv*)*y*HfNCl. The superconductors are not yet over-doped. The T_c decreases in the following order of the increasing basal spacing (*d*) with intercalation of Sr, Ba (*d* = ~10 Å) > Ca*x*(NH_3)*y*, Ba*x*(NH_3)*y*, Sr*x*(NH_3)*y* (*d* = 12 Å) > Ca*x*(THF)*y* (*d* = 15 Å), although a minimum level of doping is certainly necessary for the superconductivity. The as-prepared compounds are cointercalated with ammonia used as a solvent, which can be replaced with THF. With varying electron-doping concentrations and interlayer spacings, the highest T_c of 26.0 K was obtained for the Ca and THF cointercalated compound $Ca_{0.11}$(THF)*y*HfNCl, a new record of high T_c in the electron-doped metal nitride chloride system [59].

Multivalent rare-earth metals (Eu and Yb) can be solved in liquid ammonia, and intercalated into β-*M*NCl (*M* = Zr, Hf) from the solutions [60]. The T_c values of the ammonia cointercalated compounds *REx*(NH_3)*y*HfNCl (*RE* = Eu, Yb) are comparable with those of*AEx*(NH_3)*y*HfNCl (figure 36); 24.1 and 23.0 K for $Eu_{0.13}(NH_3)_y$ and $Yb_{0.11}(NH_3)_y$, respectively. The temperature dependence of the magnetic susceptibility of *REx*(NH_3)*yM*NCl measured in a temperature range of 100–300 K suggests that the *RE* metals exist as the paramagnetic ions $Eu^{+2}$and Yb^{+3}. The anisotropic magnetoresistance measurement has evidenced that the paramagnetism of Eu^{2+} and Yb^{3+} can coexist with the superconductivity even under high magnetic fields up to 14 T, as shown in figure 37 for $Eu_{0.08}(NH_3)y$HfNCl. The anisotropic upper critical fields ($H_{c2}^{\|ab}$ and $H_{c2}^{\|ab}$ were determined with the magnetic field parallel to the *ab* plane andc axis, respectively, on a uniaxially oriented pellet sample. The anisotropy parameter γ = (d $H_{c2}^{\|ab}$ /d*T*)/(d/d*T*) = 4.1, comparable to those of other electron-doped *β*-*M*NCl superconductors, 3.7 for $Li_{0.48}$(THF)*y*HfNCl [313] and 4.5 for $ZrNCl_{0.7}$ [314].

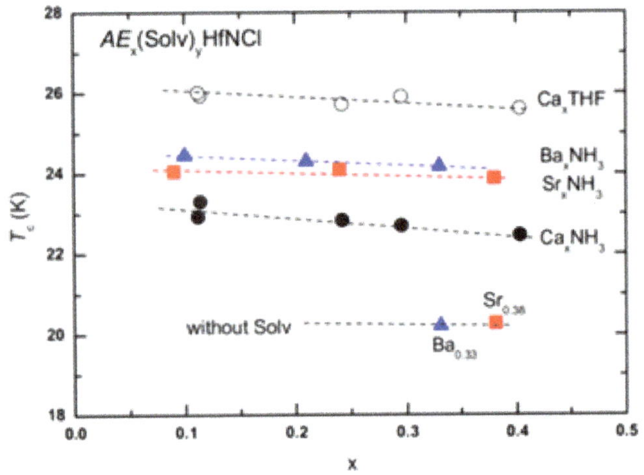

Figure 36. T_c of the as-prepared $AEx(NH_3)y$HfNCl (*AE* = Ca (●), Sr (■), Ba (▲)) as a function of concentration *x*. T_c values of the evacuated *AEx*HfNCl (*AE* = Sr, Ba) and the THF-cointercalated Ca*x*(THF) *y*HfNCl are compared in the same figure. Reprinted with permission from [59]. Copyright 2013 by IOP Publishing.

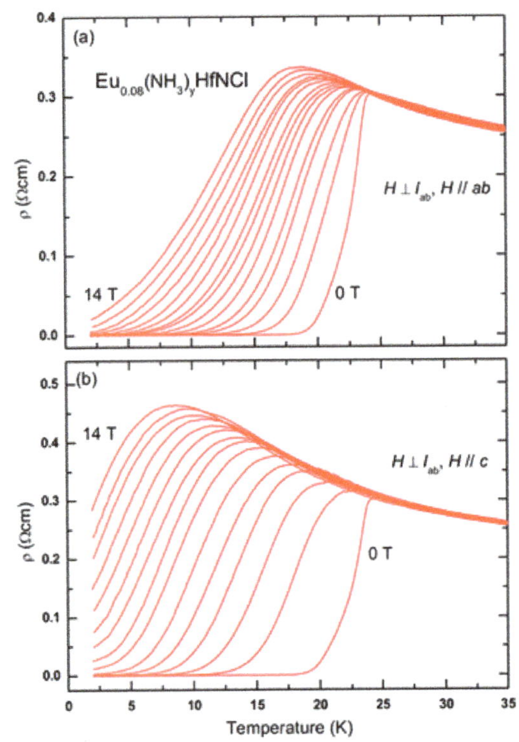

Figure 37. Temperature dependences of the resistivity of $Eu_{0.08}(NH_3)$ *y*HfNCl in various magnetic fields for (a) *H ab* ($H \perp I_{ab}$) and (b) *H c* ($H \perp I_{ab}$). The applied magnetic field was increased from 0 to 14 T with an interval of 1 T. Reprinted with permission from [60]. Copyright 2013 by IOP Publishing.

All the T_c values of the electron doped *β*-HfNCl so far determined can fit on a single curved line as a function of the basal spacing (*d*) as shown in figure 38 [59], suggesting that the alkali, alkaline-earth and rare-earth metals act as similar electron dopants, and the T_c is hardly dependent on the doping concentration. A very similar trend was also observed in the electron doped *β*-ZrNCl [315]. It should be noted that the T_c increases with increasing *d* upon cointercalation of NH_3 and THF, and then decreases gradually with the further increase of *d*upon cointercalation of PC.

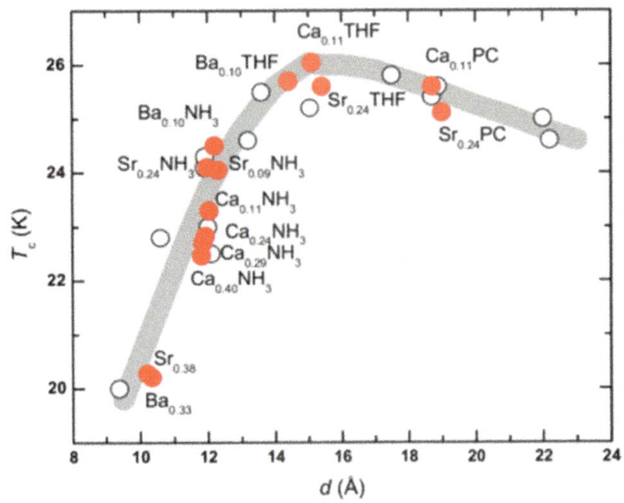

Figure 38. *Tc* dependence on the basal spacing *d* of $AE_x(Solv)_y$HfNCl (●), and from alkali metal and rare earth metal intercalated compounds (○). Reprinted with permission from [59]. Copyright 2013 by IOP Publishing.

Intercalation Compounds of TiNCl with Alkali Metals

In a previous study we prepared the alkali metal intercalation compounds of the *α*-form structured HfNBr, *Ax*(THF)*y*HfNBr (*A* = Li, Na) [316]. The intercalation was successful, and the color of the layered crystals changed from pale yellow to black. However, the resulting compounds were found to not be superconductors, but insulators with resistivity >10^7 *Ω*cm. This is quite a contrast to the fact that the electron doped *β*-HfNBr shows high-$T_c$superconductivity, like *β*-HfNCl. Later we have used *α*-form layered crystals TiNCl, which are prepared by the ammonolysis of $TiCl_4$ at elevated temperatures, followed by purification via chemical transport [317]. TiNCl can be intercalated with alkali metals as well as neutral organic molecules such as pyridine and diamines. The intercalation compounds become superconductors. The reason for the quite different behaviors of the two kinds of *α*-form crystals TiNCl and HfNBr is not clear.

(i) Alkali metal intercalation in TiNCl.

In the first attempt of the intercalation, TiNCl was subjected to reaction with various kinds of metal azides AN_3 (*A* = Li, Na, K, Rb) at elevated temperatures under vacuum [317]. The azides are thermally decomposed to metal and nitrogen, and the resulting metal is intercalated into the interlayer space between chloride layers. Part of the alkali metal is also used to extract or deintercalate chloride ions from the interlayer space, forming a metal chloride

$$TiNCl + (x + y)MN_3 \rightarrow M_xTiNCl_{1-y} + yMCl + 3/2(x + y)N_2.$$

The metal intercalated compounds show superconductivity with $T_c \sim 16$ K, although the superconducting volume fractions determined from the diamagnetic expulsion were found to be as low as 5–30%. The Rietveld analysis of the x-ray powder diffraction data revealed that the TiNCl crystalline layers are mutually shifted to accommodate the metal atoms between the chloride layers as shown in figure 39. The space groups of the resulting new polytypes are *Bmmb* for Na, and *Immm* for K or Rb intercalated compounds. The space group of Li_xTiNCl is unchanged from *Pmmn* of the pristine crystal.

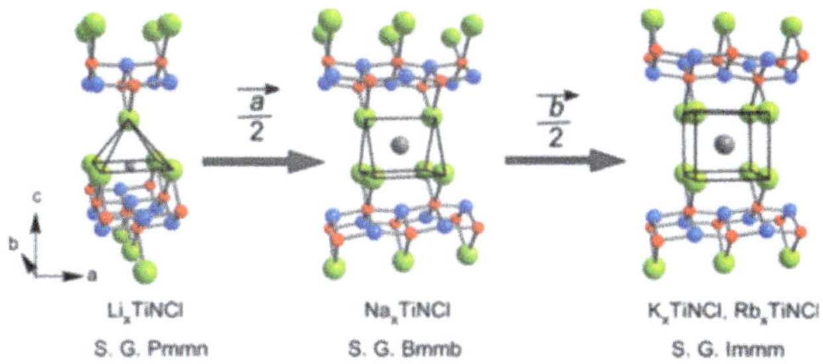

Figure 39. Schematic structures of the TiNCl polytypes with intercalated cations between the layers. Reproduced from [317] with permission of The Royal Society of Chemistry.

A mild intercalation reaction using metal naphthalene solutions in THF can produce superconductors with a much higher volume fraction >60% [56]. The as-prepared compound is co-intercalated with THF. As shown in figure 40, the basal spacing (*d*) of the compounds varies depending on the cointercalation conditions. $Na_{0.16}(THF)y$TiNCl has a basal spacing of 13.10 Å (T_c = 10.2 K), which decreases to 8.44 Å (T_c = 18.0 K) upon removal of THF by prolonged evacuation at 90 °C. The THF molecules can be replaced with larger size solvent molecules PC (propylene carbonate); the basal spacing increases to 20.53 Å with decreasing T_c to 7.4 K. Figure41 shows the T_c values as a function of 1/*d* for alkali metals and solvent cointercalated compounds. The T_c decreases with the increase of the basal spacing (*d*). The data fit on a linear line passing through the origin, suggesting the importance of the Coulomb interlayer coupling in the pairing mechanism in this system. It should also be noted that the T_c of non-cointercalated compounds *Ax*TiNCl (*A* = Na, K, Rb) also fit on this line except $Li_{0.13}$TiNCl. The as-prepared sample $Li_{0.13}(THF)y$TiNCl has a basal spacing of 13.1 Å similar to that of $Na_{0.16}$(THF) *y*TiNCl, and a T_c = 10.2 K. The non-cointercalated compound $Li_{0.13}$TiNCl was obtained by evacuation at 150 °C, which has the smallest basal spacing of 7.8 Å, the same as that of the pristine TiNCl. Unexpectedly, T_c was found to be ~6.0 K, much lower than the value expected for the small basal spacing of figure 41. Li ions are small enough in size to penetrate into chlorine layers, forming double LiCl layers between $[TiN]_2$ layers, $[TiN]_2(Li_{0.13}Cl)(ClLi_{0.13})[TiN]_2$, in which Li ions are located close to TiN superconducting layers in parallel with Cl atoms. On the other hand, in the THF cointercalated compound, Li ions can be coordinated with THF molecules between chlorine layers $[ClTi_2N_2Cl]Li_{0.26}(THF)y[ClTi_2N_2Cl]$. The low T_c of $Li_{0.13}$TiNCl against the small *d* suggests that the location of positive centers may also influence the Coulomb interlayer coupling for superconductivity. The linear relation shown in figure 41 appears to be applied to the structure where the positive centers are located between the chlorine layers as shown in figure 40 for the $Na_{0.16}(THF)y$ cointercalated compound. Another systematic study on the relation between the basal spacing and T_c has been performed on TiNBr, and again a similar linear relation was found [57].

(ii) Anisotropic superconducting properties.

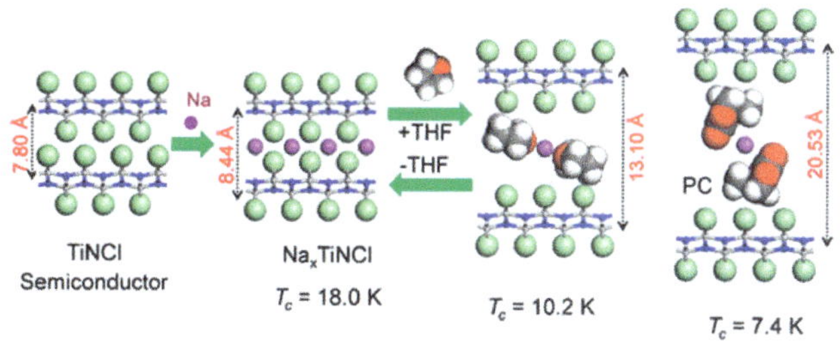

Figure 40. Schematic illustration showing the expansion of the basal spacing of TiNCl upon intercalation of Na and cointercalation of solvent molecules. Reprinted from [303] with permission from Elsevier.

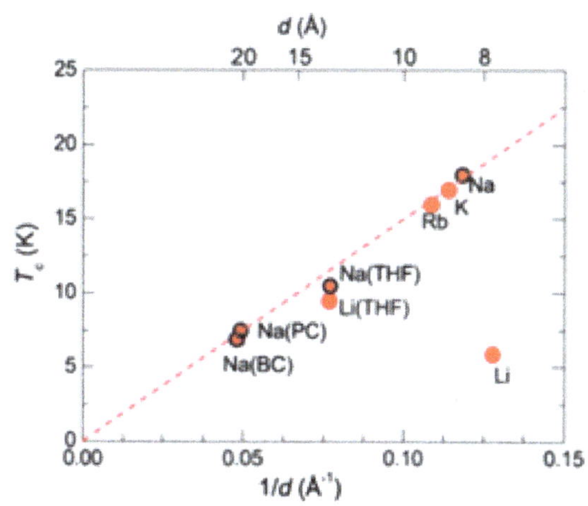

Figure 41. T_c versus 1/*d* for *Ax*TiNCl (*A* = Li, Na, K, Rb) superconductors with and without cointercalation [56]. Reprinted with permission from [56]. Copyright 2012 by the American Physical Society.

The anisotropic magnetic susceptibility of the layered superconductor $Na_{0.16}(THF)y$TiNCl was measured on the highly oriented pellet sample with the magnetic field parallel and perpendicular to the *c* and *ab* plane [56]. Figure 42 shows the anisotropic Hc_2 thus determined as a function of temperature. The anisotropy parameter was calculated to be γ = 1.5 and 1.2 for *Nax*(THF)*y*TiNCl and *Nax*TiNCl, respectively. Note that the TiNCl superconductors exhibit rather isotropic or 3D character. The expansion of the basal spacing from 8.44 to 13.10 Å by cointercalation of THF has little effect on the anisotropy. The superconductor derived from TiNBr also shows a similar γ value [57]. The characteristic superconducting parameters of theα- and the β-structured layered nitride superconductors are compared in table 12. The anisotropy parameter $\gamma = \xi ab/\xi c$ (ratio of the coherence lengths in the *ab* plane and along the *c*-axis) for $K_{0.21}$TiNBr is calculated to be ~1.3, close to that found in electron-doped TiNCl; $Na_{0.16}$TiNCl (γ = 1.2, T_c = 18.1 K) and $Na_{0.16}$(THF) *y*TiNCl (γ = 1.5, T_c = 10.2 K) [56]. In contrast, the β-structured nitrides show the anisotropy parameter γ as large as 3.7–4.5

[59, 60,313, 314]. The small anisotropy parameter γ appears to be one of the characteristic features of the α-structured superconductors. The coherence length along the c-axis (ξc) of β-$Li_{0.48}(THF)y HfNCl$ is about 16 Å, comparable with the basal spacing 17.8 Å, i.e., the separation of the superconducting layers. This suggests that the superconducting β-form layers may be weakly Josephson coupled. On the other hand, in the α-form TiNBr and TiNCl, the ξc are more than three times larger than the basal spacing, implying that the nitride layers are more strongly coupled.

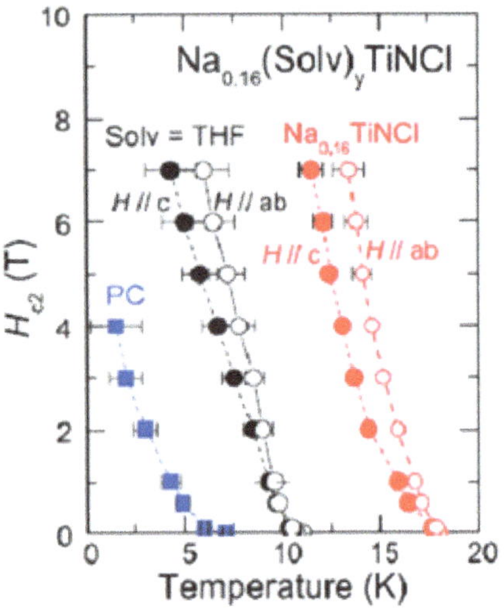

Figure 42. Hc_2–T phase diagram of Na-intercalated compounds with and without cointercalation. Reprinted with permission from [56]. Copyright 2012 by the American Physical Society.

Table 12. Characteristic superconducting parameters of the α- and β-structured layered nitride superconductors [57].

Compound	T_c (K)	d (Å)	ξ_{ab} (Å)	ξ_c (Å)	λ_{ab} (Å)	γ	Reference
α-$K_{0.21}TiNBr$	17.2	9.5	53	41	3045	1.3	[57]
α-$Na_{0.16}TiNCl$	18.1	8.4	33	28	4746	1.2	[56]
α-$Na_{0.16}(THF)_yTiNCl$	10.2	13.1	55	35	—	1.5	[56]
β-$ZrNCl_{0.7}$	13	9.8	71	16	—	4.5	[314]
β-$Li_{0.48}(THF)_yHfNCl$	25.5	18.7	60	16	4630	3.7	[313]
β-$Eu_{0.08}(NH_3)_yHfNCl$	24.3	11.9	61	15	—	4.1	[60]
β-$Ca_{0.11}(THF)_yHfNCl$	26.0	15.0	47	12	—	4.1	[59]

High-T_c layered cuprate superconductors have large anisotropy parameters γ on H_{c2}, varying in the range of 3–30 [318]. Iron pnictide superconductors LnFeAs(O, F) (Ln = La, Sm, Nd) recently discovered also show a large anisotropy of γ = 4–9 [319–323]. The large anisotropy parameters of the layered compounds have been considered to be important to realize high-T_c superconductivity. The electron-doped β-ZrNCl and β-HfNCl are also classified into this category. However, the Ba-122 superconductors such as (Ba, K)Fe_2As_2 with T_c = 38 K have been developed, which have a small isotropic parameter γ = 1.5–1.9 [231, 319, 324–327]. The layer coupling through intervening (Ba, K) atoms seems to be stronger than those of the 1111 pnictides coupled through metal oxide layers. It is interesting to note that β-$ZrNCl_{0.7}$ (table 12), which is electron-doped by a partial deintercalation of chlorine atoms from the interlayer space, has an anisotropy parameter as large as 4.5. In β-$ZrNCl_{0.7}$ with d = 9.8 Å, the nitride layers should be directly coupled without intervening alkali atoms. Nevertheless, the anisotropy parameter is comparable to, or even larger than, that of the cointercalated compound β-$Li_{0.48}(THF)y HfNCl$. It is evident that the interlayer separation is not a decisive parameter for the anisotropy on H_{c2}. The small anisotropy on H_{c2} and the Coulomb coupling between the superconducting layers should be the relevant nature of the superconductivity of the α-form layered nitrides [57]. For more discussion on the superconducting mechanisms and the anisotropy of the two different kinds of layered nitrides, a theoretical study including the electric band structure is required [328].

Intercalation Compounds of TiNCl with Neutral Amines

In the formation of the intercalation compound of β-form layered compounds, organic solvent molecules are cointercalated with metal atoms; organic molecules alone cannot be intercalated. In contrast, α-form TiNCl can intercalate neutral organic molecules without metal atoms. TiNCl can form an intercalation compound with pyridine from liquid and gas phases, $Py_{0.25}TiNCl$. The basal spacing increases to 13.5 Å with the molecular plane oriented perpendicular to the layers as schematically shown in figure 43 [317]. The compound becomes a superconductor with T_c = 8.6 K. The T_c is different from those of the alkali metal intercalated compounds. It is interesting to develop different kinds of organic compounds which can be intercalated to obtain high-T_c superconductors. Although the doping mechanism is not yet clear in the intercalation compound with pyridine, it would be reasonable to estimate that the lone pair electrons of nitrogen atoms in pyridine may act as electron donors to the TiNCl layers. It was reported that FeOCl isotypic with TiNCl forms an intercalation compound with pyridine, and the electrical conductivity increases by about seven orders due to the charge transfer from the organic Lewis base to the FeOCl layers [329]. In that study, various kinds of aliphatic amines have been intercalated into TiNCl to develop new superconductivity.

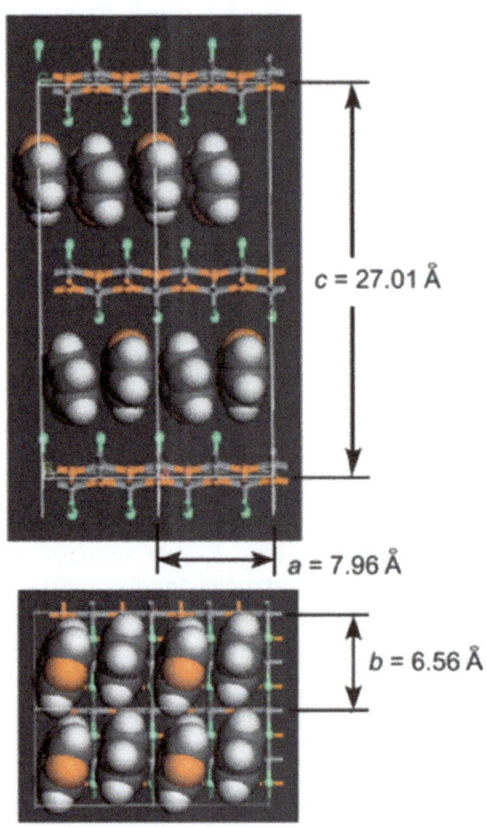

Figure 43. The arrangements of pyridine (Py) molecules in $Py_{0.25}TiNCl$

obtained by geometrical optimization. A 2 × 2 × 1 supercell is used; the views along the *b*-axis (top) and along the *c*-axis (bottom). Nitrogen atoms are orange-colored. Reproduced from [317] with permission of The Royal Society of Chemistry.

n-alkyl monoamines ($CnH_{2}n_{+1}NH_{2}$, $3 \leq n \leq 12$) can form intercalation compounds with TiNCl, expanding the basal spacing to a value in the range of 12.0 to 37 Å, with the alkyl chains oriented in various ways. All of the compounds with *n*-alkyl monoamines are not superconductors down to 2 K [55]. It is interesting that ethylene diamine ($NH_2CH_2CH_2NH_2$) can form a similar intercalation compound with TiNCl with a basal spacing of 11.12 Å, which shows superconductivity with T_c = 10.5 K [55]. Systematic studies have been done using alkylene diamines with different numbers of carbon atoms, $NH_2C_nH_{2n}NH_2$ ($2 \leq n \leq 12$). The results are shown in figure 44 [55]. Most of the diamine intercalation compounds are superconductors. The basal spacings are near 12.5 Å irrespective of the chain length of diamines, suggesting that the alkylene chains are aligned with the molecular axis parallel to the layers, and oriented along the *b*-axis as shown in figure 45. Diamines with an even number of carbon atoms appear to have larger superconducting volume fractions, and the diamines with longer alkylene chains are suitable for higher T_c. The compound with *n* = 10 (decamethylene diamine, DMDA) shows a large volume fraction > 50%, and T_c = 17.1 K, which is comparable with T_c = 18.1 K of $Na_{0.16}TiNCl$. Mechanisms for the superconductivity are not clear, and remain open problems for physicists as well as chemists.

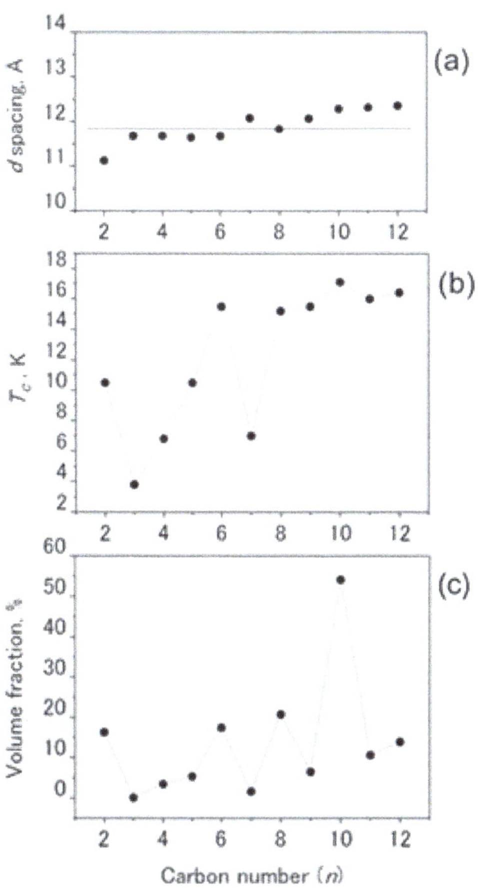

Figure 44. (a) Basal spacing (*d*), (b) superconducting transition temperature (*Tc*) and (c) superconducting volume fraction as a function of the number of carbon atoms (*n*) in the alkylene chains of diamine in the ($NH_2CnH_2nNH_2$ ($2 \leq n \leq 12$))*x*-TiNCl compounds. Reproduced from [55] with permission of The Royal Society of Chemistry.

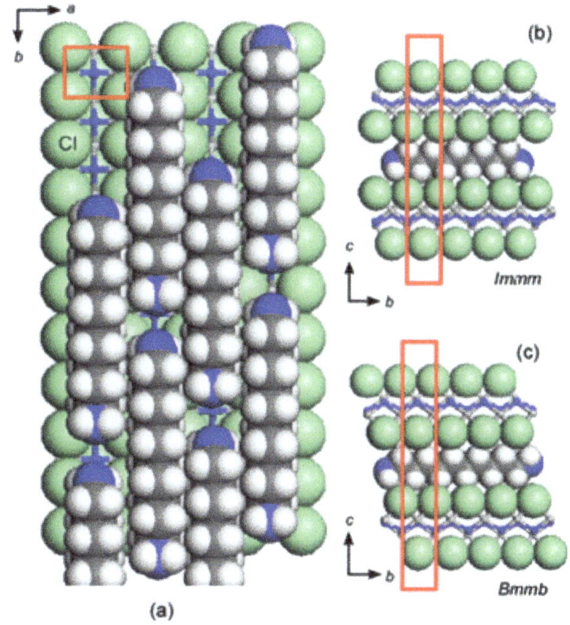

Figure 45. Schematic illustration showing the arrangements of alkylene diamine molecules between TiNCl layers; (a) *ab* projection: the linear alkylene diamine molecules are oriented parallel to the layers and aligned along the *b*-axis, (b) the *bc* projection for the arrangement of nonamethylene diamine (NMDA) molecules in the interlayer space of the TiNCl host structure with the space group *Immm*, and (c) the *bc* projection for the arrangement of decamethylene diamine (DMDA) molecules in the interlayer space of the host structure with the space group *Bmmb*. The unit cells are shown in red color in the projections. Reproduced from [55] with permission of The Royal Society of Chemistry.

Silicon Clathrates and Related Compounds with Cage-like Structures

Using high pressure and high temperature (HPHT) conditions we have prepared the barium-containing silicon clathrate compound Ba_8Si_{46}, which shows superconductivity with T_c = 8.0 K [330]. This is the first superconductor with a clathrate structure. The application of HPHT conditions is favorable for synthesizing silicon-rich binary phases such as $LaSi_5$, $LaSi_{10}$, $BaSi_6$, $Ba_{24}Si_{100}$, Ba_8Si_{46}, $NaSi_6$, Na_8Si_{46}, and Na_xSi_{136} [98, 331–334]. The silicon rich compounds are generated under high pressure by obeying Le Chatelier's principle; the molar volume of the reactant of the system decreases in the product by forming covalent networks and high coordination environments. Electrons are doped from metals into the covalent networks. Most compounds are found to become superconductors. In this study of silicon clathrates and related compounds, ternary systems of ubiquitous (commonly found) elements have been developed using HPHT conditions.

Ternary System Ca-Al-Si under HPHT Conditions

Ca–Al–Si ternary metal mixtures were pre-melted using an arc furnace, followed by remelting at 1000–1200 °C using radio frequency induction heating under an Ar atmosphere in an *h*-BN crucible. The cooled ternary mixtures were supplied for the HPHT treatment up to 5–13 GPa and 600–1000 °C by using a Kawai-type multianvil apparatus. A new ternary compound $Ca_2Al_3Si_4$ was obtained above 650 °C under a pressure of 5

GPa. It crystallizes with the space group $Cmc2_1$ and the lattice parameters a = 5.8846(8), b = 14.973(1), and c = 7.7966(8) Å [45]. The structure is composed of an aluminum silicide framework [Al_3Si_4] and layer structured [Ca_2] network interpenetrating with each other as shown in figure 46. The [Ca_2] subnetwork has an isomorphous structure with black phosphorus. The compound shows superconductivity with T_c of 6.4 K.

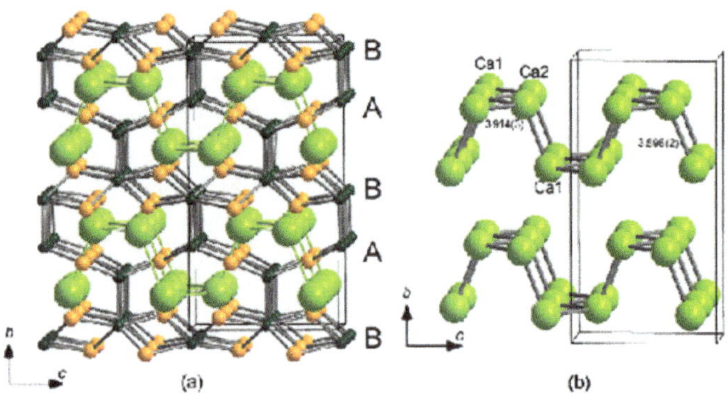

Figure 46. Schematic representation of the crystal structure of (a) $Ca_2Al_3Si_4$ composed of A [$AlSi_2$] and B [Al_2Si_2] layers with Ca atoms: small black balls, Si; orange balls, Al; green balls, Ca. (b) [Ca_2] layer which is isomorphous with black phosphorus. Reprinted from [45] with permission from Elsevier.

Under a higher pressure of 13 GPa at 1000 °C, solid solutions $Ca(Al_{1-x}Si_x)_2$ ($0.35 \leq x \leq 0.75$) isomorphous with the cubic Laves phase were obtained [71]. As shown in figure 47 the structure can be regarded as a kind of clathrate compound composed of face-sharing truncated tetrahedral cages with Ca atoms at the center, $Ca@(Al,Si)_{12}$. The compound with a stoichiometric composition CaAlSi shows superconductivity with T_c of 2.6 K [71]. This is the first superconducting Laves phase compound composed solely of commonly found elements.

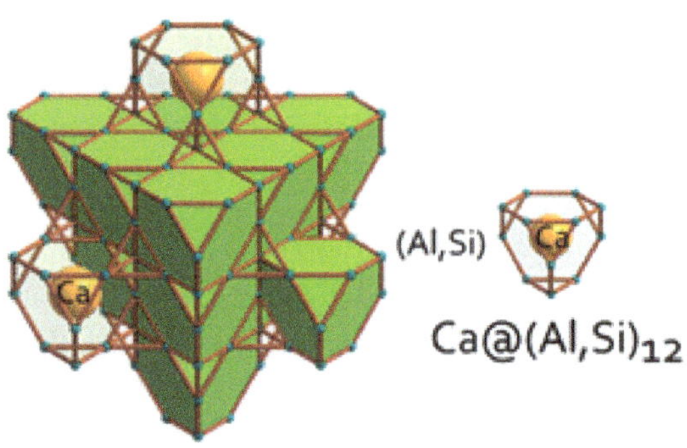

Figure 47. Crystal structure of the Laves phase $Ca(Al,Si)_2$ composed of face-sharing truncated $Ca@(Al,Si)_{12}$ tetrahedra. Reprinted with permission from [71]. Copyright 2013 by the American Chemical Society.

Ternary System Al–Mg–Si under HPHT conditions

Although the ternary system Al–Mg–Si is an important subject in the development of commercial Al alloys [335], Mg_2Si with antifluorite structure is the only compound known as bulk phase in the ternary system under ambient pressure. However, various kinds of fine coherent precipitates are formed in the Al matrix during low temperature aging, which are considered to play an important role for the hardening of a commercial Al alloy. It is reasonable to assume that the ternary and binary precipitates found in Al-based alloys are formed under high pressure generated by Al matrix of the alloy. New binary and ternary compounds in the ternary system Al–Mg–Si have been prepared using HPHT conditions, and the structures are determined using single crystals. Some of them become superconductors.

Ternary compounds $Mg(Mg_{1-x}Al_x)Si$ ($0.3 < x < 0.8$) have been prepared under HPHT conditions of 5 GPa at 800–1100 °C. The single crystal study revealed that the compound ($x = 0.45$) is isomorphous with the anticotunnite, or the TiNiSi structure, and crystallizes with space group *Pnma*, with lattice parameters a = 6.9242(2), b = 4.1380(1), c = 7.9618(2) Å, and $Z = 4$. The compound with $x > 0.5$ shows superconductivity with a T_c ~6 K [70]. The compound is a peritectic solid solution associated with other phases such as Mg_9Si_5, Al, and Si, depending on cooling protocols in the preparation. The band structure calculation on the composition of MgAlSi suggests that the Al and Mg orbitals mainly contribute to the density of states near the Fermi level, and the substitution of Mg with Al favors superconductivity.

Two kinds of magnesium-based compounds Mg_9Si_5 and Mg_4AlSi_3 have been prepared under a similar HPHT condition. Single crystal study revealed that Mg_9Si_5 crystallizes in space group $P6_3$ (no. 173) with the lattice parameters a = 12.411(1) Å, c = 12.345(1) Å, and Z = 6 [46]. The structure can be derived from the high pressure form Mg_2Si with anticotunnite structure; excess Si atoms of Mg_9Si_5 form Si–Si pairs in the prismatic cotunnite columns running along the c-axis. Mg_4AlSi_3 is obtained by a rapid cooling of a ternary mixture Mg:Al:Si = 1:1:1 from ~800 °C to room temperature under a pressure of 5 GPa. The compound crystallizes in space group *P4/ncc* (no. 130) with the lattice parameters a = 6.7225(5) Å, c = 13.5150(9) Å, and Z = 4 [46]. As shown in figure 48, the structure consists of an alternate stacking of [$AlSi_2$] layers having a Cairo pattern and [Mg_4Si] antitetragonal prismatic layers. It can be viewed as composed of hexa-Si-capped tetragonal prismatic cages Mg_8Si_6 with an Al atom at the center of each cage, $Al@Mg_8Si_6$. The compound shows superconductivity with a transition temperature T_c = 5.2 K. The formation regions of the two kinds of new magnesium-based compounds have been proposed.

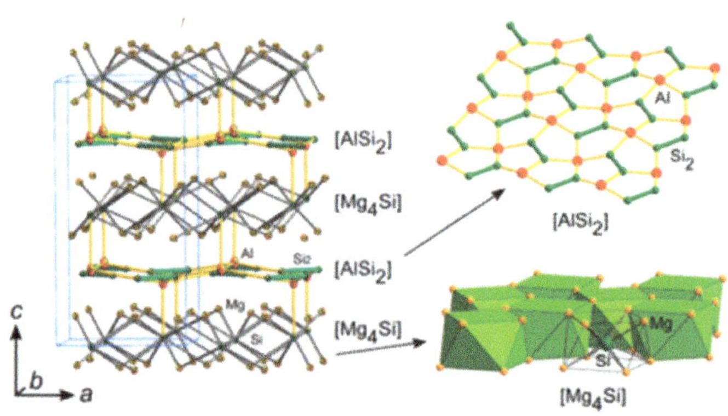

Figure 48. Structure of Mg_4AlSi_3 composed of alternate stacking of [Mg_4Si] and [$AlSi_2$] layers. The [$AlSi_2$] layers form an almost coplanar Cairo pattern with a short Si–Si bond distance (2.43 Å). The [Mg_4Si] layers consist of Mg antitetragonal prism polyhedra surrounding Si atoms. Reprinted with permission from [46]. Copyright 2013 by the American Chemical Society.

An attempt to prepare carbon analogs for silicon clathrate compounds

Silicon and carbon chemistries are often discussed comparatively from the viewpoint of belonging to the same group in the periodic table [336, 337]. It is interesting to prepare metal doped carbon clathrate compounds, in which carbon forms a sp^3 clathrate network-like open diamond framework. A hole-doped diamond was found to become a superconductor [338]. If the synthesis of an electron doped carbon clathrate is realized, it is expected that the carbon sp^3framework with a high Debye temperature should exhibit high-T_c superconductivity [339].

We have already obtained 3D carbon frameworks with cages by polymerization of C_{60} crystals under HPHT conditions [340, 341]. An attempt has been made to prepare a Ba doped clathrate-like structure from Ba-doped fulleride, Ba_3C_{60}, using HPHT conditions.

The powder Ba_3C_{60} sample was compressed using a Kawai-type multianvil press at 5–15 GPa and 500–1150 °C [97]. The x-ray powder diffraction (XRD) pattern showed that the Ba_3C_{60}compressed under 15 GPa at 900 °C is changed into an amorphous solid, which was found to be chemically stable in air and even in water. The high Vickers micro-hardness (1700 kg mm^{-2}) and the Raman spectra of the solids suggest that the C_{60} molecules are collapsed to form an amorphous 3D polymer encapsulating the Ba atoms. As shown in figure 49, the solid obtained by compression under 15 GPa at 900 °C shows a semi-metallic conductivity. It is interesting to note that the Hall coefficient of this sample is positive in the whole temperature range <300 °C, indicating that the dominant carriers are holes. As shown in figure 50, the covalent diameter of a Ba atom is too large to substitute one C atom in the carbon matrix. The diameter is rather comparable with the covalent diameter of a six-membered carbon ring. If a six-membered carbon ring with 12 electrons in the sp^3 hybridized orbitals is substituted by a Ba atom with a similar size having only two valence electrons, the carbon matrix should be efficiently hole-doped. The electrical characteristics such as conductivity, Hall coefficient, carrier density, and mobility of the carbon matrix encapsulating Ba atoms were found to be comparable with those of B-doped diamond [97]. The Ba-encapsulated carbon matrix obtained in this study is amorphous. It is interesting to prepare crystalline carbon analogs for silicon clathrate compounds.

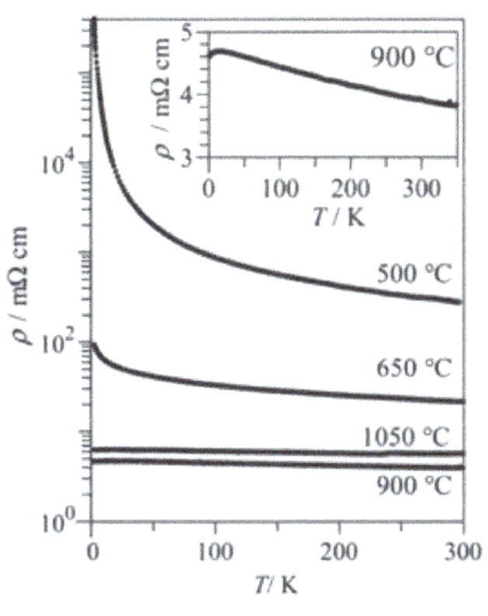

Figure 49. Temperature dependences of the electrical resistivity of Ba_3C_{60} samples obtained by treatment at various temperatures under 15 GPa. The inset shows the resistivity in an enlarged linear scale for the sample obtained at 900 °C under 15 GPa. Reprinted from [97] with permission from Elsevier.

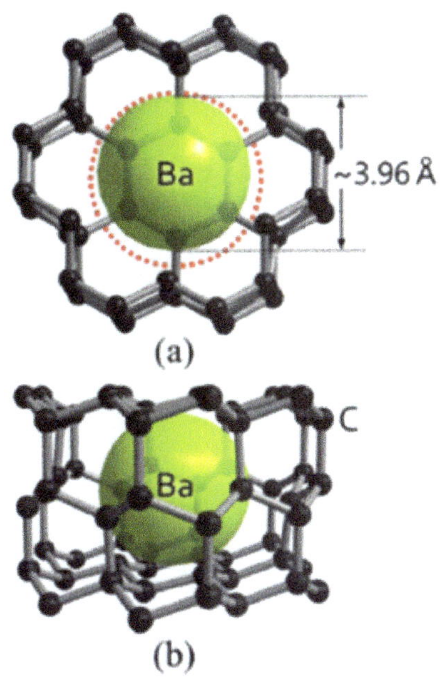

Figure 50. Schematic illustrations of a local arrangement of carbon network encapsulating a Ba atom in Ba_3C_{60} collapsed by HPHT treatment, (a) top and (b) side views; the Ba atom (covalent diameter 3.96 Å) replaces carbon atoms in a six-membered ring with comparable size. The covalent diameter (4.2 Å) of a six-membered carbon ring (red circle in (a)) includes the diameter of the six-membered ring (2.9 Å) and a fringe with a width of a half length of the C–C bonds formed with carbon atoms surrounding the ring. Reprinted from [97] with permission from Elsevier.

Other New Superconductors

Besides superconductors described in the preceding sections, many new superconductors have been discovered in this project. In this section, some of them are overviewed.

AM_2X_2-type (122-type, $M \neq$ Fe) Superconductors

The 122-type iron arsenide superconductors $AEFe_2As_2$ (AE=Ba, Ca, Sr and Eu) have been investigated intensively by many groups including the FIRST Project team (see the preceding section). These compounds have the $ThCr_2Si_2$-type structure and non-iron 122-superconductors with the same structure have also been studied widely; this class of compounds includes $SrNi_2P_2$ [342], $BaNi_2P_2$ [343], $BaRh_2P_2$, $BaIr_2P_2$ [344], $LaRu_2P_2$ [345], $LiCu_2P_2$ [346], $SrNi_2As_2$ [347], $BaNi_2As_2$ [348], etc. The AEM_2X_2 compounds sometimes crystallize in different polymorphs with $CaBe_2Ge_2$-type structure [349]. The two structures of $ThCr_2Si_2$-type and $CaBe_2Ge_2$-type are deeply concerned with each other as compared in figure 51 [350]. In $ThCr_2Si_2$-type AEM_2X_2, A, M and X atom planes are stacked in a sequence of A–(X–M_2–X)-... along the c-axis of the tetragonal cell, forming an MX_4 coordination tetrahedron. In the $CaBe_2Ge_2$-type structure, on the other hand, the sequence of the atom planes is AE–(X–M_2–X)–AE–(M–X_2–M)... where the XM_4 tetrahedron is formed as well as the MX_4tetrahedron. Fewer compounds had been known for the $CaBe_2Ge_2$-type, and

$SrPt_2As_2$ with T_c = 5.2 K is a rare example of the superconductor having this structure [350] (strictly speaking, its structure is an incommensurate orthorhombic variant of the $CaBe_2Ge_2$-type structure [351]). Band structure calculations for $CaBe_2Ge_2$-type $SrPt_2As_2$ revealed that it has two two-dimensional (2D) like Fermi surfaces as well as two three-dimensional (3D) like ones in contrast to the strong 2D character in the electronic structure of $ThCr_2Si_2$-type AFe_2As_2 [352]. Several 122-type superconductors with $ThCr_2Si_2$-type, $CaBe_2Ge_2$-type or related structures were discovered in the FIRST Project as overviewed below.

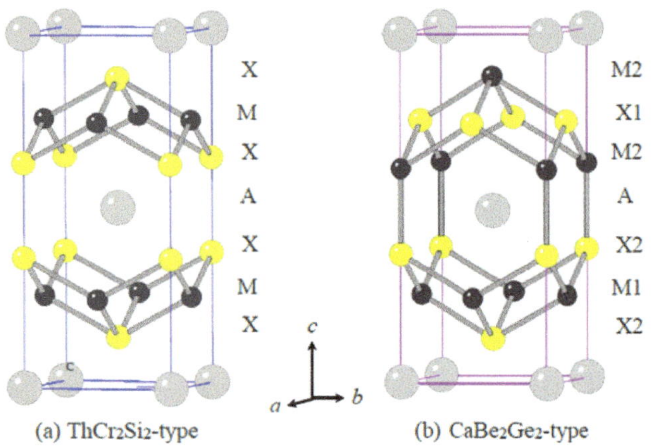

Figure 51. Comparison of the crystal structures of (a) $ThCr_2Si_2$-type and (b) $CaBe_2Ge_2$-type AM_2X_2 compounds. In the latter structure, two nonequivalent sites are present for both the M and X atoms as denoted by the suffixes 1 and 2. Reprinted with permission from [350]. Copyright 2010 by the Physical Society of Japan.

122-type antimonides and related compounds

The $SrPt_2Sb_2$ had been known to have the $CaBe_2Ge_2$-type structure with tetragonal cell of a = 4.603 and c = 10.565 Å [351] but any physical properties had not been reported. In the FIRST Project, this material was revisited to elucidate its physical properties [41]. Samples were synthesized starting from Sr, Pt and Sb in two-step procedures: arc melting and re-melting of the arc-melted specimen. Though electron probe microanalysis confirmed the composition of $SrPt_2Sb_2$ (122) for the major part of the ingot obtained, the powder x-ray pattern was not consistent with the $CaBe_2Ge_2$-type tetragonal lattice. Thus, $SrPt_2Sb_2$ has a different structure which may be some derivative of the $CaBe_2Ge_2$-type. As shown in figure 52, electrical resistivity, magnetization and specific heat measurements confirmed a bulk superconducting transition at T_c = 2.1 K for $SrPt_2Sb_2$. It is a type-II superconductor with a lower critical field (H_{c1}) of 6 Oe and upper critical field (H_{c2}) of 1 kOe at 1.8 K. Debye temperature (Θ_D) and electronic specific heat coefficient (γ) were derived from specific heat data to be Θ_D = 183 K and γ = 9.2 mJ (mol K^2)$^{-1}$. The normalized specific heat jump at T_c was calculated as $\Delta C(T_c)/\gamma T_c$ = 1.29, consistent with the BCS weak coupling limit of 1.43. Normal state electrical resistivity of $SrPt_2Sb_2$ exhibited anomalies around 250 K with thermal hysteresis which corresponded to a certain structural transition, though its detail has not yet been elucidated. $SrPt_2Sb_2$ is a 122-type superconducting antimonide discovered for the first time, and detailed studies for the structure and the phase transition are greatly desired.

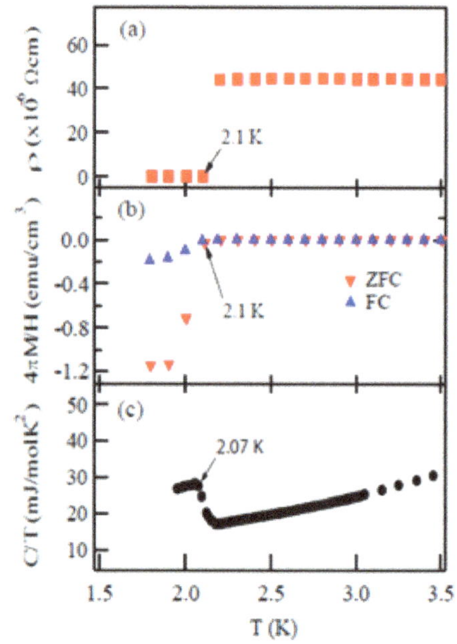

Figure 52. (a) Electrical resistivity (b) magnetization in zero field cooling (ZFC) and field cooling (FC) and (c) specific heat as functions of temperature for $SrPt_2Sb_2$. Reprinted with permission from [41].

The Ba-derivative of the aforementioned compound, $BaPt_2Sb_2$ had not been reported, and was prepared first in the FIRST Project by the arc melting method for a Ba, Pt and Sb mixtures [42]. The x-ray pattern of the $BaPt_2Sb_2$ sample was consistent with a monoclinic lattice having parameters of a = 6.702 Å, b = 6.752 Å, c = 10.47 Å and β = 91.23°. The structure of $BaPt_2Sb_2$ is shown in figure 53, which can be interpreted as a monoclinic variant of the $CaBe_2Ge_2$-type structure (note that the a-axis and b-axis of $BaPt_2Sb_2$ correspond to diagonal of the a_0-axis and b_0-axis of the original $CaBe_2Ge_2$-type lattice with the relationship of $a(b) \approx \sqrt{2}a_0$). Figure 54 gives the temperature dependency of the resistivity of $BaPt_2Sb_2$, which confirms superconducting transition at 1.8 K. Specific heat data gave Θ_D = 146 K and γ = 8.6 mJ (mol K^2)$^{-1}$, deriving $\Delta C(T_c)/\gamma T_c$ = 1.37, which is comparable with the BCS weak coupling limit. Magnetization measurements revealed type-II superconductivity with $\mu_0 H_{c2}(0)$ = 0.27 T and Ginzburg–Landau (GL) coherent length $\xi_{GL}(0)$ = 350 Å. Band structure calculations revealed that Fermi surfaces of $BaPt_2Sb_2$ resemble those of $SrPt_2As_2$ with two 2D-like Fermi surfaces and two 3D-like ones. Thus, it has more 3D-like character compared with the AFe_2As_2 system, which may account for the relatively lower T_c of $BaPt_2Sb_2$.

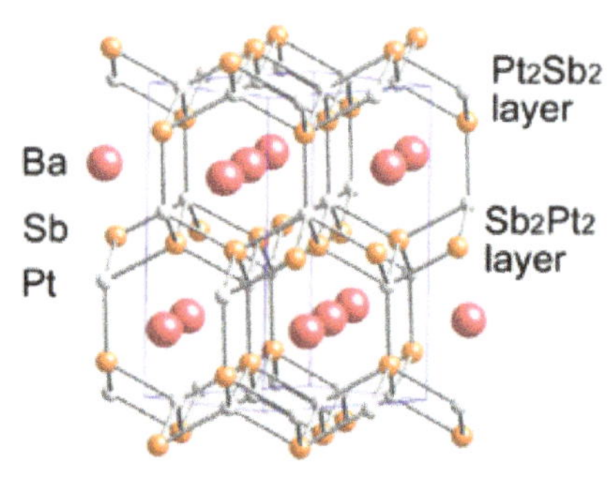

Figure 53. Crystal structure of $BaPt_2Sb_2$. Reprinted with permission from [42]. Copyright 2015 by American Physical Society.

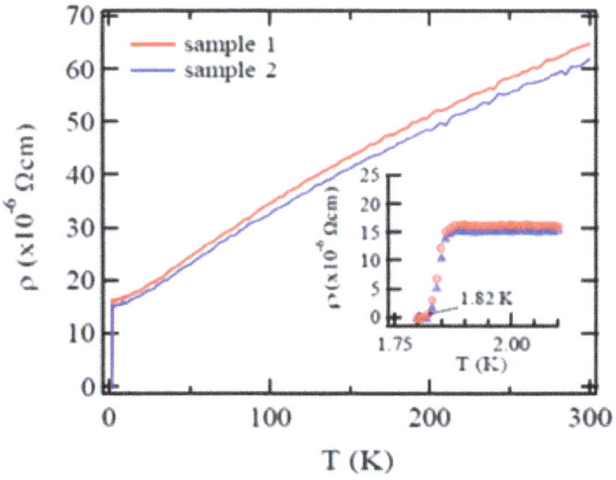

Figure 54. Temperature dependences of the electrical resistivity of two $BaPt_2Sb_2$ samples. The inset is an expanded view of the resistivity at low temperatures ranging from 1.8 to 2.2 K. Reprinted with permission from [42]. Copyright 2015 by American Physical Society.

Pd-based $CaBe_2Ge_2$-type antimonides such as $LnPd_2Sb_2$ (Ln = La, Ce, Pr, Nd, Eu) [353] were known to exist but superconductivity was not reported for this class of compounds. In the FIRST Project, $LaPd_2Sb_2$ was selected among them and reinvestigated in detail to find the superconducting transition [39]. The $LaPd_2Sb_2$ sample was prepared at 900 °C starting from La, Sb and Pd in an evacuated silica tube. The resulting specimen was a single phase of the $CaBe_2Ge_2$-type compound though its tetragonal lattice parameters (a = 4.568 Å, c = 10.266 Å) were slightly different from the previous report [353]. Figure 55 shows the temperature dependence of resistivity for $LaPd_2Sb_2$ under zero and various magnetic fields. The superconducting transition occurs at onset temperature of 1.4 K and zero resistivity temperature of 1.2 K at H = 0. Magnetization measurements revealed type-II nature superconductivity with $\mu_0H_{c2}(0)$ = 0.86 T and $\xi_{GL}(0)$ = 233 Å. Bulk superconductivity was confirmed by specific heat measurements which gave Θ_D = 210 K, γ = 6.89 mJ (mol K^2)$^{-1}$ and $\Delta C(T_c)/\gamma T_c$ = 1.325, consistent with the BCS weak coupling limit. Figure 56 gives the total density of states for La, Pd1, Pd2, Sb1 and Sb2 (see figure 51 on the atom sites with the suffixes of 1 and 2) calculated by DFT.

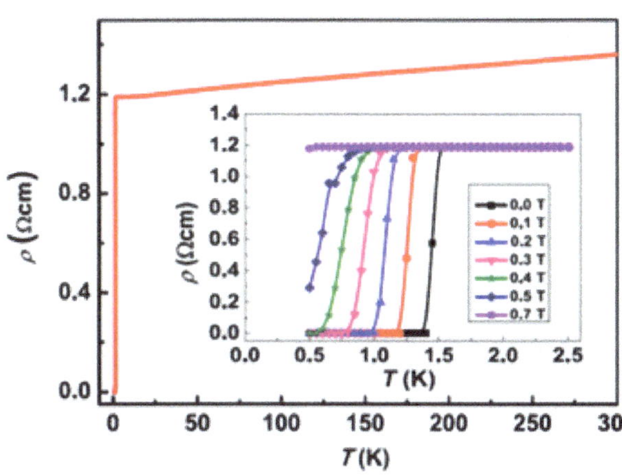

Figure 55. Temperature dependence of electrical resistivity for $LaPd_2Sb_2$. The inset shows the magnetic field dependence. Reprinted from [39] with permission from Elsevier.

It is seen that Pd contributes the most to the total DOS at the Fermi level, consistent with a general tendency that the DOS at the Fermi level are dominated by the d-band of the M atom in $CaBe_2Ge_2$-type AM_2X_2. Figure 56 reveals hybridization between Pd 4d of Pd1 (Pd2) and Sb 5p of Sb2 (Sb1) as demonstrated by synchronized modulation in partial DOS of Pd1 (Pd2) and Sb2 (Sb1). The total DOS at the Fermi level is nearly 40% of that of $SrPt_2As_2$, which may account for relatively lower T_c of $LaPd_2Sb_2$ compared with $SrPt_2As_2$.

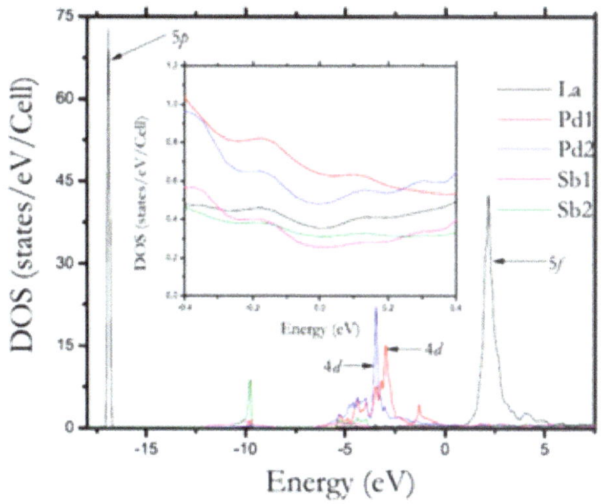

Figure 56. Calculated total density of states for La, Pd1, Pd2, Sb1 and Sb2 (see figure 51 on the atom sites with the suffixes of 1 and 2). The inset shows the contribution to Fermi level from each atom. Reprinted from [39] with permission from Elsevier.

The As-derivative of the aforementioned compound, $LaPd_2As_2$, which was known to have a $ThCr_2Si_2$-type structure [353], was also studied in the FIRST Project [40]. In the $ThCr_2Si_2$-typeAEM_2X_2, interlayer X–X bonding is sometimes formed between the adjacent M_2X_2 layers to give the collapsed tetragonal (cT) structure with a 3D network. Superconductivity in the collapsed Fe_2As_2 planes is often attained under high pressure taking well known examples of$AEFe_2As_2$ (AE = Ca, Sr, Ba, Eu) which undergo both superconducting and collapsed transitions under high pressure [354–357]. Superconductivity under ambient pressure in the cT structure is rather rare and has been reported for $AEPd_2As_2$ (AE = Ca, T_c = 1.27 K; AE = Sr, T_c = 0.92 K) [358]. In the FIRST Project, it was elucidated that $LaPd_2As_2$ has the cT structure under ambient pressure with the interlayer As–As distance of 2.318 Å, slightly shorter than the covalent single bond of 2.38 Å for As. Moreover, it was confirmed for the first time that this cT phase shows type-II superconductivity below 1 K. The superconducting and physical parameters obtained for $LaPd_2As_2$ are $\mu_0H_{c2}(0)$ = 0.402 T (by the Werthamer–Helfand–Hohenberg (WHH) theory using the $\rho_{90\%}$ point as T_c), $\xi_{GL}(0)$ = 137 Å, Θ_D = 261 K, γ = 5.56 mJ (mol K^2)$^{-1}$ and $\Delta C(T_c)/\gamma\ T_c$ = 1.17. The density of states at the Fermi level calculated from the specific heat data is as small as ~0.84 states per eV per formula unit (fu) which may explain the relatively low T_c of this cT phase.

Co-based superconductor $LaCo_2B_2$

Only few reports are available for Co-based superconducting compounds [359, 360]. In the FIRST Project, $LaCo_2B_2$, which was reported first in 1973 [361], was re-visited and was found to be superconducting after isovalent or aliovalent substitution for the constituent cations [34]. Polycrystalline samples of $(La_{1-x}Y_x)Co_2B_2$, $La(Co_{1-x}Fe_x)_2B_2$ and $LaCo_2(B_{1-x}Si_x)_2$ as well as the mother compound were prepared by the arc-melting method for La, Co, B, Y, Fe mixtures. Powder x-ray diffraction of the arc-melted $LaCo_2B_2$ sample was consistent with the tetragonal $ThCr_2Si_2$-type structure as reported previously [361], having lattice parameters of $a = 3.610$ Å and $c = 10.20$ Å. As shown in figure 57, the 10% Y-doped sample showed superconductivity below 4.4 K. Magnetic susceptibility data revealed Pauli paramagnetism for the normal state of this compound and the *M–H* curve (figure 57(b) inset) at 2 K reveals type-II nature of superconductivity. Superconductivity with $T_c \sim 4$ K was also seen in the Fe-doped system of $La(Co_{1-x}Fe_x)_2B_2$ for $x \geq 0.1$ while the Si-doped samples did not show superconductivity.

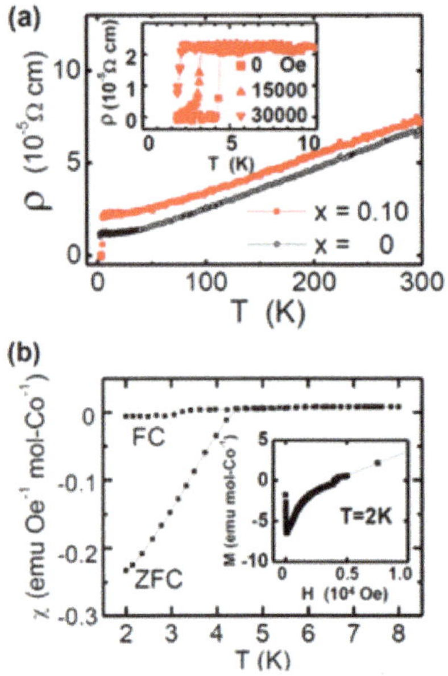

Figure 57 (a) Temperature dependence of electrical resistivity for $(La_{1-x}Y_x)Co_2B_2$. The inset shows temperature dependence of resistivity for $x = 0.1$ in magnetic field. (b) Temperature dependence of the magnetic susceptibility for the $x = 0.1$ sample under ZFC and FC conditions at 10 Oe. The inset shows the field dependence of magnetization at 2 K. Reprinted with permission from [34]. Copyright 2011 by American Physical Society.

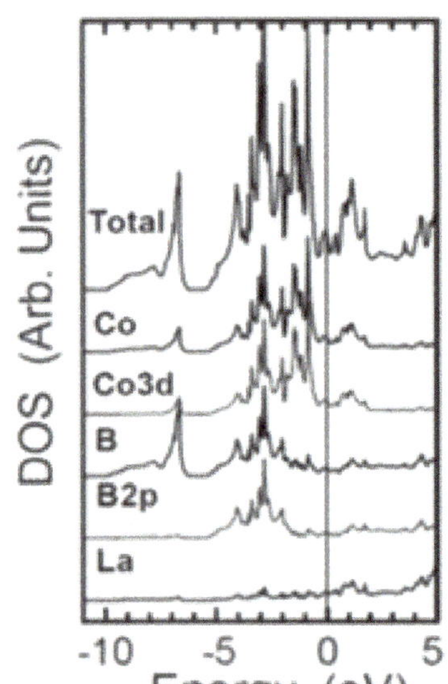

Figure 58 Calculated DOS of $LaCo_2B_2$, with the PDOS for La, Co, Co 3*d*, B, and B 2*p*. Reprinted with permission from [34]. Copyright 2011 by American Physical Society.

Electronic structure of $LaCo_2B_2$ was investigated theoretically by the DFT method; calculated DOS is shown in figure 58. From the DFT calculation, it was confirmed that La ions take +3 state and metallic conduction occurs in the CoB layer composed of highly covalent Co and B. This strong covalency suppresses the spin moment of the Co ion resulting in the Pauli paramagnetic state. Such a situation is caused by the relatively shallow B 2*p* level compared with the As 4*p* level in LaFeAsO where antiferromagnetic state is realized rather than the paramagnetic one.

Enhancement of Superconductivity by Phosphorus Doping in $BaNi_2As_2$

$BaNi_2As_2$ has a tetragonal $ThCr_2Si_2$-type structure at room temperature [348, 362, 363] but it undergoes a structural transition at ~130 K to a triclinic form where alternate Ni–Ni bonds are formed in the Ni plane (see figure 60(a)) [362]. Below 0.7 K, the triclinic phase shows superconductivity, which is believed to be of conventional BCS-type [348, 363–365]. In the FIRST Project, $BaNi_2As_2$ was reinvestigated because of this unique structural transition, i.e., this material was expected to offer a stage for studying chemical tuning of soft phonons by elemental substitution and its effects on superconductivity [35].

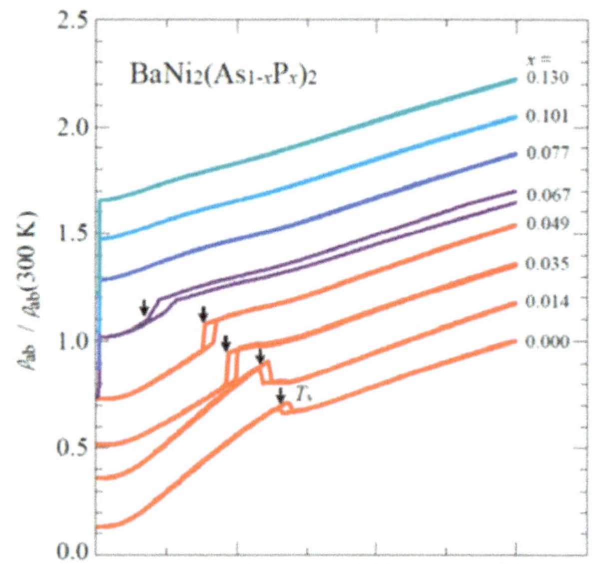

Figure 59. Temperature dependence of electrical resistivity parallel to the *ab* plane, *ρab*, normalized by the value at 300 K for $BaNi_2(As_{1-x}P_x)_2$. The data measured upon heating and cooling are plotted. For the sake of clarity, *ρab*/*ρab*(300 K) is shifted by 0.175 with respect to all data. T_s is the phase transition temperature at which the tetragonal-to-triclinic phase transition occurs. Reprinted with permission from [35]. Copyright 2012 by American Physical Society.

Single crystals of $BaNi_2(As_{1-x}P_x)_2$ were grown by a self-flux method starting from a mixture of Ba, NiAs, Ni and P [35]. The solubility limit was determined to be $x = 0.13$ and powder x-ray diffraction confirmed the single-phase nature for all specimens with $x \leq 0.13$. Figure 59 shows the temperature dependence of the electrical resistivity parallel to the *ab* plane of the $BaNi_2(As_{1-x}P_x)_2$ crystal. The tetragonal-to-triclinic transition is clearly seen in this figure as the resistivity anomaly with thermal hysteresis. The transition temperature decreases with increasing x and finally the triclinic phase disappears for $x \geqq 0.07$. Enhancement of superconductivity by phosphorus doping is striking; T_c, which is below 0.7 K for the triclinic phase with $x < 0.07$, is suddenly increased to 3.33 K in the tetragonal phase with $x = 0.077$. Figure 60 shows the phase diagram of the $BaNi_2(As_{1-}xPx)_2$ system. Triclinic phase formation is suppressed by phosphorus doping and instead superconductivity is enhanced drastically following the disappearance of the triclinic phase. The Debye frequency ω_D and logarithmic averaged phonon frequency ω_{ln} calculated from the specific heat of the tetragonal phase exhibit significant softening near the tetragonal-to-triclinic phase boundary. The low-lying soft phonons seem to play, being strongly coupled with acoustic modes, an important role in the enhancement of superconductivity in the tetragonal phase. Indeed, the normalized specific heat jump $\Delta C(T_c)/\gamma T_c$ is increased from 1.3 in the triclinic phase to 1.9 in the tetragonal phase with $x = 0.077$, i.e., the system is transformed from the weak coupling regime to the strong coupling one by phosphorus doping. Such a mechanism of T_c enhancement is common to other systems of CaC_6 [366–368] and Te [369] under high pressure.

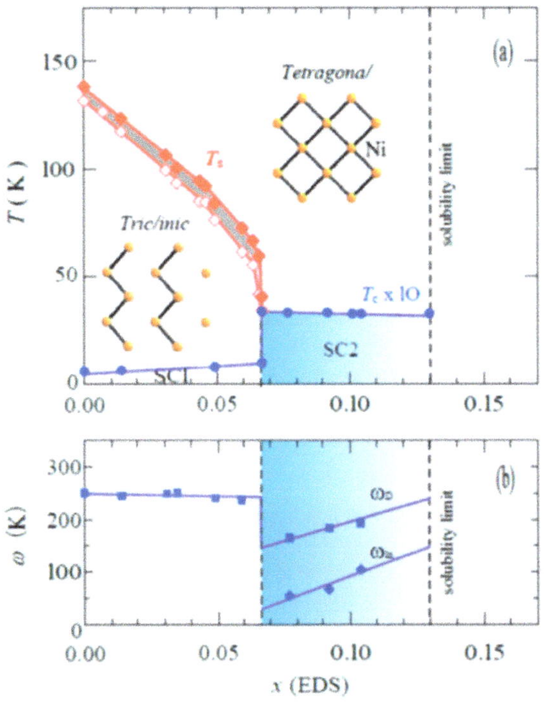

Figure 60. (a) Electronic phase diagram of $BaNi_2(As_{1-}xPx)_2$. The closed circles represent the superconducting transition temperatures T_c. For clarity, the values of T_c have been scaled by a factor of 10. SC1 and SC2 denote the superconducting phases. The open and closed diamonds represent the tetragonal-to-triclinic structural transition temperatures T_s upon cooling and heating, respectively. The insets show schematic views of Ni planes in the triclinic and tetragonal phase. (b) Debye frequency ω_D and logarithmic averaged phonon frequency ω_{ln} as a function of phosphorus content x. Reprinted with permission from [35]. Copyright 2012 by American Physical Society.

Transition metal dichalcogenides

The transition metal dichalcogenides, MX_2 have renewed interest recently and were studied intensively in this project. In this section, two topics are presented; one is the CdI_2-type telluride family and the other is the pyrite (FeS_2)-type chalcogenide family.

CdI_2-type tellurides

$IrTe_2$ has a trigonal CdI_2-type structure (see figure 61) with the space group of $P\bar{3}m1$. The Ir atom is octahedrally coordinated by the Te atoms and the edge-sharing of the $IrTe_6$ octahedron forms the Te–Ir–Te composite layers which are stacked along the c-axis of the trigonal lattice. Both the Ir and Te atoms form 2D regular triangular lattices within the ab-plane with three equivalent Ir–Ir (Te–Te) bonds. $IrTe_2$ undergoes a first-order transition at ~250 K transforming to a low temperature phase for which a monoclinic structure was first proposed [370]. Recently, a CDW-like superlattice modulation with wave vector of $\boldsymbol{q} = (1/5,0,-1/5)$ was observed in electron diffraction patterns for the low temperature phase [371]. Considering the partly filled d-orbitals of the Ir atoms, the orbital degree of freedom is believed to play an important role in this transition [372, 373]. In this project, the low temperature structure was analyzed in detail using x-ray data for single crystals [374]. The structure at 20 K was found to be triclinic (space group $P\bar{1}$) as shown in figure 62. In this triclinic structure, one out of five Ir–Ir bonds along the trigonal a-axis shrink considerably forming Ir–Ir dimers as illustrated by the yellow hatch in figure 62. The plane of the dimers propagates with the vector of $\boldsymbol{q} = (1/5, 0, -1/5)$ consistent with the aforementioned electron diffraction data. This dimerization seems to affect the physical properties of the system seriously as the first-principles band calculations indicated that tilted two-dimensional Fermi surfaces emerge in the triclinic phase with a possible switching of the conduction plane from the vassal (ab) plane in the trigonal phase to tilted plane normal to q in the triclinic phase [374].

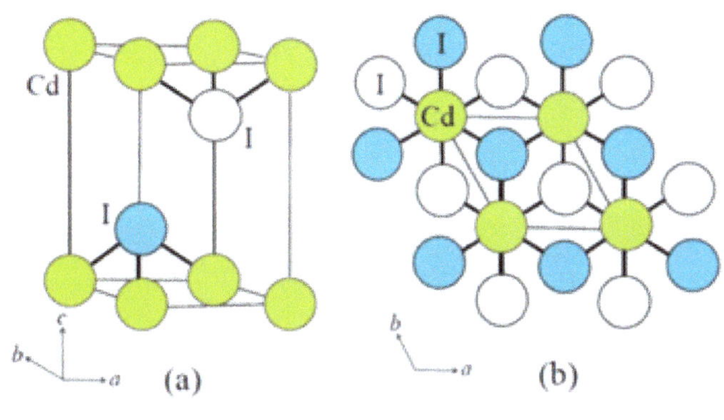

Figure 61. Crystal structure of CdI_2: (a) unit lattice and (b) composite plane consisting of the CdI_6 octahedra projected on the ab plane where the I atom, represented by the white (blue) circle, is located below (above) the Cd plane.

It was found in 2011 that partial substitution of Pt for Ir in $IrTe_2$ suppresses the formation of the low temperature phase resulting in the appearance of superconductivity [375, 376]. A phase diagram of the $Ir_{1-}xPtxTe_2$ system was first determined for the polycrystalline samples [63] and then by single crystal data in the FIRST Project [377]. Figure 63 is the phase diagram based on single crystal data; superconductivity appears for $\sim 0.04 < x < \sim 0.14$ in $Ir_{1-}xPtxTe_2$ with the highest T_c of ~3.2 K for $x = 0.04$ at the phase boundary of the trigonal and triclinic phases (the monoclinic phase in this figure corresponds to the triclinic phase in figure 62, representing the simplified symmetry). For the polycrystalline sample with $x = 0.04$, type-II superconductivity with $\mu_0 H_{c2}(0) = 0.17$ T and $\Delta C(T_c)/\gamma T_c = 1.5$ has been reported [63]. The Pt substitution works as the electron doping shifting the Fermi level upward and affecting the DOS near the Fermi level (it causes an increase of DOS near the Fermi level in the triclinic phase, and a decrease of DOS in the trigonal phase). Another effect of the substitution, which seems to be more essential, is the suppression of the triclinic phase by breaking the Ir–Ir dimers.

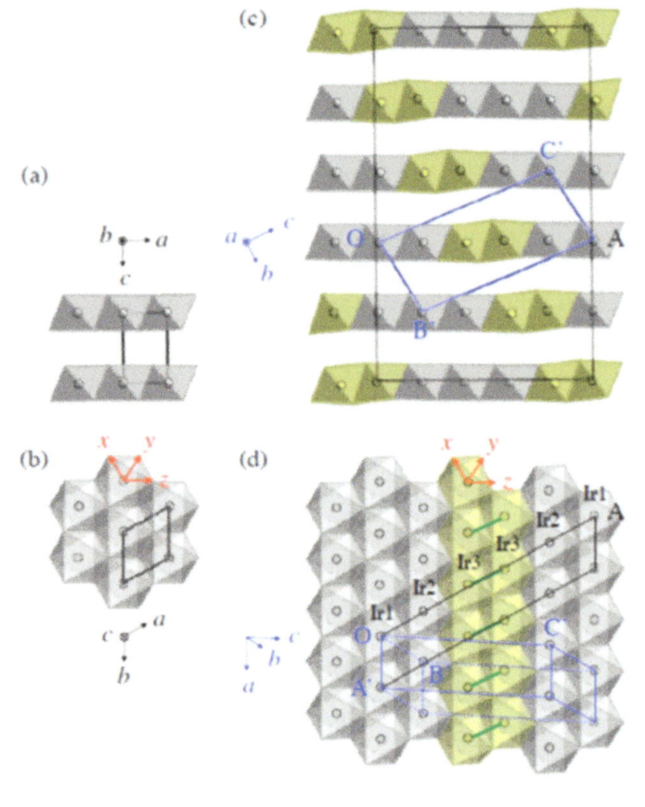

Figure 62. (a), (b) The crystal structure of $IrTe_2$ at 300 K. Black lines represent the unit cell of the trigonal lattice. (c), (d) The crystal structure of $IrTe_2$ at 20 K. Blue lines represent the unit cell of the triclinic lattice. Black lines represent the $5a \times b \times 5c$ supercell (a, b, and c are the high-temperature trigonal-cell parameters). Reprinted with permission from [374]. Copyright 2014 by the Physical Society of Japan.

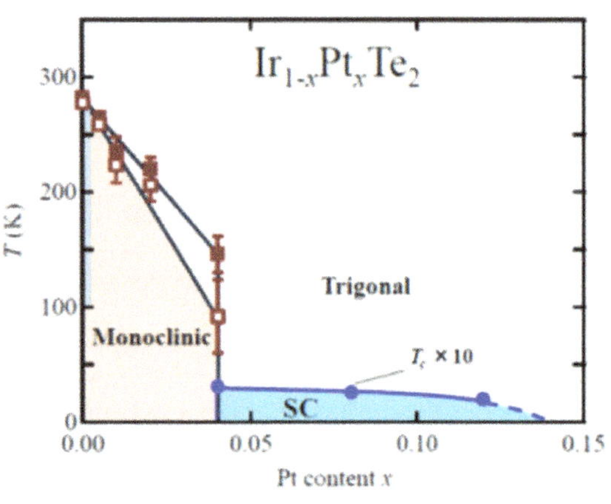

Figure 63. Phase diagram of $Ir_{1-}xPtxTe_2$ determined using single-crystal samples. Closed and open squares respectively represent the trigonal–monoclinic transition temperatureT_s upon heating and cooling (the 'monoclinic' phase corresponds to the triclinic phase in figure 62, representing the simplified symmetry). Closed circles represent the *Tc* and SC represents the bulk superconducting region. The hatched area represents a temperature range of broad hysteresis in resistivity. Reprinted with permission from [63]. Copyright 2012 by the Physical Society of Japan.

Similar results on suppression of the phase transition and appearance of superconductivity have been reported for several systems such as $Ir_{1-}xPdxTe_2$ [371], $PdxIrTe_2$ [371] and $CuxIrTe_2$ [378], where substitution or intercalation of Pd or Cu brings about essentially the same effect as the Pt substitution. In the FIRST Project, two more advances were made on $IrTe_2$ and related systems. The first is the isovalent Rh doping for $IrTe_2$ [64]. The $Ir_{1-}xRhxTe_2$ system shows similar features to $Ir_{1-}xPtxTe_2$, i.e., triclinic phase formation is suppressed by Rh doping, resulting in the appearance of superconductivity with $T_c \sim 2.6$ K, despite that band filling is unchanged by the isovalent Rh doping as long as a rigid-band picture is concerned. A distinct difference was, however, seen between $Ir_{1-}xRhxTe_2$ and $Ir_{1-}xPtxTe_2$; in the former phase, a doping of $x \sim 0.1$ is needed for the complete suppression of the triclinic phase, which is three times larger than$x = 0.03$ in the latter. This difference seems to be caused by the lower volume expansion and the aforementioned unchanged band filling in the case of Rh doping

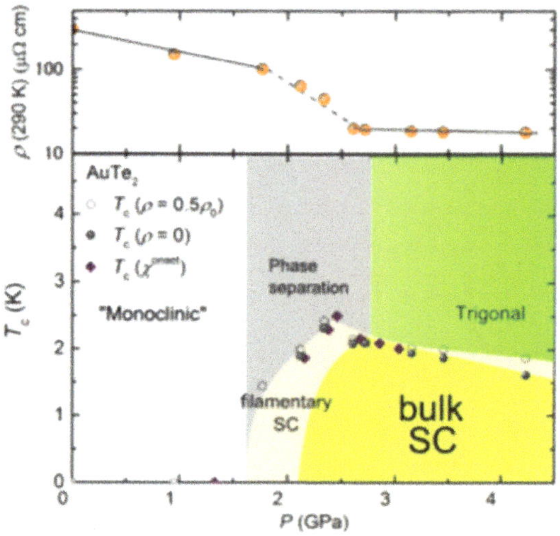

Figure 64. (Top) Pressure dependence of resistivity at 290 K, ρ(290 K) (since the applied pressure decreases by approximately 0.2 GPa on cooling, the actual pressure at 290 K is larger than the displayed pressure). (Bottom) *T–P* phase diagram for $AuTe_2$. Open (filled) circles represent *Tc* determined using $\rho_{50\%}$ point and filled diamonds indicate *Tc*determined by the onset of the diamagnetic shielding signal [61]. (The 'monoclinic' phase corresponds to the triclinic phase in figure 62, representing the simplified symmetry.) Reprinted with permission from [61]. Copyright 2013 by the Physical Society of Japan.

Another advance was attained from the study of a related system of $Au_{1-}xPtxTe_2$ [62]. $AuTe_2$ has a monoclinically distorted CdI_2-type average structure with a space group of *C*2/*m* where Te–Te zigzag chains run along the *a*-axis [379]. In actuality, the structure is associated with incommensurate modulation with a wave vector $\boldsymbol{q} = -0.4076a^* + 0.4479c^*$ and due to this modulation, Te–Te dimers with a short distance of 2.88 Å exist in the real structure instead of the zigzag chains [380, 381]. It was found that Pt doping brings about structural change from the distorted CdI_2-type for $x = 0$ to distortion-free CdI_2-type without Te–Te dimers for $x = 0.35$, via the two-phase mixed region for $x = 0.1$ and 0.15 [62]. Type-II superconductivity was observed for the distortion-free $x = 0.35$ sample with $T_c = 4.0$

K, $H_{c2}(0)$ = 12.9 kOe and $\xi_{GL}(0)$ = 160 Å. The $\Delta C(T_c)/\gamma T_c$ is 1.57, exceeding the BCS weak coupling limit. It should be noted that superconductivity appears by breaking of the Te–Te dimers, which shows striking similarity with the $Ir_{1-x}Pt_xTe_2$ system where superconductivity appears by breaking of the Ir–Ir dimers. It was also found that superconductivity is induced in $AuTe_2$ by application of pressure instead of Pt doping [61]. As shown in the phase diagram of figure 64, application of mechanical pressure causes a structural change from the distorted CdI_2-type to the distortion-free CdI_2-type via the two-phase mixed region. Superconductivity appears above 2.12 GPa with the highest T_c of 2.3 K at 2.34 GPa.

Pyrite-type Chalcogenide Family

Iridium dichalcogenides IrX_2 (X = Se and Te) sometimes take pyrite-type (FeS_2-type) structures after introduction of vacancies for the Ir sites [382]. It is also known that application of high pressure is effective to stabilize the pyrite-type form against the Cd_2I_2-type one [383]. In the Ir-deficient phase Ir_xX_2, Ir vacancies are distributed randomly forming a cubic pyrite-type structure with a space group of $Pa\bar{3}$ (figure 65). This structure can be interpreted as the NaCl-type constructed by the face centered cubic sublattice of Ir and the X–X dimers located at the center of each edge as well as at the body center of the cubic lattice. In the Ir_xX_2 phase with the particular value of x = 0.25, Ir vacancies tend to be distributed in an ordered way in which one of four Ir atoms are regularly removed [382]. The vacancy ordering results in a stoichiometric phase of Ir_3X_8 having a rhombohedral structure with the space group of $R\bar{3}$ (in Ir_3X_8, the rhombohedral distortion is far less pronounced and the vacancy ordering may not be perfect compared with the corresponding phase of Rh_3X_8 [382]).

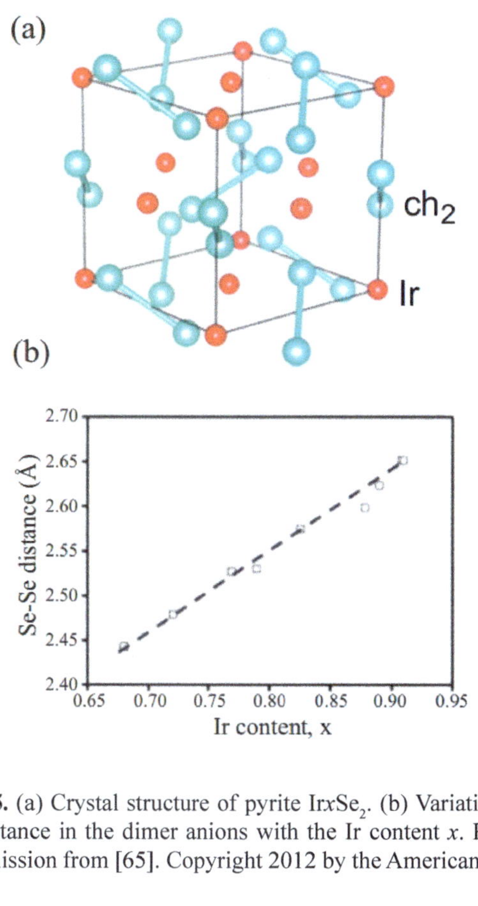

Figure 65. (a) Crystal structure of pyrite Ir_xSe_2. (b) Variation of the Se–Se distance in the dimer anions with the Ir content x. Reprinted with permission from [65]. Copyright 2012 by the American Physical Society.

In the FIRST Project, the pyrite-type Ir_xX_2 samples were synthesized under high pressure and their superconductivity was studied for the first time [65]. Figure 66 gives the electronic phase diagram obtained for the Ir_xSe_2 system. By increasing x, the system changes from insulating to metallic nature and superconductivity starts to emerge at the Ir content of $x \sim 0.75$ with the highest T_c of ~6.4 K at x = 0.91 (0.91 is the highest Ir content attained experimentally). The Se–Se distance of the Se dimer increases linearly with increasing x as shown in figure 65. A similar phase diagram was obtained for the Ir_xTe_2 system, with the highest T_c of ~4.7 K for x = 0.93. DFT calculations were carried out for the vacancy-ordered structural model of Ir_3Se_8, where there exist one long Se_1–Se_1 dimer (the edge site dimer with $r_{Se1-Se1}$ = 2.61 Å) and three short Se_2–Se_2 dimers (the body center site dimer with $r_{Se2-Se2}$ = 2.50 Å). The DFT calculations revealed that the band crossing the Fermi level consists mainly of the σ^* (anti bonding) orbital of the Se_1–Se_1 dimer and d_{z2} orbitals of the nearest Ir atoms with far less contribution from the σ^* orbitals of the Se_2–Se_2 dimers. Such a nature of the electronic structure results in the half-filled narrow conduction band which easily becomes insulating by the electron–electron correlation, electron–lattice interactions and/or the disordered Ir vacancies. DFT calculations were also performed for the vacancy-free pyrite-type structure which is composed of equivalent Se–Se dimers. In this situation, the σ^* orbitals of the Se–Se dimers contribute equally to form a wider conduction band. It should be noted that the Ir vacancy introduction causes a linear increase of the Se–Se distance of the Se dimer in correlation with the monotonous increase of T_c. In the CdI_2-type $IrTe_2$, breaking of the Te–Te dimer is essential for the appearance of superconductivity, while in pyrite-type Ir_xX_2, control of the bonding state of the X–X dimers by elongation and equalization is indispensable for inducing superconductivity. Superconducting parameters obtained for $Ir_{0.91}Se_2$ are $\mu_0H_{c2}(0)$ = 14.3 T (type-II superconductor), $\xi_{GL}(0)$ = ~48Å and $\Delta C(T_c)/\gamma T_c$ = 3.1 (strong coupling superconductor).

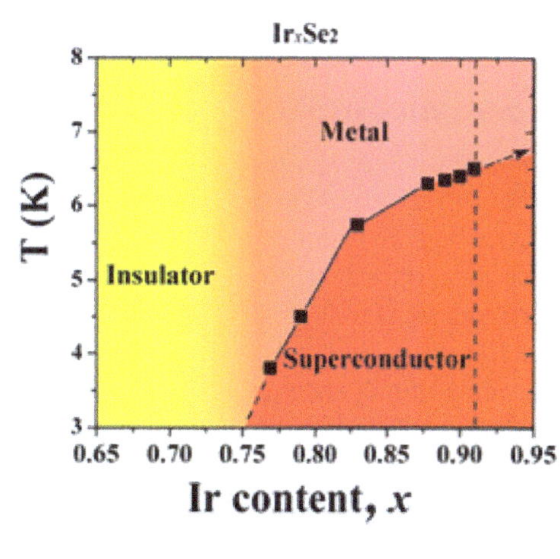

Figure 66. Electronic phase diagram of Ir_xSe_2. Reprinted with permission from [65]. Copyright 2012 by the American Physical Society.

The strong correlation between the Se–Se distance and T_c was also confirmed for the $Ir_{0.94-x}Rh_xSe_2$ system in the FIRST Project [66]. Figure 67 indicates the phase diagram of the Rh-doped system in question. With increasing Rh content, the system undergoes changes from the non-metal state to the normal-metal state with T-square resistivity via the strange-metal state with T-linear resistivity. Accompanied by this alteration, the Se–Se distance first increases then decreases, taking the maximum at $x \sim 0.4$. The striking correlation of the Se–Se distance and T_c, Θ_D and $\Delta C(T_c)/\gamma T_c$ is worth noting; both T_c and $\Delta C(T_c)/\gamma T_c$ have maximum values, while Θ_D has the minimum value when the Se–Se distance is longest. This suggests strengthening of the electron–phonon coupling and softening of phonon due to the structural instability at the edge of weak dimer states when the Rh content is $x \sim 0.4$.

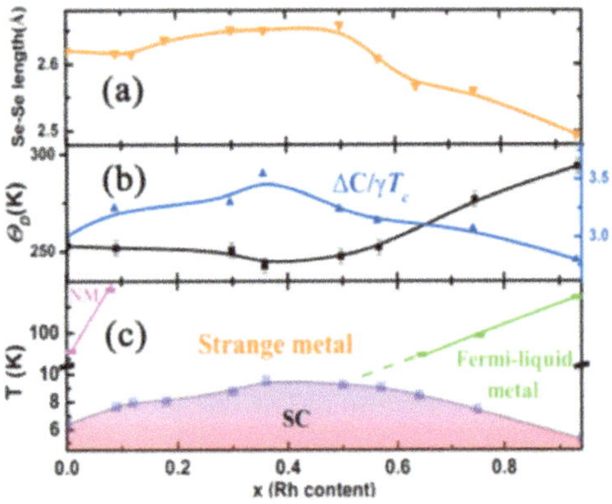

Figure 67. Properties of $Ir_{0.94-}xRhxSe_2$. (a) Variation of the Se-Se dimer bond length. (b) Debye temperature Θ_D and $\Delta C(Tc)/\gamma T_c$ as a function of the rhodium content x. (c) Electronic phase diagram of $Ir_{0.94-}xRhxSe_2$. Reprinted with permission from [66]. Copyright 2012 by the American Chemical Society.

A similar study has been carried out for the corresponding telluride system of $Ir_{0.95-}xRhxTe_2$[67]. In this system, the Te–Te distance of the Te-dimer shows a parabolic change taking the maximum between $x = 0.2$ and 0.3. Nevertheless, aforementioned correlation was not observed between the Te–Te distance and T_c nor $\Delta C(T_c)/\gamma T_c$; both parameters decrease almost linearly with x. This result may be related to the reduced electron–electron repulsion in the telluride system compared with the selenide system, which is caused by the wider width of the conduction band in the telluride system consisting of the more spatially spread Te 5p orbital.

Noncentrosymmetric Superconductors

Inversion symmetry of a crystal structure is deeply linked to the superconducting state induced in it. Usual superconductors are centrosymmetric with inversion symmetry and, in that case, the state of the Cooper pair can be described by the multiplication of a spin part and an orbital part. In order to satisfy the odd symmetry of the wave function for the exchange of pairing electrons, the spin-singlet state (asymmetric spin part) of the Cooper pair should be combined with the symmetric orbital of the s-wave or d-wave. In contrast, the spin-triplet (symmetric spin part) state should be combined with the asymmetric p-wave. Such a simple picture is not applicable for noncentrosymmetric superconductors; the paring state is no longer classified as singlet or triplet, caused by asymmetric spin–orbit coupling. Theories predicted that mixing of singlet and triplet states (a mixed parity state) occurs for a noncentrosymmetric superconducting system [384–386]. In addition to this striking feature, an anisotropic gap structure with line nodes is predicted for a noncentrosymmetric superconductor [384–386]. The first example of this class of superconductors, $CePt_3Si$, was found in 2004 [387] followed by discoveries of a variety of compounds. Among them, Ce-containing heavy-fermion materials with $BaNiSn_3$-type structure such as $CeCoGe_3$ [388], $CeIrSi_3$ [389] and $CeRhSi_3$ [390] are worthy of special mention because of their unusually large upper critical fields far beyond the Pauli limit, which seem to be concerned with the mixed parity state.

Noncentrosymmetric Silicides

A wide variety of silicide superconductors have been known and most of them crystallize in ceontrosymmetric structures showing conventional s-wave superconductivity. Some exceptions are the Ce-containing heavy-fermion superconductors described above. In the FIRST Project, we found two new noncentrosymmetric silicide superconductors, $SrAuSi_3$ [49] and Li_2IrSi_3[50]. Here, their structural and physical properties are overviewed.

$SrAuSi_3$ is the first noncentrosymmetric superconductor containing Au, which is a heavy element and may cause strong spin–orbit coupling. The $SrAuSi_3$ is stable only under high pressure and its polycrystalline sample was prepared under high temperature–high pressure conditions [49]. In figure 68, crystal structure of $SrAuSi_3$ is shown; it is a $BaNiSn_3$-type tetragonal structure in the space group of $I4mm$. The structure can be interpreted as sequence of the atom planes along the c-axis as Sr–(Au–Si_2–Si)–Sr–(Au–Si_2–Si)–Sr... which has a close relationship with the $ThCr_2Si_2$-type and $CaBe_2Ge_2$-type structures (see figure 51). Figure 69shows the temperature dependence of electrical resistivity measured by varying the magnetic field (H). T_c of $SrAuSi_3$ is 1.6 K at $H = 0$ and decreases with increasing magnetic field, giving$H_{c2}(0) \sim 2.2$ kOe which is much lower than the Pauli limit ($H_p(0) \sim 30$ kOe). This result suggests that H_{c2} is governed by the orbital pair breaking mechanism. Indeed, the orbital limit estimated from the WHH theory is ~1.5 kOe, comparable with the experimental value. However, H_{c2} of $SrAuSi_3$ increases almost linearly with temperature deviated substantially from the WHH convex upward curve. This deviation may suggest a nonspherical Fermi surface or gap anisotropy in $SrAuSi_3$. The GL coherent length and penetration depth were estimated to be$\xi_{GL}(0) = 390$ Å and $\lambda(0) = 4400$ Å giving the GL parameter $\kappa_{GL} = 11$ consistent with the type-II nature of superconductivity.

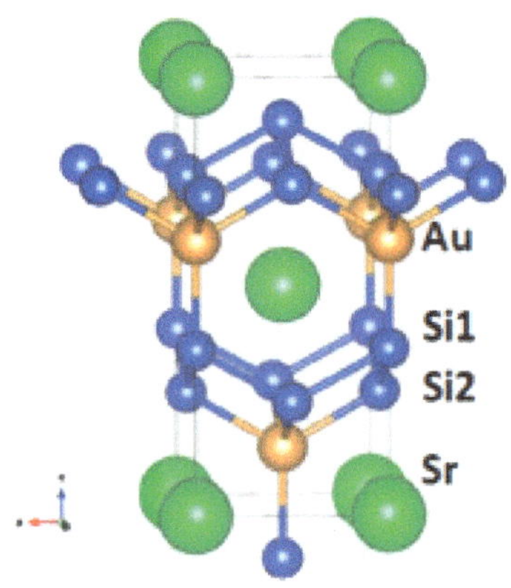

Figure 68. Crystal structure of $SrAuSi_3$. Reprinted with permission from [49]. Copyright 2014 by the American Chemical Society.

Specific heat data gave $\gamma = 6.0$ mJ (mol $K^2)^{-1}$ and $\Theta_D = 410$ K. The normalized specific heat jump is calculated to be $\Delta C(T_c)/\gamma T_c = 1.92$ and the electron–phonon coupling to be $\lambda_{ep} \approx 0.97$ using density of states at the Fermi level, 1.3 states (eV f.u.)$^{-1}$ from the band calculation. These parameters indicate that $SrAuSi_3$ is a moderately strong coupling superconductor. T_c was estimated by the McMillan formula [391], $T_c = (\Theta_D/1.45) \times \exp\{-1.04(1+\lambda_{ep})/[\lambda_{ep} - \mu^*(1 + 0.62\,\lambda_{ep})]\}$, using the standard value of 0.13 for the Coulomb repulsion parameter μ^*, resulting in ~19 K. The large difference between this value and experimental T_c of 1.6 K is striking; it may be caused by the parity mixing or the gap anisotropy in the noncentrosymmetric superconductivity. The DFT band calculation for $SrAuSi_3$ revealed that two-types of carrier exist on multiple Fermi surfaces. The major carriers conduct in the Si layers while other types of carriers conduct through the 3D network in the structure. Some of the latter carriers looked to have asymmetric spin–orbit coupling, suggesting that they may cause the unusual behaviors in $SrAuSi_3$.

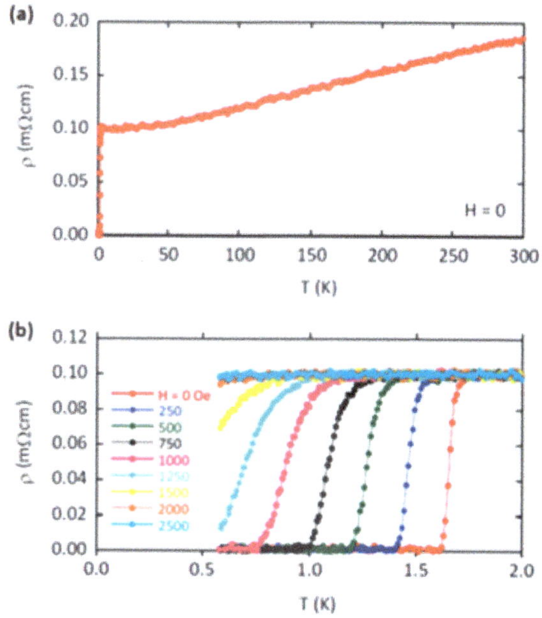

Figure 69. Temperature dependence of the electrical resistivity for a $SrAuSi_3$ polycrystalline sample: (a) data taken in the wide temperature range of 0.6–300 K at zero magnetic field ($H = 0$) and (b) data below 2 K at various magnetic field values ($H = 0$–2500 Oe). Reprinted with permission from [49]. Copyright 2014 by the American Chemical Society.

$IrSi_3$ crystallizes in a hexagonal structure with the space group $P6_3mc$ [392] (figure 70) where planar layers of four-fold Si (a distorted kagome network) are stacked along the *c*-axis sandwiching the Ir atoms. The Ir atoms are placed with unequal distances from the upper and lower neighboring Si planes leading to a polar structure with no central symmetry. In the FIRST Project, Li atoms were intercalated to this structure for the first time to obtain Li_2IrSi_3 [50]. Li_2IrSi_3 has a trigonal structure with the space group of $P31c$ as given in figure 70 where the essential nature of the Si plane stacking is preserved. In Li_2IrSi_3, the positions of Ir and Li are not symmetric concerning the distances from the upper and lower neighboring Si planes, leading to the nonpolar asymmetry. However, displacement from the symmetrical equivalent position is $\Delta z/c \sim 0.007$, which is one order of magnitude smaller than that of the mother compound of $IrSi_3$.

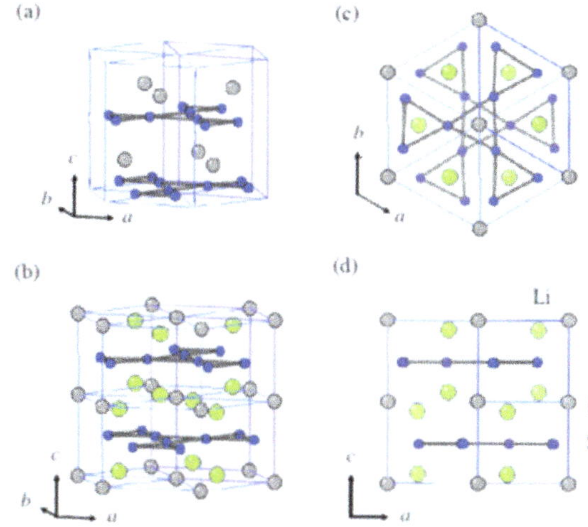

Figure 70. Crystal structures of (a) $IrSi_3$ (hexagonal, space group $P6_3mc$) and (b) Li_2IrSi_3(trigonal, space group $P31c$). Blue circles, yellow circles, and gray circles denote Si, Li, and Ir, respectively. (c) and (d) show the top and side views of the crystal structure of Li_2IrSi_3. Reprinted with permission from [50]. Copyright 2014 by the Physical Society of Japan.

Figure 71 shows resistivity in magnetic field as a function of temperature. Li_2IrSi_3 exhibits type-II superconductivity with T_c of 3.8 K. Temperature dependence of μ_0H_{c2} gave $\mu_0H_{c2}(0) = 0.3$ T and $\xi_{GL}(0) = 330$ Å (see the inset in figure 71) while the specific heat data gave $\gamma = 5.3$ mJ (mol $K^2)^{-1}$ and $\Theta_D = 484$ K, which are worth comparing with $\gamma = 0.73$ mJ (mol $K^2)^{-1}$ and $\Theta_D = 516$ K for $IrSi_3$. The normalized specific heat jump at T_c is $\Delta C(T_c)/\gamma T_c = 1.41$, consistent with the BCS weak coupling limit. All these experimental data suggest the conventional nature of superconductivity in Li_2IrSi_3; in particular, $\mu_0H_{c2}(0)$ (0.3 T) is much lower than the Pauli limit of $\mu_0H_P(0) = 6.9$ T. The weakened inversion symmetry breaking seems to account for the less noticeable unique nature of superconductivity. The appearance of superconductivity in Li_2IrSi_3 may be concerned with the enhancement of γ, i.e., a seven-fold increase of the electronic DOS at the Fermi level after the intercalation of Li for $IrSi_3$.

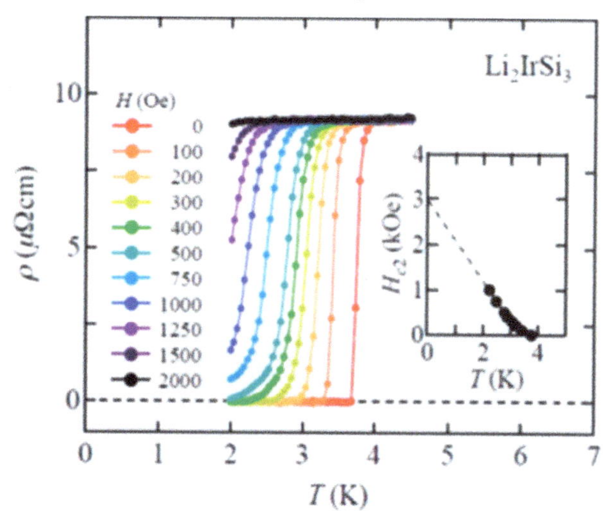

Figure 71. Temperature dependence of electrical resistivity for Li_2IrSi_3 at magnetic field H up to 2000 Oe. The inset shows the temperature dependence of the upper critical field H_{c2}. The dashed line represents the linear extrapolation of H_{c2}(T). Reprinted with permission from [50]. Copyright 2014 by the Physical Society of Japan.

Ternary equiatomic pnictides, LaMPn (M=Ir, Rh; Pn=P, As)

The ternary equiatomic pnictides with the general formula, *M'MPn* (*M'* is a large sized electropositive transition metal, *M* is a smaller transition metal, *Pn* is a pnictogen) form a large family of compounds. In the FIRST Project, the equiatomic pnictides with 4*d* and 5*d* transition metals for the *M* site were studied because previous studies have been rather confined to the systems with 3*d* transition metals for *M*. Consequently, type-II superconducting transitions were observed for the first time in LaIrP, LaIrAs and LaRhP, which were prepared using the high-pressure synthesis technique [51]. Lattice parameters for LaRhAs and LaIrP were in good agreement with previous reports [393, 394]. In figure 72, the crystal structure is shown for LaRhP, which has a tetragonal lattice with the space group of $I4_1md$ and is an ordered ternary derivative of the α-$ThSi_2$-type structure. In the structure, the Rh and P atoms are linked forming a 3D network with a trigonal planar coordination, and the La atoms are placed in the cavities of the network. There are two sets of Rh–P zigzag chains running toward the perpendicular directions.

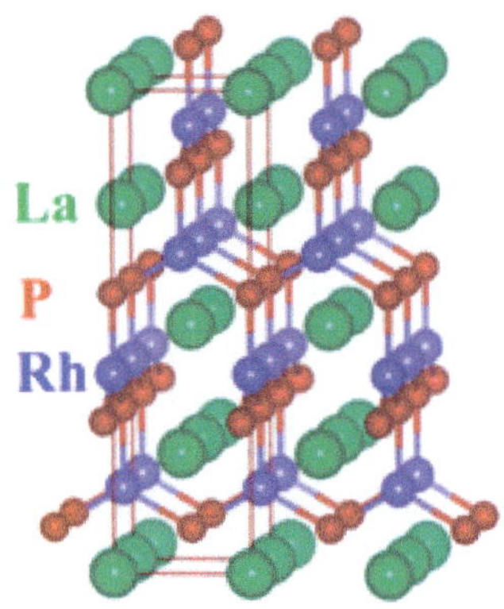

Figure 72. Crystal structure of LaRhP. Reprinted with permission from [51]. Copyright 2014 by the American Physical Society.

In table 13, superconducting parameters and the lattice parameters are shown for LaIrP, LaIrAs and LaRhP. Among them, LaIrP has the highest T_c of 5.3 K and the highest $H_{c2}(0)$ of 13.8 kOe (WHH value) or 16.4 kOe (GL value) corresponding to the highest value of electron–phonon coupling, $\lambda_{ep} = 0.67$. The $\Delta C(T_c)/\gamma T_c$ values are less than unity for all three compounds indicating the weak electron–phonon coupling regime. For every compound, the experimental $H_{c2}(0)$ is much lower than the Pauli limit of field, suggesting that the asymmetric spin–orbital coupling induced by the lack of central symmetry is not significant in the La*MPn* system.

Table 13. Superconducting and lattice parameters for the LaMX compounds [51].

Compound	LaIrP	LaIrAs	LaRhP
Superconducting parameters			
T_C(K)	5.3	3.1	2.5
$H_{C2}(0)$ (WHH) (kOe)	13.8	5.5	2.1
$H_{C2}(0)$ (GL) (kOe)	16.4	6.4	2.7
$\xi_{GL}(0)$ (Å)	14.2	22.7	34.9
$\Delta C(T_C)/\gamma T_C$	0.65	0.84	0.73
λ_{ep}	0.67	0.58	0.52
Lattice parameters (tetragonal $I4_1md$)			
a (Å)	4.2065	4.1505	4.1846
b (Å)	14.9379	14.3277	14.9358

Miscellaneous Non-iron Superconductors

Besides the superconductors described in the preceding sections, various new non-iron superconductors were discovered through the FIRST Project: they include NbSiAs ($T_c = 8.2$ K) [30], $CeNi_{0.8}Bi_2$ ($T_c = 4$ K) [37], LaNiBN ($T_c = 4.1$ K), LaPtBN ($T_c = 6.7$ K), $La_3Ni_2B_2N_3$($T_c = 15$ K) [38], La_2Sb ($T_c = 5.3$ K) [43], $Ban_{+2}Ir_4nGe_{12}n_{+4}$ ($T_c = 6.1$ K for $n = 1$; $T_c = 3.2$ K for $n = 2$) [73], Nb_4NiSi ($T_c = 7.7$ K) [72], and Ca_2InN ($T_c = 0.6$ K) [74].

4. Thin films and wires of iron-based superconductors

Epitaxial thin Film Deposition

Deposition of $Ba(Fe,Co)_2As_2$ *epitaxial films*

Fabrication of high-quality epitaxial films is essential for the study of new superconductors such as IBSCs, from the viewpoints not only of the investigation of their anisotropic physical properties but also their application to superconducting wires or tapes as well as electronic devices. Among a variety of IBSCs, the 122 type compounds have been expected to be the most promising candidates for wire application, because of their anisotropy which is substantially smaller than that of the 1111 compounds and comparable to that of MgB_2 [395]. For thin-film growth, carrier doping by the substitution of Co or P seems easier than F doping in the 1111 compounds.

Actually, the first superconducting epitaxial films of the IBSCs were realized in the 122 type compound $Sr(Fe,Co)_2As_2$ (Sr-122:Co) films with a T_c^{onset} and T_c^{zero} of approximately 20 and 15 K on LSAT (001) single-crystal substrates by Hiramatsu *et al* of Tokyo Institute of Technology (Tokyo Tech) [396]. They employed pulsed laser deposition (PLD) using a second-harmonic neodymium-doped yttrium aluminum garnet (Nd:YAG) laser ($\lambda = 532$ nm). Later Katase *et al*succeeded in fabricating Co-doped Ba-122 (Ba-122:Co) epitaxial films on LSAT substrates by using the same PLD technique and demonstrated that they had much higher stability against water vapor than Sr-122 films [397]. However, their rather large resistive transition width (ΔT_c) of approximately 3 K and low critical current density (J_c) at 5 K of 10^5 A cm^{-2} indicated that their film quality was not good enough to apply them to electronic devices and superconducting tapes.

In order to obtain higher-quality Ba-122:Co epitaxial films, Katase *et al* of Tokyo Tech Hosono group tried improving the purity of the Ba-122:Co target and homogeneity of substrate temperature [398]. A key point to obtain a high-purity Ba-122:Co target was use of fine Ba metal pieces to synthesize the BaAs precursor. Figure 73 shows x-ray diffraction patterns for Ba-122:Co films on LSAT obtained before and after such improvement. The diffraction peak from Fe impurities observed in the previous film almost disappears in the improved film. The in-plane alignment of the 122 grains was also found to be improved. The improved Ba-122:Co film exhibited a substantially higher T_c^{zero} above 20 K, narrower transition width of about 1 K, and a higher self-field J_c up to 4 MA cm^{-2} at 4 K, as shown in figure 74.

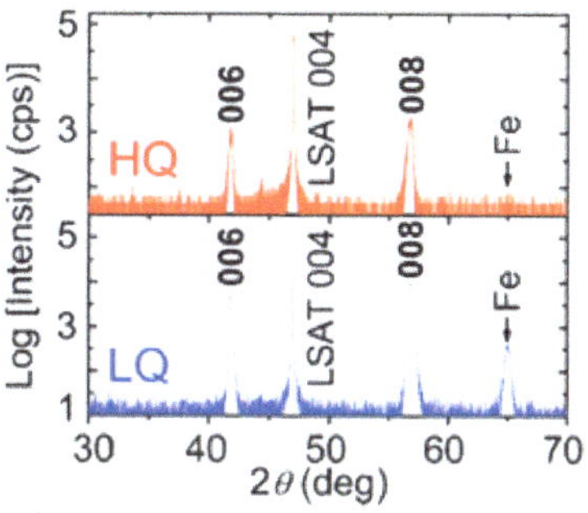

Figure 73. X-ray diffraction patterns for high-quality (HQ) and low-quality (LQ) $Ba(Fe,Co)_2As_2$ (Ba-122:Co) epitaxial films on LSAT substrates fabricated by PLD. High-quality films were obtained by improving the purity of targets and homogeneity of substrate temperature. Reprinted with permission from [398]. Copyright 2010 by the Japan Society of Applied Physics.

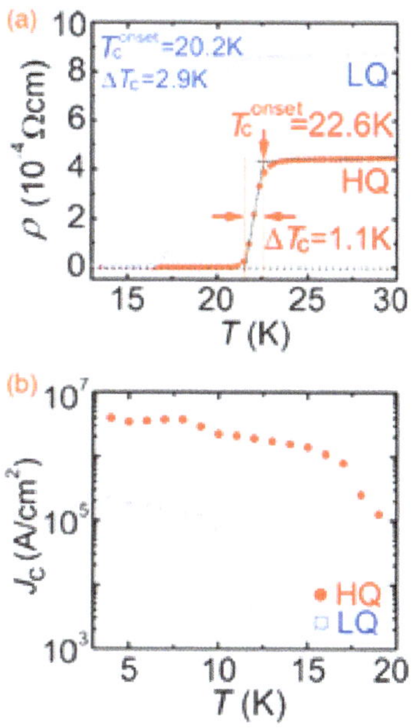

Figure 74. Temperature dependences of (a) resistivity and (b) J_c for high-quality (HQ) and lower-quality (LQ) Ba-122:Co epitaxial thin films on LSAT substrates. Reprinted with permission from [398]. Copyright 2010 by the Japan Society of Applied Physics.

They also comprehensively studied the deposition conditions for Ba(Sr)122 thin films [238] and found that Ba-122 epitaxial films could be grown in a wider temperature range of 700–900 °C than that for Sr-122. In particular, Ba-122 epitaxial films with a two-dimensional growth mode, high crystallinity, and a high self-field J_c at 4 K above 1 MA cm^{-2} could be obtained in a temperature range of 800–850 °C at a growth rate between 0.28 and 0.33 nm s^{-1}. Investigation of the Co content dependence of the transport properties for $Ba(Fe_{1-x}Co_x)_2As_2$ (Ba-122:Co) epitaxial films revealed that the highest T_c of approximately 25.5 K was obtained for $x = 0.075$ [238]. They also demonstrated that Ba-122:Co epitaxial films with a self-field J_c above 1 MA cm^{-2} could be directly prepared on MgO (100) single-crystal substrates without using a conducting buffer layer, such as the Fe buffer reported by Iida *et al* [399]. Figure 75 shows the field angular dependence of J_c for a high-quality Ba-122:Co epitaxial film on LSAT [400]. A broad J_c peak around the *c*-axis direction is observed at 6 T. This *c*-axis peak becomes less prominent at higher fields but still exists even at a high field of 15 T. This result suggested that naturally formed defects along the *c*-axis such as dislocations work as rather strong pinning centers in the Ba-122:Co epitaxial film on LSAT. This is quite different from the result reported for the Ba-122:Co epitaxial films on a Fe buffer layer where no *c*-axis peak was observed [401].

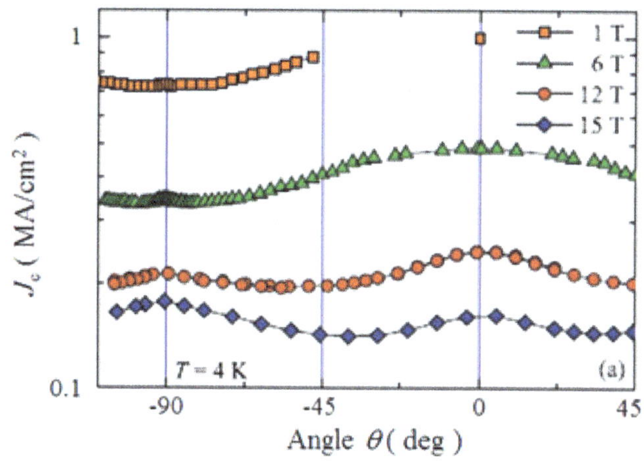

Figure 75. Angular dependence of J_c at 4 K in fields of 1–15 T for a high-quality Ba-122:Co epitaxial film on LSAT. Reprinted with permission from [400]. Copyright 2011 by IOP Publishing.

Deposition of $BaFe_2(As,P)_2$ Epitaxial Films

Superconducting Properties of $BaFe_2(As_{0.6}P_{0.4})_2$ Epitaxial Films

As described in the previous section, the 122 compounds seem the most suitable for application to superconducting wires or tapes among IBSCs. The 122 compounds exhibit superconductivity by substituting Co for the Fe site, K for the Ba(Sr) site, or P for the As site. K-doped Ba(Sr)122 has the highest T_c and upper critical field (H_{c2}) as summarized in table 14. Although synthesis of K-doped Ba-122 (Ba-122:K) epitaxial films by a molecular beam epitaxy (MBE) method was reported [402], the existence of volatile K makes it difficult to fabricate its films by a PLD method. It was also reported that the films were not stable in ambient atmosphere [402]. P-doped Ba-122 (Ba-122:P) has a higher T_c than Ba-122:Co of approximately 30 K [403] and is expected to be stable in ambient atmosphere. Adachi *et al* of ISTEC group chose this compound as a material candidate for production of superconducting tapes by a PLD method using a second-harmonic Nd:YAG laser, and examined the film preparation conditions on MgO single crystal substrates [404].

Table 14. Superconducting properties of representative iron-based superconductors.

Material	T_c (K)	$\mu_0 H_{c2}^{//c}(0)^*$ (T)	γ_H
*R*FeAs(O, F) (*R*=Nd, Sm)	47–55	80–100	5–10
$(Ba, K)Fe_2As_2$	37–38	70–135	1.5–2
$Ba(Fe, Co)_2As_2$	22–25	47–50	1.5–1.9
$BaFe_2(As, P)_2$	30–31	~60	1.5–1.9
Fe(Se, Te)	14–16	~50	1.1–1.9

High-purity targets with a nominal composition of $BaFe_2(As_{0.6}P_{0.4})_2$ were carefully synthesized by a conventional solid-state reaction method. Epitaxial films were obtained at a substrate temperature of approximately 800 °C. The energy density on the target was relatively high, approximately 10 J cm^{-2}, leading to a deposition rate of 5 nm s^{-1} at the repetition rate of 10 Hz and the substrate–target distance of 7 cm. The average FWHM value of the peaks in the ϕ-scan was about 1.5°. The obtained film exhibited a T_c^{onset} and T_c^{zero} of 26.5 and 24.0 K, as seen in the resistive transition curve of figure 76. An even higher T_c of 27.0 K was also observed for a film without patterning. The first synthesis of Ba-122:P epitaxial films was previously achieved by an MBE method [405]. The observed T_c values of the PLD films are comparable to those reported for the MBE films [405] and the single crystals [403]. Figure 77 shows the dependence of J_c on the applied field along the *c*-axis at different temperatures for the Ba-122:P epitaxial film with the T_c^{onset} of 26.5 K and a self-field J_c value at 4.2 K of 3.5 MA cm^{-2}. The film exhibited rather high in-field J_c values, for example, approximately 1 MA cm^{-2} at 4.2 K, 3 T and 10 K, 1 T, which are higher than those for Ba-122:Co films [406].

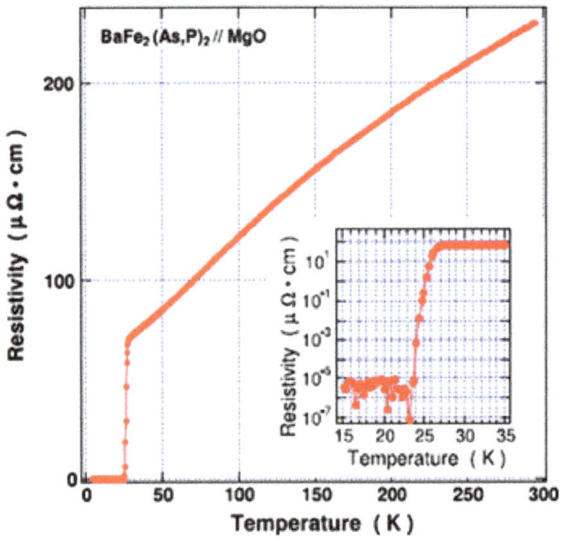

Figure 76. Temperature dependence of resistivity for a $BaFe_2(As_{0.6}P_{0.4})_2$ (Ba-122:P) epitaxial film on MgO substrate fabricated by PLD. The inset magnifies the curve near *Tc*. Reprinted with permission from [404]. Copyright 2012 by IOP Publishing.

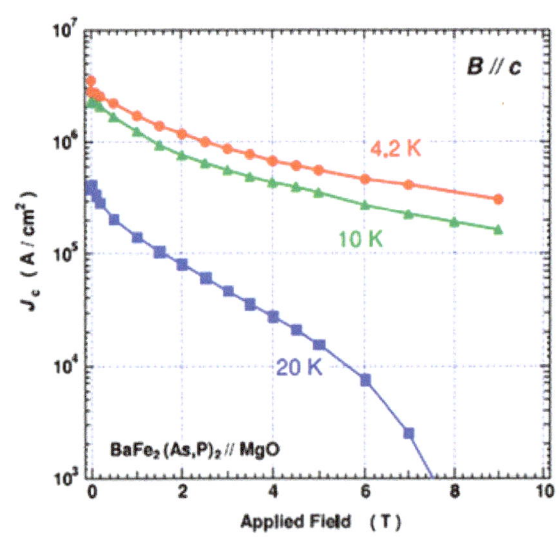

Figure 77. Magnetic field dependence of J_c at 4.2, 10, and 20 K for a Ba-122:P epitaxial film on MgO substrate fabricated by a PLD method. Magnetic field was applied parallel to the *c*-axis. Reprinted with permission from [404]. Copyright 2012 by IOP Publishing.

Superconducting Properties of $BaFe_2(As_{1-x}P_x)_2$ Epitaxial Films

The dependence of superconducting properties of $BaFe_2(As_{1-}$ $xPx)_2$ (Ba-122:P) epitaxial films on the P content was systematically investigated by Miura *et al* by using Ba-122:P targets with nominal P content *x* of 0.25, 0.33, 0.40 and 0.50 [407]. The *x* values in the films analyzed using an electron probe micro analyzer (EPMA) were found to be slightly smaller by approximately 0.05 than the nominal *x* values in bulk targets. Figure 78 shows the temperature dependence ofH_{c2} for Ba-122:P films with various analyzed *x* values in magnetic fields up to 12 T. The *x* = 0.28 film exhibits a maximum T_c^{zero} of 26.5 K, and a further increase in *x* leads to a reduction in T_c^{zero} while the *x* = 0.19 film showed a broad transition and T_c^{zero} of about 12 K. This *x* dependence ofT_c is similar to that observed in single crystals [403]. The *x* = 0.28 film also exhibits the highestH_{c2} in both field directions. The inset shows the anisotropy of H_{c2}, $\gamma_H = H_{c2}^{ab}/H_{c2}^{c}$ for the films. The *x* = 0.28 film with the optimal T_c shows the smallest γ_H value of 1.54. This is different from the case of cuprate superconductors [408], and preferable from the viewpoint of wire or tape application. Figure 79 shows the angular dependence of J_c for the *x* = 0.28 film at 10 K under different magnetic fields. This film also exhibits high in-field J_c values. It is also found that $J_c$shows minimum values in the *c*-axis direction at all fields from 0.5 to 7 T, indicating that no *c*-axis-correlated pinning centers are included in this film.

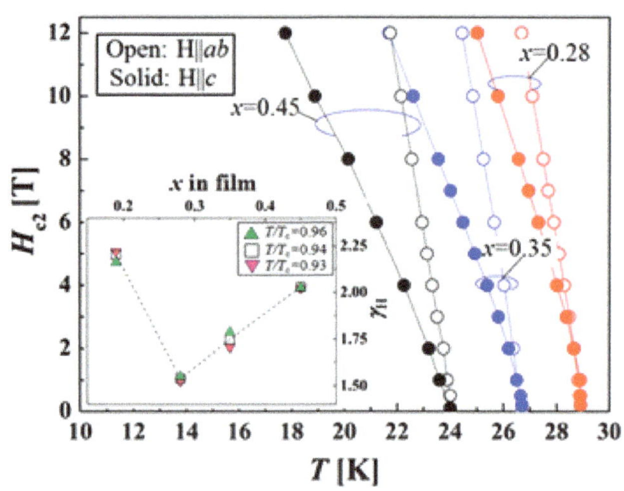

Figure 78. Temperature dependence of upper critical fields (H_{c2}) for Ba-122:P epitaxial films on MgO with analyzed P content *x* of 0.28, 0.35, and 0.45. The inset shows the *x*dependence of anisotropy parameter γ_H at *T*/*Tc* = 0.93, 0.94, and 0.96. Reprinted with permission from [407]. Copyright 2013 by the Japan Society of Applied Physics.

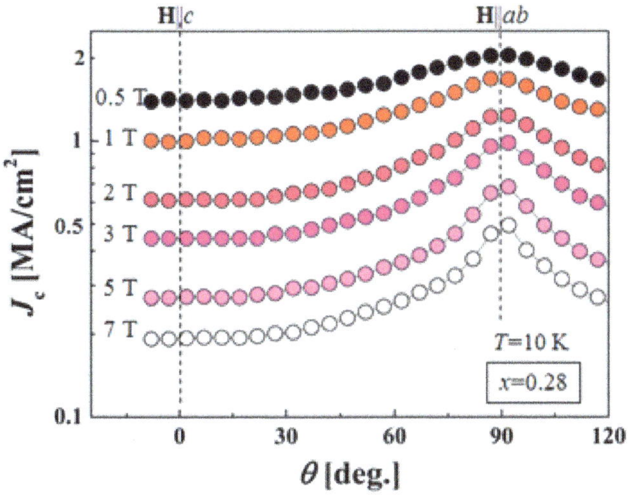

Figure 79. Angular dependences of in-field J_c at 10 K in fields of 0.5–7 T for the Ba-122:P film with x = 0.28. Reprinted with permission from [407]. Copyright 2013 by the Japan Society of Applied Physics.

$BaFe_2(As,P)_2$ films with *Strong Pinning Centers*

Addition of artificial pinning centers consisting of $BaZrO_3$ *nanoparticles*

The Ba-122:P epitaxial films on MgO substrates fabricated by a PLD method showed relatively high in-field J_c values. However, these values do not seem to be high enough for high-field application. Since the Ba-122 superconductors have very high $H_{c2}(0)$ values over 60 T, a remarkable improvement in in-field J_c would be expected, if effective vortex pinning centers could be introduced into the materials. Actually, amorphous tracks induced in Ba-122:K single crystals by heavy-ion irradiation significantly enhanced their in-field J_c and resulted in a matching field as high as 21 T without T_c degradation [409], although heavy-ion irradiation is not practical for fabrication of long wires or tapes. For the case of Ba-122 epitaxial films, naturally formed defects in Ba-122:Co films enhanced the J_c in the fields parallel to the c-axis to some extent [400], as described in the previous section. It was also reported that naturally formed nanopillars along the c-axis in the Ba-122:Co thin films prepared by PLD using a $SrTiO_3$ buffer layer and oxygen-rich targets significantly enhanced the in-field J_c around the c-axis direction [410, 411]. However, the J_c around the ab plane direction was not much improved.

For application to superconducting tapes, it is desirable to find a controllable and practical way of enhancing vortex pinning in Ba-122 films in an isotropic way. In cuprate superconductor films such as $REBa_2Cu_3O_y$ (REBCO; RE = Y or rare-earth elements) films, the addition of nanoparticles consisting of second oxide phases has been found to be effective in enhancing their in-field J_c in an isotropic way [412–414]. However, in order to apply a similar technique to Ba-122 epitaxial films, the second phases need to be chemically stable and crystallographically compatible with the Ba-122 phase, and its size should be small enough to avoid blocking of the current path.

Miura *et al* tried introducing $BaZrO_3$ (BZO) nanoparticles, which are known to be effective pinning centers in REBCO films, into Ba-122:P epitaxial films on MgO by a PLD method [415]. Approximately 80 nm thick BZO-doped Ba-122:P epitaxial films were grown from 1 mol.% and 3 mol.% BZO-doped Ba-122:P targets with the nominal P content of 0.33. The deposition conditions were similar to those employed by Adachi *et al* for the synthesis of undoped Ba-122:P films [404]. The cross-sectional elemental maps and x-ray diffraction pattern shown in figure 80(a) of the film grown from the 3 mol.% BZO-doped target indicated the presence of homogenously dispersed BZO nanoparticles. The average nanoparticle size and the average spacing were found to be 8 nm and 24 nm, respectively, leading to a density n of approximately 6.8 × 10^{22} m^{-3}. The cross-sectional TEM image of a typical BZO nanoparticle is shown in figure80(b). The size of the nanoparticle is about 5 and 10 nm parallel and perpendicular to the c-axis, respectively. The periodicity of the Ba-122 planes around the nanoparticles is only perturbed by the creation of stacking faults. However, nano-beam diffraction patterns for the nanoparticle and the Ba-122:P matrix indicated that the BZO nanoparticles are not epitaxially oriented along the Ba-122:P matrix. The T_c^{zero} values for the undoped and doped films are 26.3 and 25.0 K, respectively, indicating that introduction of the BZO nanoparticles does not induce significantT_c degradation.

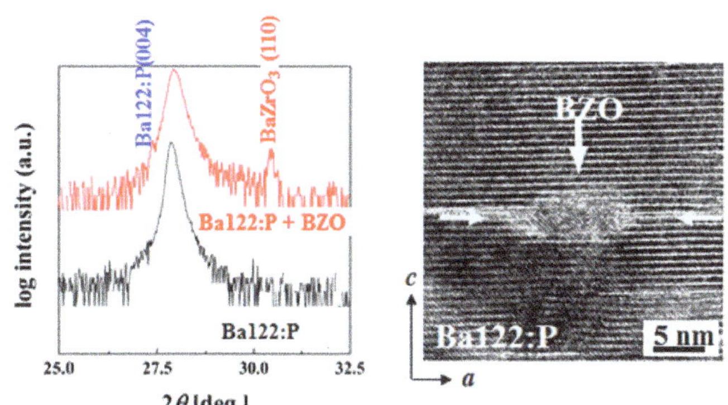

(a) (b)

Figure 80. (a) X-ray diffraction patterns for a Ba-122:P and a Ba-122:P + 3 mol% $BaZrO_3$(BZO) film prepared on MgO substrates. (b) Cross-sectional high-resolution TEM (HRTEM) image for a BZO nanoparticle in the Ba-122:P + 3 mol% BZO film. Reprinted with permission from Macmillan Publishers Ltd: [415], Copyright 2013.

Figure 81(a) shows the field dependence of J_c ($H//c$) at 5 K for the undoped Ba-122:P and BZO added Ba-122:P films. The self-field J_c ($J_c^{sf.}$) increases monotonically with the amount of BZO additive and the field decay of J_c is greatly reduced within the measured field range. The Ba-122:P film shows a characteristic crossover field H^* (90% of $J_c(H)$/ $J_c^{sf.}$ which is indicated by the arrows, followed by a power-law regime ($J_c \propto H^{-\alpha}$) with $\alpha \sim 0.40$ at intermediate fields. A more rapid decay of J_c is observed as H approaches the irreversibility field, H_{irr}. For the Ba-122:P + BZO films, H^* also increases with the BZO content. At intermediate fields, we find that both films with BZO show a slower decay of $J_c(H)$, indicating the importance of BZO nanoparticles to enhance $J_c(H)$ in magnetic fields. The non-power-law dependence observed for the Ba-122:P + BZO films is similar to that observed in RE123 films with strong pinning coming from uniformly dispersed nanoparticles [413, 414, 416].

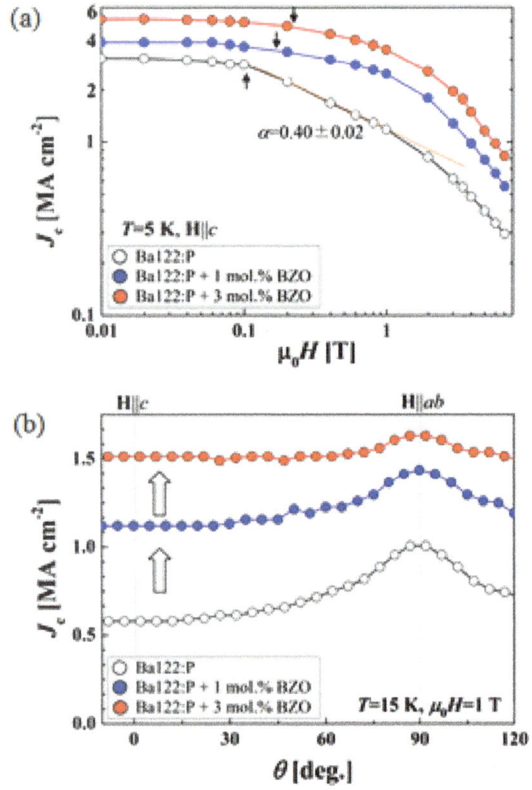

Figure 81. (a) Magnetic field dependence of J_c at 5 K for Ba-122:P, Ba-122:P + 1 mol% BZO and Ba-122:P + 3 mol% BZO films on MgO. A magnetic field was applied parallel to the *c*-axis. (b) The angular dependence of in-field J_c at 15 K, 1 T for the Ba-122:P, Ba-122:P + 1 mol% BZO and Ba-122:P + 3 mol% BZO films. Reprinted with permission from Macmillan Publishers Ltd: [415], Copyright 2013.

Figure 81(b) shows the angular dependence of J_c, $J_c(\theta)$, curves measured for the three films at 1 T, 15 K. By adding BZO the J_c increases for all orientations with respect to the J_c of the Ba-122:P film. In particular, the Ba-122:P + 3 mol% BZO film exhibits an almost isotropic J_c with the value for *H*//*c* 2.6 times higher than that of the Ba-122:P film. The minimum value of $J_c(\theta)$,$J_{c,min}$, of 1.5 MA cm^{-2} at 1 T, 15 K is over 28 times and 7 times higher than that of Ba-122:Co films with *c*-axis columnar defects [410] and Ba-122:Co films with super-lattice structures [417], respectively, in very similar field and temperature conditions, indicating strong isotropic pinning by the BZO nanoparticles.

In figure 82, the pinning force, $F_p = J_c(H) \times \mu_0 H$ is compared with that of several superconductor materials. At 15 K, F_p for the Ba-122:P + 3 mol% BZO film is over 3 times higher than that for the Ba-122:P film and higher than NbTi [418] at 4.2 K at all magnetic fields. Comparing with MgB_2 data at 15 K [419], F_p is clearly higher for $\mu_0 H > 0.5$ T. At $T = 5$ K, the F_p of the Ba-122:P + 3 mol.% BZO film, which is the minimum value in all field directions, reaches ~59 GN m^{-3} for $\mu_0 H > 3$ T up to the highest field we measured (9 T), a 50% increase over Nb_3Sn at 4.2 K [420].

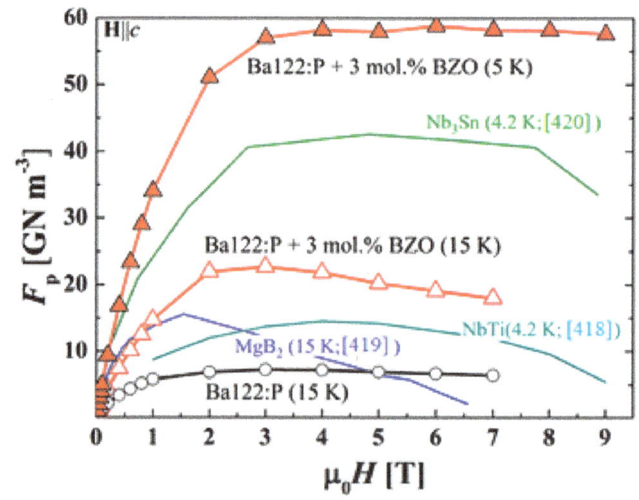

Figure 82. Pinning force F_p at magnetic fields applied parallel to the *c*-axis for the Ba-122:P + 3 mol% BZO film at 5 K and 15 K and the Ba-122:P film at 15 K. For comparison, the data for NbTi at 4.2 K [418], MgB_2 (*H//ab*) at 15 K [419], Nb_3Sn at 4.2 K [420] are included. Reprinted with permission from Macmillan Publishers Ltd: [415], Copyright 2013.

Miura *et al* also found that the characteristic magnetic field, where the maximum relative J_c enhancement by BZO addition was observed, increased monotonically by increasing the nanoparticle density. From the comparison with the previous results of REBCO films with BZO nanoparticles [421], it was deduced that the effective way to optimize the J_c performance of IBSCs as well as cuprates is dispersing nanoparticles with the average size smaller than ~3*(2$\xi(T)$) and with densities such that the nanoparticle spacing matches the intervortex distance. These results suggested the possibility of further enhancing in-field J_c properties, in particular at higher fields, by optimizing the landscape of nanoparticles or nanoscale defects in Ba-122:P epitaxial films.

Intentionally Grown c-axis Pinning Centers

Sato *et al* carefully investigated the properties of Ba-122:P epitaxial films on MgO (001) single-crystal substrates prepared by PLD using the second harmonic of an Nd:YAG laser and found that films with very high and less anisotropic in-field J_c could be obtained at certain deposition conditions [422]. They employed a semiconductor infrared diode ($\lambda = 975$ nm, and maximum power = 300 W) for substrate heating and achieved high substrate temperature (T_s) up to 1400 °C. High-purity Ba-122:P targets with the nominal P content *x* of 0.30 were used. It was found that 150–200 nm thick epitaxial films with high crystallinity ($\Delta\phi$ and $\Delta\omega$ well below 0.8°) were grown at rather high T_s of 1000–1100 °C (optimum at 1050 °C) and low growth rate of 0.2–0.4 nm s^{-1} for laser fluence of 3.0–3.5 J cm^{-2}. Their lattice parameters slightly different from the single crystal data [403] indicated existence of tensile strain in the films. The optimum film exhibited a T_c of 26.5 K, a narrow transition width ΔT_c of 1.5 K, and a very high self-field J_c up to 7 MA cm^{-2}. This J_c value is comparable to the recently reported high value for a film grown by an MBE method [423].

Figure 83 shows the angular dependence of J_c at 12 K, 3 T for Ba-122:P epitaxial films grown at the optimum T_s. The $J_c(\theta_H)$ curves exhibit a broad peak around the *c*-axis direction ($\theta_H = 0°$) in addition to the intrinsic J_c peak at $\theta_H = 90°$, indicating existence of pinning centers along the *c*-axis. With decreasing growth rate from 0.39 nm s^{-1} to 0.22 nm s^{-1}, the $\theta_H = 0°$ peak becomes more prominent and the J_c values in all directions become remarkably higher, resulting in less anisotropic angular dependence. These results indicate that the vortex pinning properties and J_c anisotropy of Ba-122:P epitaxial films can be controlled by tuning the growth rate.

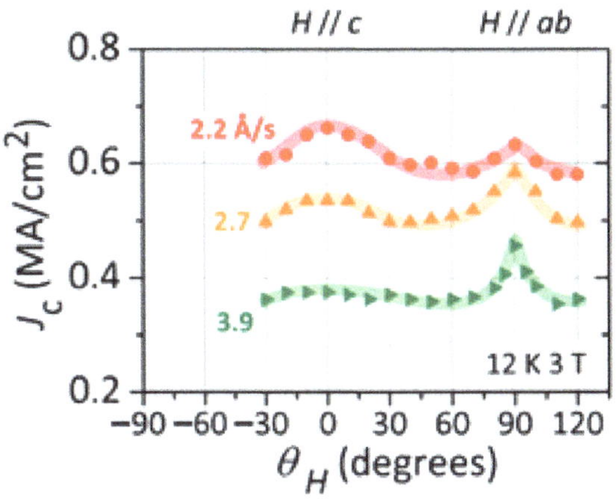

Figure 83. Angular dependence of in-field J_c at 12 K, 3 T for high-quality Ba-122:P epitaxial films grown on MgO by PLD at relatively high substrate temperatures and low deposition rates of 0.22, 0.27, and 0.39 nm s^{-1}. Reprinted with permission from [422]. Copyright 2014 by AIP Publishing LLC.

In figure 84, the magnetic field dependence of $J_c^{H//ab}$ (closed symbols) and (open symbols) of the optimum Ba-122:P epitaxial films is compared with those reported for other Ba-122 epitaxial films with high J_c [415, 417, 424–426] as well as $SmFeAsO_{1-x}F_x$ and Fe(Se,Te) films [427, 428]. The optimum Ba-122:P film exhibits an even higher $J_c^{H//c}$ than the Ba-122:P film with BZO nanoparticles at $H > 4$ T and a $J_c^{H//c}$ as high as 0.8 MA cm^{-2} at 4 K, 9 T, resulting in the highest pinning force of 72 GN m^{-3}. It would also be the noteworthy that the $J_c^{H//ab}$ value of 1.1 MA cm^{-2} at 9 T giving a pinning force of 99 GN m^{-3} is the highest obtained for IBSC films.

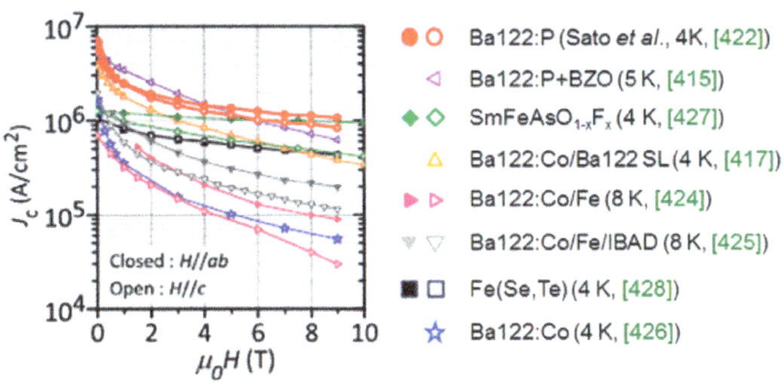

Figure 84. Comparison of magnetic field dependence of J_c for the optimum Ba-122:P epitaxial film with those reported for other Ba-122 films as well as $SmFeAsO_{1-x}F_x$ and Fe(Se,Te) films with high J_c values. The closed and open symbols represent J_c for fields parallel to the *a,b*-axes and *c*-axis, respectively. Reprinted with permission from [422]. Copyright 2014 by AIP Publishing LLC.

Figure 85 shows the cross-sectional bright-field scanning TEM (STEM) images for Ba-122:P epitaxial films grown at the growth rate of 0.22 nm s^{-1} and 0.39 nm s^{-1}. As indicated by the vertical white arrows in figure 85, there are many vertical defects with a substantially higher density than that observed in the Ba-122:Co epitaxial films by Katase *et al.* It is also found that most of the defects in the film grown at 0.22 nm s^{-1} start appearing at mid-thickness and are oriented parallel to the *c*-axis, while the defects in the latter film originate just at the substrate surface and are tiled with respect to the *c*-axis. Other planar or line defects in the *ab* plane, such as stacking faults, were not observed. Energy dispersive x-ray spectroscopy (EDX) combined with STEM revealed that the chemical composition of the defects is the same as that of the matrix region and that the impurity oxygen concentration in the films is less than the detection limit. These results indicated that the defects are not an impurity phase such as $BaFeO_2$ [429] but may be edge or threading dislocations and/or domain boundaries. The stronger vortex pinning along the *c*-axis for the film grown at the lower rate can be explained by a high density of straight defects with a lateral size (4 nm) close to the double of the ξab of Ba-122:P at 4 K [407]. The very high in-field J_c observed in these epitaxial films indicates a high potential of Ba-122:P epitaxial films for application to superconducting tapes or wires, though the very high T_s would not be favorable to film fabrication on metal substrates.

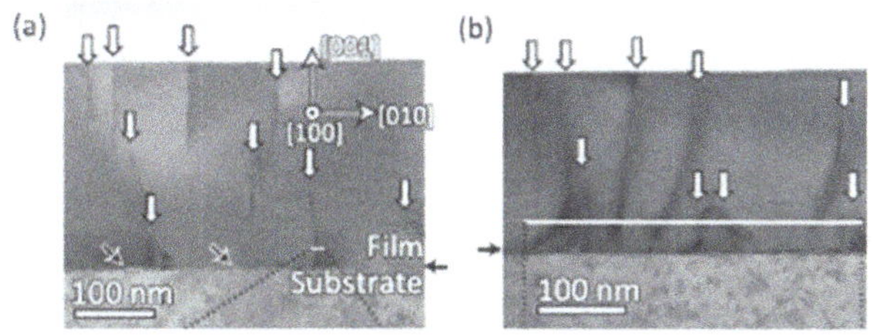

Figure 85. Cross-sectional bright-field STEM images for Ba-122:P epitaxial films grown at the growth rate of (a) 0.22 nm s^{-1} and (b) 0.39 nm s^{-1}. The arrows indicate the positions of vertical defects which are considered as strong pinning centers. Reprinted with permission from [422]. Copyright 2014 by AIP Publishing LLC.

Fabrication of Superconducting thin film Devices

Bi-crystal Josephson junctions using Ba(Fe,Co)$_2$As$_2$ epitaxial films

The Josephson junction is the most important basic element in electronic application of superconductors. For low temperature application using liquid He or a cryocooler, Josephson junctions using Nb thin films and a thin Al oxide layer as a tunnel barrier [430] have been in practical use. This sandwich-type junction exhibits ideal superconductor–insulator–superconductor (SIS) type current–voltage (*I–V*) characteristics and is very reliable. For cuprate superconductors such as $YBa_2Cu_3O_{7-x}$ (YBCO), it is very difficult to fabricate SIS type Josephson junctions because of their peculiar physical properties such as very short coherence length and *d*-wave symmetry of the superconducting gap. The very high substrate temperature of 700–800 °C for fabrication of their epitaxial films also makes it very difficult to realize a sharp film–barrier interface required for SIS junctions. However, Josephson junctions exhibiting weak-link-type or superconductor–normal metal–superconductor (SNS) type *I–V* characteristics can be readily obtained by utilizing weak links naturally formed at high-angle grain boundaries (GBs) of their thin films [431]. This weak-link behavior at a high-angle GB comes from carrier depletion due to local structural disorder near the GB. Actually, GB Josephson junctions using an epitaxial film on a bicrystal substrate (bicrystal junction) or a substrate with an artificially formed step (step-edge junction) have been applied to electronic devices such as superconducting quantum interference devices (SQUIDs) operating at the liquid-nitrogen temperature [432].

IBSCs also exhibit superconductivity only when a proper amount of charge carrier is doped, though their parent materials are not antiferromagnetic insulators but antiferromagnetic metals. Their short coherence length and rather high film growth temperature would make fabrication of SIS Josephson junctions difficult.

Katase *et al* tried fabricating Josephson junctions using high-quality Ba-122:Co epitaxial films and bicrystal substrates [433]. Ba-122:Co thin films were deposited on LSAT bicrystal substrates with a symmetrical [001]-tilt boundary having a misorientation angle θ_{GB} of 30°. Figure 86shows the I–V curve and magnetic field dependence of I_c (I_c–B curve) at 10 K for a 10 μm-wide bridge patterned across the bicrystal GB (BGB). The bridge clearly shows a resistively-shunted-junction (RSJ) type curve. The magnitude of I_c modulation, which is here defined as $[I_c(0) - I_c(0.9\text{ mT})]/I_c(0)$, is approximately 95%, indicating that most of the supercurrent originates from the Josephson current. This is actually the first demonstration of a thin film Josephson junction in IBSCs. However, the hysteresis observed in the I_c–B curves suggests that flux trapping occurs in the BGB regions, probably due to inhomogeneity in their microstructures. A 10 μm-wide bridge patterned in a single grain region of the same substrate showed an I_c of 40 mA at 10 K, implying that the I_c across the GB is suppressed to less than 1/20 of the film I_c.

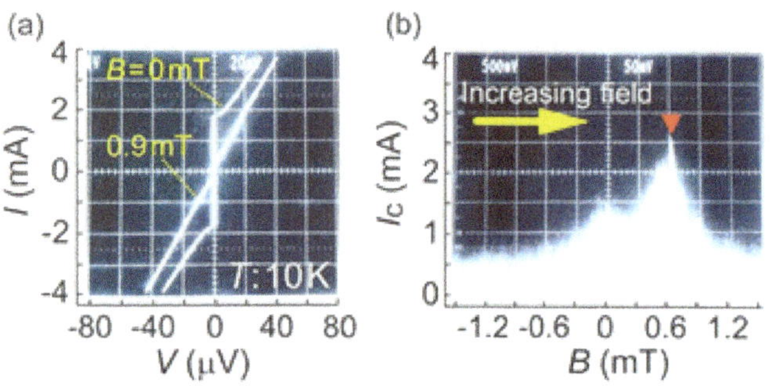

Figure 86. (a) Current–voltage (I–V) curve under B = 0 and 0.9 mT and (b) magnetic field dependence of I_c (I_c–B) curve at 10 K for a 10 μm-wide bridge patterned across the bicrystal GB (BGB) in a Ba-122:Co epitaxial film on an LSAT [001]-tilt bicrystal substrate with the misorientation angle θ_{GB} = 30°. Reprinted with permission from [433]. Copyright 2010 by AIP Publishing LLC.

Katase *et al* also fabricated Ba-122:Co epitaxial films on [001]-tilt MgO bicrystal substrates with various misorientation angles and examined the GB transport properties [434]. Though the details will be described in section 4.2.4, I–V curves for BGB junctions with θ_{GB} = 30° and 45° were found to be well fitted by the Ambegaokar–Halperin (AH) model [435], or the RSJ model taking account of thermal fluctuation, confirming that the excess current ratio is minimal for these high-angle BGB junctions. In contrast, BGB junctions with lower θ_{GB} showed I–V curves containing flux–flow behavior. Figure 87 shows the I–V curves for BGB junctions withθ_{GB} = 16°, 24°, 30° and 45°. The dotted lines are the fits by the AH model, while the solid lines are fits by a phenomenological model previously proposed to explain the fractions of flux–flow and RSJ behaviors [436]. The latter fits indicate that the fractions of the RSJ current are approximately 70%, 90%, 100% and 100% for the BGB junctions with θ_{GB} = 16°, 24°, 30° and 45°, respectively.

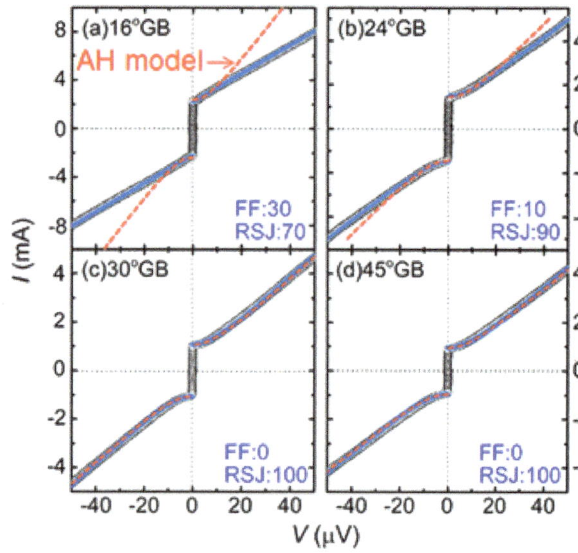

Figure 87. I–V curves at 12 K for Ba-122:Co BGB junctions with θ_{GB} = 16°, 24°, 30°, and 45° grown on MgO bicrystal substrates. The red dotted lines indicate the fits by the AH model, and the blue lines show the fits to the I–V curves with the phenomenological model combining the RSJ behavior and the FF behavior.

The BGB junctions using Ba-122:Co epitaxial films were found to exhibit the specific resistanceAR_N (A is the junction area and R_N is the junction's normal resistance) of $10^{-10} - 10^{-9}$ Ωcm², which is more than one order of magnitude lower than that for YBCO [001]-tilt bicrystal junctions [431], and metallic temperature dependence. Figure 88 shows the temperature dependence of J_c for the BGB junctions with θ_{GB} = 16°, 24°, 30°, and 45° [434]. Clear quadratic temperature dependence is observed for these high-angle BGB junctions. The solid lines in the figure are the fits to the de Gennes theory based on a conventional proximity effect in the dirty limit [437]

$$I_c = I_0\left(1 - \frac{T}{T_c}\right)^2 \frac{\kappa d}{\sinh(\kappa d)}$$

$$= \frac{V_0}{R_N}\left(1 - \frac{T}{T_c}\right)^2 \frac{\kappa d}{\sinh(\kappa d)},$$

where d is the barrier thickness, κ^{-1} is the decay length for a normal metal, and V_0 is the characteristic voltage, which is approximately proportional to the energy gap near the barrier. By using the junction resistance R_N estimated from the I–V characteristics, all the curves can be well fitted, confirming that the Ba-122:Co BGB junctions are SNS type junctions. This is in contrast to the case of the YBCO BGB junctions, which are basically SIS junctions as indicated by their quasi-linear temperature dependence of I_c and small hysteresis at low temperatures [431, 438].

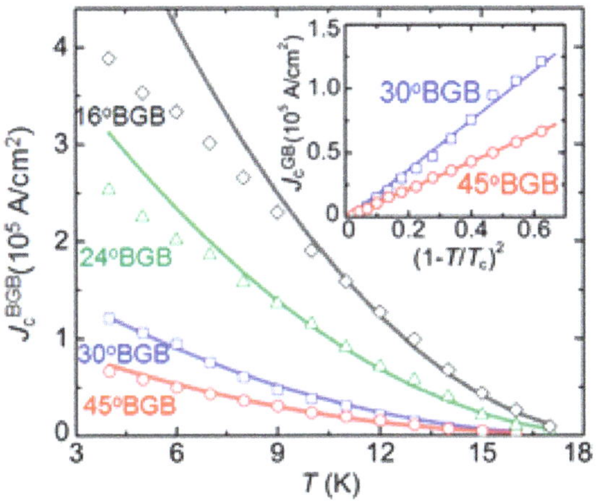

Figure 88. Temperature dependences of J_c for the BGB junctions with θ_{GB} = 16°, 24°, 30°, and 45° grown on MgO bicrystal substrates. The solid lines show the temperature dependences of J_c predicted from the de Gennes theory. The inset shows a linearized plot of the quadratic temperature dependences for the θ_{GB} = 30° and 45° junctions.

Fabrication of DC SQUIDs using bi-Crystal Junctions Shown in Section 4.2.1

Katase *et al* fabricated DC SQUIDs using Ba-122:Co BGB junctions on LSAT [001]-tilt bicrystal substrates with θ_{GB} = 30° and demonstrated their operation for the first time [439]. As schematically shown in figure 89(a), the SQUIDs have a loop with 18 × 8 μm^2 size containing two 3 μm-wide BGB junctions. Figure 89(b) shows the voltage–flux (V–Φ) characteristics at 14 K. A clear voltage modulation with ΔV = 1.4 μV is seen, though its magnitude is one order of magnitude smaller than that typically observed in practical SQUIDs [432].

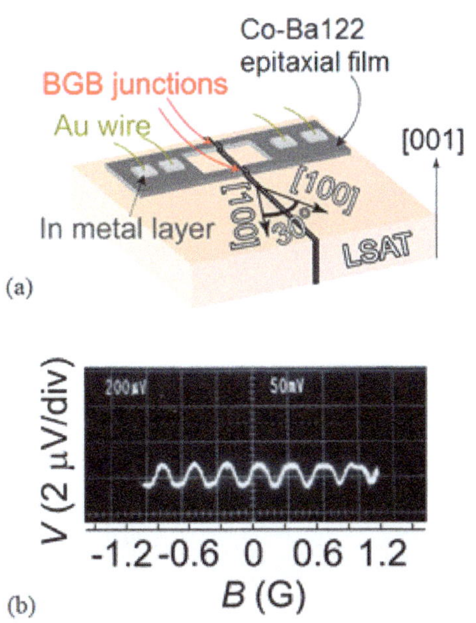

Figure 89. (a) Schematic of DC SQUID structure fabricated using a Ba-122:Co film on LSAT bicrystal substrate with θ_{GB} = 30°. (b) Voltage–flux (V–Φ) characteristics of the DC SQUID measured at 14 K. Reprinted with permission from [439]. Copyright 2010 by IOP Publishing.

Figure 90(a) shows the flux noise spectrum for the SQUID at 14 K measured using a commercial flux-locked-loop (FLL) circuit. The white noise level is 1.2 × 10^{-4} Φ_0/Hz$^{1/2}$ and the onset of 1/*f* noise is observed at about 20 Hz. The intrinsic white noise level was estimated to be 9.1 × 10^{-5} Φ_0/Hz$^{1/2}$ by subtracting the contribution of the amplifier noise. This is about one order of magnitude larger than the white noise typically observed for YBCO SQUIDs at 77 K [432, 440, 441]. As shown in figure 90(b), the SQUID or the BGB junctions exhibit a rather steep quadratic temperature dependence of I_c due to the metallic nature of the barrier. The SQUID shows I_c less than 100 μA, which is required for operation using the FLL circuit, only at temperatures very close to the junction T_c. However, I_cR_N at 14 K for instance is as small as 10μV. The observed small voltage modulation and the large white noise level for the Ba-122: Co SQUIDs can be explained by this small I_cR_N as well as the small R_N. Thus improvement of BGB junction properties seems necessary to make a SQUID based on doped Ba-122 a practical one.

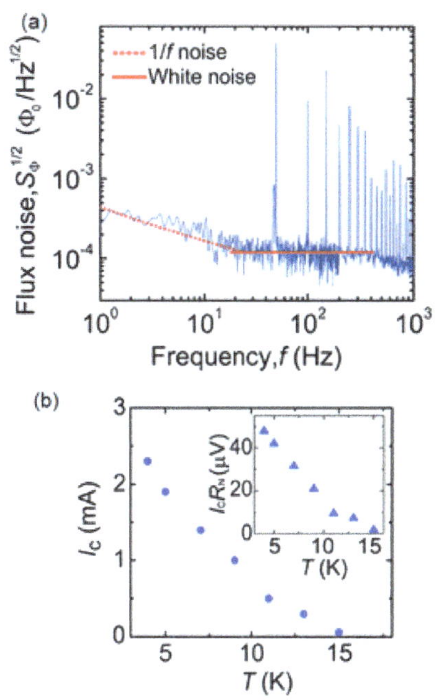

Figure 90. (a) Flux noise ($S^{1/2}{}_\Phi$) spectrum, including the equivalent input noise in the FLL circuit, measured at 14 K for a Ba-122:Co DC SQUID. (b) Temperature (T) dependence of the critical current (I_c) for the SQUID. The inset shows the temperature dependence of the I_cR_N product. Reprinted with permission from [439]. Copyright 2010 by IOP Publishing.

Fabrication of BaFe$_2$(As,P)$_2$ step-edge Junctions

Ishimaru *et al* tried fabricating step-edge junctions using Ba-122:P epitaxial films [442]. One motivation is the fact that larger I_cR_N products than BGB junctions were reported in REBCO step-edge junctions [443, 444], although this was more or less attributed to the dx^2–y^2 symmetry of the order parameter in REBCO. Another motivation is that the GB properties in Ba-122:P could be different from those for the Ba-122:Co BGB junctions. Actually substantially higher BGB J_c has been reported in Ba-122:P epitaxial films fabricated by an MBE method [423].

A step structure with height of about 70 nm and an angle of 20° was fabricated on MgO (100) substrates by an Ar ion milling technique. An approximately 90 nm thick Ba-122:P epitaxial film deposited on the substrate exhibited a self-field J_c as high as 6.7 MA cm^{-2} at 4.2 K. Figure91 shows the I–V curves for a 10 μm-wide bridge across the substrate step. The I–V curve without microwave irradiation is RSJ-type. The junction I_c and R_N are 0.41 mA and 1.35 Ω, respectively, resulting in an I_cR_N product of 0.55 mV. This I_cR_N product is one order of magnitude larger than that for the Ba-122:Co BGB junctions [433, 439]. Upon irradiation of 10.02 GHz microwave, clear Shapiro steps are observed. By changing the irradiation frequency and power, the junction I_c could be suppressed to zero, while Shapiro steps were observed up to the voltage of 0.7 mV. This confirms that the observed supercurrent has Josephson current origin. As shown in figure 92, the junction I_c could be observed up to the temperature of approximately 30 K, which is very close to the film T_c. However, magnetic field modulation of I_c was not observed for this step-edge junction. Cross-sectional TEM observation revealed that the Ba-122:P film on the step-slope did not have a single orientation and complicated grain structures, suggesting that Josephson junction is formed not along a single GB but at a GB with a very small area between certain grains. These results indicate the possibility of fabricating GB junctions with better performance for IBSCs, though control of microstructure of the junction region is required.

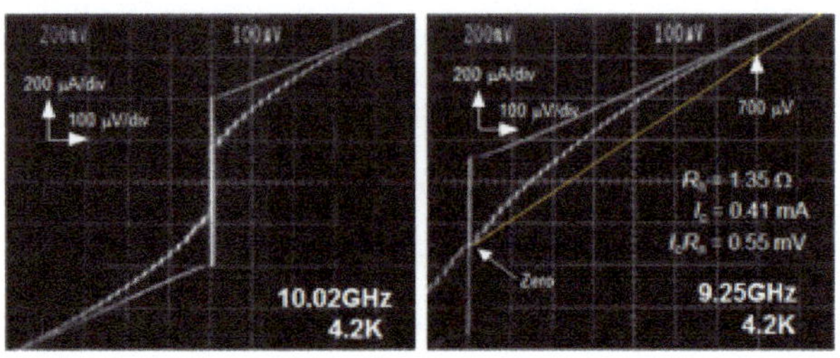

Figure 91. I–V characteristics at 4.2 K with and without microwave irradiation for a Ba-122:P step-edge junction fabricated on an MgO substrate.

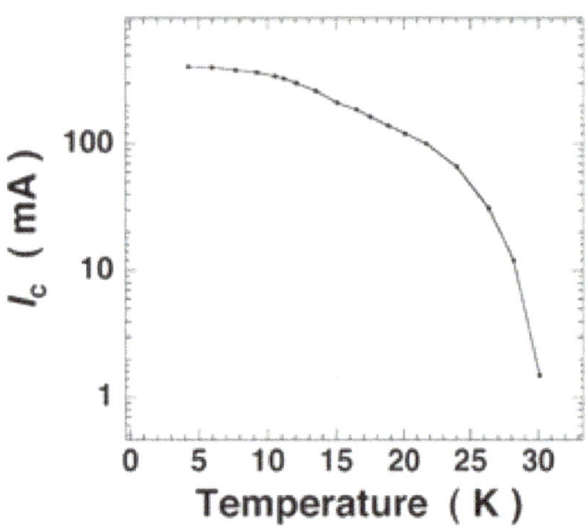

Figure 92. Temperature dependence of I_c for a Ba-122:P step-edge junction fabricated on an MgO substrate.

Effect of grain Boundary Angle on Inter-grain Transport Properties

The properties of GBs are of great importance for the application of superconducting materials, in particular, to wires and tapes [445]. It is well known that GBs block the supercurrent in cuprate superconductors such as REBCO. For REBCO, J_c across the grain boundary starts to decrease at a critical angle (θ_c) of approximately 3–5° and shows nearly exponential rapid decay with further increasing θ_{GB} [431, 446]. Because of this weak-link behavior, strict control of the in-plane orientation by employing a biaxially textured buffer layer, which is realized, for example, by an ion-beam-assisted deposition (IBAD) technique [447], is required to fabricate REBCO superconducting tapes on flexible metal substrates or coated conductors with a high J_c.

The first study on the GB properties of IBSCs using Ba-122:Co epitaxial thin films on STO [001]-tilt bicrystal substrates was reported by Lee *et al* [448]. They found that J_c across the BGB, even with a low misorientation angle of 6°, was strongly suppressed in their low-temperature laser scanning microscope imaging and transport J_c measurements in a magnetic field of 0.2–0.5 T. Katase *et al* performed a more systematic study on the transport properties of BGBs with θ_{GB} = 3–45° using high-quality Ba-122:Co epitaxial thin films with a self-field J_c of well above 1 MA cm^{-2} at 4 K prepared on both MgO and LSAT bicrystal substrates [434]. Figure 93(a) shows the self-field J_c of BGB ($J_{c\,BGB}$) at 4 and 12 K as a function of θ_{GB}. Nearly exponential decay of $J_{c\,BGB}$ for θ_{GB} above approximately 10° at both temperatures is observed, indicating that high-angle BGBs act as weak links. However, the exponential decay is more gradual than that for YBCO BGBs. Figure 93(b) shows the θ_{GB} dependence of $J_{c\,BGB}$ normalized by the film J_c at 4 K for lower-angle BGBs. The ratio starts to decrease at a critical angle θ_c of approximately 9–10°, while there is no reduction for $\theta_{GB} < \theta_c$, indicating that BGBs in this region are strong links. This critical angle is substantially larger than the value of 3–5° reported for YBCO BGBs [431].

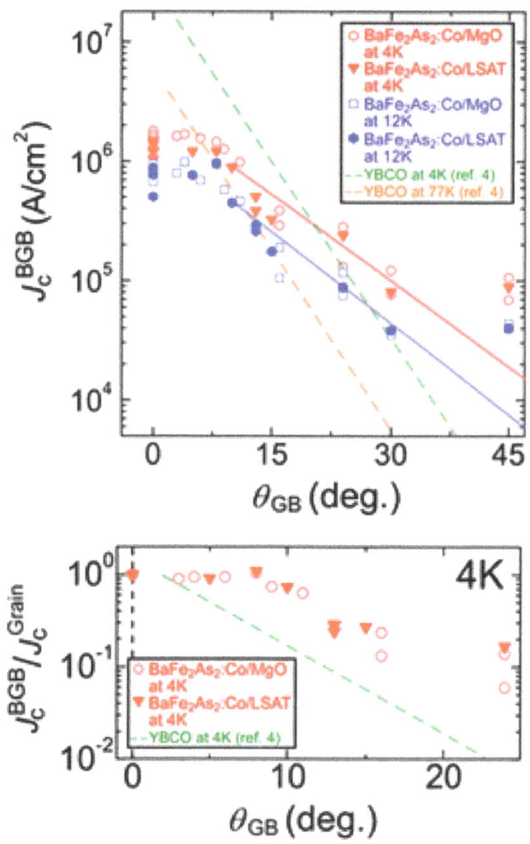

Figure 93. (a) Intergrain transport critical current density J_c^{BGB} at 4 and 12 K in a self-field as a function of misorientation angle θ_{GB} for Ba-122:Co BGB junctions grown on [001]-tilt bicrystal substrates of MgO and LSAT. The red and blue solid lines are fits to the empiri-

cal equation $J_c^{BGB} = J_{c0}\exp(-\theta_{GB}/\theta_0)$. The average data for the YBCO BGB junctions taken at 4 and 77 K [431] are also indicated by the green and orange dashed lines, respectively, for comparison. (b) Ratio of J_c^{BGB} to the intragrain J_c (J_c^{Grain} in the range θ_{GB} = 0–25° at 4 K. The dashed green line shows the result for the YBCO BGB junctions.

Figure 94 shows [001] plan-view HR-TEM images of the Ba-122:Co BGB junctions on MgO bicrystal substrates. The BGBs with θ_{GB} = 4° and 24° clearly indicated an array of misfit dislocations with the periodic distance of approximately 5.0 nm for θ_{GB} = 4° and 1.2 nm for θ_{GB} = 24°. Using a geometric tilted boundary model, the grain boundary dislocation spacing *D* is given by $D = (|b|/2)/\sin(\theta_{GB}/2)$ where $|b|$ is the norm of the corresponding Burgers vector. With the lattice constant a = 0.396 nm of Ba-122, *D* is estimated to be 5.7 nm and 1.0 nm for θ_{GB} = 4° and 24°, respectively. The estimated *D* values are very similar to the *D* values observed above. Energy dispersive spectroscopy (EDS) line spectra across the BGBs and parallel to the BGBs confirmed that the chemical compositions of the BGBs and the film region are homogeneous, and no secondary phase was observed in the BGB regions.

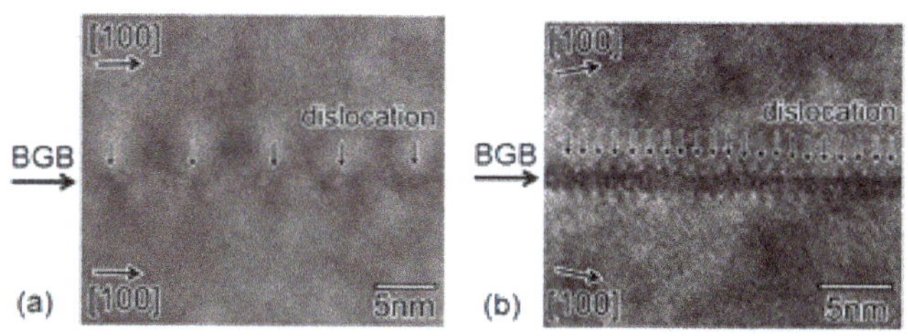

Figure 94. [001] plan-view HRTEM images of the Ba-122:Co BGB junctions on MgO bicrystal substrates with θ_{GB} = (a) 4°, and (b) 24°. Misfit dislocations are marked by the down-pointing arrows [434]. Reprinted with permission from Macmillan Publishers Ltd: [434], Copyright 2011.

The critical angle of 9° observed in Ba-122:Co BGBs corresponds to the spacing between the misfit dislocations of approximately 2.8 nm, which is comparable to or slightly larger than the coherence length ξab(T) of 2.6 nm at 4 K [319]. This is consistent with the notion that strong supercurrent channels still remain between the dislocations for the case of $\theta_{GB} < \theta_c$, while a coherent supercurrent cannot pass through the BGBs at $\theta_{GB} > \theta_c$, giving rise to the weak-link behavior. For the case of YBCO, the critical angle of 5° corresponds to D = 4.5 nm, which is much larger than ξab(T) of 1.6 nm at 4 K. It was previously pointed out that strain near the dislocation cores induces a local transition to an antiferromagnetic phase and forms insulating regions near dislocation cores in cuprate superconductors [446]. The smaller critical angle for YBCO could be explained by the formation of such insulating regions.

For higher-angle BGBs, the distance between the dislocations becomes smaller and eventually dislocations overlap each other. For the case of YBCO, the GB regions become carrier-depleted and thus insulating, though several origins for this, such as shift of chemical potential due to excess ion charge [446], band bending [431], and local structural distortion [449], were proposed. Since carrier-depleted IBSC materials are antiferromagnetic metals (or semi-metals) not insulators, a different nature of high-angle GBs is expected. Actually, as indicated in figure88, the Ba-122:Co high-angle BGB junctions exhibit normal-metal-like behavior of the junction barrier. The more gradual exponential decay of J_c^{BGB} in Ba-122:Co BGB junctions than that for YBCO BGB junctions in figure 93 can be attributed to this metallic nature of the GB region. The metallic nature of the GB region could also explain the critical angle θ_c which is substantially larger than that of YBCO and consistent with the dislocation distance *D*.

The observed larger critical angle appears to afford a great advantage for application to superconducting tapes since less strict control of the in-plane alignment for buffer layers would be required to obtain high-J_c IBSC films on flexible metal substrates. Moreover, this would be advantageous for the fabrication of superconducting wires, for example, based on the powder-in-tube (PIT) method where grains are not highly oriented.

Approach to Electrostatic Field Controlled Device

The modulation of electronic properties by applying electric fields is a commonly used technique in semiconducting materials. The electric double-layer transistor (EDLT) type device composed of an atomically flat film of an insulator/semiconductor and an ionic liquid (or a polymer electrolyte) as a gate electrode (shown in figure 95 as an example) is one of the efficient electric-field devices because such devices can accumulate extremely high currier density up to 10^{15} cm^{-2} and insulator-to-metal transitions have been demonstrated for various materials such as organic polymers, InO*x* polycrystalline films, and ZnO single-crystal films [450–453].

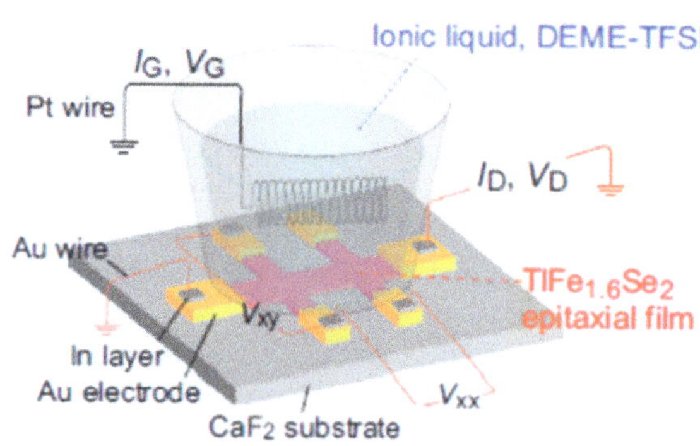

Figure 95. Schematic of the EDLT using $TlFe_{1.6}Se_2$ epitaxial film with a six-terminal Hall bar structure on a CaF_2 substrate. V_G was applied via a Pt counter electrode through the ionic liquid, DEME-TFSI, contained in a silica glass cup. Electrical contacts were formed using Au wires and In/Au pads. Reprinted with permission from [464]. Copyright 2014 by the National Academy of Sciences.

In 2008, Ueno *et al* [454] reported that the superconductivity of pristine $SrTiO_3$ single-crystal surface emerged using an EDLT structure. The critical temperature (T_c = 0.4 K) was comparable to the maximum value for a chemically doped bulk crystal [455]. This success proposed a new route to evolve superconductivity. Inspired by this result, insulator–superconductor transitions using EDLT devices have been reported in $KTaO_3$ (T_c = 0.047 K) [456], ZrNCl (T_c = 15.2 K) [457], and MoS_2 (T_c = 9–11 K) [458, 459]. The similar device has been applied on cuprate superconductors. In the case of epitaxial film of $La_{2-x}Sr_xCuO_4$, its underdoped and insulating film changed into a superconductor (T_c = 30

K) by applying −4.5 V to the gate electrode [460]. In the case of YBCO, the enhancement of T_c (from 83 to 134 K by applying −3 V (hole accumulation) of gate voltage (V_G)) was observed but its decrease in resistivity from onset T_c to lower temperature (i.e., superconducting transition) was very broad [461]. When the positive V_G was applied (i.e., electron accumulation), T_c decreased with increasing V_G [461–463], whereas the resistivity changes seriously by keeping the device under the constant V_G. From this phenomenon, it is considered that the modulation of T_c in YBCO devices is due not only to the carrier accumulation but also the changing in the defect concentration (possibly oxygen) by the electro-chemical effect.

Though the parent materials of IBSCs are generally antiferromagnetic metals, only $A_{0.8}Fe_{1.6}Se_2$(A: K, Rb, Cs, Tl) is an AFM Mott insulator, where iron vacancy (V_{Fe}) order forms a $\sqrt{5} \times \sqrt{5} \times 1$ supercell [155]. This material is called the 245 type of IBSC of which the crystal structure of the fundamental cell (tetragonal $ThCr_2Si_2$-type) is the same as that of the 122 type (see section 3.1.1). By decreasing V_{Fe}, it changes to metal and subsequently reveals superconductivity (maximum T_c = 32 K) [154].

The FIRST Project attempted to tune the transport property of $TlFe_{1.6}Se_2$ by carrier accumulation using EDLT structure [464]. The reason to select $TlFe_{1.6}Se_2$ is its chemical stability compared to alkali metal-based (K, Rb and Cs) 245-type compounds. Employing CaF_2 substrate, $TlFe_{1.6}Se_2$ film was deposited by PLD method with epitaxial relationships of [001] $TlFe_{1.6}Se_2$//[001] CaF_2 (out of plane) and [310] $TlFe_{1.6}Se_2$// [100] CaF_2 (in plane). The ordering of V_{Fe}, which indicates the insulating phase, was clearly observed in the high-angle annular dark field STEM images. The EDLT type device was fabricated using a 20 nm thick $TlFe_{1.6}Se_2$ film. The ionic liquid, N,N-diethyl-N-methyl N-(2-methoxyethyl)-ammonium bis-(trifluoromethylsulfonyl) imide, covered the EDLT device and a Pt coil electrode was inserted into the ionic liquid to act as a gate electrode. Figure 96 shows the R–T curves under applying V_G at 0, 2 and 4 V and activation energy in the high temperature region. Unfortunately, a superconductivity transition did not emerge but the electrostatic carrier doping has been controlled successfully by the EDLT structure. Moreover, the EDLT structure induced the phase transition (resistance humps marked by the arrow in figure 96(a)) assignable to a magnetic phase transition or the formation of an orbital-sensitive Mott phase, which has been commonly observed as a precursory phenomenon of transition to a superconducting state. This demonstration of carrier doping of the Mott insulator by the electrostatic method offers a way to extend the exploration of high-T_c superconductors to even insulating materials.

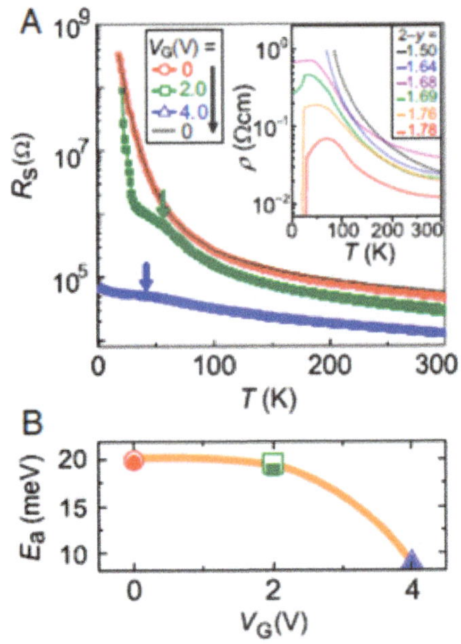

Figure 96. (A) T dependences of R_s for the $TlFe_{1.6}Se_2$ EDLT measured with increasing T(open symbols) and decreasing T (filled symbols) at $V_G = 0 \rightarrow +2.0 \rightarrow +4.0 \rightarrow 0$ V. The arrows indicate the positions of resistance humps. The reported ρ–T curves of (Tl,K) $Fe_{2-y}Se_2$ bulk materials [156] are shown for comparison (inset). A resistance hump appears at $2-y \geq 1.68$, and superconductivity emerges at $2-y \geq 1.76$. (B) The E_a estimated from the high T region of the ρ–T curve shown in (A) as a function of V_G. Reprinted with permission from [464]. Copyright 2014 by the National Academy of Sciences.

Fabrication of Superconducting tapes by Deposition

Fabrication of a Short $Ba(Fe,Co)_2As_2$ *tape*

Superconducting tapes or coated conductors which are fabricated by depositing superconducting films on flexible metal tapes with proper buffer layers have been demonstrated to be practical conductors with high critical current for the case of REBCO cuprate superconductors [465]. In order to realize a high-J_c REBCO film layer on a biaxially textured oxide buffer layer, the IBAD technique [447] or rolling-assisted biaxially textured substrate (RABiTS) technique [466] has been employed. For the case of MgB_2 with no significant GB problem, the PIT technique enables production of round superconducting wires which are more favorable for magnet application [467], though they can only be used at low temperatures below 20 K. Since IBSCs also exhibit weak-link behavior at GBs with the misorientation angle larger than the critical angle of approximately 9°, as described in the previous section [434], the coated conductor technique would be one promising candidate for production of practical conductors.

The first trial fabrication of iron-based 122 compound films on flexible metal substrates with biaxially textured buffer layers was reported by Iida *et al* [468]. By employing the Fe buffer architecture, they realized the biaxially textured growth of Ba-122:Co thin films on IBAD–MgO-buffered Hastelloy substrates, which are typically used for the fabrication of REBCO coated conductors. The films exhibited in-plane misorientation $\Delta\phi$ of about 5°, which was slightly smaller than that of the homoepitaxial MgO/IBAD–MgO layer. They also showed a broader transition width and a substantially lower self-field J_c than those for films on MgO single-crystal substrates, although a substantial improvement in J_c has recently been reported [425].

Katase *et al* succeeded in preparing biaxially textured Ba-122:Co thin films directly on IBAD–MgO-buffered Hastelloy substrates [406]. Figure 97(a) schematically shows the film on the structure. They used 10 × 10 mm^2 substrates with moderate in-plane alignment for the epitaxial-MgO layer on IBAD MgO $\Delta\Phi_{MgO}$ of 5.5–7.3°. X-ray diffraction revealed that the films had a substantially smaller $\Delta\Phi$ of approximately 3°, irrespective of the $\Delta\Phi_{MgO}$ value. As shown in figure 97(b), the Ba-122:Co films exhibited a resistive transition as sharp as that for films on MgO single crystal substrates and high self-field J_c values of 1.2–3.6 MA cm^{-2} at 2 K. Figure 98shows the magnetic field dependence of J_c (J_c–H) at 4–18 K for the film on the substrate with $\Delta\Phi_{MgO}$ = 6.1°. At 4 K, J_c(*H*//*ab*) is larger than J_c(*H*//*c*) in almost the whole field range, while crossovers are observed at 5.0 T at 12 K and 2.9 T at 16 K, respectively. At a higher *T* of 18 K,J_c(*H*//*c*) is greater than J_c(*H*//*ab*) in the whole *H* region. These results suggest the existence of naturally-formed relatively strong vortex pinning centers along the *c*-axis in the Ba-122 film on the metal substrate, as observed in the Ba-122:Co epitaxial films on LSAT substrates [400]. The pinning force density F_p was found to show the largest value of ~8 GN m^{-3} at 4 K which is smaller than those of Ba-122:Co epitaxial films on LSAT substrates. This suggests that much higher in-field J_c would also be realized on IBAD substrates by further optimizing or introducing the *c*-axis correlated pinning centers. These results demonstrated the possibility of fabricating high-J_c coated conductors with Ba-122 compounds by a rather simple low-cost process using less textured templates with a large $\Delta\Phi$.

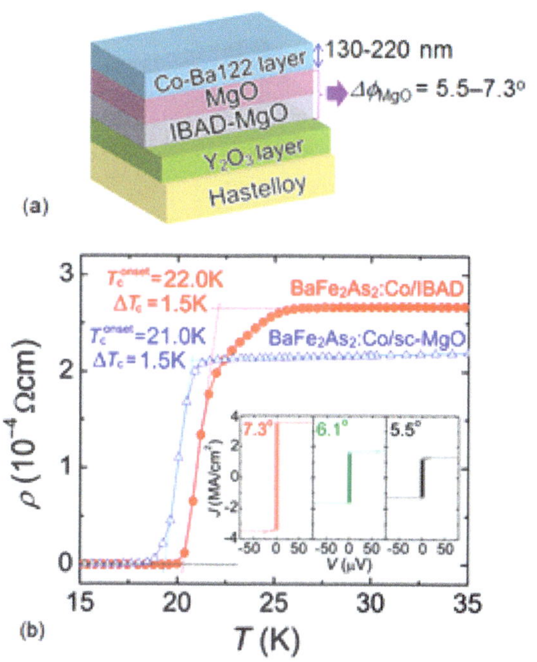

Figure 97. (a) Schematic cross-section and (b) ρ–T curves for Ba-122:Co thin films on IBAD–MgO substrate (circles) and single-crystal MgO (triangles). The inset shows the$J$–V characteristics at 2 K of the films on IBAD–MgO with (left) $\Delta\Phi_{MgO}$ = 7.3°, (middle) 6.1°, and (right) 5.5°. Reprinted with permission from [406]. Copyright 2011 by AIP Publishing LLC.

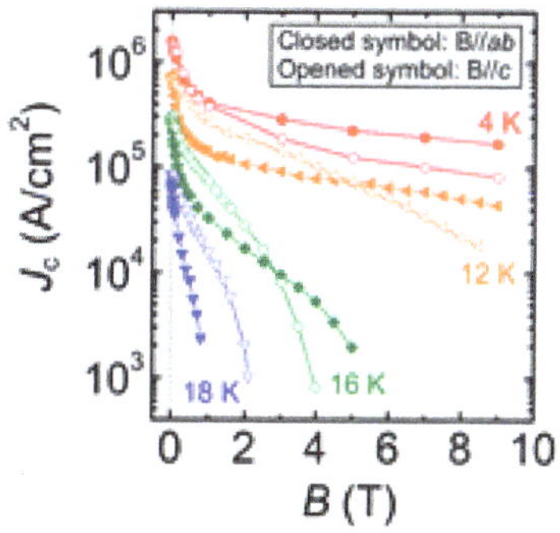

Figure 98. Magnetic field dependence of J_c at 4, 12, 16, and 18 K for a Ba-122:Co thin films on IBAD–MgO substrate. The closed and open symbols represent J_c for fields parallel to the *a*, *b*-axes and *c*-axis, respectively. Reprinted with permission from [406]. Copyright 2011 by AIP Publishing LLC.

Fabrication of a Long BaFe$_2$(As, P)$_2$ tape

As described in sections 4.1.2 and 4.1.3, Ba-122:P epitaxial films exhibited a higher T_c than Ba-122:Co films and a substantially higher in-field J_c by introducing artificial pinning centers consisting of oxide nanoparticles or tuning naturally formed pinning centers along the *c*-axis [415, 422]. Thus Miyata *et al* [469] tried fabricating long coated conductors using Ba-122:P films on flexible metal tapes using a PLD system shown in figure 99. The system is equipped with a reel-to-reel tape feeding mechanism that enables deposition on a tape longer than 1 m. Two second-harmonic Nd:YAG lasers can simultaneously generate two laser plumes on a target. The target–tape distance is variable between 32 and 57 mm. The energy density of one plume on the target was 2–3 J cm^{-2} and the repetition rate was 2.5–20 Hz. The target could be changed to a new one during a short interval of deposition without breaking vacuum.

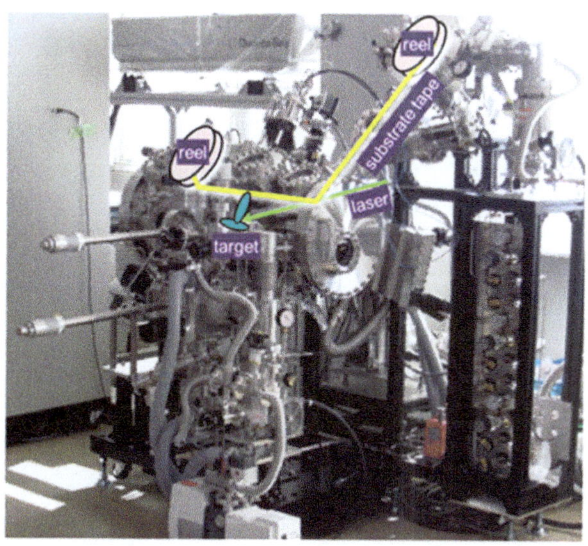

Figure 99. Photograph of a reel-to-reel PLD system designed for fabrication of long Ba-122:P coated conductors.

Two types of IBAD–MgO-buffered Hastelloy tapes, commercial and home-made, were used. The commercial tape employed homoepitaxial-MgO (epi-MgO)/IBAD-MgO/solution-deposited (SDP) Y_2O_3/Hastelloy architecture, while the latter employed epi-MgO/IBAD-MgO/sputtered

$Gd_2Zr_2O_7$ (GZO)/Hastelloy architecture. In either case, 50–100 nm-thick epi-MgO layers exhibited $\Delta\phi$ of ~5°. As the first step, deposition was performed using a Ba-122:P target with the nominal P content of 0.40 on a fixed tape without travelling at a substrate temperature of approximately 850 °C. The deposited film exhibited biaxial texture and a resistive transition with a T_c^{zero} of 17.6 K, which was substantially lower that for the film on single-crystal (sc) MgO (~24 K) [404]. In figure 100(a), x-ray diffraction patterns for the film on IBAD–MgO and sc-MgO are compared. A clear shift of the (008) diffraction peak to a lower angle side for the former film is observed. This peak position is close to that for Ba-122:P with x = 0.2, suggesting a loss of P in the film. Cross-sectional elemental map analysis suggested the existence of a phase consisting of Ba–P–O which grew along the *c*-axis. It was also found that the (004) as well as (008) peak position was sensitive to the background pressure before deposition, as shown in figure 100(b). Since a more than 1 m-long metal tape was loaded between two reels even for deposition of a short-length sample, careful degassing of the whole the tape was required to obtain low residual gas pressure.

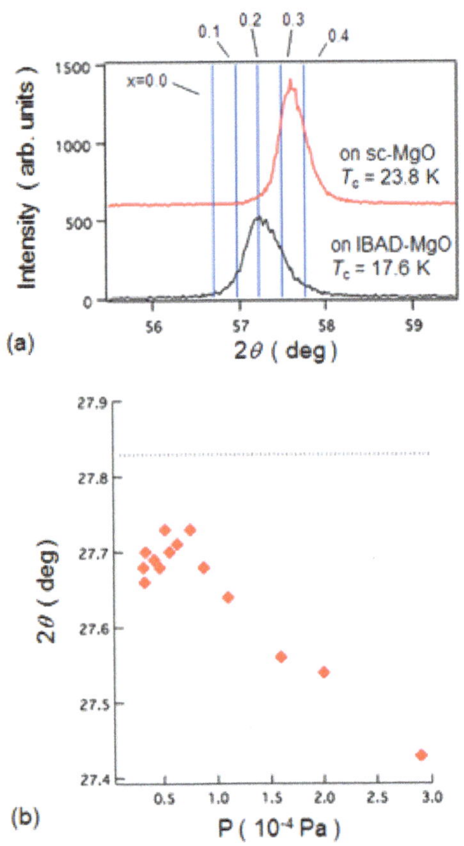

Figure 100. (a) Comparison of x-ray diffraction patterns for typical Ba-122:P films prepared on IBAD–MgO and single crystal (sc-) MgO substrates using a PLD target with the nominal P content x of 0.40. (b) Correlation between the position of the (004) diffraction peak with the residual pressure before deposition for Ba-122:P films on IBAD–MgO substrates.

Figure 101(a) shows the temperature dependence of resistance for films deposited on fixed tapes under low residual pressure below 10^{-4} Pa at heater setting temperatures of 1150–1250 °C. The films deposited at a temperature of 1200 °C exhibited the highest T_c (R 10%) of 20.7 K, while the film I_c tends to decrease with increasing the temperature, as shown in figure 101(b), suggesting a change in film morphology.

Figures 102(a) and (b) show the R–T curve and the I–V curve at 4.2 K, respectively, for a 2 mm-wide rectangular piece cut from a 5 cm-long coated conductor which was fabricated at a tape travelling speed of 6 mm min^{-1} using another Ba-122:P (x = 0.40) target. Although its T_c (R 10%) (=17.8 K), which means the temperature where resistivity down to 10% of normal state by superconductivity transition, is lower than that for the film deposited on a fixed tape, the observed self-field I_c of 0.55 mA corresponds to a J_c of 1.1 × 10^5 A cm^{-2}. Figure 103(a) shows a picture of a 15 cm-long coated conductor fabricated at a tape travelling speed of 6 mm min^{-1}. The coated conductor showed a T_c (R 10%) of 18.7 K and an overall self-field I_c at 4.2 K for 1cm width and 10 cm length of 0.47 mA, as shown in figure 103(b). This I_c corresponds to a J_c of 4.7 × 10^4 A cm^{-2}. This lower overall J_c than that for the shorter sample indicates the inhomogeneity of the film properties along the length and/or possibly across the width.

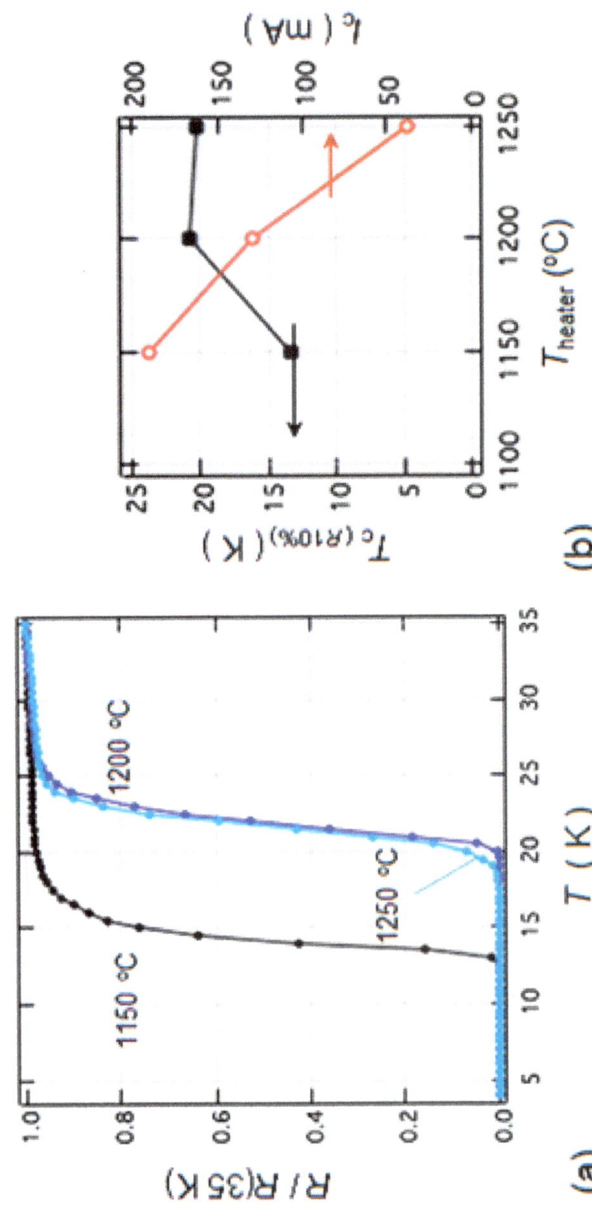

Figure 101. (a) Temperature dependences of resistance for Ba-122:P films deposited on fixed IBAD–MgO-buffered metal tapes under low residual pressure below 10^{-4} Pa at heater setting temperatures of 1150–1250 °C. (b) Dependence of T_c (R 10%) and I_c on the heater setting temperature for 2 mm-wide Ba-122:P films on IBAD–MgO-buffered metal tapes.

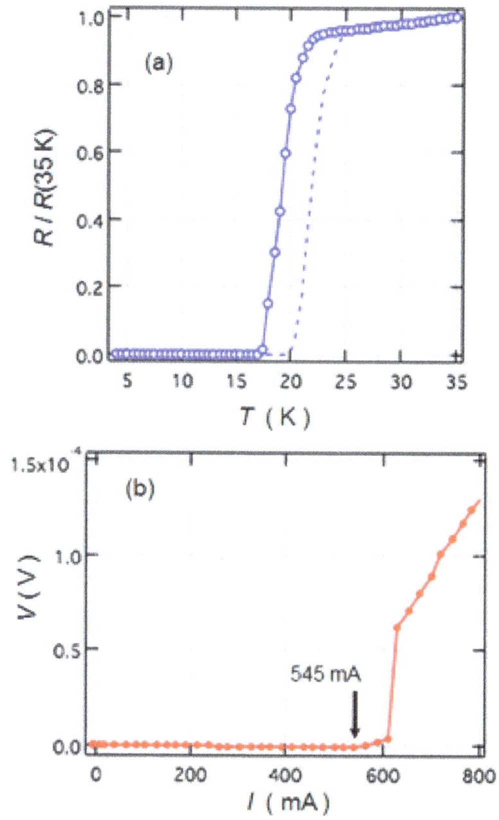

Figure 102. (a) *R–T* curve (open circle) and (b) *I–V* curve at 4.2 K for a 2 mm-wide rectangular piece cut from a 5 cm long Ba-122:P coated conductor which was fabricated at a tape travelling speed of 6 mm min^{-1} using a Ba-122:P ($x = 0.40$) target. The dashed line in (a) shows the *R–T* curve of the tape fabricated under static conditions.

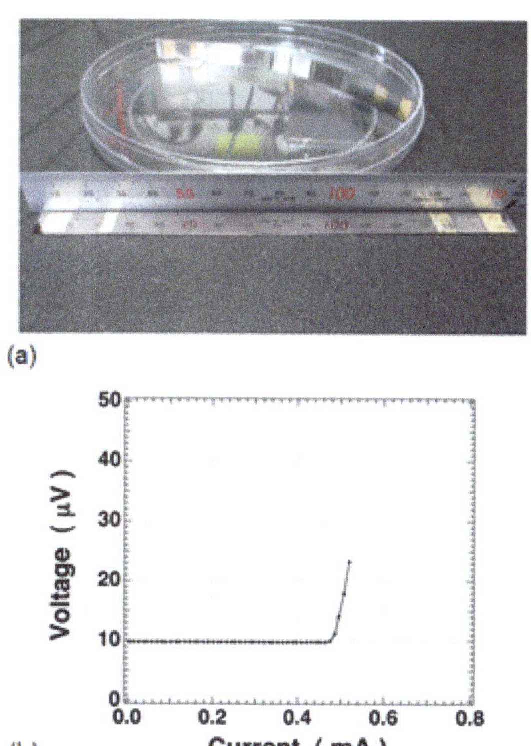

Figure 103. (a) Photograph and (b) *I–V* curve measured at 4.2 K for a 15 cm-long Ba-122:P coated conductor fabricated at a tape travelling speed of 6 mm min^{-1}. The width of the tape is 1 cm and the Au electrodes of the four probe method are attached where the distance between the inner electrodes is 10 cm.

Although the fabrication conditions have not yet been optimized, the superconducting properties of the trial-fabricated coated conductors are inferior to those for the Ba-122:P films on sc-MgO. One reason for the inferior properties could be a difference in film morphology. Scanning electron microscopy (SEM) observation of the surface of a coated conductor fabricated on a moving tape revealed a rather rough surface with bumps. This is clearly different from the surface for the high-J_c films on sc-MgO with average surface roughness R_a smaller than 4 nm.

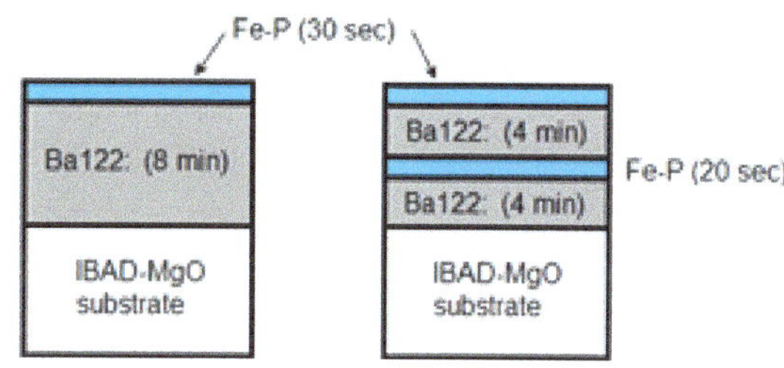

Figure 104. Schematic cross-section of Ba-122:P/Fe–P multilayer films on IBAD–MgO substrates.

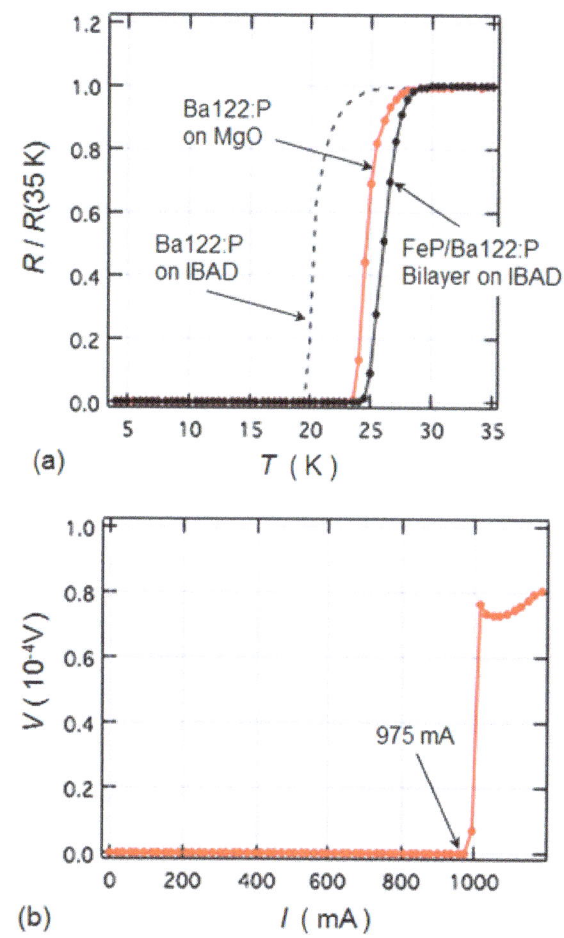

Figure 105. (a) *R–T* curves and (b) *I–V* curve measured at 4.2 K for 280 nm thick Fe_3P/Ba122:P bilayer films.

The lower T_c values for the coated conductors also suggest that the actual P content is still lower than the optimal P content in the target. In order to examine the influence of the P content, Ishimaru *et al* [470] fabricated multilayer films on the fixed IBAD–MgO-buffered tapes, as schematically shown in figure 104, using Ba-122:P ($x = 0.33$) and Fe_3P targets. Figure105 shows the *R–T* curve and the *I–V* curve at 4.2 K for a 280 nm thick Fe_3P/Ba-122:P bilayer film. The film exhibits a T_c (R 10%) of approximately 24.0 K, which is closer to the T_c for the film on sc-MgO. The self-field I_c corresponds to a J_c of 1.75 × 10^5 A cm^{-2}, which is substantially improved as compared with the single-layer film. The four-layer film showed an even higher T_c(R 10%) of 29.5 K, though its self-field J_c was decreased (0.88 × 10^5 A cm^{-2}). These results confirm that the films on IBAD–MgO-buffered tapes are actually P-deficient. Since the EDS results for the films fabricated under higher residual gas pressure indicated the formation of a Ba–P–O phase, the surface oxygen on the epi-MgO layer might react with the Ba-122:P films to some extent, leading to the P-deficient composition. The superconducting properties of Ba-122:P coated conductors could be further improved by employing a more stable thin buffer layer, for example, BZO, in addition to optimization of deposition parameters.

Fabrication of Tapes and Wires by the Powder-in-tube Method

Among IBSCs, K-doped (Ba,K)Fe_2As_2(Ba-122) and (Sr,K)Fe_2As_2(Sr-122) are most potentially useful for high field applications due to their high critical temperature (T_c) value of ~39 K and high upper critical field (H_{c2}) of over 50 T [319]. The relatively smaller anisotropy of Ba-122 compounds compared to those of cuprate superconductors [194] is also attractive for magnet applications because it is expected to bring about a higher irreversibility field, H_{irr}. Furthermore, the critical angle (θ_c) of the transition from a strong link to a weak link for Ba-122 is substantially larger than that for YBCO-based conductors [434]. For the high magnetic field magnet applications, we have to develop a high performance superconducting tape or wire conductor. In order to evaluate the potential for tape and wire applications, the development of a tape and wire processing technique is essential.

For Ba-122 and Sr-122 superconductors, PIT and coated conductor processes have been developed for tape and wire fabrications. Iron-based coated conductors have been grown by several groups utilizing existing YBCO coated conductor technology and have been found to have a self-field J_c of over 1 MA cm^{-2} [425, 471, 472]. Although, at an early stage of development, the transport J_c in IBSCs reported was disappointingly low due to the weak link grain boundary problem [448, 473–485], astonishing progress has been made for Ba(Sr)-122 tapes in the past several years. J_c for Ba(Sr)-122 tapes and wires approaches 10^4 A cm^{-2} at 4.2 K and 10 T through metal addition plus the rolling induced texture process, hot isostatic press method, cold press method, hot press method, and so on [486–495]. These results demonstrated that mechanical deformation is critical for producing high quality superconducting tapes and wires, which plays an important role in densifying the conductor core and aligning the grains of the superconducting phase. An understanding of the influence of mechanical deformation on the microstructure and superconducting properties will accelerate the development of the appropriate processes and further improve the transport J_c of Ba(Sr)-122 tapes and wires. In this section, we report our recent development of Ba(Sr)-122 tape conductors by applying a conventional PIT method. Emphasis was placed on the relation between the microstructure and critical current properties. The influence of superconducting core density, grain alignment, and microstructure on J_c in the tapes was systematically investigated.

Fabrication of Tape Conductors

Ba-122 and Sr-122 tapes were fabricated by applying a standard *ex situ* PIT process [496]. First we prepared the precursor powders of Ba-122 and Sr-122 as shown in figure 106 [495]. Ba or Sr filings, K plates, Fe powder and As pieces were mixed to the nominal composition of $Ba_{0.6}K_{0.4}Fe_2As_{2.1}$ and $Sr_{0.6}K_{0.4}Fe_2As_{2.1}$ in an Ar gas atmosphere using a ball-milling machine and the materials were put into a Nb tube of 6 mm outer diameter and 5 mm inner diameter. The Nb tube was put into a stainless steel tube, both ends of which were pressed and sealed by arc welding in an Ar-gas atmosphere. In order to compensate for loss of elements, the starting mixture contained 10–20% excess K. The stainless steel tube was heat treated at 900 °C for 10 h and then cooled to a room temperature in a box furnace. After the heat treatment, the precursor was removed from the Nb tube and ground into powder with an agate mortar in a glove box filled with high purity argon gas. Figure 107 shows the XRD pattern of the precursor powder [495]. This analysis together with the magnetization measurement indicate that the precursor powder obtained by this method has fairly good quality.

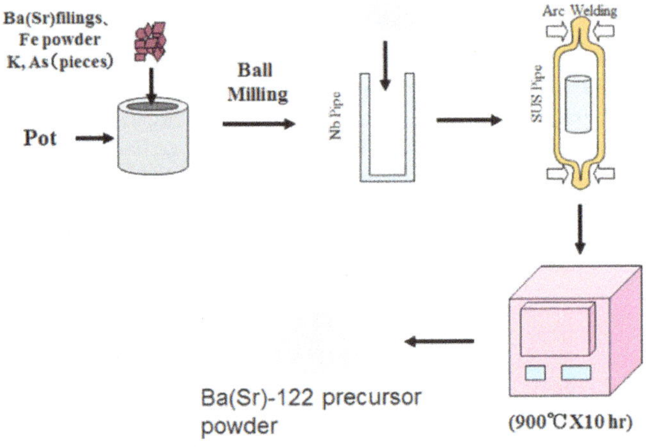

Figure 106. Fabrication of the Ba(Sr)-122 precursor powder.

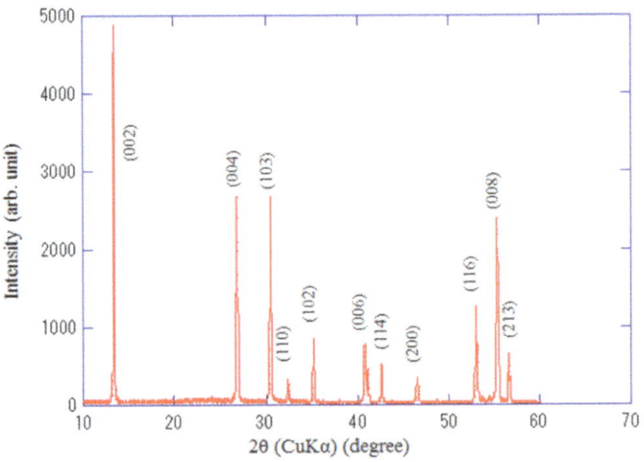

Figure 107. X-ray diffraction pattern of the Ba-122 precursor powder. Reprinted with permission from [495]. Copyright 2013 by IOP Publishing.

The precursor powder was packed into an Ag tube of outer diameter: 6–8 mm, inner diameter: 3.5–4 mm and length: 50 mm, and the Ag tube was groove rolled into a wire with a rectangular cross section of ~2 × ~2 mm. The wire was then cold rolled into a tape using a flat rolling machine, initially into 0.8 mm in thickness followed by intermediate annealing at 800 °C for 2 h and then into 0.40 ~ 0.20 mm in thickness. For some of the tapes we applied uniaxial pressing. For this uniaxial pressing, the flat rolled tape was cut into samples of 35 mm in length and the sample was sandwiched between two hardened steel dies and pressed. The uniaxial pressure was widely changed from ~0.1 to ~4 GPa. The flat rolled and uniaxially pressed tapes were subjected to a final sintering heat treatment at 850 °C for 2 ~ 10 h to obtain Ba-122/Ag and Sr-122/Ag superconducting tapes. All the sintering heat treatments were carried out by putting the tapes into a stainless steel tube, both ends of which were pressed and sealed by arc welding in an Ar atmosphere. We also fabricated seven-filamentary tapes. A cold worked mono-filamentary wire with 1.3 mm diameter was cut into seven short wires with length of 40 mm. The seven wires were bundled together and put into another Ag tube, and the assemblage was cold worked into tape with intermediate annealing, uniaxially pressed, and finally heat treated under the same condition as mono-filamentary tapes.

Critical Current and Microstructure

The transport critical current I_c at 4.2 K in magnetic fields were measured by the standard four-probe resistive method in liquid helium (4.2 K) using a 12 T superconducting magnet. The criterion voltage to measure I_c was 1 μV cm^{-1}. The transport critical current density, J_c was estimated by dividing I_c by the cross sectional area of the Ba-122 or Sr-122 superconducting core. Magnetic fields up to 12 T were applied parallel to the tape surface and perpendicular to the tape axis.

Figure 108 shows J_c versus magnetic field curves at 4.2 K of two uniaxially pressed Ba-122/Ag tapes with final tape thicknesses of 0.40 and 0.47 mm [495] and Sr-122/Ag tape with 0.40 mm thickness [497]. For comparison, the J_c–B curves of a groove rolled Ba-122/Ag wire with a rectangular cross-section (~2 × ~2 mm) and a flat rolled tape (0.4 mm thickness) with no uniaxial pressing are also shown in the figure. All the wires and tapes were subjected to a final heat treatment of 850 °C for 10 h. The figure clearly indicates that the J_c increases with the process from groove rolling to uniaxial pressing. The J_c of the grooved rolled wire is as low as ~10^3 A cm^{-2} in applied magnetic fields. Similar low J_c values were reported for Ba-122 and Sr-122 round wires fabricated by the conventional PIT process [486, 487, 498]. A substantial J_c increase was obtained by the application of the cycles of flat rolling and subsequent heat treatment [494]. However, the most remarkable result in this figure is that another large J_c enhancement was obtained by the application of uniaxial pressing. All the pressed tapes show J_c over 10^4 A cm^{-2} at 10 T, indicating that a high J_c is obtained with good reproducibility. The data in figure 108 suggest that the application of higher uniaxial pressure will result in higher J_c values. This can be realized as will be discussed later.

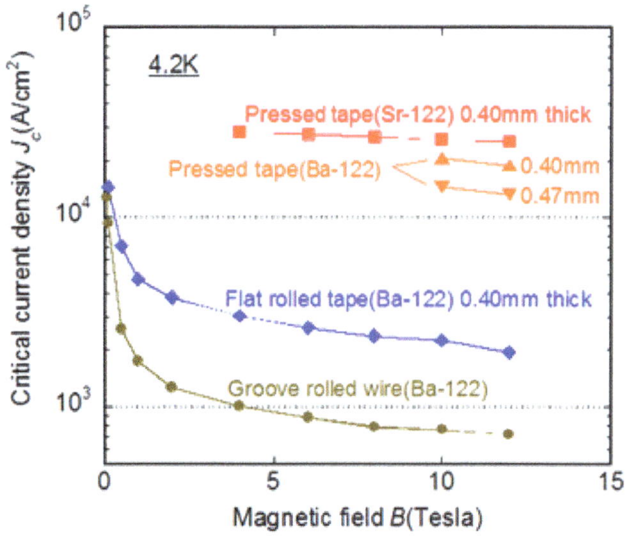

Figure 108. J_c–B curves of uniaxially pressed Ba-122/Ag and Sr-122/Ag tapes [495, 497]. The data of groove rolled wire and flat rolled tape are also shown for comparison. Reprinted with permission from [495]. Copyright 2013 by IOP Publishing.

In order to investigate the mechanism of J_c enhancement, we observed the microstructure change during the process. Fractured cross-sections of the wires and tapes were observed by SEM. Figures 109(a)–(c) show the grain structures of the Ba-122 core for the groove rolled wire, the flat rolled tape and uniaxially pressed tape, respectively [495]. All of the wires and tapes were finally heat treated at 850 °C for 10 h. The microstructure of the groove rolled wire shows non-uniformity in the grain size distributed widely from ~1 to ~10 μm. The flat rolled and pressed tapes show more uniform grain structure, whose average size is a few micrometers. We consider that the cycles of deformation and subsequent heat treatment break up larger grains into smaller grains, resulting in the more uniform grain structure of the Ba-122 core. The difference of grain structure between the flat rolled tape and pressed tape is not significant, although a large J_c difference is obtained for rolled and pressed tapes. This large J_c difference can be explained by the difference of Ba-122 core density as will be discussed later. Figures110(a) and (b) show XRD patterns of the Ba-122 core surface of the flat rolled and pressed tapes, respectively [495]. For the XRD observation the Ag sheath was peeled off. It should be noted that the relative intensities of the (00*l*) peaks of both tapes are not as high as those observed for the pressed precursor powder in figure 106 and show more random orientation. This is different from PIT processed Bi-2223/Ag tapes, in which the cycles of flat rolling and subsequent heat treatment produce stronger *c*-axis grain alignment due to the larger anisotropic morphology of the Bi-2223 crystal [499]. Our results for Ba-122/Ag tapes are also in contrast to the Fe-sheathed PIT Sr-122 tape [484, 489, 491], in which much stronger grain orientation was observed, similar to the Bi-2223 tape. This difference seems to be caused by the different sheath material and different processing parameters such as reduction ratio. Almost no difference of relative (00*l*) peaks was observed between rolled and pressed tapes, indicating that the grain orientation of the pressed tape is as low as the flat rolled tape. This is consistent with the grain structure shown in figures 109(b) and (c). From the almost same grain morphology and XRD pattern between the rolled and pressed tapes, the grain orientation can be ruled out as a possible origin of the large enhancement of J_c by the uniaxial pressing.

Figure 109. Grain structures of the Ba-122 core for the (a) groove rolled wire, (b) flat rolled tape and (c) uniaxially pressed tape. Reprinted with permission from [495]. Copyright 2013 by IOP Publishing.

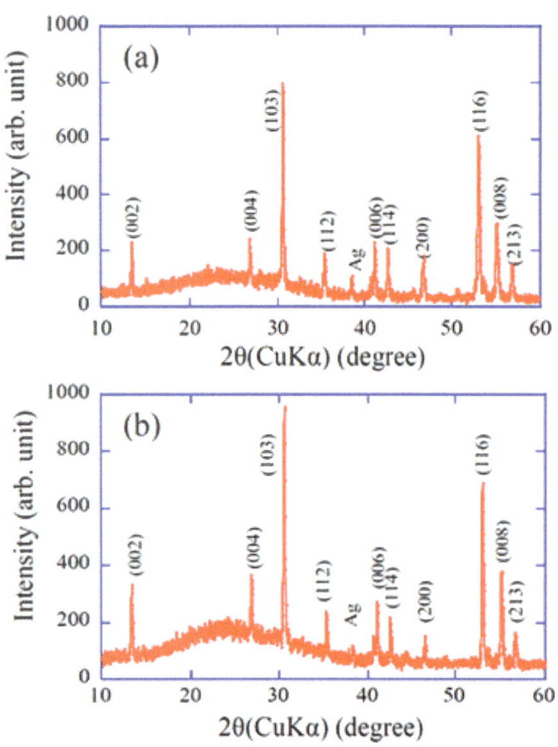

Figure 110. X-ray diffraction patterns of (a) flat rolled and (b) uniaxially pressed tapes. Reprinted with permission from [495]. Copyright 2013 by IOP Publishing.

For the PIT processed superconducting wires and tapes, the density of the superconducting core seems to be one of the major factors that governs the J_c values of tapes and wires. Because it is difficult to directly measure the density of the thin superconducting core, researchers sometimes use Vickers hardness as an indicator of the density of the superconducting core [500,501]. In this study of the superconducting wires and tapes fabricated by the PIT method, we performed the Vickers micro-hardness (*HV*) measurements of the Ba-122 core in order to investigate the influence of the core density on the J_c of our wire and tapes. Figure 111 shows the *HV* of the groove rolled wire, flat rolled tape and uniaxially pressed tape [495]. All the wires and tapes were finally heat treated at 850 °C for 10 h. The *HV* measurements were made on a polished transverse cross-section of each sample with 0.05 kg load and 10 s duration in a row at the center of the cross section. For the rolled tape, the *HV* at two different cross-sections was measured, because *HV* varies more widely in the rolled tape depending on the position. Although the scattering of *HV* is large, the figure clearly shows that the average *HV* increases with the progression of deformation process from groove rolling to uniaxial pressing. The average *HV* values of the groove-rolled wire, flat-rolled tape and uniaxially pressed tape are 87.1, 94.0 and 117, respectively. This increase of *HV* should be attributed to the increase of the Ba-122 superconducting core density. Thus, the increase of J_c with the deformation process in figure 108 can be explained by the increase of Ba-122 core density in the wires and tapes.

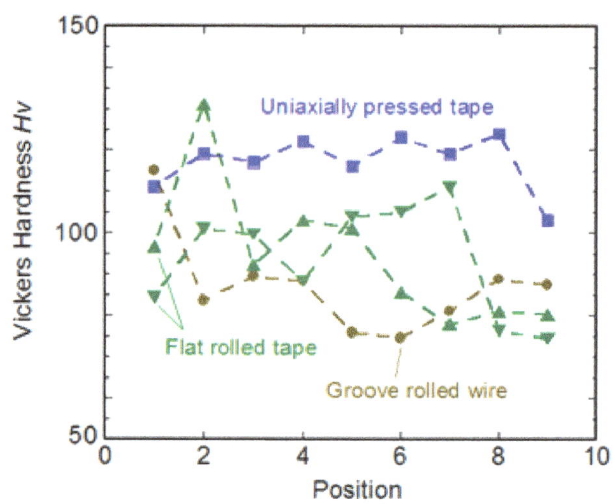

Figure 111. Vickers micro-hardness (*HV*) of the groove rolled wire, flat rolled tape and uniaxially pressed tape. Reprinted with permission from [495]. Copyright 2013 by IOP Publishing.

In the case of PIT processed Bi-2223/Ag tapes, it is reported that cold uniaxial pressing is more effective in enhancing transport J_c than flat rolling [502]. This can be attributed to a change in the micro-crack structure and more uniform deformation achieved by pressing rather than rolling. The uniaxial pressing of the Bi-2223/Ag tape introduces micro-cracks along the longitudinal direction of the tape, while in the flat rolled tape cracks appear along the transverse direction of the tape. During rolling, pressure varies along the arc of contact between the two rolls and the tape. The stress induced by the inhomogeneous pressure in the tape promotes the alignment of cracks transverse to the length direction [495]. However, in the case of uniaxial pressing, inhomogeneous deformation of the superconducting core occurs along the width of the tape, resulting in cracks along the direction of the tape length. In addition, the forces applied by uniaxial pressing are uniformly distributed perpendicular to the surface of the tape, thus resulting in higher and homogeneous compression. The higher uniform pressure reduces voids, improves texture formation, and thus further improves J_c. It is supposed that the longitudinal cracks have no influence on the transport superconducting currents along the tape, while the transverse cracks should be barriers of the superconducting currents. This influence of cracks on the superconducting current flow along the tape was directly evidenced by magneto-optical imaging.

It is expected that cracks are also introduced in our Ba-122 tape and that the different direction of residual cracks is another important reason for the difference of J_c in the rolled and pressed tapes in addition to the difference of densification of superconducting cores. An apparent difference in crack structures between the rolled and pressed tapes was also observed. Figures112(a) and (b) show SEM images of the cracks in the flat rolled and pressed tapes, respectively. The observation was carried out on the tape plane of the as-rolled or as-pressed tapes without subsequent heat treatment. The photographs were taken of the core surface after the Ag sheath was peeled off. As shown in figure 112(a), all cracks observed in the as-rolled tape run transverse to the tape length, while they run parallel to the tape length in the as-pressed tape as shown in figure 112(b). We consider that the observed difference in crack direction provides another key to elucidating the mechanism of the positive influence of applying uniaxial pressing [495]. It is speculated that the stress situation induced by the pressing, together with the heat treatment after pressing, acts so as to heal the transverse cracks produced by the previous rolling process, which run transverse to the tape length and reduce the effective cross-sectional area for superconducting current flow along the tape length.

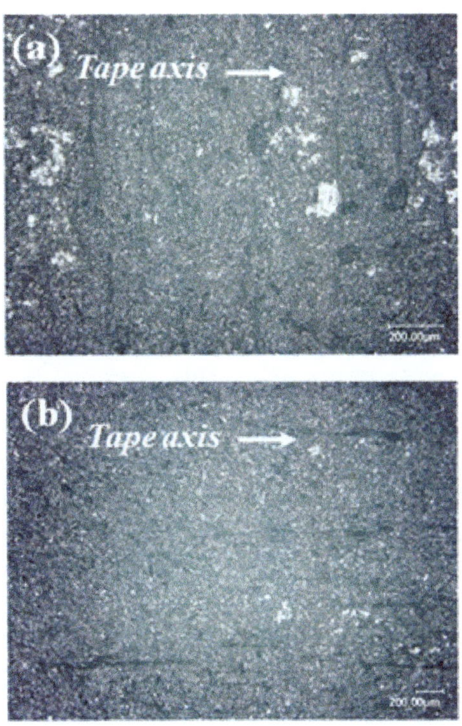

Figure 112. Optical microscopy images of (a) flat rolled and (b) uni-axially pressed tapes.

Effects of High Uniaxial Pressure and Degree of Flat Rolling

As mentioned before, J_c significantly enhanced by the uniaxial pressing under 0.4 GPa. This suggests that much higher J_c values can be obtained by increasing the uniaxial pressure. Thus, we increased the uniaxial pressure up to 4 GPa [503]. Figure 113 shows J_c–B curves of the single and seven filamentary Ba-122/Ag tapes uniaxially pressed under 2–4 GPa before final heat treatment at 850 °C for 2–4 h [503]. The J_c–B curves of the Ba-122 tape pressed under 0.4 GPa, commercial Nb–Ti and Nb_3Sn wires are also shown for comparison. Much higher J_c values than the tape pressed under 0.4 GPa are obtained by increasing the uniaxial pressure. J_c values of all the pressed Ba-122 tapes show very small field dependence as observed for the tapes shown in figure 108, and J_c values well above 5.0×10^4 A cm^{-2} in 10 T are obtained, indicating that a high J_c values are obtained with good reproducibility. It should be noted that a high J_c exceeding the practical level of 10^5 A cm^{-2} at 4.2 K is obtained in magnetic fields up to 6 T for mono-filamentary tape and J_c maintains a high value of 8.6×10^4 A cm^{-2} in 10 T. The seven-filamentary tape also sustains J_c as high as 5.3×10^4 A cm^{-2} at 10 T. These high J_c values highlight the importance of uniaxial pressing for enhancing the J_c of Ba-122 tape conductors.

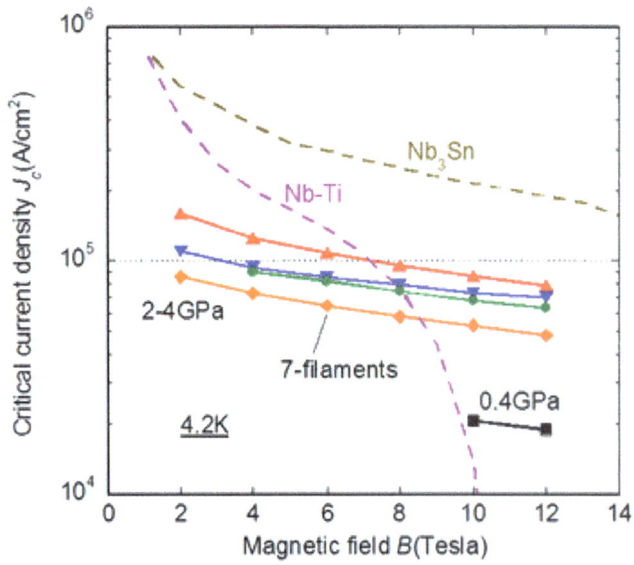

Figure 113. J_c–B curves at 4.2 K of uniaxially pressed Ba-122/Ag tapes. Pressure was changed from 2 to 4 GPa. For comparison the data of Ba-122 tape (figure 108) and Nb–Ti and Nb_3Sn commercial wires are also included in the figure. Reprinted with permission from Macmillan Publishers Ltd: [503], Copyright 2014.

The J_c values of the Ba-122/Ag tapes exceed the values of the Nb–Ti conductor in magnetic fields higher than ~8 T. Furthermore, in comparison with the Nb_3Sn, the J_c of Ba-122 tapes show very small magnetic field dependence. As will be described later, Ba-122 surpasses the J_c performance of Nb_3Sn in high magnetic fields. Our previous study [486, 487] indicates that the magnetic field dependence of J_c at 4.2 K of the flat rolled Ba-122 tape is very small, comparable to that of the melt-textured Bi-2212/Ag tape [504]. These results suggest that the Ba-122 superconducting wires will be competitive with well-established Nb-based superconductors and Bi-based oxide for high magnetic field applications in the near future.

The large reduction of tape thickness only by applying flat rolling is also found to be effective in enhancing J_c values [503]. Figure 114 shows the J_c–B curves of the flat rolled Ba-122 tapes. These J_c values are much higher than that of the flat rolled tape shown in figure 108. Figure 114clearly indicates that J_c significantly increases when the tape thickness is reduced by flat rolling. J_c reached a maximum value of 4.5×10^4 A cm^{-2} at 10 T when the thickness of the tape reduced to 0.26 mm. But when the thickness is reduced further, degradation in J_c was observed. However, further improvement in J_c values could be achieved for thinner tapes when we applied uniaxial pressing instead of flat rolling, as shown in figure 113. SEM observation of these thin tapes indicates that the transverse micro-cracks observed in figure 112 appear to decrease in number with decreasing tape thickness. With decreasing tape thickness, the stress and strain distribution in the rolling process seems to become similar to those in uniaxial pressing when the rolling diameter is constant, leading to the disappearance of transverse micro-cracks. This is one of the reasons why J_c increases with decreasing tape thickness in the flat rolling process.

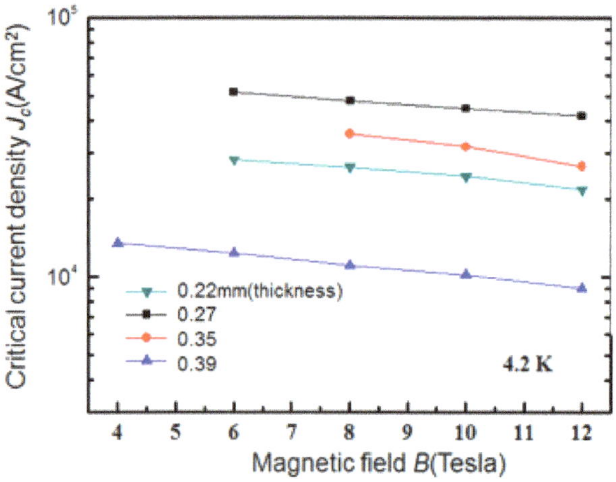

Figure 114. J_c–B curves at 4.2 K of flat rolled Ba-122/Ag tapes. Reprinted with permission from Macmillan Publishers Ltd: [503], Copyright 2014.

Figure 115 shows XRD patterns of the flat rolled and uniaxially pressed Ba-122/Ag tapes shown in figures 113 and 114 [503]. The data of randomly orientated Ba-122 precursor powder are also included in the figure for comparison. All the tapes consist of a main phase, $Ba_{1-x}K_xFe_2As_2$, however, Ag peaks from the sheath material are also detected. The relative intensities of the (00*l*) peaks with respect to that of the (103) peak in all tapes are strongly enhanced, when compared to the randomly oriented powder, indicating that a well-defined *c*-axis grain orientation is obtained by flat rolling and pressing. The relative intensities of the (00*l*) peak are higher than those of the flat rolled tape in figure 110(a), indicating that a large degree of flat rolling is effective in enhancing the *c*-axis grain orientation. However, it remains at almost the same level when decreasing the tape thickness below 0.39 mm, as shown in figure 115, suggesting that the grain orientation is hardly further improved by any further flat rolling process. Therefore, grain orientation can be ruled out as a possible mechanism of J_c enhancement with decreasing tape thickness in figure 114. In contrast, a higher relative intensity of (00*l*) peaks was observed in the pressed tape. This relative intensity in figure 115 is much higher than that of the uniaxially pressed tape in figure 110(b). Thus, the high uniaxial pressure is effective in enhancing the *c*-axis grain orientation. These results indicate that high uniaxial pressure is more effective in improving the *c*-axis grain orientation than the large degree of flat rolling. However, it should be noted that the degree of *c*-axis grain orientation in our pressed tapes is still lower than that in the Fe sheathed PIT processed Sr-122 tapes [489,491]. This suggests that the grain orientation could be further enhanced by optimizing the processing parameters of the tape conductors or by using harder sheath materials.

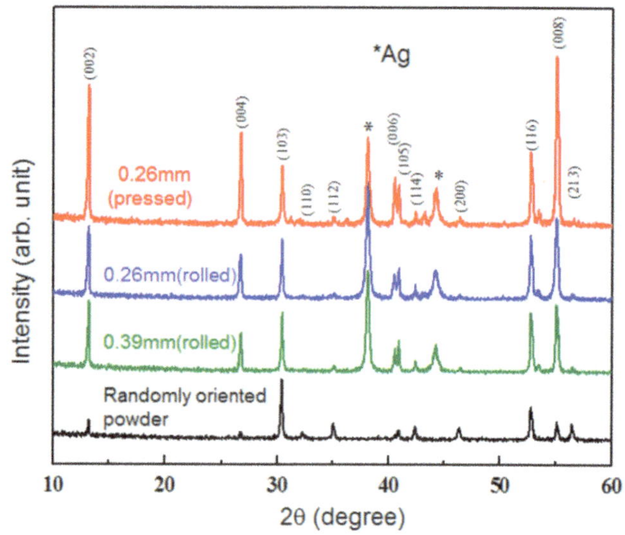

Figure 115. X-ray diffraction patterns of rolled and pressed Ba-122/Ag tapes. The data of Ba-122 powder with random grain orientation are also shown for comparison. Reprinted with permission from Macmillan Publishers Ltd: [503], Copyright 2014.

We evaluated the effect of flat rolling on *HV* and J_c. Both the *HV* and J_c clearly increase with decreasing tape thickness. This indicates that in addition to the disappearance of transverse cracks, the increase of Ba-122 density might be another reason for *Jc* enhancement by the hard flat rolling. Figure 116 shows J_c (10 T, 4.2 K) as a function of *HV* for both rolled and pressed tapes [503]. J_c increases with increasing *HV* and, hence, with increasing Ba-122 core density. A strong correlation between the hardness of the tapes and J_c was observed. With increasing hardness, the J_c of the Ba-122 core increased, however the hardness and J_c of flat rolled tapes do not surpass the hardness and J_c of the uniaxially pressed tapes, respectively. This suggests that uniaxial pressing of tapes yields much better J_c–H characteristics than flat rolling [495]. However, it seems that there is no discontinuity in the J_c–HV curve between flat rolling and uniaxial pressing. This suggests that there is no essential difference in the superconducting current limiting mechanism between flat rolling and uniaxial pressing. Thus, we can expect that high J_c values comparable to or higher than pressed tapes can be obtained only by applying a high degree of flat rolling. As is well known, flat rolling is more convenient and useful than uniaxial pressing for the fabrication of practical level long tape conductors.

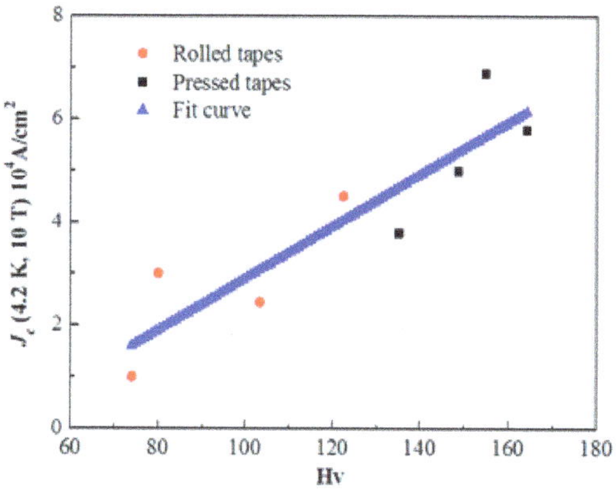

Figure 116. J_c at 4.2 K and 10 T as a function of Vickers hardness *HV* for the flat rolled and uniaxially pressed Ba-122/Ag tapes. Reprinted with permission from Macmillan Publishers Ltd: [503], Copyright 2014.

Figures 117(a) and (b) show typical SEM images of the polished surface for the flat rolled and uniaxially pressed Ba-122/Ag tapes. These observations were carried out on the tape plane of the tapes. Although rolling can reduce voids and improve the density of the Ba-122 core, the microstructures are still porous and quite inhomogeneous. In contrast, the pressed tapes with higher hardness and J_c apparently have a denser and more uniform microstructure than the rolled tapes with lower hardness and J_c. This result is consistent with the *HV* analysis in figure116. Generally speaking, the pressure introduced by the uniaxial pressing is higher and more uniform than that introduced by the rolling. Thus, the microstructure in the pressed sample is denser and more uniform than that in the rolled one.

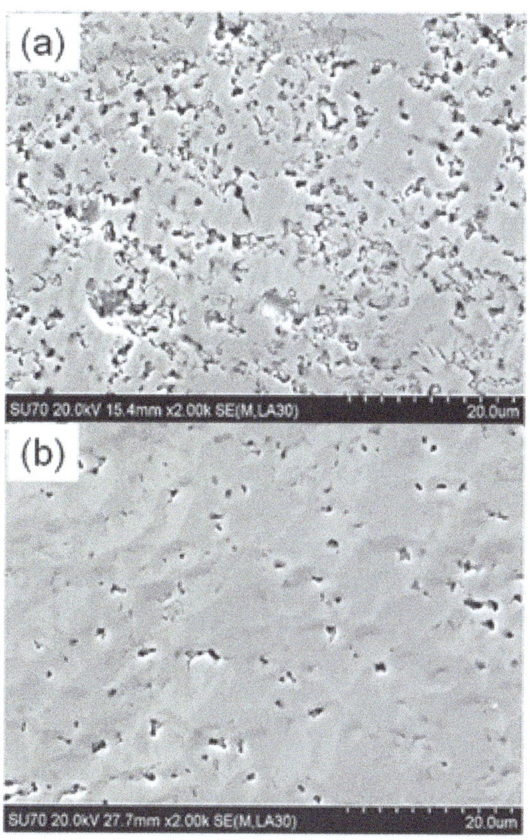

Figure 117. Scanning electron microscopy images of (a) flat rolled and (b) uniaxially pressed Ba-122/Ag tapes.

In the previous sections, we used an Ag tube as the metal sheath of Ba-122 tapes. The Ag sheath was completely annealed by the heat treatment and became very soft after the heat treatment. Thus, the Ag sheathed Ba-122 tape is not practical from the aspect of mechanical strength. In order to solve this problem we fabricated Ba-122 tape conductors applying a new sheath structure of double sheaths; stainless steel as an outer sheath and Ag as an inner sheath [505]. The inner Ag sheath was used to avoid the reaction between stainless steel and Ba-122. Figure118 shows J_c–B curves at 4.2 K of the double sheathed Ba-122 tapes fabricated with flat rolling and uniaxial pressing. The inset shows transverse cross sections of flat rolled (top) and pressed (bottom) tapes. For comparison, data of commercial superconductors, Nb–Ti and Nb_3Sn, are also shown in the figure. We found that the rolled tapes show J_c values of 7.7×10^4 A cm^{-2} at 4.2 K and 10 T with high homogeneity. These J_c are the highest values reported so far for IBSC tapes and wires fabricated by a scalable rolling process. It should be noted that the use of hard stainless steel as the outer sheath increases not only the mechanical strength of the tape but also the density of Ba-122 core. The application of uniaxial pressing to the double sheathed tape further increased J_c at 4.2 K and 10 T up to 9.0×10^4 A cm^{-2}. The transport J_c–B curves for both rolled and pressed tapes show extremely small magnetic field dependence and the J_c values exceed 3×10^4 A cm^{-2} in 28 T, which are much higher than those of low-temperature commercial superconductors. The microstructure investigations indicate that such high J_c values were achieved by the higher density of the core, more uniform deformation of the Ba-122 core and higher degree of grain orientation [505]. These results indicate that the combination of the double sheath and rolling is very promising for fabricating long Ba-122 tape conductors for a high magnetic field magnet which can generate fields higher than 20 T.

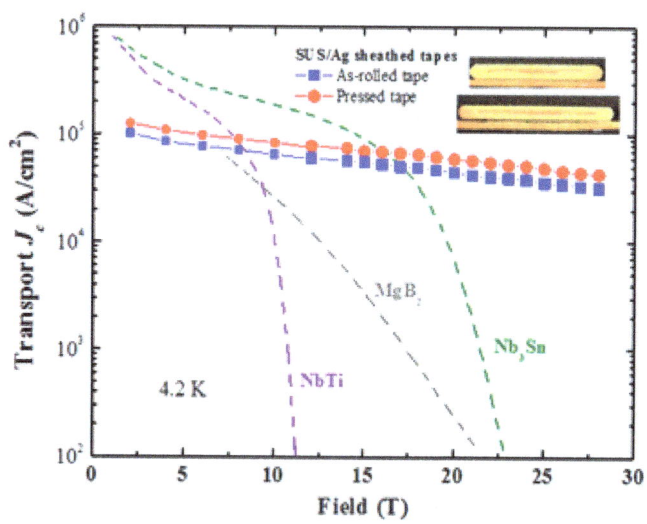

Figure 118. J_c versus magnetic field curves at 4.2 K of double sheathed Ba-122 tapes fabricated with rolling and pressing. For comparison, MgB_2 wires are shown in the figure. Reprinted with permission from [505]. Copyright 2015 by IOP Publishing.

Temperature Dependence of Critical Current Density

The temperature dependence of the transport critical current was measured using a 10 T split type superconducting magnet with a variable temperature insert cryostat that can control the temperature between 10 and 40 K [497]. We also investigated the magnetic field orientation dependence of J_c for Sr-122/Ag tapes to evaluate the J_c anisotropy of the tape conductors.

Figure 119 shows J_c–B curves of a uniaxially pressed Sr-122/Ag wire at four different temperatures of 10, 20, 25 and 30 K in external magnetic fields applied parallel and perpendicular to the tape surface [497]. J_c was estimated by dividing the I_c by the cross sectional area of the Sr-122 superconducting core of the tape conductor. The uniaxial pressure was 0.4 GPa, which brought a thickness reduction ratio of 20%. At 20 K, J_c value was almost 10^4 A cm^{-2}at 0 T. This J_c decreases gradually with increasing magnetic field applied parallel to the tape surface. J_c values in perpendicular fields are lower than those in parallel fields; however, field dependence is still small. The J_c–B curves at temperatures below 25 K show small slopes, nearly equal to those at 10 K, suggesting that the *ex situ* Sr-122/Ag tape has a high potential in high magnetic fields at temperatures below 25 K. Furthermore, even at 30 K, which is close to the $T_c$of this superconducting material, J_c was still observed in fields up to around 7 T. These results are due to the high B_c2 values of Sr-122. The J_c value in a magnetic field applied perpendicular to the tape surface was about half the value in the magnetic field applied parallel to the tape surface at temperatures below 20 K. This J_c anisotropy of the Sr-122/Ag tape is smaller than those of Bi-2212/Ag and Bi-2223/Ag tape conductors. This result also suggests that the Sr-122/Ag tape shows lower anisotropy in J_c and is useful for practical magnet applications.

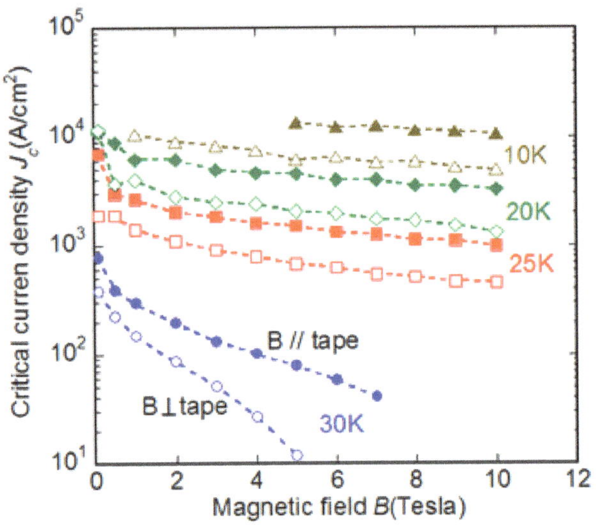

Figure 119. Temperature dependence of J_c–B curves of the Sr-122/Ag tape. A magnetic field was applied parallel (perpendicular to the tape axis) and perpendicular to the tape surface. Reprinted with permission from [497]. Copyright 2014 by IOP Publishing.

Figures 120(a) and (b) show J_c of a pressed Sr-122/Ag tape as a function of external magnetic field directions to the tape surface at 20 and 30 K, respectively [497]. Zero and 90° in the figure correspond to the magnetic field directions perpendicular and parallel to the tape surface, respectively. Maximum and minimum J_c values were obtained at around 0° and 90° in every magnetic field. This angular dependence is similar to the anisotropy of high-T_c oxide superconducting tapes, Y-123(coated conductor), 2212/Ag and B-2223/Ag tapes. The ratio between maximum and minimum J_c values (J_c anisotropy) is almost two at every magnetic field up to 3 T. However, the J_c anisotropy increased with increasing magnetic field as shown in figure 119.

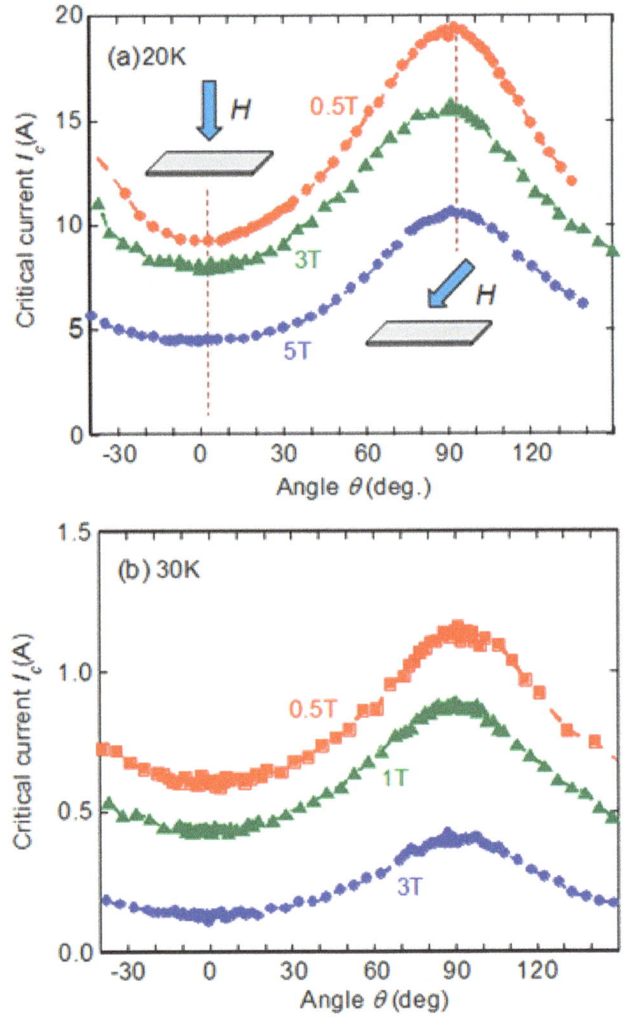

Figure 120. I_c at 20 K and 30 K as a function of strength and direction of external magnetic field. Reprinted with permission from [497]. Copyright 2014 by IOP Publishing.

Summary and Future Prospects

Among the many fabrication stages of superconducting tapes, flat rolling has been commonly used to densify and realize the grain orientation of the superconducting core [494, 496, 505]. Our results showed a large increase in J_c due to the improvement of the Ba(Sr)-122 core density and preferred orientation in the initial step by the rolling process. Upon further rolling to smaller tape thicknesses, degradation of critical current density was observed due to the introduction of transverse micro-cracks. However, a larger reduction of tape thickness by flat rolling tends to decrease the transverse micro-cracks and to enhance J_c values. This suggests that higher J_c can be realized by only applying flat rolling.

Furthermore, when the tape was uniaxially pressed, J_c values were significantly increased by another improvement of the core density and grain orientation. Excellent transport J_c values of ~10^5 A cm^{-2} under magnetic fields up to 6 T were obtained in uniaxially pressed Ba-122/Ag tapes. A higher core density and more grain orientation are responsible for this high $J_c$performance in the pressed samples. The use of a hard metal as the outer sheath of a double sheathed tape is also effective in increasing the core density and J_c values. However, it should be emphasized that practical applications of uniaxial pressing for the manufacture of long tapes require specialized machines for continuous pressing of the tape. Fortunately, there have been successful attempts at producing long Bi-2223/Ag wires by periodic pressing [506] and eccentric rolling [507], which might be also applied to the production of long Ba-122 wires with high transport J_c.

Ba(Sr)-122/Ag tape conductors show very small field dependence of J_c with small J_c anisotropy. With further improvements in the critical current density and wire fabrication technology, Ba(Sr)-122 superconductors will have a very promising future in high-field applications.

NEW FUNCTIONAL MATERIALS AND DEVICES FOUND WITHIN THIS PROJECT

Exploration of novel superconductors needs a non-conventional approach in various aspects, such as the material system and synthetic processes as high pressure and electric field effect. As a result, there should be a much higher probability than other materials research to find new functionalities or to encounter new phenomena during the research [508]. A well-known example is the discovery of high-performance thermoelectric properties in $NaCo_2O_4$ in the course of comparative study of high-T_c cuprates with layered cobaltites [509]. This feature comes from the richness of materials in which we are engaging.

In this project, this team proceeded exploration of new superconductors along with seeking new functional materials. Some of representative achievements are briefly described in this section.

Catalytic activity for Ambient Pressure NH_3 Synthesis

Ammonia is the simplest active nitrogen molecule and one of the most commercially produced chemicals, reaching 160 million tons per year. While industrial ammonia synthesis from N_2 and H_2 is conducted using the Haber–Bosch process using iron-based catalysts [510] at 400–600 °C and 20–40 MPa, such high reaction temperatures are disadvantageous with respect to the equilibrium and exothermic reaction (46.1 kJ mol^{-1}) of ammonia synthesis. The rate-determining step of ammonia synthesis is cleavage of the N≡N bond, because the bond energy is extremely large (945 kJ mol^{-1}). Transition metals such as Fe or Ru are indispensable for the promotion of N≡N bond cleavage, in addition to electron donors that provide electrons to the transition metals. An N_2 molecule is fixed to form a bond with a transition metal by donating electrons from its bonding orbitals and accepting electrons to the antibonding π orbitals (back-donation) [511]. This back-donation is effectively enhanced by electron donors, which further weakens the N≡N bond and results in the cleavage of N_2. We assume electron injection is essentially indispensable to enhancing the efficiency of ammonia synthesis using Fe or Ru-catalysts. However, the situation is not so simple, i.e., it is extremely difficult to realize a low work function and chemical and thermal stability; both are generally incompatible. Although the catalytic activity of Ru is drastically enhanced by adding alkali or alkaline earth metals with small work functions [512], these metals are practically inapplicable for ammonia synthesis because these metals are so chemically active that the reaction with the produced ammonia and/or N_2 forms metal nitrides and amides.

We examined the catalytic activity of a stable electride, C12A7:e$^-$ as an efficient promoter with high electron donating power and chemical stability for a Ru catalyst utilizing unique properties of this material, i.e., low work function (2.4 eV) comparable metal potassium but chemical inertness [513]. Electride is an ionic crystal in which electrons serve as anions. 12CaO . $7Al_2O_3$(C12A7), which is a constituent of commercial alumina cement, works as complexant to electrons and the resulting material became the first electride that is stable at temperatures above room temperature and at ambient atmosphere [514, 515]. As shown figure 121, the unit cell of C12A7 has a positively charged framework structure composed of 12 sub-nanometer-sized cages which are connected to each other by sharing a mono-oxide layer to embrace $2O^{2-}$in the two cages as the counter anions to compensate the electro-neutrality. Chemical reduction processes are used to inject four electrons into four of the 12 cages by extracting two O^{2-} ions accommodated in the cavities as counter anions to compensate for the positive charge on the cage wall. The resultant chemical formula is represented by $[Ca_{24}Al_{28}O_{64}]^{4+}(e^-)_4$. The injected electrons occupy a unique conduction band called 'the cage conduction band' (CCB), which is derived from the 3-dimensionally connected cages by sharing an oxide monolayer, and can migrate through the thin cage wall by tunneling, which leads to metallic conduction (about 1500 S cm^{-1} at room temperature). This electron-trapped cage structure of the bulk is retained up to the top surface if the sample is appropriately heated. Such electrons encapsulated in the cages of C12A7:e$^-$ can be readily replaced with a hydride ion (H$^-$) by heating in H_2 gas. The incorporated H$^-$ ions desorb as H_2 molecules at about 400 °C, leaving electrons in the positively charged framework of C12A7: the incorporation and release of H$^-$ ions on C12A7:e$^-$ are entirely reversible. Please note that the electride formation and reversible storage ability of H$^-$totally originate from the unique crystal structure of C12A7 described above. Such formation was impossible for other oxides bearing Al_2O_3 and CaO.

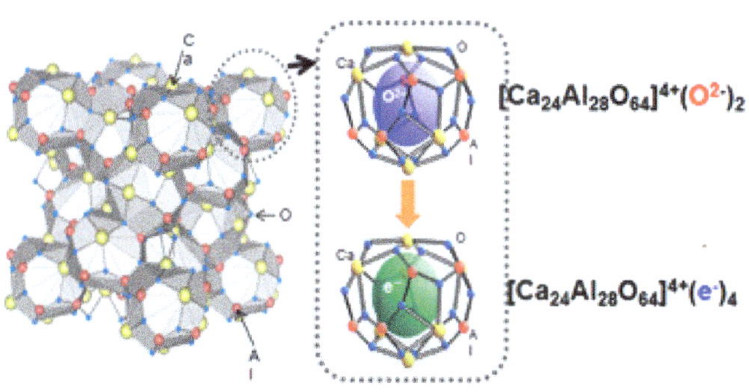

Figure 121. Crystal structure of 12CaO . $7Al_2O_3$(C12A7). C12A7 has a cubic symmetry and two formula units with the chemical composition of $Ca_{24}Al_{28}O_{64}$, and is an insulator with a bad gap of ~7 eV. Stoichiometric C12A7 (C12A7:O^{2-}) accommodates two O^{2-} ions in two cages out of 12 cages constituting a unit cell to preserve the electro-neutrality of the positively charged cage walls and its chemical formula may be expressed as $[Ca_{24}Al_{28}O_{63}]^{4+}(O^{2-})_2$. All of these in-cage oxygen ions may be replaced with electrons keeping the original case structure by an appropriate chemical treatment. The resulting material with a formula $[Ca_{24}Al_{28}O_{63}]^{4+}(4e^-)$ may be regarded as an electride.

C12A7 electride, abbreviated as C12A7:e, is a metallic conductor and exhibits a metal–superconductor transition at 0.2–0.4 K.

We found that C12A7:e$^-$ exhibits a superconductor transition at 0.2–0.4 K depending on the carrier concentration under an ambient atmosphere [516]. Although T_c is very low, this is the first *s*-metal superconductor under ambient pressure. Alkali and alkaline-earth metals do not show superconductivity under ambient pressure but some high pressure phases exhibit T_c. We found the crystal structure of C12A7 is similar to that of superconducting phases and considered that the admixture of Ca *d*-orbitals to the *s*-orbital of the cage electrons is the origin of the superconductivity of C12A7:e$^-$ [517].

We deposited nano-sized Ru particles to C12A7:e$^-$ powders using the chemical vapor deposition method to enhance the amount of captured N_2 on their surfaces. Figure 122 shows a comparison in catalytic activity for NH_3 synthesis [108]. The activation energy over Ru-loaded C12A7:e$^-$ is reduced to almost half of that over other Ru-catalysts and the turn-over-frequency (TOF, the measure of catalytic activity per active site) is larger by an order of magnitude than that of the latter including the best Ru-based catalyst (Ru/Cs-loaded MgO). The excellent activity of C12A7:e$^-$ demonstrates the electron-donating effect plays an essential role in NH_3synthesis. Very recently, it has been clarified by examining the kinetics of the isotope exchange reaction between ^{14}N and ^{15}N with DFT calculations that the rate determining step of NH_3synthesis over Ru-loaded C12A7:e$^-$ shifts from dissociation of the N≡N bond to the N–H bond formation process [518].

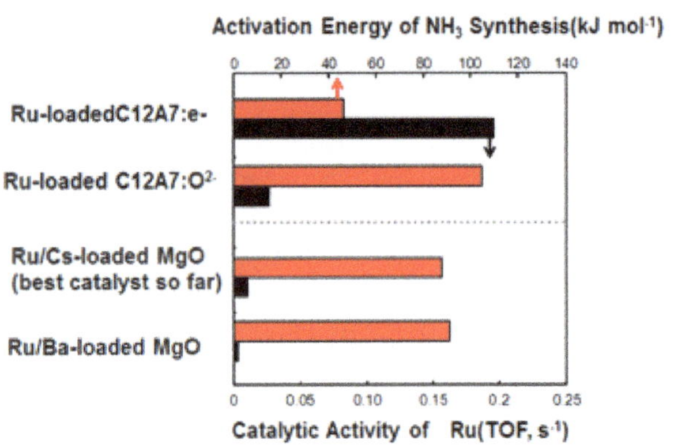

Figure 122. Comparison of Ru-loaded catalysts for ammonia synthesis. The reaction conditions; catalyst 0.3 g, flow rate of H_2/N_2 (3:1) gas 60 ml min^{-1}, temperature 400 °C, and pressure 0.1 MPa. Reprinted with permission from Macmillan Publishers Ltd: [108], Copyright 2012.

Another surprising finding is that Ru-loaded C12A7:e$^-$ does not exhibit hydrogen-poisoning [108] as shown in figure 123. An obstacle for industrial ammonia synthesis with a Ru-loaded catalyst is hydrogen-poisoning in high H_2 pressure. Because NH_3 synthesis over Ru catalysts in general degrades by hydrogen adatoms formed on Ru surfaces, the reaction order for H_2 on Ru catalysts often approaches −1 (i.e., the reaction rate *decreases* with the partial pressure of H_2). Such H_2 poisoning on Ru catalysts is a serious obstacle for industrial ammonia production, which requires high pressure conditions to collect the resulting NH_3 in the form of liquid (NH_3easily becomes liquid above 0.85 MPa at room temperature). The chemical industry is therefore currently searching for a supported Ru catalyst that promotes N_2 dissociation but suppresses H_2poisoning. Figure 124 shows a tentative mechanism for ammonia synthesis over Ru-loaded C12A7:e$^-$. The robustness of Ru-loaded C12A7:e$^-$ to H_2-poisoning comes from its ability of reversible H-storage and release. C12A7:e$^-$ reacts with H_2 to form C12A7:H$^-$ in which H is incorporated into the cage as H$^-$. Although this reaction over Ru-free C12A7:e$^-$ is irreversible in temperature range below ~450 °C, the H-release temperature is drastically reduced to ~300 °C [519]. Desorption of H$^-$ to react with activated nitrogen species on Ru surfaces to form N–H bonds, leaving an electron in the cage. We consider that hydrogen poisoning on Ru-surfaces is suppressed by preferential entrapping of a H adatom in the cage to form H$^-$ and the incorporated H$^-$ in the cage is activated by a strong driving force for stable N–H bond formation.

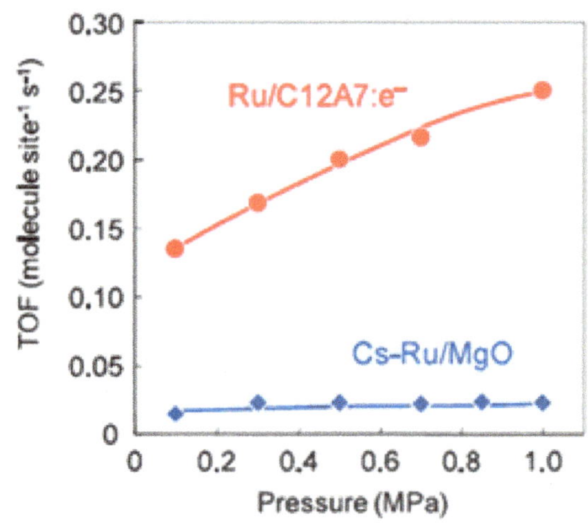

Figure 123. Catalytic activity of Ru-loaded C12A7:e$^-$ as a function of partial hydrogen pressure. Data of a representative Ru-catalyst are shown for comparison. Note that the activity is proportional to partial H_2 pressure. There is no serious H_2-poisoning which is a common drawback of Ru-catalysts. Reprinted with permission from Macmillan Publishers Ltd: [108], Copyright 2012.

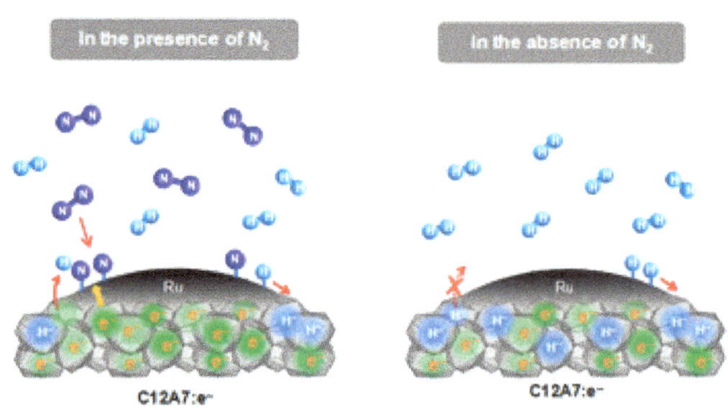

Figure 124. Tentative reaction mechanism to explain suppression of H_2-poisoning in Ru-loaded C12A7:e$^-$. Reprinted with permission from Macmillan Publishers Ltd: [108], Copyright 2012.

Bipolar oxide thin Film Transistor

Thin film transistor (TFT) is a fundamental building block of modern integrated circuits. Oxide semiconductor-based TFTs are an emerging technology to take over the current Si TFT technology because they have superior properties such as larger electron mobility, lower defect density, low temperature fabrication and high optical transparency [520]. In particular, progress in TFTs using transparent amorphous oxide semiconductors—such as InGaZnO*x* (IGZO) [521,522]—is so remarkable that the application of IGZO-TFTs has started to drive energy-saving and high resolution displays for mobile phones, tablet PCs, PC monitors and large-screen TVs using organic light-emitting diodes. The next challenge in oxide TFTs is application to logic circuits. So far, oxide TFTs work only as unipolar devices mostly *n*-type, or recently realized *p*-type [523], but do not exhibit inversion/ambipolar operation; i.e., complementary circuits cannot be made from TFTs of the same oxide material. Therefore, the next challenge is to realize oxide-based ambipolar TFTs and complementary circuits using only a single oxide semiconductor channel.

In this project, we succeeded in fabricating an ambipolar oxide TFT using an SnO channel, and demonstrated operation of a complementary-like inverter [111]. It is of interest to note that SnO has the same crystal structure as FeSe, which is a parent compound of IBSCs, and exhibits a insulator–metal–superconductor transition under high pressure [524]. This is the first success in oxide TFTs. We achieved clear ambipolar operation with saturation mobilities of ~0.81 for the *p*-channel and ~10^{-3} $cm^2\ (Vs)^{-1}$ for *n*-channel modes in the SnO TFTs. The maximum voltage gain of ~2.5 was obtained in the complementary-like inverter circuit as shown in figure125. This is the first demonstration of a complementary-like circuit using a single oxide semiconductor channel and would provide an important step toward practical oxide electronics. Further, the low temperature process (250 °C at maximum) is compatible with emerging flexible electronics technology.

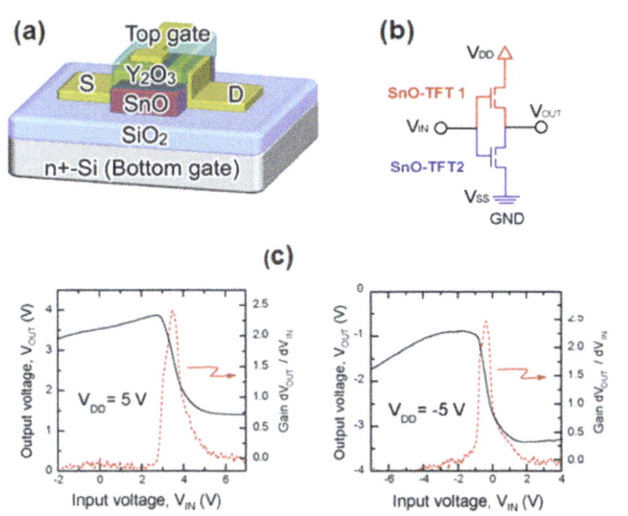

Figure 125. Ambipolar oxide TFTs using SnO. (a) TFT structure, (b) equivalent circuit of SnO complementary metal-oxide semiconductor (CMOS) inverter, and (c) their characteristics [111]. Reprinted with permission from [111]. Copyright 2011 by John Wiley and Sons.

Metallic Ferroelectric Materials: Slater Insulators

In 1995, Anderson and Blount [525] predicted the possibility of a ferroelectric metal in which a ferroelectric-like structural transition occurs in the metallic state. They were motivated by the BCS theory and considered the possibility of such a material which would improve T_c utilizing strong electron–phonon coupling. It is almost a consensus that metals do not exhibit ferroelectricity because static internal electric fields are screened by conduction electrons. In fact, no clear example of such a material had been reported up to 2013. In this project, we could identify this type of material. One is $LiOsO_3$ [106] and another is $LaFeAsO_{0.5}H_{0.5}$ [211]. Since the latter was described in section 3.1, we focus on the former hereafter.

Yamaura (a member of this project) and his collaborators found [106] that the high-pressure-synthesized material $LiOsO_3$ exhibits a structural transition at a temperature (T_s) of 140 K as shown in figure 126. A complementary usage of XRD and neutron diffraction revealed that structural transition at 140 K is due to centrosymmetric ($R\bar{3}c$) to non-centrosymmetric ($R3c$) phase in metallic $LiOsO_3$. This transition is structurally equivalent to the ferroelectric transition of $LiNbO_3$ [526], involving a continuous shift in the mean position of Li^+ ions on cooling below 140 K. Its discovery realizes the scenario suggested by Anderson and Blount, i.e., the existence of ferroelectric-like soft phonons could stabilize non-centrosymmetric superconductivity at enhanced temperatures.

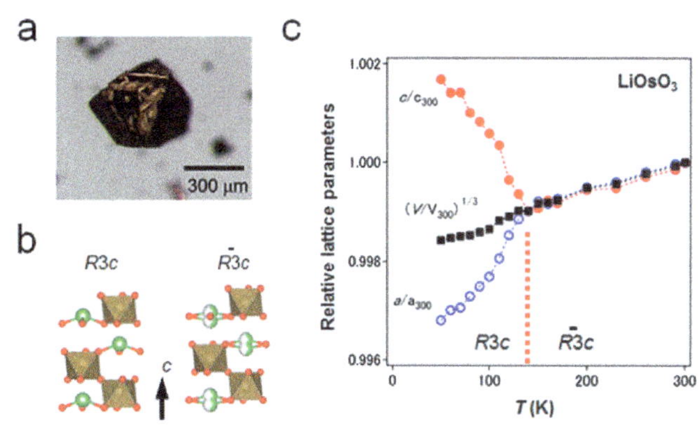

Figure 126. Slater insulator $LiOsO_3$. (a) Photo of a flex-grown single crystal, (b) schematic views of centrosymmetric (right) and non-centrosymmetric (left) $LiNbO_3$-type structures, (c) thermal evolution of the rhombohedral lattice parameters of $LiOsO_3$measured by neutron diffraction. The parameters are normalized by the 300 K values of$a$ = 5.0699(1) Å and c = 13.2237(2) Å [106].

PERSPECTIVE ON SUPERCONDUCTING TAPES AND WIRES

Among recently discovered new superconductors, IBSCs exhibit the properties most attractive for tape or wire applications. Although their T_c of 30–56 K is below liquid nitrogen temperature, their upper critical fields (H_{c2}) of 60–100 T are comparable to those for cuprate superconductors with T_c values over 90 K [319]. In particular, 122 compounds, such as Ba-122 doped with K, P, or Co, have very low anisotropy comparable to or lower than that of MgB_2[319, 395, 407], implying that high irreversibility fields close to H_{c2} or high in-field J_c are expected. Moreover, 122 compounds were found to have the critical angle for the transition from strong-link to weak-link behavior approximately twice of the value for REBCO, suggesting less sensitivity to grain misorientation at GBs [434].

Actually, the in-field J_c of Ba(Sr)122:K wires having superconducting cores without biaxial textures which are fabricated by an *ex situ* PIT method has recently been improved greatly, as shown in figure 127. J_c at 4.2 K and

10 T for Ba(Sr)122:K wires has reached 10^4 A cm^{-2} through metal addition plus a rolling induced texture process such as hot isostatic or cold press methods [490, 491, 493, 495]. Such mechanical deformation was found to be effective for densifying the conductor core and aligning the grains of the superconducting phase, resulting in high quality superconducting wires. By applying uniaxial pressure of 2–4 GPa, J_c has been further improved and approached 10^5 A cm^{-2}, which is a practical level as a superconducting wire [503]. In section 4.4, we showed the magnetic field dependence of J_c at 4.2 K for Ba-122:K wires with a Ag sheath (figure 113), which were fabricated by NIMS group using a uniaxial pressure process and a more practical flat rolling process. The wires fabricated by employing the former process exhibit a J_c at 10 T close to 10^5 A cm^{-2} for a single core wire and a high J_c of 5 × 10^4 A cm^{-2}even for a 7-filament wire. Though J_c of the wires through the latter process is still lower, 4 × 10^5A cm^{-2} at 10 T, these J_c values are well over that of the Nb–Ti practical conductor. Moreover, in comparison with the Nb_3Sn conductor, the J_c of Ba-122:K tapes through both processes show very small magnetic field dependence, indicating that Ba-122 has great potential to surpass the J_c performance of Nb_3Sn in high magnetic fields.

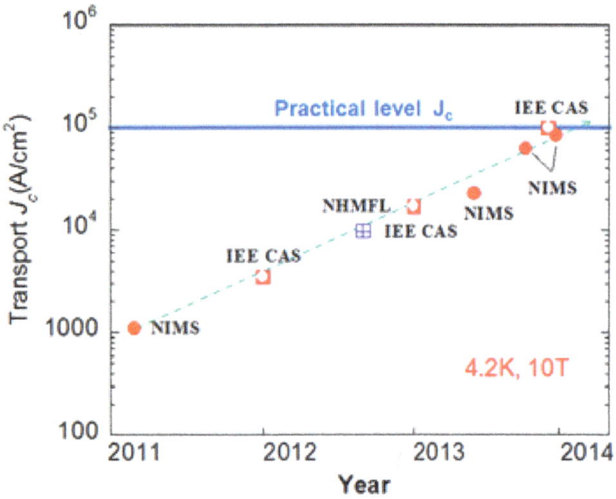

Figure 127. Recent evolution of transport J_c at 4.2 K, 10 T for Ba(Sr)122:K PIT wires or tapes reported by groups of groups from the National Institute for Materials Science (NIMS), Japan, Institute of Electrical Engineering of the Chinese Academy of Science (IEECAS) and National High Magnetic Field Laboratory (NHMFL), USA.

The superior J_c properties of Ba-122 at higher fields have been demonstrated more clearly in thin film works. In figure 128, J_c–H properties at 4.2 K for various Ba-122 films on single-crystal and IBAD–MgO-buffered metal substrates are compared with those for Nb–Ti and Nb_3Sn conductors. Nb_3Sn exhibits a steep decrease of J_c at fields near 20 T, which is close to its H_{c2}. The Ba-122:P film on MgO fabricated by an MBE method shows J_c ($H//c$) over 10^5 and 10^4A cm^{-2} at 20 and 35 T, respectively [527]. The Ba-122:P films with dense *c*-axis-correlated pinning centers (line dislocations in the mother phase) [422] or $BaZrO_3$ nanoparticles [415] by a PLD method exhibit even higher J_c values at fields below 9 T and a rather slow decay, suggesting that their J_c values at higher fields would be higher than that for the MBE film. Another significance of these results is that the in-field performance of IBSCs, in particular Ba-122, can be remarkably improved by introduced nanometer-size vortex pinning centers, as already demonstrated in REBCO. This would be true not only for films or coated conductors but also for PIT wires. Although the in-field J_c performance of Ba-122 coated conductors, in particular longer-length ones, is still substantially lower than that for high-quality films on single crystal substrates, this difference could be mainly attributed to technical problems and should be overcome in the near future. In fact, for the case of 11 compound Fe(Se,Te) of which films can be grown at lower substrate temperatures, coated conductors, not long, with J_c over 10^5 A cm^{-2} at 30 T have been demonstrated [528].

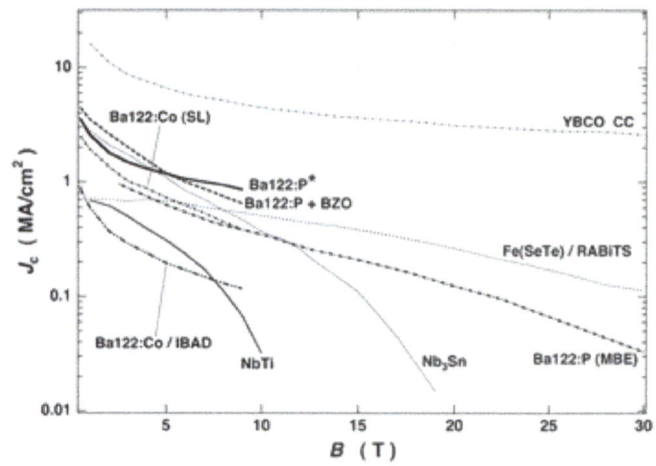

Figure 128. J_c–$B(//c)$ curves at 4.2 K reported for Ba-122 and Fe(Se,Te) films on single-crystal substrates and some technical substrates such as IBAD and RABiTS (rolling assisted biaxially textured substrates). The data of Nb–Ti and Nb_3Sn commercial wires and YBCO coated conductors are also shown for comparison. Ba-122:P*; Tokyo Institute of Technology (TIT) [422], Ba-122:P + BZO; ISTEC [415], Ba-122:Co (SL); Wisconsin Univ. and NHMFL [417], Ba-122:P (MBE); Nagoya Univ. and IFW Dresden [527], Ba-122:Co/IBAD; IFW Dresden [425], Fe(Se,Te)/RABiTS; Brookhaven National Laboratory (BNL), USA and NHMFL [528].

The superior J_c performance of IBSCs such as Ba-122 mentioned above indicates that their most promising application would be wires or tapes for magnets used at low temperatures and high fields well over 20 T. The competitive candidate for such applications is REBCO coated conductors, which exhibit even higher in-field J_c at low temperatures. However, one of the current critical issues of REBCO coated conductors for this application is suppression of shielding current, which is required to generate stable and highly precise fields [529]. In order to overcome this difficulty, the fabrication technique for narrow multi-filamentary structures is being developed [530]. Another problem is the mechanical strength against delamination which originates from its multilayer structure [531]. If the J_c performance of iron-based multifilamentary PIT wires is further improved by refining the fabrication process and reaches a practical level even at fields of 20–30 T, high J_c in higher magnetic fields of IBSCs could be a concrete advantage over REBCO coated conductors, though there are some further issues such as the sheath material with higher mechanical strength and lower cost. Concerning iron-based coated conductors, more technical challenges still remain to demonstrate an advantage over practical conductors, especially REBCO coated conductors. These include high in-field J_c on less-textured buffer layers, narrow and thicker conductors with a lower aspect ratio, higher mechanical strength, and so on.

Although coated conductors have been fabricated using the same oxide buffer-layer technology as for REBCO, special buffer-layer materials more compatible with IBSCs would be necessary to demonstrate their own potential.

Another challenge common to PIT wires and coated conductors is the development of IBSCs with less toxic elements, which would further stimulate their application. Considering the history that it needed more than 20 years to develop commercial REBCO coated conductors, further continuing research and development is definitely required to realize practical wires or tapes based on IBSCs discovered only seven years ago. Fortunately, the material variety of IBSCs is the largest among all the superconductor families, and the discovery of new types of superconducting materials has been continuing to date. The intrinsic nature of this materials system [508, 532] is to have a Fermi surface composed of multi-pockets arising from multi-orbitals of Fe *3d*. This nature provides a wide opportunity in which various degrees of freedom can contribute to the emergence of superconductivity. Recent research has clarified the contribution of orbital [533], charge [534] and phonons [535] in addition to spin [159, 160], which was initially proposed. The current status may be regarded as 'actors are ready!'. These actors may work cooperatively to raise T_c, which has been experimentally suggested recently. We may expect discoveries of new IBSCs which exhibit higher J_c keeping less anisotropy from the background. *Iron is still hot*!

ACKNOWLEDGEMENTS

This work was executed under JSPS FIRST (Funding Program for World-Leading Innovative R&D on Science and Technology) Program initiated by CSTP. The results described herein were obtained by many collaborators and graduate students participated. One of the authors (HH) thanks the administrative staff of the project at TIT and a funding from MEXT Element Strategy Initiative to form a core research center for harvesting achievements.

REFERENCES

1. Rogalla H and Kes P H (ed) 2012 100 Years of Superconductivity (London: CRC Press)
2. A calculation method of Tc using DFT was developed by Gross et al: Lüders M, Marques M A L, Lathiotakis N N, Floris A, Profeta G, Fast L, Continenza A, Massidda S and Gross E K U 2005 Phys. Rev. B 72 024545 Marques M A L, Lüders M, Lathiotakis N N, Profeta G, Floris A, Fast L, Continenza A, Gross E K and Massidda S 2005 Phys. Rev. B 72 024546
3. Kamihara Y, Hiramatsu H, Hirano H, Kawamura R, Yanagi H, Kamiya T and Hosono H 2006 J. Am. Chem. Soc. 128 10012
4. Kamihara Y, Watanabe T, Hirano M and Hosono H 2008 J. Am. Chem. Soc. 130 3296
5. Watanabe T, Yanagi H, Kamiya T, Kamihara Y, Hiramatsu H, Hirano M and Hosono H 2007 Inorg. Chem. 46 7719
6. Hosono H 2007 Thin Solid Films 515 5745
7. Hosono H 2009 Physica C 469 314
8. Hosono H 2013 Japan. J. Appl. Phys. 52 090001
9. Hanna H, Muraba Y, Matsuishi S, Igawa N, Kodama K, Shamoto S and Hosono H 2011 Phys. Rev. B 84 024521
10. Matsuishi S, Hanna T, Muraba Y, Kim S W, Kim J E, Takata M, Shamoto S, Smith R I and Hosono H 2012 Phys. Rev. B 85 014514
11. Iimura S, Matsuishi S, Sato H, Hanna T, Muraba Y, Kim S W, Kim J E, Takata M and Hosono H 2012 Nat. Commun. 3 943
12. Matsuishi S, Maruyama T, Iimura S and Hosono H 2014 Phys. Rev. B 89 094510
13. Ishida J, Iimura S, Matsuishi S and Hosono H 2014 J. Phys.: Condens. Matter. 26 435702
14. Guo Y F, Wang X, Li J, Sun Y, Tsujimoto Y, Belik A A, Matsushita Y, Yamaura K and Takayama-Muromachi E 2012 Phys. Rev. B 86 054523
15. Guo Y F et al 2010 Phys. Rev. B 82 054506
16. Muraba Y, Matsuishi S and Hosono H 2014 Phys. Rev. B 89 094501
17. Muraba Y, Matsuishi S and Hosono H 2014 J. Phys. Soc. Japan 83 033705
18. Hanna T, Muraba Y, Matsuishi S and Hosono H 2013 Appl. Phys. Lett. 103 142601
19. Muraba Y, Matsuishi S, Kim S-W, Atou T, Fukunaga O and Hosono H 2010 Phys. Rev. B 82 180512(R)
20. Katase T, Iimura S, Hiramatsu H, Kamiya T and Hosono H 2012 Phys. Rev. B 85 140516(R)
21. Katase T, Hiramatsu H, Kamiya T and Hosono H 2013 New J. Phys. 15 073019
22. Hiramatsu H, Katase T, Kamiya T and Hosono H 2013 IEEE Trans. Appl. Superconduct. 23 7300405
23. Kudo K, Iba K, Takasuga M, Kitahama Y, Matsumura J, Danura M, Nogami Y and Nohara M 2013 Sci. Rep. 3 1478
24. Guo Y, Wang X, Li J, Zhang S, Yamaura K and Takayama-Muromachi E 2012 J. Phys. Soc. Japan 81 064704
25. Katayama N et al 2013 J. Phys. Soc. Japan. 82 123702
26. Kudo K, Mizukami T, Kitahama Y, Mitsuoka D, Iba K, Fujimura K, Nishimoto N, Hiraoka Y and Nohara M 2014 J. Phys. Soc. Japan. 83 025001
27. Kudo K, Kitahama Y, Fujimura K, Mizukami T, Ota H and Nohara M 2014 J. Phys. Soc. Japan. 83 093705
28. Kudo K et al 2013 Sci. Rep. 3 3101
29. Guo J G, Lei C H, Hayashi F and Hosono H 2014 Nat. Commun. 5 4756
30. Ryu G, Kim S W, Mizoguchi H, Matsuishi S and Hosono H 2012 Europhys. Lett. 99 27002
31. Yajima T, Nakano K, Takeiri F, Ono T, Hosokoshi Y, Matsushita Y, Hester J, Kobayashi Y and Kageyama H 2012 J. Phys. Soc. Japan 81 103706
32. Yajima T, Nakano K, Takeiri F, Nozaki Y, Kobayashi Y and Kageyama H 2013 J. Phys. Soc. Japan 82 033705
33. Nakano K, Yajima T, Takeiri F, Green M A, Hester J, Kobayashi Y and Kageyama H 2013 J. Phys. Soc. Japan 82 074707
34. Mizoguchi H, Kuroda T, Kamiya T and Hosono H 2011 Phys. Rev. Lett. 106 237001

35. Kudo K, Takasuga M, Okamoto Y, Hiroi Z and Nohara M 2012 Phys. Rev. Lett. 109 097002
36. Matsuishi S, Nakamura A, Muraba Y and Hosono H 2012 Supercond. Sci. Technol. 25 084017
37. Mizoguchi H, Matsuishi S, Hirano M, Tachibana M, Takayama-Muromachi E, Kawaji H and Hosono H 2011 Phys. Rev. Lett. 106 057002
38. Imamura N, Mizoguchi H and Hosono H 2012 J. Am. Chem. Soc. 134 2516
39. Ganesanpotti S, Yajima T, Tohyama T, Li Z, Nakano K, Nozaki Y, Tassel C, Kobayashi Y and Kageyama H 2014 J. Alloy. Compd 583 151
40. Ganesanpotti S, Yajima T, Nakano K, Nozaki Y, Yamamoto T, Tassel C, Kobayashi Y and Kageyama H 2014 J. Alloy. Compd 613 370
41. Imai M, Emura S, Nishio M, Matsushita Y, Ibuka S, Eguchi N, Ishikawa F, Yamada Y, Muranaka T and Akimitsu J 2013 Supercond. Sci. Technol. 26 075001
42. Imai M, Ibuka S, Kikugawa N, Terashima T, Uji S, Kageyama H, Yajima T and Hase I 2015 Phys. Rev. B 91 014513
43. Mizoguchi H and Hosono H 2011 Chem. Commun. 47 3778
44. Lei H and Hosono H 2013 Europhys. Lett. 104 17003
45. Tanaka M, Zhang S, Tanaka Y, Inumaru K and Yamanaka S 2013 J. Solid State Chem. 198 445
46. Ji S, Imai M, Zhu H and Yamanaka S 2013 Inorg. Chem. 52 3953
47. Ibuka S, Imai M, Naka T and Nishio M 2014 Supercond. Sci. Technol. 27 025012
48. Kudo K, Fujimura K, Onari S, Ota H and Nohara M 2015 Phys. Rev. B at press
49. Isobe M, Yoshida H, Kimoto K, Arai M and Takayama-Muromachi E 2014 Chem. Mater. 26 2155
50. Pyon S et al 2014 J. Phys. Soc. Japan 83 093706
51. Qi Y, Guo J, Lei H, Xiao Z, Kamiya T and Hosono H 2014 Phys. Rev. B 89 024517
52. Mizugichi Y et al 2012 Phys. Rev. B 86 220510
53. Demura S et al 2013 J. Phys. Soc. Japan 82 033708
54. Yajima T, Takeiri F, Nozaki Y, Li Z, Tohyama T, Green M A, Kobayashi Y and Kageyama H 2014 J. Phys. Soc. Japan 83 073705
55. Yamanaka S, Umemoto K, Zheng Z, Suzuki Y, Matsui H, Toyota N and Inumaru K 2012 J. Mater. Chem. 22 10752
56. Zhang S, Tanaka M and Yamanaka S 2012 Phys. Rev. B 86 024516
57. Zhang S, Tanaka M, Watanabe E, Zhu H, Inumaru K and Yamanaka S 2013 Supercond. Sci. Technol. 26 122001
58. Zheng Z and Yamanaka S 2011 Chem. Mater. 23 1558
59. Zhang S, Tanaka M, Zhu H and Yamanaka S 2013 Supercond. Sci. Technol. 26 085015
60. Zhang S, Tanaka M, Onimaru T, Takabatake T, Isikawa Y and Yamanaka S 2013 Supercond. Sci. Technol. 26 045017
61. Kitagawa S, Kotegawa H, Tou H, Ishii H, Kudo K, Nohara M and Harima H 2013 J. Phys. Soc. Japan 82 113704
62. Kudo K, Ishii H, Takasuga M, Iba K, Nakano S, Kim J, Fujiwara A and Nohara M 2013 J. Phys. Soc. Japan 82 063704
63. Pyon S, Kudo K and Nohara M 2012 J. Phys. Soc. Japan 81 053701
64. Kudo K, Kobayashi M, Pyon S and Nohara M 2013 J. Phys. Soc. Japan 82 085001
65. Qi Y, Matsuishi S, Guo J, Mizoguchi H and Hosono H 2012 Phys. Rev. Lett. 109 217002
66. Guo J, Qi Y, Matsuishi S and Hosono H 2012 J. Am. Chem. Soc. 134 20001
67. Guo J, Qi Y and Hosono H 2013 Phys. Rev. B 87 224504
68. Sathish C I et al 2012 J. Solid State Chem. 196 579
69. Sathish C I, Shirako Y, Tsujimoto Y, Feng H L, Sun Y, Akaogi M and Yamaura K 2014 Solid State Commun. 177 33
70. Ji S, Tanaka M, Zhang S and Yamanaka S 2012 Inorg. Chem. 51 10300
71. Tanaka M, Zhang S, Inumaru K and Yamanaka S 2013 Inorg. Chem. 52 6039
72. Ryu G, Kim S W, Matsuishi S, Kawaji H and Hosono H 2011 Phys. Rev. B 84 224518
73. Guo J, Yamaura J, Lei H, Matsuishi S, Qi Y and Hosono H 2013 Phys. Rev. B 88 140507 (R)
74. Jeong S, Matsuishi S, Lee K, Toda Y, Kim S W and Hosono H 2014 Supercond. Sci. Technol. 27 055005
75. Kudo K, Kobayashi M, Kakiya S, Danura M and Nohara M 2012 J. Phys. Soc. Japan 81 035002
76. Yajima T et al 2013 J. Phys. Soc. Japan. 82 013703
77. Park S W, Mizoguchi H, Kodama K, Shamoto S, Otomo T, Matsuishi S, Kamiya T and Hosono H 2013 Inorg. Chem. 52 13363
78. Liu X, Matsuishi S, Fujitsu S and Hosono H 2012 Phys. Rev. B 85 104403
79. Liu X, Matsuishi S, Fujitsu S and Hosono H 2012 Phys. Rev. B 84 214439
80. Liu X, Matsuishi S, Fujitsu S, Ishigaki T, Kamiyama T and Hosono H 2012 J. Am. Chem. Soc. 134 11687
81. Mizoguchi H and Hosono H 2011 J. Am. Chem. Soc. 133 2394
82. Lei H, Yamaura J, Guo J, Qi Y, Toda Y and Hosono H 2014 Inorg. Chem. 53 5684
83. Lee K, Kim S-W, Toda Y, Matsuishi S and Hosono H 2013 Nature 494 336
84. Anzai A, Fuchigami M, Yamanaka S and Inumaru K 2012 Mater. Res. Bull. 47 2062
85. Shirako Y, Shi Y G, Aimi A, Mori D, Kojitani H, Yamaura K, Inaguma Y and Akaogi M 2012 J. Solid State Chem. 191 167
86. Shirako Y et al 2011 Phys. Rev. B 83 174411
87. Wang X X, Guo Y F, Shirako Y, Yamaura K and Takayama-Muromachi E 2011 Physica C 471 763
88. Shirako Y, Kojitani H, Oganov A R, Fujino K, Miura H,

Mori D, Inaguma Y, Yamaura K and Akaogi M 2012 Am. Mineralogist 97 159

89. Zhang S, Yoshikawa M, Inumaru K and Yamanaka S 2013 Inorg. Chem. 52 10571
90. Yajima T 2015 submitted
91. Sun Y S, Guo Y F, Wang X X, Tsujimoto Y, Matsushita Y, Shi Y G, Wang C, Belik A A and Yamaura K 2012 Appl. Phys. Lett. 100 161907
92. Feng H L, Guo Y, Sathish C I, Wang X, Yuan Y-H and Yamaura K 2014 JPS Conf. Proc. 1 012002
93. Feng H L, Tsujimoto Y, Guo Y, Sun Y, Sathish C I and Yamaura K 2013 High Pressure Res. 33 221
94. Mizoguchi H, Kamiya T, Matsuishi S and Hosono H 2011 Nat. Commun. 2 470
95. Yajima T, Kitada A, Kobayashi Y, Sakaguchi T, Bouilly G, Kasahara S, Terashima T, Takano M and Kageyama H 2012 J. Am. Chem. Soc. 134 8782
96. Tassel C, Goto Y, Kuno Y, Hester J, Green M, Kobayashi Y and Kageyama H 2013 Angew. Chem. Int. Edn 53 10377
97. Tanaka M, Zhang S, Onimaru T, Takabatake T, Inumaru K and Yamanaka S 2014 Carbon 73 125
98. Yamanaka S, Komatsu M, Tanaka M, Sawa H and Inumaru K 2014 J. Am. Chem. Soc. 136 7717
99. Nishikubo Y, Nakano S, Kudo K and Nohara M 2012 Appl. Phys. Lett. 100 252104
100. Kitada A, Kasahara S, Terashima T, Kobayashi Y, Yoshimura K and Kageyama H 2011 Appl. Phys. Exp. 4 035801
101. Sakurai H, Kolodiazhnyi T, Michiue Y, Takayama-Muromachi E, Tanabe Y and Kikuchi H 2012 Angew. Chem. Int. Edn 51 6653
102. Wang X X et al 2011 Phys. Rev. B 83 100410
103. Wang X X, Guo Y G, Shi Y G, Li J J, Zhang S B and Yamaura K 2012 J. Phys.: Conf. Ser. 400 032109
104. Lei H, Yin W-G, Zhong Z and Hosono H 2014 Phys. Rev. B 89 020409
105. Wang X et al 2012 Inorg. Chem. 51 6868
106. Shi Y et al 2013 Nat. Mat. 12 1024
107. Shi Y, Guo Y, Yu S, Arai M, Sato A, Belik A A, Yamaura K and Takayama-Muromachi E 2010 J. Am. Chem. Soc. 132 8474
108. Kitano M et al 2012 Nat. Chem. 4 934
109. Toda Y, Hirayama H, Kuganathan N, Torrisi A, Sushko P V and Hosono H 2013 Nat. Commun. 4 2378
110. Kobayashi Y et al 2012 Nat. Mater. 11 507
111. Nomura K, Kamiya T and Hosono H 2011 Adv. Mater. 23 3431
112. Takahashi H, Igawa K, Arii K, Kamihara Y, Hirano M and Hosono H 2008 Nature 453 376
113. Liu R H et al 2008 Phys. Rev. Lett. 101 087001
114. Chen X H, Wu T, Wu G, Liu R H, Chen H and Fang D F 2008 Nature 453 761
115. Ren Z-A et al 2008 Chin. Phys. Lett. 25 2215
116. Fujioka M et al 2013 Supercond. Sci. Technol. 26 085023
117. Ren Z-A et al 2008 Europhys. Lett. 83 17002
118. Wang C et al 2008 Europhys. Lett. 83 67006
119. Rotter M, Tegel M and Johrendt D 2008 Phys. Rev. Lett. 101 107006
120. Hosono H, Matsuishi S, Nomura N and Hiramatsu H 2009 Butsuri. 64 807 (in Japanese)
121. Aswathy P M, Anooja J B, Sarum P M and Syamaprasad U 2010 Supercond. Sci. Technol. 23 073001
122. Johnston D C 2010 Adv. Phys. 59 803
123. Peglione J and Greene R L 2010 Nat. Phys. 6 645
124. Johrendt D 2011 J. Mat. Chem. 21 13726
125. Fujitsu S, Matsuishi S and Hosono H 2012 Int. Mater. Rev. 57 311
126. Wang N L, Hosono H and Dai P C 2013 Iron-Based Superconductors—Materials, Properties and Mechanism (Singapore: Pan Stanford Publishing)
127. Johnson P D, Xu G and Yin W G (ed) 2015 Iron-Based Superconductivity (New York: Springer)
128. Pottgen R and Johrendt D 2008 Z. Naturforsch. 63b 1135
129. Johnson V and Jeitschko W 1974 J. Solid State Chem. 11 161
130. Zimmer B I, Jeitschko W, Albering J H, Glaum R and Reehuis M 1995 J. Alloys Compd 229 238
131. Yanagi H, Watanabe T, Kodama K, Iikubo S, Shamoto S, Kamiya T, Hirano M and Hosono H 2009 J. Appl. Phys. 105 093936
132. Yanagi H, Kawamura R, Kamiya T, Kamihara Y, Hirano M, Nakamura T, Osawa H and Hosono H 2008 Phys. Rev. B 77 224431
133. Watanabe T, Yanagi H, Kamihara Y, Kamiya T, Hirano M and Hosono H 2008 J. Sol. State Chem. 181 2117
134. Kayanuma K, Kawamura R, Hiramatsu H, Yanagi H, Hirano M, Kamiya T and Hosono H 2008 Thin Solid Films 516 5800
135. Kayanuma K, Hiramatsu H, Hirano M, Kawamura R, Yanagi H, Kamiya T and Hosono H 2007 Phys. Rev. B 76 195325
136. Quebe P, Terbüchte L J and Jeitschko W 2000 J. Alloys Compd 302 70
137. Matsuishi S, Inoue Y, Nomura T, Yanagi H, Hirano M and Hosono H 2008 J. Am. Chem. Soc. 130 14428
138. Just G and Paufler P 1996 J. Alloys Compd 232 1
139. Wang X C, Liu Q Q, Lv Y X, Gao W B, Yang L X, Yu R C, Li F Y and Jin C Q 2008 Solid State Commun. 148 538
140. Parker D R, Pitcher M J, Baker P J, Franke I, Lancaster T, Blundell S J and Clarke S J 2009 Chem. Commun. 2189
141. Ogino H, Matsumura Y, Katsura Y, Ushiyama K, Horii S, Kishio K and Shimoyama J 2009 Supercond. Sci. Technol. 22 075008
142. Ogino H, Katsura Y, Horii S, Kishio K and Shimoyama J 2009 Supercond. Sci. Technol. 22 085001
143. Ogino H, Sato S, Kishio K, Shimoyama J, Tohei T and Ikuhara Y 2010 Appl. Phys. Lett. 97 072506

144. Chen G F, Xia T-L, Yang H X, Li J Q, Zheng P, Luo J L and Wang N L 2009 Supercond. Sci. Technol. 22 072001
145. Zhu X, Han F, Mu G, Zeng B, Cheng P, Shen B and Wen H-H 2009 Phys. Rev. B 79 024516
146. Shirage P M, Kihou K, Lee C H, Kito H, Eisaki H and Iyo A 2011 J. Am. Chem. Soc. 133 9360
147. Hsu F-C et al 2008 Proc. Natl Acad. Sci. USA 105 14262
148. Medvedev S et al 2009 Nat. Mater. 8 630
149. Wang Q-Y et al 2012 Chin. Phys. Lett. 29 037402
150. He S L et al 2013 Nat. Mater. 12 605
151. Tan S Y et al 2013 Nat. Mater. 12 634
152. Liu D et al 2012 Nat. Commun. 3 931
153. Ge J-F, Liu Z-L, Liu C, Gao C-L, Qian D, Xue Q-K, Liu Y and Jia J-F 2014 Nat. Mater. 14 285
154. Guo J, Jin S, Wang G, Wang S, Zhu K, Zhou T, He M and Chen X 2010 Phys. Rev. B 82 180520
155. Zavalij P et al 2011 Phys. Rev. B 83 132509
156. Fang M-H, Wang H-D, Dong C-H, Li Z-J, Feng C-M, Chen J and Yuan H Q 2011 Europhys. Lett. 94 27009
157. Ivanovskii A L 2011 Physica C 471 409
158. Malaeb W et al 2008 J. Phys. Soc. Japan. 77 093714
159. Mazin I I, Singh D J, Johannes M D and Du M H 2008 Phys. Rev. Lett. 101 057003
160. Kuroki K, Onari S, Arita R, Usui H, Tanaka Y, Kontani H and Aoki H 2008 Phys. Rev. Lett. 101 087004
161. Chubukov A V, Efremov D V and Eremin I 2008 Phys. Rev. B 78 134512
162. Graser S, Maier T A, Hirschfeld P J and Scalapino D J 2009 New J. Phys. 11 025016
163. Ikeda H, Arita R and Kuneš J 2010 Phys. Rev. B 81 054502
164. Daghofer M, Moreo A, Riera J A, Arrigoni E, Scalapino D J and Dagotto E 2008 Phys. Rev. Lett. 101 237004
165. Thomale R, Platt C, Hanke W and Bernevig B A 2011 Phys. Rev. Lett. 106 187003
166. Wang F, Zhai H, Ran Y, Vishwanath A and Lee D-H 2009 Phys. Rev. Lett. 102 047005
167. Qian T et al 2011 Phys. Rev. Lett. 106 187001
168. Kontani H and Onari S 2010 Phys. Rev. Lett. 104 157001
169. Suzuki K, Usui H, Iimura S, Sato Y, Matsuishi S, Hosono H and Kuroki K 2014 Phys. Rev. Lett. 113 027002
170. Nomura T, Kim S W, Kamihara Y, Hirano M, Sushko P V, Kato K, Takata M, Shluger A L and Hosono H 2008 Supercond. Sci. Technol. 21 125028
171. de la Cruz C et al 2008 Nature 453 899
172. Malavasi L, Artioli G A, Ritter C, Mozzati M C, Maroni B, Pahari B and Canesch A 2010 J. Am. Chem. Soc. 132 2417
173. Hess C, Kondrat A, Narduzzo A, Hamann-Borrero J E, Klingeler R, Werner J, Behr G and Büchner B 2009 Europhys. Lett. 87 17005
174. Kamihara Y et al 2010 New J. Phys. 12 033005
175. Margadonna S, Takabayashi Y, McDonald M T, Brunelli M, Wu G, Liu R H, Chen X H and Prassides K 2009 Phys. Rev. B 79 014503
176. Drew A J et al 2009 Nat. Mater. 8 310
177. Kito H, Eisaki H and Iyo A 2008 J. Phys. Soc. Japan 77 063707
178. Kodama K, Ishikado M, Esaka F, Iyo A, Eisaki H and Shamoto S 2011 J. Phys. Soc. Japan 80 034601
179. Zhigadlo N D, Katrych S, Weyeneth S, Puzniak R, Moll P J W, Bukowski Z, Karpinski J, Keller H and Batlogg B 2010 Phys. Rev. B 82 064517
180. Sefat A S, Huq A, McGuire M A, Jin R, Sales B C, Mandrus D, Cranswick L M D, Stephens P W and Stone K H 2008 Phy. Rev. B 78 104505
181. Dong X L et al 2010 Phys. Rev. B 82 212506
182. Cao G et al 2009 Phys. Rev. B 79 174505
183. Maroni B, Malavasi L, Mozzati M C, Grandi M S, Hill A H, Chermisi D, Dore P and Postorino P 2010 Phys. Rev. B 82 104503
184. Canfield P C and Bud'ko S L 2010 Annu. Rev. Condens. Matter Phys. 1 27
185. Wen H-H, Mu G, Fang L, Yang H and Zhu X 2008 Europhys. Lett. 82 17009
186. Sefat A S, Jin R, McGuire M A, Sales B C, Singh D J and Mandrus D 2008 Phys. Rev. Lett. 101 117004
187. Canfield P C, Bud'ko S L, Ni N, Yan J Q and Kracher A 2009 Phys. Rev. B 80 060501
188. Ni N, Thaler A, Kracher A, Yan J Q, Bud'ko S L and Canfield P C 2009 Phys. Rev. B 80 024511
189. Jiang S, Xing H, Xuan G, Wang C, Ren Z, Feng C, Dai J, Xu Z and Cao G 2009 J. Phys.: Condens. Matter. 21 382203
190. Sharma S, Bharathi A, Chandra S, Reddy V R, Paulraj S, Satya A T, Sastry V S, Gupta A and Sundar C S 2010 Phys. Rev. B 81 174512
191. Tarascon J M, Greene L H, Barboux P, McKinnon W R, Hull G W, Orlando T P, Delin K A, Foner S and McNiA E J Jr 1987 Phys. Rev. B 36 8393
192. Sato M, Kobayashi Y, Lee S C, Takahashi H, Satomi E and Miura Y 2010 J. Phys. Soc. Japan. 11 014710
193. Ni N, Tillman M E, Yan J-Q, Kracher A, Hannahs S T, Bud'ko S L and and Canfield P C 2009 Phys. Rev. B 78 214515
194. Tanatar M A et al 2009 Phys. Rev. B 79 094507
195. Moll P J W, Puzniak R, Balakirev F, Rogacki K, Karpinski J, Zhigadlo N D and Batlogg B 2010 Nat. Mater. 9 628
196. Nagamatsu J, Nakagawa N, Muranaka T, Zenitani Y and Akimitsu J 2001 Nature 410 63
197. Karpinski J, Zhigadlo N D, Katrych S, Puzniak R, Rogacki K and Gonnelli R 2007 Phys. C 456 3
198. Eltsev Yu Nakao K, Lee S, Masui T, Chikumoto N, Tajima S, Koshizuka N and Murakami M 2003 J. Low Temp. Phys. 131 1069
199. Wu M K, Ashburn J R, Torng C J, Hor P H, Meng R L, Gao L, Huang Z J, Wang Y Q and Chu C W 1987 Phys. Rev. Lett. 58 908
200. Maeda H, Tanaka Y, Fukutomi M and Asano T 1988 Japan. J. Appl. Phys. 27 L209

201. Schilling A, Cantoni M, Guo J D and Ott H R 1993 Nature 363 56
202. Asada Y 1991 Bull. Japan. Inst. Metals 30 832 (in Japanese)
203. Tozer S W, Kleinsasser A W, Penney T, Kaiser D and Holtzberg F 1987 Phys. Rev. Lett. 59 1768
204. Iye Y, Tamegai T, Sakakibara T, Goto T, Miura N, Takeya H and Takei H 1988 Physica C 153–155 26
205. Hagen S J, Jing T W, Wang Z Z, Horvath J and Ong N P 1988 Phys. Rev. B 37 7928
206. Farrell D E, Bonham S, Foster J, Chang Y C, Jiang P Z, Vandervoort K G, Lam D J and Kogan V G 1989 Phys. Rev. Lett. 63 782
207. Köhler A and Behr G 2009 J. Supercond. Novel Magn. 22 565
208. Artioli G A, Malavasi L, Mozzati M C and Fernandez Y D 2009 J. Am. Chem. Soc. 131 12044
209. Lee C-H, Iyo A, Eisaki H, Kito H, Fernandez-Diazi M T, Ito T, Kohou K, Matsuhata H, Brenden M and Yamada K 2008 J. Phys. Soc. Japan 77 083704
210. Iimura S et al 2013 Phys. Rev. B 88 060501
211. Hiraishi M et al 2014 Nat. Phys. 10 300
212. Sunagawa M et al 2014 Sci. Rep. 4 4381
213. Harnagea L et al 2011 Phys. Rev. B 83 094523
214. Kumar N, Chi S, Chen Y, Rana K G, Bigam A K, Thamizhavel A, William R II, Dhar S K and Lynn J W 2009 Phys. Rev. B 80 144524
215. Leithe-Jasper A, Schnelle W and Rosner H 2008 Phys. Rev. Lett. 101 207004
216. Saha S R, Butch N P, Kirshenbaum K and Paglione J 2009 Phys. Rev. B 79 224519
217. Li L J et al 2009 New J. Phys. 11 025008
218. Danura M, Kudo K, Oshiro Y, Araki S, Kobayashi T C and Nohara M 2011 J. Phys. Soc. Japan. 80 103701
219. Kudo K, Matsumura J, Danura M and Nohara M (unpublished data)
220. Schnelle W, Leithe-Jasper A, Gumeniuk R, Burkhardt U, Kasinathan D and Rosner H 2009 Phys. Rev. B 79 214516
221. Han F et al 2009 Phys. Rev. B 80 024506
222. Qi Y, Gao Z, Wang L, Zhang X, Wang D, Yao C, Wang C, Wang C and Ma Y 2011 Europhys. Lett. 96 47005
223. Nishikubo Y, Kakiya S, Danura M, Kudo K and Nohara M 2010 J. Phys. Soc. Japan 79 095002
224. Wang X L, Shi H Y, Yan X W, Yuan Y C, Lu Z-Y, Wang X Q and Zhao T-S 2010 Appl. Phys. Lett. 96 012507
225. Saha S R, Drye T, Kirshenbaum K, Butch N P, Zavalij P Y and Paglione J 2010 J. Phys.: Condens. Matter. 22 072204
226. Shirage P M, Miyazawa K, Kito H, Eisaki H and Iyo A 2008 Appl. Phys. Express 1 081702
227. Cortes-Gil R and Clarke S J 2011 Chem. Mat. 23 1009
228. Sasmal K, Lv B, Lorenz B, Guloy A M, Chen F, Xue Y-Y and Chu C-W 2008 Phys. Rev. Lett. 101 107007
229. Lv B, Gooch M, Lorenz B, Chen F, Guloy A M and Chu C W 2009 New J. Phys. 11 025013
230. Aswartham S et al 2012 Phys. Rev. B 85 224520
231. Bukowski Z, Weyeneth S, Puzniak R, Moll P, Katrych S, Zhigadlo N D, Karpinski J, Keller H and Batlogg B 2009 Phys. Rev. B 79 104521
232. Wang C et al 2009 Phys. Rev. B 79 054521
233. Wu G et al 2008 Europhys. Lett. 84 27010
234. Saha S R, Butch N P, Drye T, Magill J, Ziemak S, Kirshenbaum K, Zavalij P Y, Lynn J W and Paglione J 2012 Phys. Rev. B 85 024525
235. Gao Z, Qi Y, Wang L, Wang D, Zhang X, Yao C, Wang C and Ma Y 2011 Europhys. Lett. 95 67002
236. Lv B, Deng L, Gooch M, Wei F, Sun Y, Meen J K, Xu Y-Y, Lorenz B and Chu C-W 2011 Proc. Natl Acad. Sci. USA 108 15705
237. Qi Y, Gao Z, Wang L, Wang D, Zhang X, Yao C, Wang C, Wang C and Ma Y 2012 Supercond. Sci. Technol. 25 045007
238. Katase T, Hiramatsu H, Kamiya T and Hosono H 2012 Supercond. Sci. Technol. 25 084015
239. Yakita H et al 2014 J. Am. Chem. Soc. 136 846
240. Sala A et al 2014 Appl. Phys. Express 7 073102
241. Cheng P, Shen B, Mu G, Zhu X, Han F, Zeng B and Wen H-H 2009 Europhys. Lett. 85 67003
242. Kakiya S, Kudo K, Nishikubo Y, Oku K, Nishibori E, Sawa H, Yamamoto T, Nozaka T and Nohara M 2011 J. Phys. Soc. Japan 80 093704
243. Nohara M, Kakiya S, Kudo K, Oshiro Y, Araki S, Kobayashi T C, Oku K, Nishibori E and Sawa H 2012 Solid State Commun. 152 635
244. Zhou W, Zhuang J, Yuan F, Li X, Xing X, Sun Y and Shi Z 2014 Appl. Phys. Express 7 063102
245. Rutzinger D, Bartsch C, Doerr M, Rosner H, Neu V, Doert T and Ruck M 2010 J. Solid State Chem. 183 510
246. Kasahara S, Shibauchi T, Hashimoto K, Nakai Y, Ikeda H, Terashima T and Matsuda Y 2011 Phys. Rev. B 83 060505(R)
247. Löhnert C, Stürzer T, Tegel M, Frankovsky R, Friederichs G and Johrendt D 2011 Angew. Chem. Int. Edn 50 9195
248. Ni N, Allred J M, Chan B C and Cava R J 2011 Proc. Natl Acad. Sci. USA 108 E1019
249. Stürzer T, Derondeau G and Johrendt D 2012 Phys. Rev. B 86 060516(R)
250. Hieke C, Lippmann J, Stürzer T, Friederichs G, Nitsche F, Winter F, Pöttgen R and Johrendt D 2013 Phil. Mag. 93 3680
251. Katayama N, Sugawara K, Sugiyama Y, Higuchi T, Kudo K, Mitsuoka D, Mizokawa T, Nohara M and Sawa H 2014 J. Phys. Soc. Japan 83 113707
252. Sawada K et al 2014 Phys. Rev. B 89 220508(R)
253. Li J et al 2012 J. Am. Chem. Soc. 134 4068
254. Krzton-Maziopa A, Shermadini Z, Pomjakushina E, Pomjakushin V, Bendele M, Amato A, Khasanov R, Luetkens H and Conder K 2011 J. Phys.: Condens. Matter. 23 052203

255. Wang A F et al Phys. Rev. B 83 060512
256. Park J T et al 2011 Phys. Rev. Lett. 107 177005
257. Buffinger D R, Ziebarth R P, Stenger V A, Recchia C and Pennington C H 1993 J. Am. Chem. Soc. 115 9267
258. Ying T P, Chen X L, Wang G, Jin S F, Zhou T T, Lai X F, Zhang H and Wang W Y 2012 Sci. Rep. 2 426
259. Ying T P, Chen X L, Wang G, Jin S F, Lai X F, Zhou T T, Zhang H, Shen S J and Wang W Y 2013 J. Am. Chem. Soc. 135 2951
260. Burrard-Lucas M et al 2013 Nat. Mater. 12 15
261. Sedlmaier S J et al 2014 J. Am. Chem. Soc. 136 630
262. Zhang A M, Xia T L, Liu K, Tong W, Yang Z R and Zhang Q M 2013 Sci. Rep. 3 1216
263. Bednorz J G and Müller K A 1986 Z. Phys. B 64 189
264. Hebard A F, Rosseinsky M J, Haddon R C, Murphy D W, Glarum S H, Palstra T T M, Ramirez A P and Kortan A R 1991 Nature 350 600
265. Deslandes F, Nazzal A I and Torrance J B 1991 Physica C 179 85
266. Arita R, Yamasaki A, Held K, Matsuno J and Kuroki K 2007 Phys. Rev. B 75 174521
267. Axtell E A III, Ozawa T, Kauzlarich S M and Singh R R P 1997 J. Solid State Chem. 134 423
268. Adam A and Schuster H U 1990 Z. Anorg. Allg. Chem. 584 150
269. Wang X F, Yan Y J, Ying J J, Li Q J, Zhang M, Xu N and Chen X H 2010 J. Phys.: Condens. Matter. 22 075702
270. Sun Y L, Jiang H, Zhai H F, Bao J K, Jiao W H, Tao Q, Shen C Y, Zeng Y W, Xu Z A and Cao G H 2012 J. Am. Chem. Soc. 134 12893
271. Ozawa T C and Kauzlarich S M 2008 Sci. Technol. Adv. Mater. 9 033003
272. Ozawa T C and Kauzlarich S M 2001 Chem, Mater. 13 1804
273. Liu R H, Tan D, Song Y A, Li Q J, Yan Y J, Ying J J, Xie Y L, Wang X F and Chen X H 2009 Phys. Rev. B 80 144516
274. Doan P, Gooch M, Tang Z, Lorenz B, Moller A, Tapp J, Chu P C W and Guloy A M 2012 J. Am. Chem. Soc. 134 16520
275. Kitagawa S, Ishida K, Nakano K, Yajima T and Kageyama H 2013 Phys. Rev. B 87 060510(R)
276. Von Rohr F, Schilling A, Nesper R, Baines C and Bendele M 2013 Phys. Rev. B 88 140501(R)
277. Nozaki Y et al 2013 Phys. Rev. B 88 214506
278. Gooch M, Doan P, Tang Z, Lorenz B, Guloy A M and Chu P C W 2013 Phys. Rev. B 88 064510
279. Pachmayr U and Johrendt D 2014 Solid State Sci. 28 31
280. Von Rohr F, Nesper R and Schilling A 2014 Phys. Rev. B 89 094505
281. Shi Y G, Wang H P, Zhang X, Wang W D, Huang Y and Wang N L 2013 Phys. Rev. B 88 144513
282. Huang Y, Wang H P, Wang W D, Shi Y G and Wang N L 2013 Phys. Rev. B 87 100507
283. Xu H C et al 2014 Phys. Rev. B 89 155108
284. Fan G, Zhang X, Shi Y and Luo J 2013 Sci. China Phys., Mechan. Astron. 56 2399
285. Huang Y, Wang H P, Chen R Y, Zhang X, Zheng P, Shi Y G and Wang N L 2014 Phys. Rev. B 89 155120
286. Gooch M, Doan P, Lorenz B, Tan Z J, Guloy A M and Chu C W 2013 Supercond. Sci. Technol. 26 125011
287. Yu W, Dong X-L, Ma M-W, Yang H-X, Zhang C, Zhou F, Zhou X-J and Zhao Z-X 2014 Chinese Phys. Lett. 31 077401
288. Litvinchuk A P, Doan P, Tang Z and Guloy A M 2013 Phys. Rev. B 87 064505
289. Zhai H F et al 2013 Phys. Rev. B 87 100502(R)
290. Singh D J 2012 New J. Phys. 14 123003
291. Yan X W and Lu Z Y 2013 J. Phys.: Condens. Matter. 25 365501
292. Subedi A 2013 Phys. Rev. B 87 054506
293. Frandsen B A et al 2014 Nat. Commun. 5 5761
294. Wang G, Zhang H, Zhang L and Liu C 2013 J. Appl. Phys. 113 243904
295. Kodenkandath T A, Lalena J N, Zhou W L, Carpenter E E, Sangregorio C, Falster A U, Simmons W B Jr, O'Connor C J and Wiley J B 1999 J. Am. Chem. Soc. 121 10743
296. Tassel C et al 2010 Phys. Rev. Lett. 105 167205
297. Fleming R M et al 1991 Nature 352 701
298. Yamanaka S 2000 Annu. Rev. Mater. Sci. 30 53
299. Yamanaka S 2010 J. Mater. Chem. 20 2922
300. Yamanaka S, Hotehama K and Kawaji H 1998 Nature 392 580
301. Yamanaka S, Kawaji H, Hotehama K and Ohashi M 1996 Adv. Mater. 8 771
302. Yamanaka S and Tou H 2001 Curr. Opin. Solid State Mater. Sci. 5 545
303. Kasahara Y, Kuroki K, Yamanaka S and Taguchi Y 2015 Physica C doi:10.1016/j.physc.2015.02.022
304. Tou H, Maniwa Y, Koiwasaki T and Yamanaka S 2001 Phys. Rev. Lett. 86 5775
305. Taguchi Y, Hisakabe M and Iwasa Y 2005 Phys. Rev. Lett. 94 217002
306. Sugimoto A, Shohara K, Ekino T, Zheng Z and Yamanaka S 2012 Phys. Rev. B 85 144517
307. Tou H, Maniwa Y and Yamanaka S 2003 Phys. Rev. B 67 100509
308. Kuroki K 2010 Phys. Rev. B 81 104502
309. Bill A, Morawitz H and Kresin V Z 2002 Phys. Rev. B 66 100501
310. Bill A, Morawitz H and Kresin V Z 2003 Phys. Rev. B 68 144519
311. Chen X, Zhu L P and Yamanaka S 2002 J. Solid State Chem. 169 149
312. Shamoto S, Takeuchi K, Yamanaka S and Kajitani T 2004 Physica C 402 283
313. Tou H, Maniwa Y, Koiwasaki T and Yamanaka S 2001 Phys. Rev. B 63 020508

314. Tou H et al 2005 Phys. Rev. B 72 020501
315. Hotehama K, Koiwasaki T, Umemoto K, Yamanaka S and Tou H 2010 J. Phys. Soc. Japan 79 014707
316. Yamanaka S, Okumura H and Zhu L-P 2004 J. Phys. Chem. Solids 65 565
317. Yamanaka S, Yasunaga T, Yamaguchi K and Tagawa M 2009 J. Mater. Chem. 19 2573
318. Marouchkine A 2004 Room-Temperature Superconductivity (Cambridge: Cambridge International Science Publishing)
319. Putti M et al 2010 Supercond. Sci. Technol. 23 034003
320. Welp U, Xie R, Koshelev A E, Kwok W K, Cheng P, Fang L and Wen H H 2008 Phys. Rev. B 78 140510
321. Jaroszynski J et al 2008 Phys. Rev. B 78 174523
322. Welp U, Chaparro C, Koshelev A E, Kwok W K, Rydh A, Zhigadlo N D, Karpinski J and Weyeneth S 2011 Phys. Rev. B 83 100513
323. Hunte F, Jaroszynski J, Gurevich A, Larbalestier D C, Jin R, Sefat A S, McGuire M A, Sales B C, Christen D K and Mandrus D 2008 Nature 453 903
324. Chen G F, Li Z, Dong J, Li G, Hu W Z, Zhang X D, Song X H, Zheng P, Wang N L and Luo J L 2008 Phys. Rev. B 78 224512
325. Altarawneh M M, Collar K, Mielke C H, Ni N, Bud'ko S L and Canfield P C 2008 Phys. Rev. B 78 220505
326. Yuan H Q, Singleton J, Balakirev F F, Baily S A, Chen G F, Luo J L and Wang N L T 2009 Nature 457 565
327. Yamamoto A et al 2009 Appl. Phys. Lett. 94 062511
328. Yin Q A, Ylvisaker E R and Pickett W E 2011 Phys. Rev. B 83 014509
329. Kanamaru F, Shimada M, Koizumi M, Takano M and Takada T 1973 J. Solid State Chem. 7 297
330. Yamanaka S, Enishi E, Fukuoka H and Yasukawa M 2000 Inorg. Chem. 39 56
331. Yamanaka S, Izumi S, Maekawa S and Umemoto K 2009 J. Solid State Chem. 182 1991
332. Yamanaka S and Maekawa S 2006 Z. Naturforsch. B 61 1493
333. Fukuoka H, Ueno K and Yamanaka S 2000 J. Organomet. Chem. 611 543
334. Kurakevych O O, Strobel T A, Kim D Y, Muramatsu T and Struzhkin V V 2013 Cryst. Growth Des. 13 303
335. Andersen S J, Marioara C D, Froseth A, Vissers R and Zandbergen H W 2005 Mat. Sci. Eng. 390 127
336. Yamanaka S 2010 Dalton Trans. 39 1901
337. Yamanaka S 2014 The Physics and Chemistry of Inorganic Clathrates ed G S Nolas (Dordrecht: Springer) chapter 7
338. Ekimov E A et al 2004 Nature 428 542
339. Connetable D et al 2003 Phys. Rev. Lett. 91 247001
340. Yamanaka S et al 2006 Phys. Rev. Lett. 96 076602
341. Yamanaka S, Kini N S, Kubo A, Jida S and Kuramoto H 2008 J. Am. Chem. Soc. 130 4303
342. Ronning F, Bauer E D, Park T, Baek S-H, Sakai H and Thompson J D 2009 Phys. Rev. B 79 134507
343. Mine T, Yanagi H, Kamiya T, Kamihara K, Hirano M and Hosono H 2008 Solid State Commun. 147 111
344. Hirai D, Takayama T, Higashinaka R, Aruga-Katori H and Takagi H 2009 J. Phys. Soc. Japan 78 023706
345. Jeitschko W, Glaum R and Boonk L 1987 J. Solid State Chem. 69 93
346. Han J-T, Zhou J-S, Cheng J-G and Goodenough J B 2010 J. Am. Chem. Soc. 132 908
347. Bauer E D, Ronning F, Scott B L and Thompson J D 2008 Phys. Rev. B 78 172504
348. Ronning F, Kurita N, Bauer E D, Scott B L, Park T, Klimczuk T, Movshovich R and Thompson J D 2008 J. Phys.: Condens. Matter. 20 342203
349. Parthe E, Chabot B, Braun H F and Engel N 1983 Acta. Cryst. B 39 588
350. Kudo K, Nishikubo Y and Nohara M 2010 J. Phys. Soc. Japan 79 123710
351. Imre A, Hellmann A, Wenski G, Graf J, Johrendt D and Mewis A 2007 Z. Anorg. Allg. Chem. 633 2037
352. Shein I R and Ivanovskii A L 2011 Phys. Rev. B 83 10450
353. Hoffman W K and Jeistchko W 1985 Monatsh. Chemi. 116 569
354. Uhoya W O, Montgomery J M, Tsoi G M, Vohra Y K, McGuire M A, Sefat A S, Sales B C and Weir S T 2011 J. Phys.: Condens. Matter. 23 122201
355. Saha S R, Butch N P, Drye T, Magill J, Ziemak S, Kirshenbaum K, Zavalij P Y, Lynn J W and Paglione J 2012 Phys. Rev. B 85 24525
356. Yamazaki T, Takeshita N, Kobayashi R, Fukazawa H, Kohori Y, Kihou K, Lee C, Kito H, Iyo A and Eisaki H 2010 Phys. Rev. B 81 224511
357. Uhoya W O, Tsoi G M, Vohra Y K, McGuire M A and Sefat A S 2011 J. Phys.: Condens. Matter. 23 365703
358. Anand V K, Kim H, Tanatar M A, Prozorov R and Johnston D C 2013 Phys. Rev. B 87 224510
359. Tsutsumi K, Takayanagi S, Ishikawa M and Hirano T 1995 J. Phys. Soc. Japan 64 2237
360. Takada K, Sakurai H, Takayama-Muromachi E, Izumi F, Dilanian R A and Sasaki T 2003 Nature 422 53
361. Niihara K, Shishido T and Yajima S 1973 Bull. Chem. Soc. Japan 46 1137
362. Sefat A S, McGuire M A, Jin R, Sales B C, Mandrus D, Ronning F, Bauer E D and Mozharivskyj Y 2009 Phys. Rev. B 79 094508
363. Kurita N, Ronning F, Tokiwa Y, Bauer E D, Subedi A, Singh D J, Thompson J D and Movshovich R 2009 Phys. Rev. Lett. 102 147004
364. Subedi A and Singh D J 2008 Phys. Rev. B 78 132511
365. Shein I R and Ivanovskii A L 2009 Phys. Rev. B 79 054510
366. Kim J S, Boeri L, Kremer R K and Razavi F S 2006 Phys. Rev. B 74 214513
367. Gauzzi A, Takashima S, Takeshita N, Terakura C, Takagi H, Emery N, Hérold C, Lagrange P and Loupias G 2007 Phys. Rev. Lett. 98 067002
368. Gauzzi A et al 2008 Phys. Rev. B 78 064506

369. Mauri F, Zakharov O, de Gironcoli S, Louie S G and Cohen M L 1996 Phys. Rev. Lett. 77 1151

370. Matsumoto N, Taniguchi K, Endoh R, Takano H and Nagata S 1999 J. Low Temp. Phys. 117 1129

371. Yang J J, Choi Y J, Oh Y S, Hogan A, Horibe Y, Kim K, Min B I and Cheong S-W 2012 Phys. Rev. Lett. 108 116402

372. Ootsuki D et al 2012 Phys. Rev. B 86 014519

373. Ootsuki D et al 2013 J. Phys. Soc. Japan 82 093704

374. Toriyama T et al 2014 J. Phys. Soc. Japan 83 033701

375. Pyon S, Kudo K and Nohara M 2011 NS2 2011: Int. Workshop Novel Superconductors and Super Materials 2011

376. Pyon S, Kudo K and Nohara M 2011 ICNSCT 2011: Int. Conf. Novel Superconductivity in Taiwan 2011

377. Pyon S, Kudo K and Nohara M 2013 Physica C 494 80

378. Kamitani M, Bahramy M S, Arita R, Seki S, Arima T, Tokura Y and Ishiwata S 2013 Phys. Rev. B 87 180501(R)

379. Tunell G and Pauling L 1952 Acta Crystallogr. 5 375

380. Schutte W J and de Boer J L 1988 Acta Crystallogr. B 44 486

381. Janner A and Dam B 1989 Acta Crystallogr. A 45 115

382. Jobic S, Evain M, Brec R, Deniard P, Jouanneaux A and Rouxel J 1991 J. Solid State Chem. 95 319

383. Leger J M, Pereira A S, Haines J, Jobic S and Brec R 2000 J. Phys. Chem. Solids 61 27

384. Gor'kov L P and Rashba E I 2001 Phys. Rev. Lett. 87 037004

385. Frigeri P A, Agterberg D F, Koga A and Sigrist M 2004 Phys. Rev. Lett. 92 097001

386. Frigeri P A, Agterberg D F and Sigrist M 2004 New J. Phys. 6 115

387. Bauer E, Hilscher G, Michor H, Paul C, Scheidt E W, Gribanov A, Seropegin Y, Noël H, Sigrist M and Rogl P 2004 Phys. Rev. Lett. 92 027003

388. Settai R, Sugitani I, Okuda Y, Thamizhavel A, Nakashima M, O□nuki Y and Harima H 2007 J. Magn. Magn. Mater. 310 844

389. Settai R, Miyauchi Y, Takeuchi T, Levy F, Sheikin I and O□nuki Y 2008 J. Phys. Soc. Japan. 77 073705

390. Kimura N, Ito K, Aoki H, Uji S and Terashima T 2007 Phys. Rev. Lett. 98 197001

391. McMillan W L 1968 Phys. Rev. 167 331

392. White J G and Hockings E F 1971 Inorg. Chem. 10 1934

393. Sologub O L, Salamakha P S, Sasakawa T, Chen X, Yamanaka S and Takabatake T 2002 J. Alloys Compd 34 6

394. Loehken A, Reiss G J, Johrendt D and Mewis A 2005 Z. Anorg. Allg. Chem. 631 1144

395. Tanabe K and Hosono H 2012 Japan. J. Appl. Phys. 51 010005

396. Hiramatsu H, Katase T, Kamiya T, Hirano M and Hosono H 2008 Appl. Phys. Express 1 101702

397. Katase T, Hiramatsu H, Yanagi H, Kamiya T, Hirano M and Hosono H 2009 Solid State Commun. 149 2121

398. Katase T, Hiramatsu H, Kamiya T and Hosono H 2010 Appl. Phys. Express 3 063101

399. Iida K, Hänisch J, Hühne R, Kurth F, Kidszun M, Haindl S, Werner J, Schultz L and Holzapfel B 2009 Appl. Phys. Lett. 95 192501

400. Maiorov B, Katase T, Baily S A, Hiramatsu H, Holesinger T G, Hosono H and Civale L 2011 Supercond. Sci. Technol. 24 055007

401. Iida K, Hänisch J, Thersleff T, Kurth F, Kidszun M, Haindl S, Hühne H, Schultz S and Holzapfel B 2010 Phys. Rev. B 81 100507

402. Takeda S, Ueda S, Yamagishi T, Agatsuma S, Takano S, Mitsuda A and Naito M 2010 Appl. Phys. Express 3 093101

403. Kasahara K et al 2010 Phys. Rev. B 81 184519

404. 304. Adachi S, Shimode T, Miura M, Chikumoto N, Takemori A, Nakao K, Oshikubo Y and Tanabe K 2012 Supercond. Sci. Technol. 25 105015

405. Kawaguchi T, Sakagami A, Mori Y, Tabuchi M, Takeda Y and Ikuta H 2014 Supercond. Sci. Technol. 27 065005

406. Katase T, Hiramatsu H, Matias V, Sheehan C, Ishimaru Y, Kamiya T, Tanabe K and Hosono H 2011 Appl. Phys. Lett. 98 242510

407. Miura M, Adachi S, Shimode T, Wada K, Takemori A, Chikumoto N, Nakao K and Tanabe K 2013 Appl. Phys. Express 6 093101

408. Shimoyama J, Kitazawa K, Shimizu K, Ueda S, Horii S, Chikumoto N and Kishio K 2003 J. Low Temp. Phys. 131 1043

409. Fang L et al 2012 Appl. Phys. Lett. 101 012601

410. Lee S et al 2010 Nat. Mater. 9 397

411. Tarantini C et al 2010 Appl. Phys. Lett. 96 142510

412. Haugan T, Barnes P N, Wheeler R, Meisenkothen F and Sumption M 2004 Nature 430 867

413. Gutiérrez J et al 2007 Nat. Mater. 6 367

414. Miura M, Maiorov B, Baily S A, Haberkorn N, Willis J O, Marken K, Izumi T, Shiohara Y and Civale L 2011 Phys. Rev. B 83 184519

415. Miura M, Maiorov B, Kato T, Shimode T, Wada K, Adachi S and Tanabe K 2013 Nat. Commun. 4 2499

416. Maiorov B, Baily S A, Zhou H, Ugurlu O, Kennison J A, Dowden P C, Holesinger T G, Foltyn S R and Civale L 2009 Nat. Mater. 8 398

417. Lee S et al 2013 Nat. Mater. 12 392

418. Cooly L D, Lee P J and Larbalestier D C 1996 Phys. Rev. B 53 6638

419. Zhuang C G, Meng S, Yang H, Jia Y, Wen H H, Xi X X, Feng Q R and Gan Z Z 2008 Supercond. Sci. Technol. 21 082002

420. Godeke A 2006 Supercond. Sci. Technol. 19 R68

421. Miura M, Maiorov B, Willis J O, Kato T, Sato M, Izumi T, Shiohara Y and Civale L 2013 Supercond. Sci. Technol. 26 035008

422. Sato H, Hiramatsu H, Kamiya T and Hosono H 2014 Appl. Phys. Lett. 104 182603

423. Sakagami A, Kawaguchi T, Tabuchi M, Ujihara T,

Takeda Y and Ikuta H 2013 Physica C 494 181

424. Iida K et al 2010 Appl. Phys. Lett. 97 172507
425. Trommler S, Hänisch J, Matias V, Hühne R, Reich E, Iida K, Haindl S, Schultz L and Holzapfel B 2012 Supercond. Sci. Technol. 25 084019
426. Hiramatsu H, Katase T, Ishimaru Y, Tsukamoto A, Kamiya T, Tanabe K and Hosono H 2012 Mater. Sci. Eng. B 177 515
427. Iida K et al 2013 Sci. Rep. 3 2139
428. Braccini V et al 2013 Appl. Phys. Lett. 103 172601
429. Zhang Y et al 2011 Appl. Phys. Lett. 98 042509
430. Gurvitch M, Washington M A and Huggins H A 1983 Appl. Phys. Lett. 42 472
431. Hilgenkamp H and Mannhart J 2002 Rev. Mod. Phys. 74 485
432. Clarke J and Braginski A I 2004 The SQUID Handbook (Weinheim: Wiley)
433. Katase T, Ishimaru Y, Tsukamoto A, Hiramatsu H, Kamiya T, Tanabe K and Hosono H 2010 Appl. Phys. Lett. 96 142507
434. Katase T, Ishimaru Y, Tsukamoto A, Hiramatsu H, Kamiya T, Tanabe K and Hosono H 2011 Nat. Commun. 2 409
435. Ambegaokar V and Halperin B I 1969 Phys. Rev. Lett. 22 1364
436. Saitoh K, Ishimaru Y, Fuke H and Enomoto Y 1997 Japan. J. Appl. Phys. 36 L272
437. De Gennes P G 1964 Rev. Mod. Phys. 36 225
438. Delin K A and Kleinsasser A W 1996 Supercond. Sci. Technol. 9 227
439. Katase T, Ishimaru Y, Tsukamoto A, Hiramatsu H, Kamiya T, Tanabe K and Hosono H 2010 Supercond. Sci. Technol. 23 082001
440. Lee L P, Longo J, Vinetskiy V and Cantor R 1995 Appl. Phys. Lett. 66 1539
441. Wakana H, Adachi S, Hata K, Hato T, Tarutani Y and Tanabe K 2009 IEEE Trans. Appl. Supercond. 19 782
442. Ishimaru Y, Murai Y, Adachi S and Tanabe K 2013 Extended Abstracts of Int. Workshop on Novel Superconductors and Super Materials 2013 (Tokyo) p 104
443. Poppe U, Divin Y Y, Faley M I, Wu J S, Jia C L, Shadrin P and Urban K 2001 IEEE Trans. Appl. Supercond. 11 3768
444. Sarnelli E, Testa G, Crimaldi D, Monaco A and Navacerrada M A 2005 Supercond. Sci. Technol. 18 L35
445. Larbalestier D, Gurevich A, Feldmann D M and Polyanskii A 2001 Nature 414 363
446. Gurevich A and Pashitskii E A 1998 Phys. Rev. B 57 13878
447. Iijima Y, Tanabe N, Kohno O and Ikeno Y 1992 Appl. Phys. Lett. 60 769
448. Lee S et al 2009 Appl. Phys. Lett. 95 212505
449. Browning N D, Buban J P, Nellist P D, Norton D P, Chisholm M F and Pennycook S J 1998 Physica C 294 183
450. Dhoot A S, Yuen J D, Heeney M, McCulloch I, Moses D and Heeger A J 2006 Proc. Natl Acad. Sci. USA 103 11834
451. Panzer M J and Frisbie C D 2006 Adv. Funct. Mater. 16 1051
452. Misra R, McCarthy M and Hebard A F 2007 Appl. Phys. Lett. 90 052905
453. Shimotani H, Asanuma H, Tsukazaki A, Ohtomo A, Kawasaki M and Iwasa Y 2007 Appl. Phys. Lett. 91 082106
454. Ueno K, Nakamura S, Shimotani H, Ohtomo A, Kimura N, Nojima T, Aoki H, Iwasa Y and Kawasaki M 2008 Nat. Mater. 7 855
455. Schooley J F, Hosier W R, Ambler E and Becker J H 1965 Phys. Rev. Lett. 14 305
456. Ueno K, Nakamura S, Shimotani H, Yuan H T, Kimura N, Nojima T, Aoki H, Iwasa Y and Kawasaki M 2011 Nat. Nonotechnol. 6 408
457. Ye J T, Inoue S, Kobayashi K, Kasahara Y, Yuan H T, Shimotani H and Iwasa Y 2010 Nat. Mater. 9 125
458. Taniguchi K, Matsumoto A, Shimotani H and Takagi H 2012 Appl. Phys. Lett. 101 042603
459. Ye J T, Zhang Y J, Akashi R, Bahramy M S, Arita R and Iwasa Y 2012 Science 338 1193
460. Bollinger A T, Dubuis G, Yoon J, Pavuna D, Misewich J and Božović I 2011 Nature 472 458
461. Dhoot A S, Wimbush S C, Benseman T, MacManus-Driscoll J L, Cooper J R and Friend R H 2010 Adv. Mater. 22 2529
462. Leng X, Garcia-Barriocanal J, Bose S, Lee Y and Goldman A M 2011 Phys. Rev. Lett. 107 027001
463. Nojima T, Tada H, Nakamura S, Kobayashi N, Shimotani H and Iwasa Y 2011 Phys. Rev. B 84 020502
464. Katase T, Hiramatsu H, Kamiya T and Hosono H 2014 Proc. Natl Acad. Sci. USA 111 3979
465. Selvamanickam V et al 2009 IEEE Trans. Appl. Supercond. 19 3225
466. Goyal A et al 1996 Appl. Supercond. 4 403
467. Tomsic M, Rindfleisch M, Yue J, McFadden K, Phillips J, Sumption M D, Bhatia M, Bohnenstiehl S and Collings E W 2002 7 Int. J. Appl. Ceramic Technol. 4 250
468. Iida K et al 2011 Appl. Phys. Express 4 013103
469. Miyata S, Ishimaru Y, Adachi S, Shimode T, Murai Y, Chikumoto N, Nakao K and Tanabe K 2013 Extended Abstracts of Int. Workshop on Novel Superconductors and Super Materials 2013 (Tokyo) p 101
470. Ishimaru Y, Miyata S, Adachi S, Shimode T, Murai Y, Chikumoto N, Nakao K and Tanabe K 2015 in preparation
471. Tarantini C et al 2010 Appl. Phys. Lett. 96 142519
472. Si W, Han S J, Shi X, Ehrlich S N, Jaroszynski J, Amit Goyal A and Li Q 2013 Nature Commun. 4 2337
473. Yamamoto A et al 2008 Appl. Phys. Lett. 92 252501
474. Moore J D et al 2008 Supercond. Sci. Technol. 21 092004

475. Kametani F et al 2009 Supercond. Sci. Technol. 22 015010
476. Gao Z, Wang L, Qi Y, Wang D, Zhang X, Ma Y, Yang H and Wen H 2008 Supercond. Sci. Technol. 21 112001
477. Wang L, Gao Z, Qi Y, Zhang X, Wang D and Ma Y 2009 Supercond. Sci. Technol. 22 015019
478. Qi Y, Zhang X, Gao Z, Zhang Z, Wang L, Wang D and Ma Y 2009 Physica C 469 717
479. Mizuguchi Y, Deguchi K, Tsuda S, Yamaguchi T, Takeya H, Kumakura H and Takano Y 2009 Appl. Phys. Express 2 083004
480. Ma Y, Gao Z, Qi Y, Zhang X, Wang L, Zhang Z and Wang D 2009 Physica C 469 651
481. Wang L, Qi Y, Wang D, Zhang X, Gao Z, Zhang Z, Ma Y W, Awaji S, Nishijima G and Watanabe K 2010 Physica C 470 183
482. Wang L, Qi Y, Zhang Z, Wang D, Zhang X, Gao Z, Yao C and Ma Y W 2010 Supercond. Sci. Technol. 23 054010
483. Durrell J H, Eom C-B, Gurevich A, Hellstrom E E, Tarantini C, Yamamoto A and Larbalestier D C 2011 Rep. Prog. Phys. 74 124511
484. Wang L, Qi Y P, Zhang X P, Wang D L, Gao Z S, Wang C L, Yao C and Ma Y W 2011 Physica C 471 1689
485. Fujioka M, Kota T, Matoba M, Ozaki T, Takano Y, Kumakura H and Kamihara Y 2011 Appl. Phys. Express 4 063102
486. Togano K, Matsumoto A and Kumakura H 2011 Appl. Phys. Express 4 043101
487. Matsumoto A, Togano K and Kumakura H 2012 Supercond. Sc. Technol. 25 125010
488. Togano K, Matsumoto A and Kumakura H 2012 Solid State Commun. 152 740
489. Gao Z S, Wang L, Chao Y, Qi Y P, Wang C L, Zhang X P, Wang D L, Wang C D and Ma Y W 2011 Appl. Phys. Lett. 99 242506
490. Weiss J D, Tarantini C, Jiang J, Kametani F, Polyanskii A A, Larbalestier D C and Hellstrom E E 2012 Nat. Mater. 11 682
491. Gao Z S, Ma Y W, Yao C, Zhang X P, Wang C L, Wang D L, Awaji S and Watanabe K 2012 Sci. Rep. 2 998
492. Ma Y 2012 Supercond. Sci. Technol. 25 113001
493. Yao C, Lin H, Zhang X, Wang D, Zhang Q, Ma Y, Awaji S and Watanabe K 2013 Supercond. Sci. Technol. 26 075003
494. Togano K, Gao Z S, Taira H, Ishida S, Iyo A, Kihou K, Eisaki H, Matsumoto A and Kumakura H 2013 Supercond. Sci. Technol. 26 065003
495. Togano K, Gao Z S, Matsumoto A and Kumakura H 2013 Supercond. Sci. Technol. 26 115007
496. Kumakura H, Matsumoto A, Fujii H and Togano K 2001 Appl. Phys. Lett. 79 2435
497. Matsumoto A, Gao Z, Togano K and Kumakura H 2014 Supercond. Sci. Technol. 27 025011
498. Ma Y W, Wang L, Qi Y P, Gao Z S, Wang D L and Zhang X P 2011 IEEE Trans. Appl. Supercond. 21 2878
499. Flukiger R, Graf T, Decroux M, Groth C and Yamada Y 1991 IEEE Trans. Magn. 2 1258
500. Satou M, Yamada Y, Murase S, Kitamura T and Kamisada Y 1994 Appl. Phys. Lett. 64 640
501. Parrell J A, Dorris S E and Larbalestier D C 1994 Physica C 231 137
502. Parrell J A, Polyanskii A A, Pashitski A E and Larbalestier D C 1996 Supercond. Sci. Technol. 9 393
503. Gao Z S, Togano K, Matsumoto A and Kumakura H 2014 Sci. Rep. 4 4065
504. Kumakura H, Togano K, Maeda H, Kase J and Morimoto T 1991 Appl. Phys. Lett. 58 2830
505. Gao Z, Togano K, Matsumoto A and Kumakura H 2015 Supercond. Sci. Technol. 28 012001
506. Marti F, Grasso G, Huang Y B and Flükiger R 1998 Supercond. Sci. Technol. 11 1251
507. Kopera L, Kovác P and Husek I 1998 Supercond. Sci. Technol. 11 433
508. Hosono H and Kuroki K 2015 Physica C doi:10.1016/j.physc.2015.02.020
509. Terasaki I, Sasago Y and Uchinokura K 1997 Phys. Rev. B 56 R12685
510. Mittasch A 1950 Adv. Catal. 2 81
511. Rao C N R and Rao G R 1991 Surf. Sci. Rep. 13 221–63
512. Urabe K, Aika K and Ozaki A 1975 J. Catal. 38 430
513. TodaY Yanagi H, Ikenaga E, Kim J-J, Kobata M, Ueda S, Kamiya T, Hirano M, Kobayashi K and Hosono H 2007 Adv. Mat. 19 3564
514. Matsuishi S, Toda Y, Miyakawa M, Hayashi K, Kamiya T, Hirano M, Tanaka I and Hosono H 2003 Science 301 626
515. Kim S-W and Hosono H 2012 Phil. Mag. 92 2596
516. Miyakawa M, Kim S-W, Hirano M, Kohama Y, Kawaji H, Atake T, Ikegami H, Kono K and Hosono H 2007 J. Am. Chem. Soc. 129 7270
517. Hosono H, Kim S-W, Matsuishi S, Tanaka S, Miyake A, Kagayama T and Shimizu K 2015 Phil. Trans. R. Soc. A 373 20140450
518. Kitano M, Kanbara S, Inoue Y, Kuganathan N, Sushko P V, Yokoyama T, Hara M and Hosono H 2015 Nat. Commun. 6 7731
519. Hayashi F, Toda Y, Kanie Y, Kitano M, Inoue Y, Yokoyama T, Hara M and Hosono H 2013 Chem. Sci. 4 3124
520. Nomura K, Ohta H, Ueda K, Kamiya T, Hirano M and Hosono H 2003 Science 300 1269
521. Nomura K, Ohta H, Takagi A, Kamiya T, Hirano M and Hosono H 2004 Nature 432 488
522. Hosono H 2006 J. Non-Cryst. Sol. 352 851
523. Ogo Y, Hiramatsu H, Nomura K, Yanagi H, Kamiya T, Hirano M and Hosono H 2008 Appl. Phys. Lett. 93 032113
524. Forthaus M K, Sengupta K, Heyer O, Christensen N E, Svane A, Syassen K, Khomskii D I, Lorenz T and Abd-Elmeguid M M 2010 Phys. Rev. Lett. 105 157001
525. Anderson P W and Blount E I 1995 Phys. Rev. Lett. 14

217

526. Boysen H and Altorfer F A 1994 Acta Crystallogr. B50 405

527. Kurth F et al 2014 ASC 2014: Applied Superconductivity Conference (Charlotte)

528. Si W, Han S J, Shi X, Ehrlich S N, Jaroszynski J, Goyal A and Li Q 2013 Nat. Commun. 4 1347

529. Yanagisawa Y, Nakagome H, Uglietti D, Kiyoshi T, Hu R, Takematsu T, Takao T, Takahashi M and Maeda H 2010 IEEE Trans. Appl. Supercond. 20 744

530. Machi T, Nakao K, Kato T, Hirayama T and Tanabe K 2013 Supercond. Sci. Technol. 26 105016

531. Takematsu T, Hua R, Takao T, Yanagisawa Y, Nakagome H, Uglietti D, Kiyoshi T, Takahashi M and Maeda H 2010 Physica C 470 674

532. Aoki H and Hosono H 2015 Phys. World 28 31

533. Onari S and Kontani H 2013 Phys. Rev. Lett. 109 137001

534. Misawa T and Imada M 2014 Nat. Commun. 5 5738

535. Lee J J et al 2014 Nature 515 245

CHAPTER 13

COMPETITION BETWEEN THE SUPERCONDUCTIVITY AND NEMATIC ORDER IN THE HIGH-TC SUPERCONDUCTOR

Jing Wang and Guo-Zhu Liu

INTRODUCTION

Unconventional superconductors usually refer to those superconductors that cannot be understood within the conventional Bardeen–Cooper–Schrieffer (BCS) theory. Notable examples of unconventional superconductors are the high-T_c cuprate superconductor, the heavy fermion superconductor and the iron-based superconductor. Unlike BCS superconductors, unconventional superconductivity is generally driven by electron–electron interactions and often has a magnetic origin. Another interesting property of the unconventional superconductor is that its ground state is not unique. In addition to the defining superconducting state, unconventional superconductors also exhibit a variety of other symmetry-broken ground states, including antiferromagnetic, nematic and stripe states, upon tuning such parameters as doping and pressure [1–5]. A widely recognized notion is that the long-range superconducting order competes and, under certain circumstances, coexists with other long-range orders. The competition and possible coexistence between different orders can give rise to rich properties and hence have attracted intense theoretical and experimental interest in recent years.

The successful microscopic theory of competing orders has not yet been established to date, primarily because the pairing mechanism in most unconventional superconductors is still undetermined. A realistic and commonly used strategy is to build a low-energy effective field theory on phenomenological grounds. One can first write down the Ginzburg–Landau (GL) actions for two bosonic order parameters and then introduce certain coupling terms between these two scalar fields. Such a generalized GL model has recently been applied to describe the competing orders in a number of unconventional superconductors [6–15]. An early success of such a theoretical investigation is the prediction of a field-induced antiferromagnetic core in the superconducting vortices of high-T_c superconductors [6]. This prediction was subsequently confirmed in experiments [16, 17]. Interestingly, experiments further found that the antiferromagnetic order not only exists in the vortex cores, but also extends into the superconducting region [16] and exhibits non-trivial spatial modulation [17, 18]. A phenomenological field theory that contains a simple quadratic–quadratic coupling term between the superconducting and antiferromagnetic order parameters was put forward to understand these new findings [7, 8].

Recently, the issue of competing orders has attracted revived interest. It is found that the competitive interaction between distinct orders can drive an instability, which gives rise to a general tendency of a first order transition [11, 14]. This phenomenon may account for the first order transition observed in some unconventional superconductors [4]. In addition, non-uniform glassy electronic phases and Brazovskii type transitions are predicted to emerge due to the competition between two long-range orders [10]. Another interesting observation is that the competition between superconducting and antiferromagnetic orders can help to judge the gap symmetry of iron-based superconductors [9, 12]. Furthermore, the competition between superconducting and nematic orders might be responsible for [15] the electronic anisotropy observed in the vortex state of the FeSe superconductor [19].

The effective field theory adopted in previous analyses of the competing orders normally contains only two bosonic order parameters. The fermionic degrees of freedom are usually completely integrated out in the spirit of the Hertz–Millis–Moriya (HMM) theory [20–22]. This integration procedure is expected to be applicable in systems that do not contain gapless fermionic excitations. For instance, the iron-

Citation: Jing Wang and Guo-Zhu Liu, Competition between the superconductivity and nematic order in the high-Tc superconductor, doi:10.1088/1367-2630/15/7/073039.

based superconductors seem to have an s-wave energy gap, so the electronic excitations are fully gapped and can be safely integrated out [9, 12]. However, such integration manipulation is not always valid. Indeed, its validity has recently been questioned in several itinerant electron systems [23–26]. In the systems that exhibit gapless fermionic excitations, integrating out fermions may lead to singularities, especially in the vicinity of the quantum critical point (QCP). Actually, infrared singularities have been found on the border of several quantum phase transitions [23–26]. In order to properly describe the quantum critical behavior in these systems, it is more appropriate to maintain both bosonic order parameter and gapless fermions in the effective theory. When a long-range order competing with superconductivity also couples to gapless fermions, it would be interesting to go beyond the HMM theory and examine the role of gapless fermions. The recent analysis presented in [13, 27] did suggest non-trivial roles played by gapless fermions.

In various types of superconductors, the superconductivity may compete with several possible long-range orders. To examine the role of fermions, we wish to study a prototypical model which describes the competition between two distinct long-range orders, contains gapless fermions and in the meantime is technically controllable. In this paper, we choose to consider the competition between the superconductivity and the nematic order in the context of high-T_c superconductors. In recent years, there has been increasing experimental evidence pointing toward the existence of an electronic nematic phase in some high-T_c superconductors [1–3,28–31], especially $YBa_2Cu_3O_{6+}\delta$ and $Bi_2Sr_2CaCu_2O_{8+}\delta$. According to these experiments, a nematic order is predicted to compete and coexist with the superconductivity, which is schematically plotted in figure 1. The nematic transition and the coupling of the nematic order to gapless fermions have stimulated intense research efforts [1–3, 27, 32–42]. From a field-theoretical viewpoint, the nematic order parameter is a real scalar field and does not carry a finite wave vector, which substantially simplifies theoretical calculations.

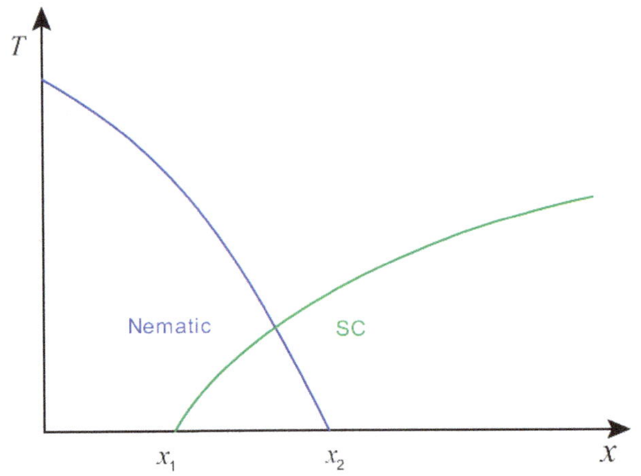

Figure 1. Schematic phase diagram on (x,T)-plane of high-T_c superconductors. X represents doping concentration. x_1 and x_2 are the QCPs of the superconducting and nematic phase transitions, respectively.

It is known that the high-T_c superconductor has a dx^2–y^2 energy gap, which vanishes at four nodes, $(\pm\frac{\pi}{2}, \pm\frac{\pi}{2})$. Therefore, gapless nodal quasiparticles (qps) are present even at the lowest energy in the superconducting phase. These nodal qps are believed to be responsible for many of the anomalous low-temperature properties of the superconducting dome. When a nematic QCP exists somewhere in the superconducting dome, as shown in figure 1, the fluctuation of the nematic order parameter couples to gapless nodal qps. This coupling can generate non-Fermi liquid behaviors and other anomalous phenomena in the vicinity of the nematic QCP [27, 36, 37, 39–42]. In particular, the ratio κ between the gap velocity v_Δ and the Fermi velocity v_F of nodal qps is driven to vanish, i.e. $\kappa = v_\Delta/v_F \rightarrow 0$, by the critical nematic fluctuation [37], leading to extreme velocity anisotropy. These unusual behaviors may have significant effects on the interplay between the superconductivity and the nematic order, which is the topic of this paper.

In this paper, we first write down an effective field theory that describes both the competitive interaction between two bosonic (superconducting and nematic) order parameters and the Yukawa-type coupling between the nematic order parameter and the nodal qps. We then carry out a detailed renormalization group (RG) analysis [43] within this effective theory. Specifically, we will derive and solve the RG flow equations of all physical parameters so as to determine the possible stable fixed points. We will demonstrate that gapless fermionic degrees of freedom can fundamentally change the basic properties of the interplay between the superconductivity and the nematic order. In case all the fermions are entirely neglected, the ordering competition may be strong enough to produce runaway behavior and turn the continuous phase transitions to first order [14]. However, once the dynamics of gapless nodal qps are properly incorporated, a stable fixed point exists with the two originally competing bosonic order parameters decoupled from each other in the vicinity of the nematic QCP. As a result, both the superconducting and nematic transitions remain continuous. In addition, the d-wave superconductivity can coexist with the nematic order homogeneously. Our results indicate that it is important to include the dynamics of gapless fermions in the theoretic description of competing orders.

In section 2, we write down the effective action which contains two bosonic order parameters and gapless nodal qps. In section 3, we make RG calculations and derive the flow equations for all parameters in the effective action. In section 4, we present numerical solutions of the flow equations and discuss the physical implications. The paper ends in section 5 with a summary and conclusion.

EFFECTIVE FIELD THEORY OF COMPETING ORDERS

We first need to write down an effective field theory to describe the competition between superconducting and nematic orders. This will be done largely on phenomenological grounds. In the phase diagram presented in figure 1, the horizonal axis is doping concentration x. The QCP of the superconducting transition is x_1, which is roughly $x_1 \approx 0.05$ in many high-T_c superconductors. The anticipated QCP for a nematic transition is represented by x_2. So far, the precise value and even the very existence of x_2 have not yet been unambiguously determined. Here, we assume that x_2 is larger than x_1, which implies a bulk coexistence of superconducting and nematic orders. In this system, there are three types of degrees of freedom: superconducting order parameter ψ, nematic order parameter ϕ and gapless nodal qps Ψ. The competition between superconducting and nematic orders can be described by a repulsive quadratic–quadratic coupling term, $\propto\psi^2\phi^2$, which is widely adopted in the description of competing orders [12–14,27]. In addition to this competitive interaction, the nematic order parameter ϕ also interacts with gapless nodal qps Ψ, which is usually described by a Yukawa-type coupling term. There is, however, no direct coupling between the superconducting order parameter and the nodal qps. Firstly, the nodal qps are excited from the $dx^2{-}y^2$ gap nodes where the superconducting order parameter vanishes. Secondly, these qps are known to have a sharp peak and a very long lifetime in the superconducting dome in the absence of competing orders [44], so their coupling to ψ must be quite weak. Furthermore, in this paper we are mainly interested in the physical properties in the close vicinity of the nematic QCP x_2, where the fluctuation of the nematic order parameter is critical. Unless the superconducting QCP x_1 coincides with, or is very close to, x_2, the fluctuation of the superconducting order parameter is not critical at x_2. Therefore, the coupling between the superconducting order parameter and the nodal qps is not as important as that between the critical nematic fluctuation and the nodal qps at x_2 and can simply be neglected.

On the basis of the above qualitative analysis, we can write down the following partition function:

$$Z = \int \mathcal{D}\psi\mathcal{D}\phi\mathcal{D}\bar{\Psi}\mathcal{D}\Psi\, e^{S}, \tag{1}$$

where the effective action is

$$S = S_\psi + S_\phi + S_\Psi + S_{\psi\phi} + S_{\Psi\phi}, \tag{2}$$

$$S_\psi = \frac{1}{2}\int \frac{d^3q}{(2\pi)^3}(-2\alpha + q^2)\psi^2 + \frac{\beta}{2}\int d^2\mathbf{r}\, d\tau\, \psi^4, \tag{3}$$

$$S_\phi = \frac{1}{2}\int \frac{d^3q}{(2\pi)^3}\left(-2r + q^2\right)\phi^2 + \frac{u}{2}\int d^2\mathbf{r}\, d\tau\, \phi^4, \tag{4}$$

$$S_\Psi = \int \frac{d^3k}{(2\pi)^3}[\Psi^\dagger_{1i}(-i\omega + v_F k_x\tau^z + v_\Delta k_y\tau^x)\Psi_{1i} + \Psi^\dagger_{2i}(-i\omega + v_F k_y\tau^z + v_\Delta k_x\tau^x)\Psi_{2i}], \tag{5}$$

$$S_{\psi\phi} = \gamma\int d^2x\, d\tau\, \psi^2\phi^2, \tag{6}$$

$$S_{\Psi\phi} = \int d^2x\, d\tau[\lambda\phi(\Psi^\dagger_{1i}\tau^x\Psi_{1i} + \Psi^\dagger_{2i}\tau^x\Psi_{2i})], \tag{7}$$

where $\tau^{x,y,z}$ are Pauli matrices and the flavor index i sums the integers from 1 to N. $\Psi^\dagger_1$ represents the nodal QPs excited from $(\frac{\pi}{2}, \frac{\pi}{2})$ and $(-\frac{\pi}{2}, -\frac{\pi}{2})$ points, while $\Psi^\dagger_2$ represents the other two. The physical flavor of the nodal qps, $N = 2$. Here, r is the tuning parameter for a nematic transition with $r = 0$ at x_2. $v_{F,\Delta}$ are the Fermi velocity and gap velocity of the nodal qps, respectively. The competitive interaction term $\psi^2\phi^2$ has a positive coefficient, $\gamma > 0$. $S_{\Psi\phi}$ represents the Yukawa coupling between the nematic order parameter and gapless nodal qps, with λ being its coupling constant. The free propagators of all the fields are shown in figure 2.

(a) (b) (c)

Figure 2. (a) Free propagator of superconducting field ψ; (b) free propagator of nematic field ϕ; and (c) free propagator of nodal qps Ψ.

In order to simplify calculations, it proves convenient to make two transformations [37]: $\phi \rightarrow \phi/\lambda$, and $r \rightarrow \lambda^2 r$. It is now easy to rewrite equation (7) as

$$S_{\Psi\phi} = \int d^2x\, d\tau\, \phi(\Psi^\dagger_{1i}\tau^x\Psi_{1i} + \Psi^\dagger_{2i}\tau^x\Psi_{2i}). \tag{8}$$

The effective action represented by equation (2) was studied recently in [27]. It was demonstrated that both the superfluid density and the critical temperature T_c are significantly suppressed at the nematic QCP x_2. However, the superconducting order parameter ψ was assumed in [27] to be classical, which is valid only when x_2 is not close to x_1. In this paper, we go beyond such an approximation and consider the quantum fluctuations of both the superconducting and nematic order parameters. As emphasized in [27], the gapless nodal qps can have an important impact on the competition between the superconductivity and the nematic order. The simplest way to include the fermionic degrees of freedom is to introduce the polarization function $\Pi(q)$, due to the nodal qps, into the effective action of the nematic order, S_ϕ. To the leading order of $1/N$-expansion, the polarization function $\Pi(q)$ is represented by the one-loop Feynman diagram shown in figure 3(a) and formally given by [37, 42]

$$\Pi(\mathbf{q}, \epsilon) = N\int \frac{d^2\mathbf{k}\, d\omega}{(2\pi)^3}\mathrm{Tr}[\tau^x G_0(\mathbf{k}, \omega)\tau^x G_0(\mathbf{k}+\mathbf{q}, \omega+\epsilon)],$$

where

$$G_0(\mathbf{k}, \omega) = \frac{1}{-\mathrm{i}\,\omega + v_\mathrm{F}k_x\tau^z + v_\Delta k_y\tau^x}$$

is the free propagator for the nodal qps Ψ_1 (the free propagator for the nodal qps Ψ_2 can be similarly written down). The polarization function $\Pi(\epsilon,\mathbf{q})$ has already been calculated previously [27, 37] and is known to have the form

$$\Pi(\mathbf{q}, \epsilon) = \frac{N}{16 v_\mathrm{F} v_\Delta}\left[\frac{\epsilon^2+v_\mathrm{F}^2 q_x^2}{\sqrt{\epsilon^2+v_\mathrm{F}^2 q_x^2+v_\Delta^2 q_y^2}}+(q_x\leftrightarrow q_y)\right]. \tag{9}$$

Including this term, the quadratic part of S_ϕ becomes

$$\left(-2r+q^2\right)\phi^2 \to \left[-2r+q^2+\Pi(q)\right]\phi^2. \tag{10}$$

From the expression of polarization $\Pi(q)$, it is easy to see that the inclusion of $\Pi(q)$ does not change the dynamical exponent $z = 1$ of ϕ. However, $\Pi(q)\propto q$, so it dominates over the kinetic term q^2 in the low energy regime. More importantly, the polarization $\Pi(q)$ introduces two important quantities, the nodal qps' Fermi velocity v_F and the gap velocity v_Δ, into the effective action of ϕ.

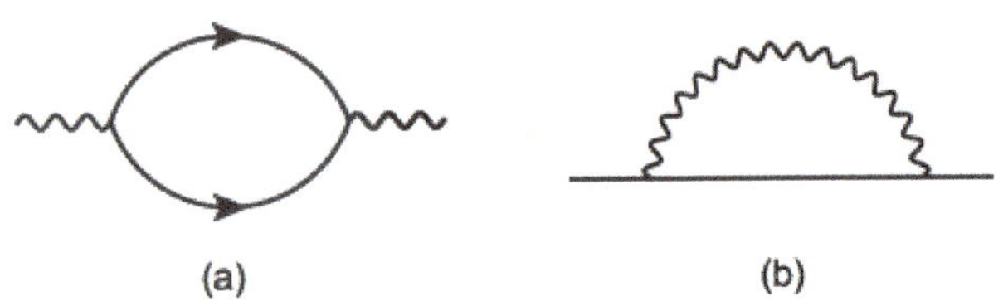

Figure 3. (a) Polarization function for nematic field ϕ; (b) Fermion self-energy correction due to nematic fluctuation.

Although the polarization $\Pi(q)$ represents the influence of the nodal qps, we cannot completely integrate the nodal qps out and drop them from the effective theory. These gapless nodal qps should be maintained for several reasons. Firstly, according to the general spirit of RG, one can safely integrate out high-energy modes at low energies. However, the nodal qps are gapless and hence exist even at the lowest energy. Integrating out the gapless fermions completely may lead to unphysical singularities [24–26]. Secondly, the coupling between the gapless fermions and the order parameter fluctuation often causes non-Fermi liquid behavior in observable quantities. In the case of nematic transition, the critical nematic fluctuation gives rise to fermion velocity renormalization and extreme anisotropy, which would be overlooked if the fermions are fully integrated out. As will be shown below, the velocity ratio $\kappa = v_\Delta/v_\mathrm{F}$ plays a non-trivial role.

The effective field theory contains seven parameters: $\alpha,r,\beta,u,\gamma,v_\mathrm{F},v_\Delta$. They are all subjected to renormalizations due to the mutual interactions among three field operators: ψ, ϕ and Ψ. We will study the flow of these seven parameters under scaling transformations and eventually obtain seven RG equations. Since we study the ψ–ϕ interaction and the Ψ–ϕ interaction on equal footings, these seven RG equations are self-consistently coupled to each other. The low-energy behaviors of these parameters and the possible fixed points can be determined by solving these coupled RG equations.

RENORMALIZATION GROUP CALCULATIONS

In this section, we make a RG analysis and obtain the flow equations of all the aforementioned parameters. In order to examine the impacts of the gapless nodal qps, we go beyond the HMM theory and maintain the nodal qps throughout our calculations. We first analyze the coupling between the nematic order and the nodal qps and derive the RG equations for the fermion velocities, $v_{\mathrm{F},\Delta}$. We then consider the competitive interaction between the superconducting and nematic order parameters and obtain the RG equations for the rest of the five parameters, $(\alpha,r,\beta,u,\gamma)$, which depend on the fermion velocities, $v_{\mathrm{F},\Delta}$. These seven equations are self-consistently coupled to each other since the fermion velocities $v_{\mathrm{F},\Delta}$ appearing in the equations of $(\alpha,r,\beta,u,\gamma)$ flow according to their own equations.

Our RG analysis will be performed within the framework presented in [43]. According to the RG theory [43], we first employ the following scaling transformations:

$$k_i = k_i' \mathrm{e}^{-l}, \tag{11}$$

$$\omega = \omega' \mathrm{e}^{-l}, \tag{12}$$

$$q_i = q_i' \mathrm{e}^{-l}, \tag{13}$$

$$\epsilon = \epsilon' \mathrm{e}^{-l}, \tag{14}$$

where $i = x,y$ and l represent the scaling parameter. Next we need to determine how all of the field operators flow under these transformations. It is easy to know that the superconducting order parameter ψ should be re-scaled as

$$\psi(\mathbf{k}, \omega) = \psi'(\mathbf{k}', \omega')\,\mathrm{e}^{5l/2} \tag{15}$$

on the basis of its action equation (3). Similarly, the rescaling behavior of the nodal qps $\Psi_{1,2}$ is found to be [37]

$$\Psi_{1,2}(\mathbf{k}, \omega) = \Psi_{1,2}'(\mathbf{k}', \omega')\,\mathrm{e}^{(4+C_1)l/2}, \tag{16}$$

where the explicit expression of constant C_1 will be defined below in equation (20).

However, determining the rescaling behavior of the nematic field ϕ is much more complicated. In the standard RG theory [43], one should regard the kinetic term of ϕ as the fixed point and obtain the rescaling behavior of ϕ by requiring such a term invariant under the scaling transformations. Nevertheless, the kinetic term of nematic order $S\phi$ is irrelevant in the low-energy region; therefore, it cannot be regarded as the fixed point. One has to choose another term to serve as the fixed point.

As already demonstrated in section 2, the polarization function $\Pi(q)$ generated by the strong interaction with gapless nodal qps dominates over the kinetic term at low energies. Naturally, one might expect to choose the polarization term, $\Pi(q)$ $\phi^2(q)$, to serve as the fixed point. However, we emphasize here that it is not appropriate to obtain the scaling of ϕ by requiring $\Pi(q)$ invariant under RG transformations. Equation (9) tells us that $\Pi(q)$ contains two fermion velocities, $v_{F,\Delta}$, which are apparently scale-dependent and flow strongly as l varies. If we insist on requiring that the polarization term is invariant under the scaling transformations, we have to assume that $v_{F,\Delta}$ are always constants and do not flow with a varying l. Under this assumption, the important property of singular fermion velocity renormalization cannot be properly taken into account in our theoretical description of competing orders. For these reasons, we believe it is not appropriate to adopt the polarization term as the fixed point of this model.

Since both the kinetic and polarization terms are not good choices for the fixed point, we have to choose another term from the effective action. It appears that the Yukawa coupling term, λ $\phi \Psi^{\dagger}\tau^{x}\Psi$, is the only available candidate. However, there is a very important problem: how to treat the coupling constant λ? In the conventional perturbation theory, usually one can make a perturbative expansion in powers of λ. Unfortunately, this scheme does not apply to this model because explicit RG calculations [2] have revealed that λ tends to diverge at the lowest energy. It was later realized that a reasonable route to access such a model is to fix λ at a certain finite value [36–38] and perform a perturbative expansion in powers of $1/N$, where N is apparently the flavor. In this formalism, λ is a constant and does not flow with running l. One can first absorb λ into ϕ [37] and then require that the Yukawa coupling term is invariant under the scaling transformations. It is now easy to know that the nematic field transforms as [37, 38]

$$\phi(\mathbf{q}, \epsilon) = \phi'(\mathbf{q}', \epsilon')\, e^{(2+C_3-C_1)l}, \tag{17}$$

where C_3 will be defined below in equation (20). This expression is to be used in the following calculations.

Flow Equations of v_F and v_Δ

The Yukawa-type interaction between the nematic order and the gapless nodal qps has recently been investigated in several papers [36, 37, 39–42]. It is well known that the Fermi velocity of nodal qps is indeed not equal to the gap velocity, i.e. $v_F \neq v_\Delta$. Experiments [44, 45] have determined that the velocity ratio $\kappa = v_\Delta/v_F \approx 0.1$. This ratio is a very important parameter since it enters into a number of observable quantities of high-T_c superconductors, including electric conductivity [46, 47], thermal conductivity [47], superfluid density [47, 48] and T_c [48]. An interesting property revealed and discussed in [36, 37, 39–42] is that the velocity anisotropy is significantly enhanced by the nematic fluctuation.

The calculation of nodal qps' self-energy function and the derivation of the flow equations have already been presented in previous publications [37, 42] and therefore are not shown here. It is only necessary to summarize the basic calculations as well as the relevant results. To the leading order, the fermion self-energy is represented by the diagram in figure 3(b) and has the form

$$\Sigma(\mathbf{k}, \omega) = \int \frac{d^2\mathbf{q}\, d\epsilon}{(2\pi)^3} G_0(\mathbf{k}+\mathbf{q}, \omega+\epsilon) \frac{1}{\Pi(q)}. \tag{18}$$

As shown in [37], it can be written as

$$\frac{d\Sigma(\mathbf{k}, \omega)}{d\ln\Lambda} = C_1(-i\omega) + C_2 v_F k_x \tau^z + C_3 v_\Delta k_y \tau^x, \tag{19}$$

where

$$C_1 = \frac{2(v_\Delta/v_F)}{N\pi^3} \int_{-\infty}^{\infty} dx \int_0^{2\pi} d\theta \frac{x^2 - \cos^2\theta - (v_\Delta/v_F)^2 \sin^2\theta}{\left[x^2 + \cos^2\theta + (v_\Delta/v_F)^2 \sin^2\theta\right]^2} \mathcal{G}(x, \theta),$$

$$C_2 = \frac{2(v_\Delta/v_F)}{N\pi^3} \int_{-\infty}^{\infty} dx \int_0^{2\pi} d\theta \frac{\cos^2\theta - x^2 - (v_\Delta/v_F)^2 \sin^2\theta}{\left[x^2 + \cos^2\theta + (v_\Delta/v_F)^2 \sin^2\theta\right]^2} \mathcal{G}(x, \theta),$$

$$C_3 = \frac{2(v_\Delta/v_F)}{N\pi^3} \int_{-\infty}^{\infty} dx \int_0^{2\pi} d\theta \frac{x^2 + \cos^2\theta - (v_\Delta/v_F)^2 \sin^2\theta}{\left[x^2 + \cos^2\theta + (v_\Delta/v_F)^2 \sin^2\theta\right]^2} \mathcal{G}(x, \theta),$$

$$\mathcal{G}^{-1} = \frac{x^2 + \cos^2\theta}{\sqrt{x^2 + \cos^2\theta + (v_\Delta/v_F)^2 \sin^2\theta}} + \frac{x^2 + \sin^2\theta}{\sqrt{x^2 + \sin^2\theta + (v_\Delta/v_F)^2 \cos^2\theta}}.$$

Using the Dyson equation, it is easy to get a renormalized fermion propagator

$$G_{\psi}^{-1}(\mathbf{k},\omega) = -\mathrm{i}\,\omega + v_{\mathrm{F}}k_x\tau^z + v_{\Delta}k_y\tau^x - \Sigma(\mathbf{k},\omega), \tag{20}$$

which leads to the following RG equations:

$$\frac{\mathrm{d}v_{\mathrm{F}}}{\mathrm{d}l} = (C_1 - C_2)v_{\mathrm{F}}, \tag{21}$$

$$\frac{\mathrm{d}v_{\Delta}}{\mathrm{d}l} = (C_1 - C_3)v_{\Delta}, \tag{22}$$

$$\frac{\mathrm{d}(v_{\Delta}/v_{\mathrm{F}})}{\mathrm{d}l} = (C_2 - C_3)(v_{\Delta}/v_{\mathrm{F}}), \tag{23}$$

where $l > 0$ is the running scale. A straightforward analysis showed that the ratio $v_{\Delta}/v_{\mathrm{F}}$ flows to zero at the lowest energy, giving rise to a novel fixed point of extreme velocity anisotropy [37]. Such a fixed point in turn leads to a number of non-trivial consequences, such as an unusual broadening of the spectral function [36], non-Fermi liquid behavior [39], an enhancement of the dc thermal conductivity [40] and a suppression of the superconductivity [27]. To analyze the influence of velocity renormalization and especially the extreme anisotropy manifested at nematic QCP on the nature of a superconducting transition, we require that the constant fermion velocities appearing in the polarization $\Pi(q)$ flow with the running scale l according to equations (21) and (22).

The extreme velocity anisotropy is a special feature of the nematic QCP, where $r = 0$ and the nematic fluctuation is critical. Away from nematic QCP, $r \neq 0$, so the nematic fluctuation leads only to a relatively unimportant renormalization of the fermion velocities. For $r \neq 0$, $v_{\mathrm{F},\Delta}$ and therefore their ratios $v_{\Delta}/v_{\mathrm{F}}$ remain finite.

Flow Equations of α, r, β, u and γ

We next consider the competitive interaction between the nematic and superconducting order parameters. Our analysis follows closely the scheme presented in a recent work of She *et al* [14]. It was assumed in [14] that all the fermionic degrees of freedom can be integrated out and their effects can be represented by the dynamical exponent z. Compared with [14], the main difference here is the inclusion of polarization $\Pi(q)$ in the effective action of nematic order ϕ, which is supposed to reflect the influence of nodal qps. The corresponding (sub) action that describes the ordering competition is

$$S_{\mathrm{com}} = S_{\psi} + S_{\phi} + S_{\psi\phi}, \tag{24}$$

where

$$S_{\psi} = \frac{1}{2}\int \frac{\mathrm{d}^2\mathbf{k}\,\mathrm{d}\omega}{(2\pi)^3}\left(-2\alpha + \mathbf{k}^2 + \omega^2\right)\psi^2 + \frac{\beta}{2}\int \prod_{m=1}^{4} \frac{\mathrm{d}^2\mathbf{k}_m\,\mathrm{d}\omega_m}{(2\pi)^3}\delta^2\left(\sum \mathbf{k}_m\right)\delta\left(\sum \omega_m\right)\psi^4,$$

$$S_{\phi} = \int \frac{\mathrm{d}^2\mathbf{q}\,\mathrm{d}\epsilon}{(2\pi)^3}\frac{1}{2}\left[-2r + \mathbf{q}^2 + \epsilon^2 + \Pi(q)\right]\phi^2 + \frac{u}{2}\int \prod_{m=1}^{4} \frac{\mathrm{d}^2\mathbf{q}_m\,\mathrm{d}\epsilon_m}{(2\pi)^3}\delta^2\left(\sum \mathbf{q}_m\right)\delta\left(\sum \epsilon_m\right)\phi^4,$$

$$S_{\psi\phi} = \gamma\int \prod_{i=1,2} \frac{\mathrm{d}^2\mathbf{k}_i\,\mathrm{d}\omega_i d^2\mathbf{q}_i\,\mathrm{d}\epsilon_i}{(2\pi)^6}\psi(\mathbf{k}_i,\omega_i)\phi(\mathbf{q}_i,\epsilon_i)\delta^2\left(\mathbf{k}_1 + \mathbf{k}_2 + \mathbf{q}_1 + \mathbf{q}_2\right)\delta\left(\omega_1 + \omega_2 + \epsilon_1 + \epsilon_2\right).$$

Before performing a standard RG analysis within this action, it is convenient to rescale the momenta and energy by Λ, i.e. $\mathbf{k} \to \mathbf{k}/\Lambda$, $\omega \to \omega/\Lambda$.

Each field operator can be separated into a slow mode and a fast mode, i.e.

$$\psi = \psi_{\mathrm{s}} + \psi_{\mathrm{f}}, \tag{25}$$

$$\phi = \phi_{\mathrm{s}} + \phi_{\mathrm{f}}. \tag{26}$$

After introducing a UV cutoff Λ, we can define the slow mode of the superconducting order parameter as $\psi_{\mathrm{s}} = \psi(k)$ with $0 < k < e^{-l}\Lambda$ and the fast mode as $\psi_{\mathrm{f}} = \psi(k)$ with $e^{-l}\Lambda < k < \Lambda$, using the formalism of [43]. Based on such modes of separation, the effective action (24) is decomposed into three parts: S^{s} that contains only the slow modes, S^{f} that contains only the fast modes and S^{sf} that contains both the slow and fast modes. More concretely, we have

$$\begin{aligned} S_{\mathrm{com}} &= S^{\mathrm{s}} + S^{\mathrm{f}} + S^{\mathrm{sf}} \\ &= \left(S_{\psi}^{\mathrm{s}} + S_{\phi}^{\mathrm{s}} + S_{\psi\phi}^{\mathrm{s}}\right) + \left(S_{\psi}^{\mathrm{f}} + S_{\phi}^{\mathrm{f}}\right) + S^{\mathrm{sf}}, \end{aligned} \tag{27}$$

where

$$S_{\psi}^{s}=\frac{1}{2}\int\frac{d^{2}\mathbf{k}\,d\omega}{(2\pi)^{3}}\left(-2\alpha+\mathbf{k}^{2}+\omega^{2}\right)\psi_{s}^{2}+\frac{\beta}{2}\int\prod_{m=1}^{4}\frac{d^{2}\mathbf{k}_{m}\,d\omega_{m}}{(2\pi)^{3}}\delta^{2}\left(\sum\mathbf{k}_{m}\right)\delta\left(\sum\omega_{m}\right)\psi_{s}^{4},$$

$$S_{\phi}^{s}=\frac{1}{2}\int\frac{d^{2}\mathbf{q}\,d\epsilon}{(2\pi)^{3}}\left[-2r+\mathbf{q}^{2}+\epsilon^{2}+\Pi(q)\right]\phi_{s}^{2}+\frac{u}{2}\int\prod_{m=1}^{4}\frac{d^{2}\mathbf{q}_{m}\,d\epsilon_{m}}{(2\pi)^{3}}\delta^{2}\left(\sum\mathbf{q}_{m}\right)\delta\left(\sum\epsilon_{m}\right)\phi_{s}^{4},$$

$$S_{\psi\phi}^{s}=\gamma\int\prod_{i=1,2}\frac{d^{2}\mathbf{k}_{i}\,d\omega_{i}d^{2}\mathbf{q}_{i}\,d\epsilon_{i}}{(2\pi)^{6}}\psi(\mathbf{k}_{i},\omega_{i})\phi(\mathbf{q}_{i},\epsilon_{i})\delta^{2}\left(\mathbf{k}_{1}+\mathbf{k}_{2}+\mathbf{q}_{1}+\mathbf{q}_{2}\right)\delta\left(\omega_{1}+\omega_{2}+\epsilon_{1}+\epsilon_{2}\right),$$

$$S_{\psi}^{f}=\frac{1}{2}\int\frac{d^{2}\mathbf{k}\,d\omega}{(2\pi)^{3}}\left(-2\alpha+\mathbf{k}^{2}+\omega^{2}\right)\psi_{f}^{2}+\frac{\beta}{2}\int\prod_{m=1}^{4}\frac{d^{2}\mathbf{k}_{m}\,d\omega_{m}}{(2\pi)^{3}}\delta^{2}\left(\sum\mathbf{k}_{m}\right)\delta\left(\sum\omega_{m}\right)\psi_{f}^{4},$$

$$S_{\phi}^{f}=\frac{1}{2}\int\frac{d^{2}\mathbf{q}\,d\epsilon}{(2\pi)^{3}}\left[-2r+\mathbf{q}^{2}+\epsilon^{2}+\Pi(q)\right]\phi_{f}^{2}+\frac{u}{2}\int\prod_{m=1}^{4}\frac{d^{2}\mathbf{q}_{m}\,d\epsilon_{m}}{(2\pi)^{3}}\delta^{2}\left(\sum\mathbf{q}_{m}\right)\delta\left(\sum\epsilon_{m}\right)\phi_{f}^{4},$$

$$S^{sf}=\int\frac{d^{2}\mathbf{k}\,d\omega}{(2\pi)^{3}}\left[(3\beta)\psi_{f}\psi_{f}\psi_{s}\psi_{s}+(3u)\phi_{f}\phi_{f}\phi_{s}\phi_{s}\right]+\int\prod_{i=1,2}\frac{d^{2}\mathbf{k}_{i}\,d\omega_{i}d^{2}\mathbf{q}_{i}\,d\epsilon_{i}}{(2\pi)^{6}}\delta^{2}\left(\mathbf{k}_{1}+\mathbf{k}_{2}+\mathbf{q}_{1}+\mathbf{q}_{2}\right)$$
$$\times\delta\left(\omega_{1}+\omega_{2}+\epsilon_{1}+\epsilon_{2}\right)\left(\gamma\psi_{s}\psi_{s}\phi_{f}\phi_{f}+\gamma\psi_{f}\psi_{f}\phi_{s}\phi_{s}+4\gamma\psi_{f}\psi_{s}\phi_{f}\phi_{s}\right).$$

After this decomposition, the partition function can be rearranged in the following way:

$$\begin{aligned}Z&=\int\mathcal{D}\psi_{s}\mathcal{D}\phi_{s}\mathcal{D}\bar{\Psi}\mathcal{D}\Psi\int\mathcal{D}\psi_{f}\mathcal{D}\phi_{f}\times\exp\left[(S_{\psi}^{s}+S_{\phi}^{s}+S_{\psi\phi}^{s})+(S_{\psi}^{f}+S_{\phi}^{f})+S^{sf}\right]\\&=\int\mathcal{D}\psi_{s}\mathcal{D}\phi_{s}\mathcal{D}\bar{\Psi}\mathcal{D}\Psi\exp\left(S_{\psi}^{s}+S_{\phi}^{s}+S_{\psi\phi}^{s}\right)\int\mathcal{D}\psi_{f}\mathcal{D}\phi_{f}\exp\left[(S_{\psi}^{f}+S_{\phi}^{f})+S^{sf}\right]\\&=\int\mathcal{D}\psi_{s}\mathcal{D}\phi_{s}\mathcal{D}\bar{\Psi}\mathcal{D}\Psi\exp\left(S_{\psi}^{\prime s}+S_{\phi}^{\prime s}+S_{\psi\phi}^{\prime s}\right)\\&\equiv\int\mathcal{D}\psi_{s}\mathcal{D}\phi_{s}\mathcal{D}\bar{\Psi}\mathcal{D}\Psi\exp\left(S_{\text{eff}}^{\prime}\right).\end{aligned}\tag{28}$$

The next step is to integrate over all the fast modes and obtain an effective action of the slow modes. The functional integration can be performed using the standard diagrammatic techniques. The propagators for the superconducting order ψ and the nematic order ϕ, shown in figure 2, are

$$G_{\psi}(\mathbf{k},\omega)=\frac{1}{-2\alpha+\mathbf{k}^{2}+\omega^{2}},\tag{29}$$

$$G_{\phi}(\mathbf{q},\epsilon)=\frac{1}{-2r+\mathbf{q}^{2}+\epsilon^{2}+\Pi(q)}.\tag{30}$$

The polarization appearing in $G_{\phi}(\mathbf{q},\epsilon)$ reflects the presence of gapless nodal qps. As already pointed out, $\Pi(q)$ dominates over the kinetic term q^2 in the low-energy regime. In order to further simplify the nematic propagator, we consider the close vicinity of the nematic QCP where r is very small. In this case, we are allowed to approximate the nematic propagator by

$$\begin{aligned}G_{\phi}(\mathbf{q},\epsilon)&=\frac{1}{-2r+\mathbf{q}^{2}+\epsilon^{2}+\Pi(q)}\\&\approx\frac{1}{-2r+\Pi(q)}\quad\text{(at low energy)}\\&\approx\frac{1}{\Pi(q)\left[1-\frac{2r}{\Pi(q)}\right]}\\&\approx\frac{1}{\Pi(q)}+\frac{2r}{\Pi^{2}(q)}.\end{aligned}\tag{31}$$

Now let us work in the spherical coordinates. We first define $\epsilon\equiv vq_0=vq\cos\theta$, $q_1=q\sin\theta\cos\phi$ and $q_2=q\sin\theta\sin\phi$, with v being the velocity of the nematic order parameter ϕ. For convenience, we could extract constant v from the energy ω of the superconducting order parameter ψ. Accordingly, the velocities of nodal qps, v_F and v_Δ, can be divided by v. After this manipulation, v_F and v_Δ become dimensionless. Now the polarization function can be written in the form

$$\Pi(q, \theta, \varphi) = qD(\theta, \varphi), \tag{32}$$

where the function

$$D(\theta, \varphi) = \frac{N}{16 v_F v_\Delta} \left(\frac{\cos^2\theta + v_F^2 \sin^2\theta \cos^2\varphi}{\sqrt{\cos^2\theta + v_F^2 \sin^2\theta \cos^2\varphi + v_\Delta^2 \sin^2\theta \sin^2\varphi}} + \frac{\cos^2\theta + v_F^2 \sin^2\theta \sin^2\varphi}{\sqrt{\cos^2\theta + v_F^2 \sin^2\theta \sin^2\varphi + v_\Delta^2 \sin^2\theta \cos^2\varphi}} \right)$$

is dimensionless. Before proceeding with the next calculations, it is helpful to define

$$F_1 \equiv F_1(v_F, v_\Delta) = \int_0^\pi d\theta \int_0^{2\pi} d\varphi \frac{\sin\theta}{(2\pi)^3 D(\theta, \varphi)}, \tag{33}$$

$$F_2 \equiv F_2(v_F, v_\Delta) = \int_0^\pi d\theta \int_0^{2\pi} d\varphi \frac{\sin\theta}{(2\pi)^3 D^2(\theta, \varphi)}, \tag{34}$$

$$F_3 \equiv F_3(v_F, v_\Delta) = \int_0^\pi d\theta \int_0^{2\pi} d\varphi \frac{\sin\theta}{(2\pi)^3 D^3(\theta, \varphi)}. \tag{35}$$

With these arrangements, we can now turn to calculate the one-loop contribution to the RG equations of all the parameters.

α,r

The diagrams contributing to α to the leading order are shown in figure 4(a). We perform the following calculations:

$$\begin{aligned} S[\alpha] &= \int^b \frac{d^3q}{(2\pi)^3} \psi_s(q)\psi_s(-q) \left[-2\alpha + 3\beta \int_b^1 \frac{d^3q'}{(2\pi)^3} G_\psi^f + \gamma \int_b^1 \frac{d^3q'}{(2\pi)^3} G_\phi^f \right] \\ &\approx \int^b \frac{d^3q}{(2\pi)^3} \psi_s(q)\psi_s(-q)(-2\alpha) \exp\left\{ -\frac{l}{2\alpha} \left[\frac{3\beta}{2\pi^2}(1+2\alpha) + \gamma(F_1 + 2rF_2) \right] \right\} \\ &= \int^1 \frac{d^3q}{(2\pi)^3} e^{-3l} \psi_s(q)\psi_s(-q) e^{5l/2} e^{5l/2} (-2\alpha) \\ &\quad \times \exp\left\{ -\frac{l}{2\alpha} \left[\frac{3\beta}{2\pi^2}(1+2\alpha) + \gamma(F_1 + 2rF_2) \right] \right\} \\ &\equiv \int^1 \frac{d^3q}{(2\pi)^3} \psi_s(q)\psi_s(-q)(-2\alpha'), \end{aligned} \tag{36}$$

where

$$\alpha' = \alpha \exp\left\{ 2l - \frac{l}{2\alpha} \left[\frac{3\beta}{2\pi^2}(1+2\alpha) + \gamma(F_1 + 2rF_2) \right] \right\}.$$

Its derivative with respect to running scale l is

$$\frac{d\alpha}{dl} = 2\alpha - \left[\frac{3\beta}{4\pi^2}(1+2\alpha) + \frac{\gamma}{2}(F_1 + 2rF_2) \right]. \tag{37}$$

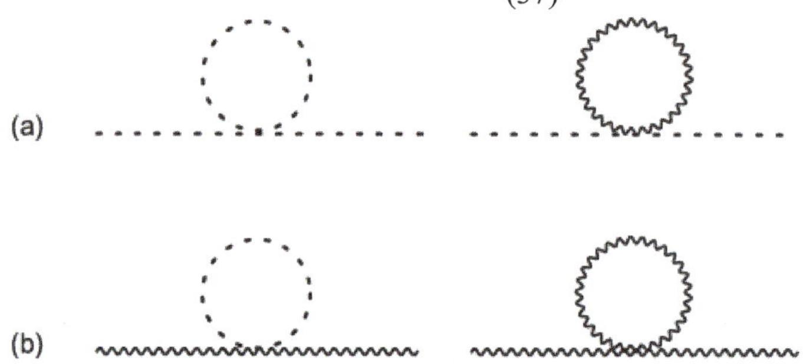

Figure 4. One-loop corrections to mass parameters α in (a) and r in (b), respectively.

By calculating the diagrams shown in figure 4(b), the flow equation for parameter r can be obtained similarly

$$\frac{\mathrm{d}r}{\mathrm{d}l} = [1 + 2(C_3 - C_1)]r - \left[\frac{3u}{2}(F_1 + 2rF_2) + \frac{\gamma}{4\pi^2}(1 + 2\alpha)\right]. \tag{38}$$

In the absence of fermionic degrees of freedom, $F_1 = F_2 = \frac{1}{2\pi^2}$ and $[1 + 2(C_3 - C_1)]r$ replaced with $2r$, then the flow equation of α is identical to that of r [14]. Such an 'exchange symmetry' is certainly broken by the gapless nodal qps via the polarization $\Pi(q)$ appearing in the effective action of nematic order ϕ.

β,u

The one-loop corrections to β are depicted in figure 5(a). By paralleling the steps performed in equation (36), we can similarly obtain

$$S[\beta] \equiv \int^{1} \prod_{m=1}^{4} \frac{\mathrm{d}^2\mathbf{k}_m\, \mathrm{d}\omega_m}{(2\pi)^3} \delta^2\left(\sum \mathbf{k}_m\right) \delta\left(\sum \omega_m\right) \frac{\beta'}{2}|\psi_s|^4,$$

where

$$\beta' = \beta \exp\left\{1 - \frac{2}{\beta}\left[\frac{9\beta^2}{2\pi^2} + \gamma^2(F_2 + 4rF_3)\right]\right\} l.$$

It leads to

$$\frac{\mathrm{d}\beta}{\mathrm{d}l} = \beta - \left[\frac{9\beta^2}{\pi^2} + 2\gamma^2(F_2 + 4rF_3)\right]. \tag{39}$$

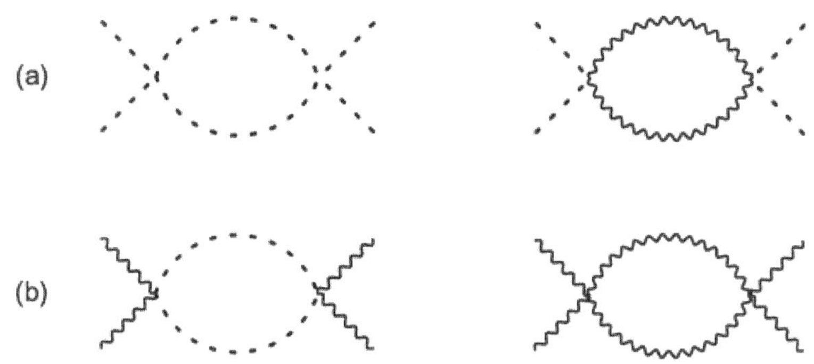

Figure 5. One-loop corrections to quartic coefficients β in (a) and u in (b), respectively.

By calculating the diagrams shown in figure 5(b), the RG equation for u is found to be

$$\frac{\mathrm{d}u}{\mathrm{d}l} = [-1 + 4(C_3 - C_1)]u - \left[\frac{\gamma^2}{\pi^2} + 18u^2(F_2 + 4rF_3)\right]. \tag{40}$$

Once again, the influence of gapless nodal qps is reflected in the functions $F_{2,3}$ and $C_{1,3}$.

γ

Now we consider the flow of γ, which characterizes the strength of the competitive interaction. The one-loop corrections to the term $\gamma\psi^2\phi^2$ have three diagrams, presented in figure 6. Following a similar procedure, we eventually obtain

$$S[\gamma] \equiv \int^{1} \prod_{m=1}^{4} \frac{\mathrm{d}^2\mathbf{k}_m}{(2\pi)^2} \frac{\mathrm{d}\omega_m}{2\pi} \delta^2\left(\sum \mathbf{k}_m\right) \delta\left(\sum \omega_m\right) \gamma' \psi_s^2 \phi_s^2,$$

where

$$\gamma' = \gamma \exp\left\{l - \left[\frac{3\beta}{\pi^2} + 6u(F_2 + 4rF_3) + 8\gamma(F_1 + 2rF_2 + 2\alpha F_1)\right] l\right\}.$$

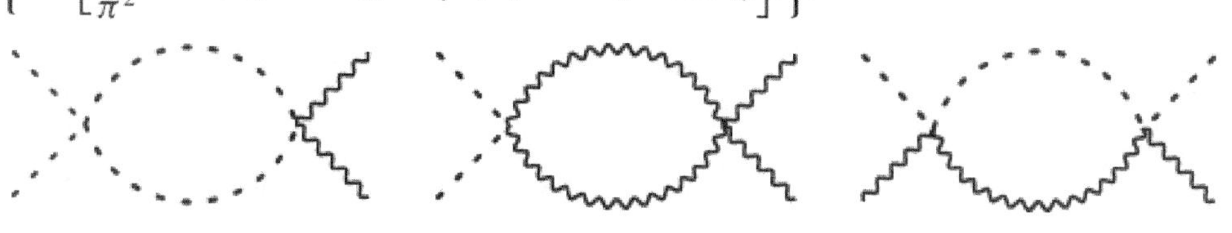

Figure 6. One-loop corrections to coupling constant γ.

The flow equation of γ is therefore given by

$$\frac{d\gamma}{dl} = \gamma\left\{2(C_3 - C_1) - \left[\frac{3\beta}{\pi^2} + 6u(F_2 + 4rF_3) + 8\gamma\,(F_1 + 2rF_2 + 2\alpha F_1)\right]\right\}. \quad (41)$$

RESULTS AND ANALYSIS

In section 3, we obtained the RG equations for a number of relevant parameters. In order to specify the possible fixed point of the system under consideration, we need to solve these RG equations.

For later reference, it is useful to list all the RG equations obtained in the last section:

$$\frac{d\alpha}{dl} = 2\alpha - \left[\frac{3\beta}{4\pi^2}(1 + 2\alpha) + \frac{\gamma}{2}(F_1 + 2rF_2)\right], \quad (42)$$

$$\frac{dr}{dl} = [1 + 2(C_3 - C_1)]r - \left[\frac{3u}{2}(F_1 + 2rF_2) + \frac{\gamma}{4\pi^2}(1 + 2\alpha)\right], \quad (43)$$

$$\frac{d\beta}{dl} = \beta - \left[\frac{9\beta^2}{\pi^2} + 2\gamma^2(F_2 + 4rF_3)\right], \quad (44)$$

$$\frac{du}{dl} = [-1 + 4(C_3 - C_1)]u - \left[\frac{\gamma^2}{\pi^2} + 18u^2(F_2 + 4rF_3)\right], \quad (45)$$

$$\frac{d\gamma}{dl} = \gamma\left\{2(C_3 - C_1) - \left[\frac{3\beta}{\pi^2} + 6u(F_2 + 4rF_3) + 8\gamma\,(F_1 + 2rF_2 + 2\alpha F_1)\right]\right\} \quad (46)$$

$$\frac{dv_F}{dl} = (C_1 - C_2)v_F, \quad (47)$$

$$\frac{dv_\Delta}{dl} = (C_1 - C_3)v_\Delta. \quad (48)$$

Compared with the case in which the action contains only two bosonic order parameters, gapless nodal qps show their existence by entering into the three functions $F_{1,2,3}$, which are all functions of the fermion velocities, $v_{F,\Delta}$. We now address the influence of these nodal qps on the fixed-point properties of the interacting system.

For finite r, the quantum fluctuation of the nematic order and its coupling to gapless nodal qps are relatively weak and hence only lead to unimportant renormalizations of fermion velocities. Actually, the nematic fluctuation is singular only in the close vicinity of the nematic QCP where $r = 0$. Here, we focus on the nematic QCP and solve the above RG equations self-consistently, after taking into account the singular velocity renormalization of the nodal qps driven by the critical nematic fluctuation.

Theoretical Analysis

Before solving the RG equations, it is useful to first make a simple theoretical analysis of the possible scaling behavior. At the nematic QCP, $r = 0$, so the RG equations can be simplified to

$$\frac{d\alpha}{dl} = 2\alpha - \left[\frac{3\beta}{4\pi^2}(1 + 2\alpha) + \frac{\gamma}{2}F_1\right], \quad (49)$$

$$\frac{d\beta}{dl} = \beta - \left[\frac{9\beta^2}{\pi^2} + 2\gamma^2 F_2\right], \quad (50)$$

$$\frac{du}{dl} = [-1 + 4(C_3 - C_1)]u - \left[\frac{\gamma^2}{\pi^2} + 18u^2F_2\right], \quad (51)$$

$$\frac{d\gamma}{dl} = \gamma\left\{2(C_3 - C_1) - \left[\frac{3\beta}{\pi^2} + 6uF_2 + 8\gamma F_1\,(1 + 2\alpha)\right]\right\}. \quad (52)$$

Here, we are particularly interested in the behavior of the competitive coupling constant γ. The initial values of two quartic coefficients, β and u, are taken to be positive to ensure that the system is stable. Numerical computations show that $F_{1,2,3}$ are all positive and driven to vanish as $l \to \infty$ due to the extreme velocity anisotropy, namely $v_\Delta/v_F \to 0$, at the nematic QCP. Moreover, since $C_3 - C_1 \to 0$ as $l \to \infty$, the right-hand side of equation (52) is actually negative in the low-energy region. As a result, the parameter γ is expected to be strongly irrelevant at low energy. It thus turns out that the superconducting and nematic orders might be decoupled, which is directly owing to the presence of gapless nodal qps.

To examine the effects of gapless nodal qps, here we present a set of RG equations

$$\frac{d\alpha}{dl} = 2\alpha - \frac{1}{8\pi^2}[3\beta(1 + 2\alpha) + \gamma], \quad (53)$$

$$\frac{d\beta}{dl} = \beta - \frac{1}{4\pi^2}(9\beta^2 + \gamma^2), \quad (54)$$

$$\frac{du}{dl} = u - \frac{1}{4\pi^2}(9u^2 + \gamma^2), \quad 55$$

$$\frac{d\gamma}{dl} = \gamma\left[1 - \frac{1}{4\pi^2}(3\beta + 3u + 4\gamma)\right], \quad (56)$$

which are obtained without considering the fermionic degrees of freedom [14]. It is easy to see that the right-hand side of equation (56) is not always negative. These equations have been analyzed in [14]. It was found that the fixed point structure depends on the initial values of $\beta$and u. For certain values of β and u, the system has a stable fixed point where the coupling parameter γ approaches a finite constant γ^*. In such a case, the superconducting and nematic orders experience a strong competitive interaction and tend to suppress each other. For other values of β and u, the system has no stable fixed point, which implies the instability of the system and the appearance of the first order transition. In any case, the properties of the system, without including the nodal qps, are significantly different from those obtained after including the nodal qps.

The above analysis strongly suggests that gapless nodal qps can have a significant influence on the fixed points of this interacting system. In particular, including the dynamics of nodal qps may fundamentally change the nature of the competitive interaction between the superconducting and nematic orders.

Numerical Results and Physical Implications

In order to confirm our qualitative analysis, we now solve the RG equations numerically. As already pointed out, the fermion velocities $v_{F,\Delta}$ are no longer constants at the nematic QCP: they become scale-dependent and their ratio v_Δ/v_F vanishes at the lowest energy. In this case, the fixed point should be obtained by self-consistently solving all the RG equations presented in equations (42)–(48). However, since the equations of $v_{F,\Delta}$ are relatively independent of all the others, they can be solved first. It is easy to get $v^*_F = 0, v^*_\Delta = 0$, in the limit $l \to \infty$. This immediately implies that $F_1 = F_2 = F_3 = 0$. Now the rest of the set of equations become much simpler, given by

$$\frac{d\alpha}{dl} = 2\alpha - \frac{3\beta}{4\pi^2}(1+2\alpha) = 0, \tag{57}$$

$$\frac{d\beta}{dl} = \beta - \frac{9\beta^2}{\pi^2} = 0, \tag{58}$$

$$\frac{du}{dl} = -u - \frac{\gamma^2}{\pi^2} = 0, \tag{59}$$

$$\frac{d\gamma}{dl} = -\gamma\frac{3\beta}{\pi^2} = 0. \tag{60}$$

These coupled equations have two solutions:

$$(1) \quad r^* = u^* = \gamma^* = 0, \quad \beta^* = \alpha^* = 0; \tag{61}$$

$$(2) \quad r^* = u^* = \gamma^* = 0, \quad \beta^* = \frac{\pi^2}{9}, \quad \alpha^* = \frac{1}{22}. \tag{62}$$

One can check that the second solution corresponds to a stable fixed point and that the first one is unstable. It is important to notice that $\gamma^* = 0$ is at the stable fixed point, which means the competitive interaction between the superconducting and nematic order parameters is actually irrelevant. These results indicate that these two long-range orders are decoupled from each other at the nematic QCP. Hence, we can draw two conclusions. Firstly, due to this decoupling, the superconducting and nematic long-range orders can coexist homogeneously. Secondly, both the superconducting and nematic phase transitions remain continuous, which is apparently different from the results of the first order transitions obtained without considering the fermionic degrees of freedom.

It is now interesting to further discuss the role played by gapless nodal qps. Currently, the quantum critical phenomena are usually investigated within HMM theory, which supposes that all the fermionic degrees of freedom can be entirely integrated out. If we use this scheme in our case, we would obtain an effective action that consists solely of two bosonic order parameters, ψ and ϕ. Gapless nodal qps only show their existence in the polarization $\Pi(q)$, which contributes a $\propto \Pi(q)\phi^2$ term to the effective Lagrangian of ϕ. The fermion velocities, $v_{F,\Delta}$, have to take bare values and cannot be renormalized, because the coupling between the nematic order and nodal qps cannot be properly accounted for once nodal qps are completely integrated out. It would not be possible to incorporate the extreme velocity anisotropy driven by a critical nematic fluctuation into the theoretical analysis. However, as demonstrated in the above calculations, such an extreme velocity anisotropy does have a significant effect on the RG trajectories. It is therefore necessary to include the dynamics of nodal qps.

In the absence of nodal qps, the propagator of the nematic order parameter behaves as $G\phi \propto 1/q^2$. This propagator is strongly singular in the small momenta limit $q \to 0$. After including gapless nodal qps, the nematic propagator becomes $G\phi \propto 1/q$, which is less singular compared with $G\phi \propto 1/q^2$ as $q \to 0$. It is clear that the coupling with nodal qps weakens the critical fluctuation of the nematic order parameter, which in turn leads to the decoupling between the nematic and superconducting orders. Apparently, gapless nodal qps do play an important role and thus should be seriously considered in the effective theory of competing orders.

SUMMARY AND DISCUSSION

In summary, we have carried out a RG analysis within an effective low-energy field theory that describes the interplay between the superconductivity and the nematic order in the context of d-wave high-T_c superconductors. Different from some previous theoretical treatments, we go beyond the HMM framework and incorporate the gapless nodal qps explicitly in our calculations. After analyzing the RG equations of a number of physical parameters, we have demonstrated that gapless nodal qps have a significant impact on the interplay between d-wave superconductivity and the nematic order. If the nodal qps are entirely neglected, the competition between the superconducting and nematic orders can result in runaway behavior, which in turn drives the first order transition [14]. However, including the dynamics of nodal qps can change this picture fundamentally and give rise to a stable fixed point with these two bosonic order parameters decoupled from each other in the vicinity of the nematic QCP. Therefore, both the superconducting and nematic phase transitions remain continuous. Moreover, the d-wave superconductivity can coexist with the nematic order homogeneously. These results indicate that it should be important to include the dynamics of gapless fermions in a theoretical description of the competing orders.

The competition between the superconductivity and the nematic order is just a simple example of the rich phenomena of ordering competition in unconventional superconductors. It would be more interesting to investigate the competition between superconductivity and antiferromagnetism, which is believed to be a fundamental issue not only in high-T_c cuprate superconductors but also in heavy fermion and iron-based superconductors. Compared with the case of the competing nematic order, the interplay between superconductivity and antiferromagnetism is more complicated and is expected to exhibit more interesting behaviors. For instance, the antiferromagnetic order parameter is a complex scalar field and carries a finite wave vector **Q**, which is often incommensurate. Furthermore, the antiferromagnetic order parameter may acquire a non-trivial dynamical exponent, $z \neq 1$, due to its coupling to gapless fermions, which would make RG calculations more involved [14]. Nevertheless, despite the technical difficulties, the general formalism presented in this paper can be applied to analyze the effects of the fermionic degrees of freedom on the interplay between superconductivity and antiferromagnetism.

Acknowledgments

We are grateful to Jian-Huang She for their very helpful communications and Jing-Rong Wang for their recent collaboration on relevant projects. GZL acknowledges support by the National Natural Science Foundation of China under grants numbers 11074234 and 11274286 and the Visitors Program of MPIPKS at Dresden.

REFERENCES

1. Kivelson S A, Bindloss I P, Fradkin E, Oganesyan V, Tranquada J M, Kapitulnik A and Howald C 2003 Rev. Mod. Phys. 75 1201
2. Vojta M 2009 Adv. Phys. 58 699
3. Fradkin E, Kivelson S A, Lawler M J, Eisenstein J P and Mackenzie A P 2010 Annu. Rev. Condens. Matter Phys. 1 153 Fradkin E 2012 Proc. Les Houches Summer School on 'Modern Theories of Correlated Electron Systems' (Les Houches, Haute Savoie, France, May 2009) (Lecture Notes in Physics vol 843) ed D C Cabra, A Honecker and P Pujol (Berlin: Springer) pp 53–106
4. Gegenwart P, Si Q and Steglich F 2008 Nature Phys. 4 186 Stockert O, Kirchner S, Steglich F and Si Q 2012 J. Phys. Soc. Japan 81 011001
5. Knebel G, Aoki D and Flouquet J 2009 arXiv:0911.5223
6. Arovas D P, Berlinsky A J, Kallin C and Zhang S-C 1997 Phys. Rev. Lett. 79 2871
7. Demler E, Sachdev S and Zhang Y 2001 Phys. Rev. Lett. 87 067202
8. Kivelson S A, Lee D-H, Fradkin E and Oganesyan V 2002 Phys. Rev. B 66 144516
9. Vorontsov A B, Vavilov M G and Chubukov A V 2009 Phys. Rev. B 79 060508
10. Nussinov Z, Vekhter I and Balatsky A V 2009 Phys. Rev. B 79 165122
11. Millis A J 2010 Phys. Rev. B 81 035117
12. Fernandes R M and Schmalian J 2010 Phys. Rev. B 82 014521
13. Moon E G and Sachdev S 2010 Phys. Rev. B 82 104516
14. She J-H, Zaanen J, Bishop A R and Balatsky A V 2010 Phys. Rev. B 82 165128
15. Chowdhury D, Berg E and Sachdev S 2011 Phys. Rev. B 84 205113
16. Lake B et al 2001 Science 291 1759
17. Lake B et al 2002 Nature 415 299
18. Hoffman J E, Hudson E W, Lang K M, Madhavan V, Eisaki H, Uchida S and Davis J C 2002 Science 295 466
19. Song C-L et al 2011 Science 332 1410
20. Hertz J 1976 Phys. Rev. B 14 1165
21. Millis A J 1993 Phys. Rev. B 48 7183
22. Moriya T 1995 Spin Fluctuations in Itinerant Electron Magnetism (Berlin: Springer)
23. Belitz D, Kirkpatrick T R and Vojta T 1997 Phys. Rev. B 55 9452
24. Chubukov A V, Pepin C and Rech J 2004 ′ Phys. Rev. Lett. 92 147003 Rech J, Pepin C and Chubukov A V 2006 ′ Phys. Rev. B 74 195126
25. Abanov A and Chubukov A V 2004 Phys. Rev. Lett. 93 255702
26. Strack P, Takei S and Metzner W 2010 Phys. Rev. B 81 125103 Thier S C and Metzner W 2011 Phys. Rev. B 84 155133
27. Liu G-Z, Wang J-R and Wang J 2012 Phys. Rev. B 85 174525
28. Ando Y, Segawa K, Komiya S and Lavrov A N 2002 Phys. Rev. Lett. 88 137005
29. Hinkov V, Haug D, Fauque B, Bourges P, Sidis Y, Ivanov A, Bernhard C, Lin C T and Keimer B 2008 Science 319 597
30. Daou R et al 2010 Nature 463 519
31. Lawler M J et al 2010 Nature 466 347
32. Kivelson S A, Fradkin E and Emery V J 1998 Nature 393 550
33. Halboth C J and Metzner W 2000 Phys. Rev. Lett. 85 5162 Metzner W, Rohe D and Andergassen S 2003 Phys. Rev. Lett. 91 066402 Dell'Anna L and Metzner W 2006 Phys. Rev. B 73 045127
34. Oganesyan V, Kivelson S A and Fradkin E 2001 Phys. Rev. B 64 195109
35. Vojta M, Zhang Y and Sachdev S 2000 Phys. Rev. B 62 6721 Vojta M, Zhang Y and Sachdev S 2000 Phys. Rev. Lett. 85 4940
36. Kim E-A, Lawler M J, Oreto P, Sachdev S, Fradkin E and Kivelson S A 2008 Phys. Rev. B 77 184514
37. Huh Y and Sachdev S 2008 Phys. Rev. B 78 064512
38. Sachdev S 2011 Quantum Phase Transitions (Cambridge: Cambridge University Press) chapter 17
39. Xu C, Qi Y and Sachdev S 2008 Phys. Rev. B 78 134507
40. Fritz L and Sachdev S 2009 Phys. Rev. B 80 144503
41. Kim E-A and Lawler M J 2010 Phys. Rev. B 81 132501
42. Wang J, Liu G-Z and Kleinert H 2011 Phys. Rev. B 83 214503
43. Shankar R 1994 Rev. Mod. Phys. 66 129
44. Orenstein J and Millis A J 2000 Science 288 468
45. Chiao M, Hill R W, Lupien Ch, Taillefer L, Lambert P, Gagnon R and Fournier P 2000 Phys. Rev. B 62 3554
46. Lee P A 1993 Phys. Rev. Lett. 71 1887
47. Durst A and Lee P A 2000 Phys. Rev. B 62 1270
48. Lee P A and Wen X-G 1997 Phys. Rev. Lett. 78 4111 Paramekanti A and Randeria M 2002 Phys. Rev. B 66 214517

CHAPTER 14

INTERPLAY AND COMPETITION BETWEEN SUPERCONDUCTIVITY AND CHARGE ORDERINGS

Konrad Kapcia

Keywords: Extended Hubbard model, Phase separation, Superconductivity, Charge orderings, Local pairing, Phase diagrams

INTRODUCTION

The interplay and competition between superconductivity and charge orderings is currently under intense investigations (among others in high temperature superconductors such as cuprates, barium bismuthates, fullerenes, and several other nonconventional superconducting materials, e.g., the Chevrel phases) [1]. They belong to a unique group of extreme type II superconductors and generally exhibit low carrier density, a small value of Fermi energy ($E_F \leq 0.1 \div 0.3$ eV), and a short coherence length ξ_0 ($\xi_0 k_F \approx 1 \div 10$). These general features are consistent with short-range, almost unretarded effective interactions response for local pairing.

In this report, we will concentrate on the intriguing problem of phase separation and the competition between superconductivity and charge orderings. The model Hamiltonian considered has the following form:

$$\hat{H} = U\sum_i \hat{n}_{i\uparrow}\hat{n}_{i\downarrow} - \mu\sum_i \hat{n}_i - 2I\sum_{\langle i,j\rangle} \hat{\rho}_i^+\hat{\rho}_j^- + \frac{W}{2}\sum_{\langle i,j\rangle} \hat{n}_i\hat{n}_j, \quad (1)$$

where $\hat{n}_i = \sum_\sigma \hat{n}_{i\sigma}$, $\hat{n}_{i\sigma} = \hat{c}_{i\sigma}^+\hat{c}_{i\sigma}$, $\hat{\rho}_i^+ = (\hat{\rho}_i^-)^\dagger = \hat{c}_{i\uparrow}^+\hat{c}_{i\downarrow}^+$. $\hat{c}_{i\sigma}^+$ $(\hat{c}_{i\sigma})$ denotes the creation (annihilation) operator of an electron with spin σ=↑,↓ at the site i, which satisfy canonical anticommutation relations:

$$\left\{\hat{c}_{i\sigma}, \hat{c}_{j\sigma'}^+\right\} = \delta_{ij}\delta_{\sigma\sigma'}, \{\hat{c}_{i\sigma}, \hat{c}_{j\sigma'}\} = \left\{\hat{c}_{i\sigma}^+, \hat{c}_{j\sigma'}^+\right\} = 0, \quad (2)$$

where δ_{ij} is the Kronecker delta. $\sum_{\langle i,j\rangle}$ indicates the sum over nearest-neighbor sites i and jindependently. z will denote the number of nearest-neighbors. U, I, and W are the interactions parameters, I_0=zI, W_0=zW. μ is the chemical potential, connected with the concentration of electrons by the formula: $n = (1/N)\sum_i \langle\hat{n}_i\rangle$, with 0≤n≤2 and N is the total number of lattice sites.

The interactions U, I, and W will be treated as the effective ones and will be assumed to include all the possible contributions and renormalizations like those coming from the strong electron-phonon coupling or from the coupling between electrons and other electronic subsystems in solid or chemical complexes [1]. In such a general case, arbitrary values and signs of U, I, and W are important to consider. In the absence of the external field conjugated with the SS order parameter $(\Delta = (1/N)\sum_i \langle\hat{\rho}_i^-\rangle)$, there is a symmetry between I>0 (s-pairing) and I<0 (η-pairing, $\Delta_Q = (1/N)\sum_i \exp(i\mathbf{Q}\cdot\mathbf{R}_i)\langle\hat{\rho}_i^-\rangle$, $\mathbf{Q}$ is half of the smallest reciprocal lattice vector) cases for model (1), thus we restrict ourselves only to the I>0 case. The CO parameter is defined as $n_Q = (1/N)\sum_i \exp(i\mathbf{Q}\cdot\mathbf{R}_i)\langle\hat{n}_i\rangle$.

We have performed extensive study of the phase diagrams (PDs) of model (1) [2]. In this report, we investigate the properties of model (1) for the case of on-site attraction (U<0, local pair or Bose condensation limit [3–8]) and intersite repulsion (W>0). In the analysis of the model atT≥0, we have adopted a variational approach (VA), which treats the on-site interaction Uexactly and the intersite interactions W and I within the mean-field approximation (MFA). The PDs of model (1) have been investigated until now for the special cases: W=0 [3–12] and I=0 [13–21] only. Some ground state results for W≠0 and I≠0 have been also obtained [5, 22, 23].

Citation: Konrad Kapcia, Interplay and Competition Between Superconductivity and Charge Orderings, DOI 10.1007/s10948-013-2152-1.

Within the VA the intersite interactions are decoupled within the MFA, which allows us to calculate the averages n, n_Q, Δ, and Δ_Q. It gives a set of four self-consistent equations (for homogeneous phases). The definitions of homogeneous phases with the values of order parameters are as follows: (i) SS—$n_Q=0$, $\Delta\neq0$, $\Delta_Q=0$; (ii) CO—$n_Q\neq0$, $\Delta=0$, $\Delta_Q=0$; (iii) M—$n_Q\neq0$, $\Delta\neq0$, $\Delta_Q\neq0$; (iv) NO—$n_Q=0$, $\Delta=0$, $\Delta_Q=0$. It is important to find a solution corresponding to the lowest energy.

Phase separation (PS) is a state in which two domains with different electron concentration: n_+ and n_- exist in the system (coexistence of two homogeneous phases). The free energies of the PS states are calculated in a standard way, using Maxwell's construction (e.g., [5, 18, 19, 24]). In model (1) for the range of parameters considered in this paper, only one PS state can occur, which is coexistence of the SS and CO phases.

In the report, we have used the following convention. A second- (first-)order transition is a transition between homogeneous phases with a (dis-)continuous change of the order parameter at the transition temperature. A transition between homogeneous phase and PS state is symbolically named as a "third-order" transition [5, 6, 17–21].

RESULTS IN THE LOCAL PAIRING LIMIT

Phase Diagrams

One should noticed that PDs obtained are symmetric with respect to half-filling (n=1) because of the particle–hole symmetry of Hamiltonian (1), so the PDs will be presented only in the range $\bar{\mu} = \mu - U/2 - W_0 \leq 0$ and $0 \leq n \leq 1$.

For any U≤0 and fixed W>0, the PDs are qualitatively similar, all (first-order, second-order and "third- order") transition temperatures decrease monotonically with increasing U and in the VA for U=0 the transition temperatures account for a half of those in the limit U→−∞, what can be symbolically written as

$$k_B T_c(U \to -\infty) = 2k_B T_c(U = 0), \qquad (3)$$

where T_c denotes the transition temperature (which can be SS–NO, SS–CO, CO–NO, PS–CO, or PS–SS).

Notice that in the U→−∞ limit model (1) is equivalent with the hard-core boson model on a lattice [25–29]. Moreover, in that limit model (1) can be derived as effective Hamiltonian at the strong-coupling limit of the extended Hubbard model by the degenerate perturbation theory [22, 23, 30].

In the range of the attractive on-site interaction, the structure of PDs of model (1) depends on the ratio W/I only (cf. Fig. 1 for $U/I_0=-1$). One can distinguish two ranges of the ratio W/I in which the system exhibits substantially different behaviors:

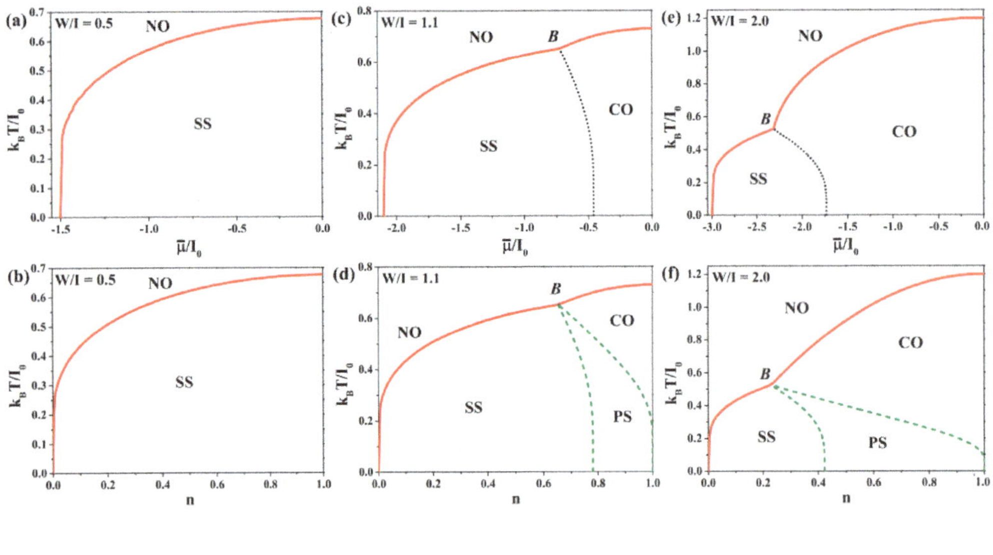

Figure 1: k_BT/I_0 vs. $\bar{\mu}/I_0$ phase diagrams (upper row) and corresponding k_BT/I_0 vs. n diagrams (lower row) for $U/I_0=-1.0$ and different values of W/I=0.5,1.1,2.0 (as labeled). Dotted, solid, and dashed lines indicate first-order, second-order, and "third-order" boundaries, respectively. B denotes bicritical points.

(i) 0≤W/I<1. In Figs. 1a,b, we present particular PDs for W/I=0.5. For 0<W/I<1 and U<0, only the second-order SS–NO transitions occur with increasing temperature. If we analyze the system for fixed n, for this range of model parameters, the PS states do not occur and the obtained PDs have the same structure as those derived in [3–5]. The transition between homogeneous SS and NO phases taking place with increasing temperature is second order for arbitrary $\bar{\mu}$ and n, and it decreases monotonically with increasing $|\bar{\mu}|/I_0|\bar{\mu}|/I_0$ and |1−n|.

(ii) 1<W/I. A few particular PDs in this regime are presented in Figs. 1c–f. For W/I=1 the SS, the CO and the M phases are degenerate at n=1. For W/I>1, three homogeneous phases (SS, CO, NO) appear. The SS–NO and CO–NO transitions are of the second order and these transition temperatures are decreasing function of $|\bar{\mu}||\bar{\mu}|$ and |1−n|. The SS–CO transition is discontinuous for fixed $\bar{\mu}$, and thus the PS state SS–CO is stable in the definite range of n. All transitions lines meet at a bicritical point B. With increasing W, the B-point moves along the boundary between SS and NO phases toward larger $|\bar{\mu}||\bar{\mu}|$ ($|1-n|$). This is due to the fact that in the VA the SS–NO transition is independent of W (for fixed n). The region of the CO phase occurrence is extended, whereas the region of the SS phase stability is reduced by increasing the ratio W/I. The first-order SS–CO as well as the "third-order" SS–PS and PS–CO transition temperatures increase with $|\bar{\mu}||\bar{\mu}|$ and $|1-n|$, respectively.

One should notice that for W_{ij} restricted to nearest neighbors ($W_2=0$) the PS state is strictly degenerated at T=0 with the M phase in the whole range of stability of both these states [22,23]. This degeneration is removed at T>0, even for $W_2=0$ and the PS state occurs on PDs. Repulsive $W_2>0$ between next-nearest neighbors destabilizes the PS state with respect to the M phase, whereas attractive $W_2<0$ extends the stability region of PS state and eliminates the M phase.

In the case of attractive W<0 (precisely for W/I<−1), the model can exhibit phase separation NO–NO (electron droplets state) at low temperatures [20, 22, 23, 31] and for U<0 the PDs as a function of n have the similar structure as those derived in [31]. We leave deeper analysis of this problem to future publications.

Order Parameters and Thermodynamic Properties

Let us focus now on the temperature dependencies of the order parameters and thermodynamic properties of the system at the sequence of transitions: SS → CO → NO for W/I=2.0, $U/I_0=-1.0$ and $\bar{\mu}/I_0=-2.0\,\bar{\mu}/I_0=-2.0$.

The temperature dependencies of the order parameters: Δ and n_Q are presented in Fig. 2. It is clearly seen that at the SS–CO transition (at $k_BT_{c1}/I_0=0.40$) the both order parameters change discontinuously. The CO–NO transition at $k_BT_{c2}/I_0=0.83$ is of the second order.

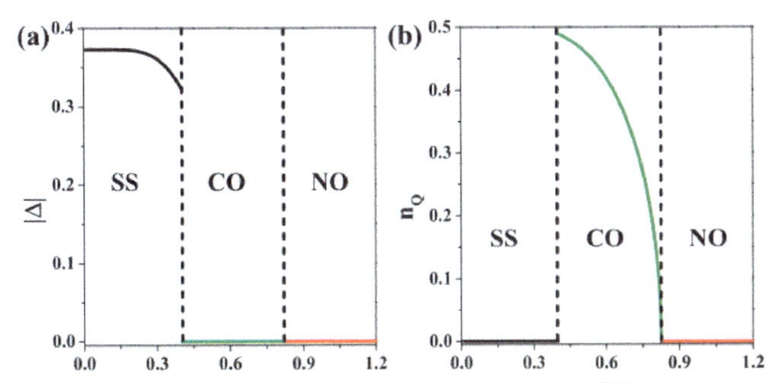

Figure 2: Temperature dependencies of (**a**) superconducting order parameter $|\Delta|$ and (**b**) charge order parameter n_Q for W/I=2.0, $U/I_0=-1.0$ and $\bar{\mu}$ /I0=$-2.0\,\bar{\mu}/I_0=-2.0$.

Calculating the grand potential per site $\omega=-1/(N\beta)\times\ln(\mathrm{Tr}[\exp(-\beta\hat{H})])$ within the VA one can obtain thermodynamic characteristics of the system for arbitrary temperature. The entropys and the specific heat c can be derived as $s=-(\partial\omega/\partial T)$ and $c=-T(\partial^2\omega/\partial T^2)$. ω, s, and c as a function of temperature are shown in Fig. 3. s increases monotonically with increasing T. It is discontinuous at T_{c1} whereas it is continuous at T_{c2}. One can notice that in the high-temperature limit $s/k_B\rightarrow\ln(4)\approx1.386$ (there are four possible configurations at each site). The peak in c(T) is associated with the first-order transition (at T_{c1}), while the λ-point behavior is typical for the second-order transition (at T_{c2}).

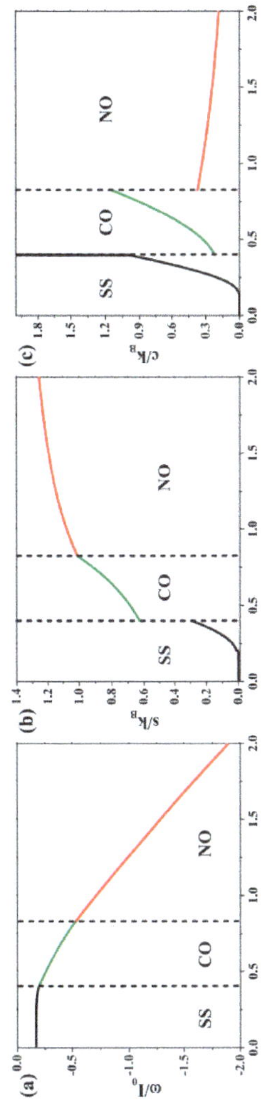

Figure 3: Temperature dependencies of thermodynamics parameters: (**a**) the grand potential ω, (**b**) the entropy s, and (**c**) the specific heat c for W/I=2.0, $U/I_0=-1.0$ and $\bar{\mu}/I_0=-2.0\,\bar{\mu}/I_0=-2.0$.

CONCLUSIONS AND FINAL REMARKS

In this report, we have presented some particular PDs of the extended Hubbard model with pair hopping and intersite density-density interactions in the zero-bandwidth limit for the case of local attraction U<0. One finds that the system considered can exhibit very interesting multicritical behaviors. Our investigations show that, depending on the values of interaction parameters (the ratio W/I), the system can exhibit three homogeneous phases: superconducting, charge-ordered and nonordered. The SS–NO and CO–NO transitions are of the second order. The SS–CO transition is discontinuous (for fixed μ), what leads to phase separation on the phase diagrams as a function of n for W/I>1. The homogeneous mixed phase (with nonzero both charge-ordered and superconducting order parameters) never occurs on PDs at T>0, at least in the absence of the next-nearest neighbors interactions. On the contrary, the PS state: SS–CO is found to be stable in definite ranges of model parameters and temperatures.

Our results are exact in the limit of infinite dimensions, where the MFA treatment of intersite terms is rigorous one [3–6, 11]. In finite dimensions due to quantum fluctuations connected with the I term, the regions of the ordered homogeneous phases occurrence are extended in comparison with the VA results [5, 6].

Acknowledgements

The author is indebted to Professor Stanisław Robaszkiewicz for very fruitful discussions during this work and careful reading of the manuscript. The work has been financed by National Science Center (NCN) as a research project in the years 2011–2013, under Grant No. DEC-2011/01/N/ST3/00413. We thank the European Commission and the Ministry of Science and Higher Education (Poland) for the partial financial support from the European Social Fund—Operational Programme "Human Capital"—POKL.04.01.01-00-133/09-00—"Proinnowacyjne kształcenie, kompetentna kadra, absolwenci przyszłości" as well as the Foundation of Adam Mickiewicz University in Poznań for the support from its scholarship programme.

REFERENCES

1. Micnas, R., Ranninger, J., Robaszkiewicz, S.: Rev. Mod. Phys. 62, 113 (1990)
2. Kapcia, K., Robaszkiewicz, S.: in preparation
3. Robaszkiewicz, S., Pawłowski, G.: Physica C 210, 61 (1993)
4. Robaszkiewicz, S.: Acta Phys. Pol. A 85, 117 (1994)
5. Kapcia, K., Robaszkiewicz, S., Micnas, R.: J. Phys. Condens. Matter
6. 24, 215601 (2012)
7. Kapcia, K., Robaszkiewicz, S.: J. Phys.: Condens. Matter 25, 065603 (2013)
8. Bułka, B., Robaszkiewicz, S.: Phys. Rev. B 54, 13138 (1996)
9. Robaszkiewicz, S., Bułka, B.: Phys. Rev. B 59, 6430 (1999)
10. Kapcia, K.: Acta Phys. Pol. A 121, 733 (2012)
11. Bari, R.A.: Phys. Rev. B 7, 2128 (1973)
12. Ho, W.-C., Barry, J.H.: Phys. Rev. B 16, 3172 (1977)
13. Wiecko, C., Allub, R.: Phys. Rev. B 35, 2041 (1987)
14. Robaszkiewicz, S.: Phys. Status Solidi (b) 59, K63 (1973)
15. Micnas, R., Robaszkiewicz, S., Chao, K.A.: Phys. Rev. B 29, 2784 (1984)
16. Mancini, F., Mancini, F.P.: Phys. Rev. E 77, 061120 (2008)
17. Mancini, F., Mancini, F.P.: Eur. Phys. J. B 73, 581 (2010)
18. Kapcia, K., Kłobus, W., Robaszkiewicz, S.: Acta Phys. Pol. A 118, 350 (2010)
19. Kapcia, K., Robaszkiewicz, S.: J. Phys. Condens. Matter 23, 105601 (2011)
20. Kapcia, K., Robaszkiewicz, S.: J. Phys. Condens. Matter 23, 249802 (2011)
21. Kapcia, K., Robaszkiewicz, S.: Acta Phys. Pol. A 121, 1029 (2012)
22. Kapcia, K., Kłobus, W., Robaszkiewicz, S.: Acta Phys. Pol. A 121, 1032 (2012)
23. Robaszkiewicz, S., Pawłowski, G.: Acta Phys. Pol. A 90, 569 (1996)
24. Pawłowski, G., Robaszkiewicz, S.: Acta Phys. Pol. A 91, 409 (1997)
25. B ̨ak, M.: Acta Phys. Pol. A 106, 637 (2004)
26. Gerisch, T., Rieckers, A.: J. Stat. Phys. 91, 759 (1998)
27. Gerisch, T., Münzner, R., Rieckers, A.: J. Stat. Phys. 91, 1021 (1998)
28. Micnas, R., Robaszkiewicz, S.: Phys. Rev. B 45, 9900 (1992)
29. Micnas, R., Robaszkiewicz, S., Kostyrko, T.: Phys. Rev. B 52, 6863 (1995)
30. Schmid, G., Todo, S., Troyer, M., Dorneich, A.: Phys. Rev. Lett.
31. 88, 167208 (2002)
32. Robaszkiewicz, S., Micnas, R., Chao, K.A.: Phys. Rev. B 23, 1447 (1981)
33. Bursill, R.J., Thompson, C.J.: J. Phys. A, Math. Gen. 26, 4497 (1993)

CHAPTER 15

SUPERCONDUCTIVITY IN FE AND AS BASED COMPOUNDS

A. Bussmann-Holder, A. Simon1, H. Keller, A. R. Bishop

Keywords: Multiband superconductivity, FeAs based compounds

The discovery of high temperature superconductivity in Fe-As type compounds [1,2] has produced a surge in solid state research since their maximum T_c value well exceeds that in MgB_2 and places them in the vicinity of cuprate superconductors. Their proximity to a magnetic phase has invoked the possibility that strong correlations dominate the superconductivity [3], and that electron-lattice interactions are too weak to account for the pairing mechanism and the high values of T_c [4]. In addition, first principles calculations of the electron-phonon coupling constant [4] together with inelastic neutron scattering studies [5, 6] showed that the coupling is small and cannot directly cause high temperature superconductivity. On the other hand, local structural probes such as extended x-ray absorption fine structure (EXAFS) measurements reveal an anomalous temperature dependence of local Fe-As lattice displacements [7] which closely resembles the ones observed in cuprate superconductors [8]. This observation suggests that local rather than global electron-lattice interaction effects are involved and that polaron formation is responsible for the anomalies observed in EXAFS data [7].

In addition to the local lattice instabilities, the complex Fermi surface topology points to multiband superconductivity [9, 10] analogous to MgB_2. Tunneling experiments [11], infrared spectroscopy [12], specific heat [13], lower critical field measurements [14], point contact spectroscopy [15], as well as penetration depth results [16 – 21] and magnetic torque studies [22] show considerable deviations from single gap behavior and have been interpreted in terms of multiple isotropic gaps. In the following we use a multiband model with polaronic couplings [23] to quantitatively demonstrate that Fe-As type superconductors are two-gap s-wave superconductors with strong local electron lattice coupling. While the former property places these systems in the company of MgB_2, the latter links them more to cuprates where polaron formation onsets at high temperatures [24]. However, opposite to cuprates, polaron coherence does not take place, i.e., a stripe like phase with preformed pairs is absent, consequently the T_c's of these compounds will probably not reach the values of cuprates. This seems to be confirmed experimentally since increasing the chemical pressure does not lead to further enhancements of T_c beyond a certain critical pressure but rather a reduction [25]. We start from an effective BCS-type Hamiltonian extended to account for two-gap superconductivity with interband coupling [26]:

$$\begin{aligned}
H &= H_0 + H_1 + H_2 + H_{12} \\
H_0 &= \sum_{k_1\sigma} \xi_{k_1} c^+_{k_1\sigma} c_{k_1\sigma} + \sum_{k_2\sigma} \xi_{k_2} d^+_{k_2\sigma} d_{k_2\sigma} \\
H_1 &= \sum_{k_1 k_1' q} V_1(k_1,k_1') c^+_{k_1+q/2\uparrow} c^+_{-k_1+q/2\downarrow} c_{-k_1'+q/2\downarrow} c_{k_1'+q/2\uparrow} \\
H_2 &= \sum_{k_2 k_2' q} V_2(k_2,k_2') d^+_{k_2+q/2\uparrow} d^+_{-k_2+q/2\downarrow} d_{-k_2'+q/2\downarrow} d_{k_2'+q/2\uparrow} \\
H_{12} &= \sum_{k_1 k_2 q} V_{12}(k_1,k_2) \{ c^+_{k_1+q/2\uparrow} c^+_{-k_1+q/2\downarrow} d_{-k_2+q/2\downarrow} d_{k_2+q/2\uparrow} + h.c. \}.
\end{aligned} \tag{1}$$

Here, c^+ , c, d^+ , d are (electron / hole) creation and annihilation operators of the As p-Fe d hybridized bands in the FeAs sheets with momentum k dependent energies ξ. The intraband pairing potentials V_i are governed by electron lattice interactions, which have a sizable strength in one channel only, whereas they are too small in the other channel to support superconductivity alone. The interband interaction V_{12} is explicitly assumed to be multiphonon mediated, stemming from strong anharmonicity, as observed experimentally [7, 27]. Including local polaronic couplings means that we modify H_0 as:

$$\overline{H}_0 = \sum_{k_i\sigma} \xi_{k_i} c^+_{k_i\sigma} c_{k_i\sigma} + \sum_q \hbar\omega_q b^+_q b_q + \frac{1}{\sqrt{N}} \sum_{q,\sigma,k_1} \gamma_i(q) c^+_{k_i+q\sigma} c_{k_i\sigma} (b_q + b^+_{-q}); i = 1,2, \tag{2}$$

Citation: A. Bussmann-Holder, A. Simon, H. Keller, A. R. Bishop, Superconductivity in Fe and As Based Compounds, DOI: 10.1007/s10948-009-0586-2.

where hω is the momentum q dependent phonon energy, b, b^+ are phonon annihilation and creation operators, and γ is the local electron-lattice coupling. By using the canonical Lang-Firsov transformation [28], electron and lattice degrees of freedom can be decoupled, whereby a rigid harmonic oscillator shift of the phonon operators takes place while electronic energies are exponentially reduced to $\tilde\xi_k = \xi_k \exp[-\gamma^2 \coth(\hbar\omega/2kT)]$. The superfluid stiffness is calculated within linear response theory through the relation between the current and the induced transverse gauge field [29]:

$$\rho_s^\alpha = \frac{1}{2N}\sum_k \left\{ \left(\frac{\partial \tilde\xi_{k_i}}{\partial k_\alpha}\right)^2 \frac{\partial f(E_{k_i})}{\partial E_{k_i}} + \frac{1}{2}\frac{\partial^2 \tilde\xi_{k_i}}{\partial k_i^2}\left[1-\frac{\tilde\xi_{k_i}}{E_{k_i}}\tanh\frac{E_{k_i}}{2kT}\right]\right\}. \tag{3}$$

Here $E_k = \sqrt{\tilde\xi^2 + \bar\Delta_i^2}$, where the values of $\bar\Delta_i$ are obtained from Eq. 1 by applying standard techniques:

$$< c_{k_1\uparrow}^+ c_{-k_1\downarrow}^+ >= \frac{\bar\Delta_{k_1}}{2E_{k_1}}\tanh\frac{\beta E_{k_1}}{2} = \bar\Delta_{k_1}\Phi_{k_1}$$

$$< d_{k_2\uparrow}^+ d_{-k_2\downarrow}^+ >= \frac{\bar\Delta_{k_2}}{2E_{k_2}}\tanh\frac{\beta E_{k_2}}{2} = \bar\Delta_{k_2}\Phi_{k_2}$$

$$\bar\Delta_{k_2} = \sum_{k_2'} V_2(k_2,k_2')\bar\Delta_{k_2'}\Phi_{k_2'} + \sum_{k_1} V_{1,2}(k_1,k_2)\bar\Delta_{k_1}\Phi_{k_1}$$
$$\bar\Delta_{k_1} = \sum V_2(k_2,k_2')\bar\Delta_{k_2'}\Phi_{k_2'} + \sum V_{2,1}(k_2,k_1)\bar\Delta_{k_1}\Phi_{k_1}. \tag{4}$$

The above coupled gap equations have to be solved simultaneously and self-consistently in order to derive their temperature dependence and T_c. Guided by experimental data [16 – 19], both gaps are taken to be isotropic s-wave. In Figs. 1 we compare the theoretically derived results for the normalized in-plane penetration depth $\lambda_{ab}^{-2}(T)/\lambda_{ab}^{-2}(0) = \sigma_{sc}(T)$ to experimental data for two Fe type superconductors, namely, $Ba_{0.6}K_{0.4}Fe_2As_2$ [16], $Ba_{1-x}K_xFe_2As_2$ [19]. (Note, that x is undefined in the latter compound). The good agreement between experiment and theory supports the present approach. Also shown in the figures are the individual contributions $\omega\rho_{sc}^s$ and $(1-\omega)\rho_{sc}^l$ to the total superfluid densities: $\rho_{sc}(T) = \lambda_{ab}^{-2}(T)/\lambda_{ab}^{-2}(0). = \omega\rho_{sc}^s + (1-\omega)\rho_{sc}^l$, where ρ_{sc}^s and ρ_{sc}^l are the individual normalized s-wave superconducting densities of the small, large gap, respectively, and ω is the weighting factor. The ratios ω /(1−ω) from the small ρ_{sc}^s and the large density ρ_{sc}^l contributions vary considerably with the transition temperature: While for the smaller T_c system, $Ba_{1-x}K_xFe_2As_2$, ρ_{sc}^l dominates the temperature dependence of the penetration depth, this is reversed for the higher T_c compound, $Ba_{1-x}K_xFe_2As_2$, where the major contribution stems from ρ_{sc}^s. In both cases, however, an inflection point at low temperatures is present which demonstrates that the two contributing gaps must have substantially different gap values, i.e., a small one coexists with a large one where their difference in size is a factor of three or more. This is in full agreement with experimental findings [16 – 21]. In Fig. 1c we show the variations of the temperature dependence of the normalized penetration depth versus $t=T/T_c$ for various ratios of the two superfluid density contributions. Obviously, the temperature dependence varies strongly with this ratio and may thus explain the non-exponential behavior at low temperatures of the superfluid density observed experimentally [20].

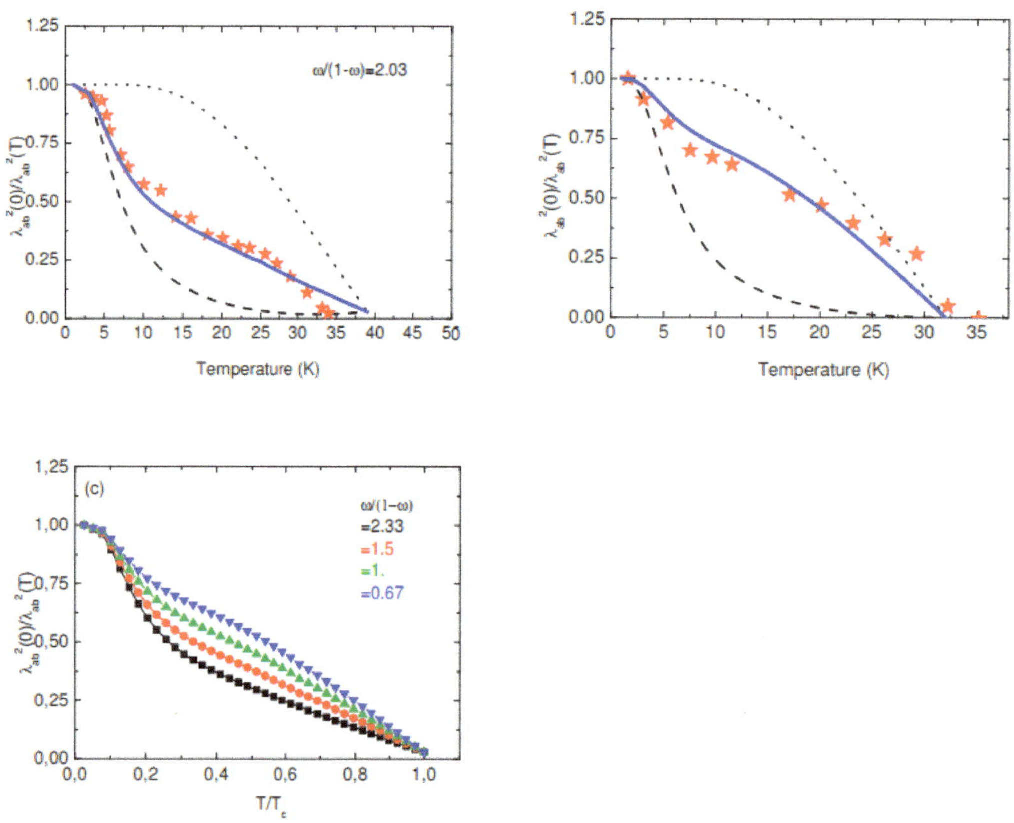

Figure 1: (a, b) Comparison of experimental (full stars) and theoretical (full lines) temperature dependencies of the normalized in-plane penetration depths together with the individual contributions from the two components ρ_{sc}^s (dashed lines) and ρ_{sc}^l (dotted lines). In (a) the

experimental data have been taken from Ref. 16 for $Ba_{0.6}K_{0.4}Fe_2As_2$. In (b) the data from Ref. 19 for $Ba_{1-x}K_xFe_2As_2$ have been plotted. The model parameters for both systems are given in Table 1 together with theoretically derived ones shown in Fig. 2. (c) Calculated temperature dependencies for varying ω /(1−ω) to the total $\rho_{sc}(T) = \lambda_{ab}^{-2}(T)/\lambda_{ab}^{-2}(0). = \omega\rho_{sc}^{s} + (1-\omega)\rho_{sc}^{l}$.

From the present analysis it appears that for small T_c an almost single superfluid density dominates the penetration depth. With increasing T_c the contributions from the individual superfluid densities approach each other, to be about equally contributing at optimal values of T_c thus achieving the maximum enhancement mechanism of the two-band model. Beyond this value the trend reverses and the leading gap changes its character, analogous to Al doped MgB_2 [30], to approach again the nearly single superfluid density behavior. In order to demonstrate this evolution, we have calculated the penetration depth for values of T_c smaller than and intermediate between those displayed in Figs. 1 and near optimal doping where T_c=50K. Experimental data for this regime are currently absent.

If the above conclusions regarding the individual superfluid density contributions to the penetration depth are correct then we would expect that the small to large superfluid density ratio ω /(1−ω) crucially influences the value of T_c. This trend is shown in Fig. 2, which can be tested experimentally by systematically measuring the superfluid density as a function of doping. Note, that the maximum value of T_c is predicted to be 56K for ω /(1−ω) = 1. 55

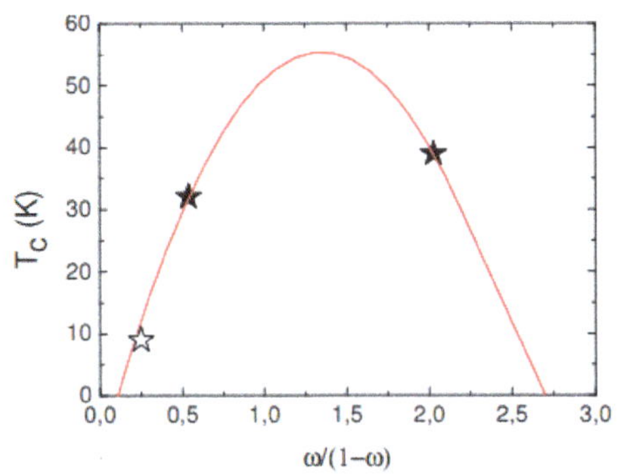

Figure 2: Calculated dependence of T_c on the ratio ω /(1−ω) . The full line is our prediction of how T_c should depend on this ratio. The full stars refer to the values used in Figs. 1a, b and as derived for $FeSe_x$ (open star) from Ref. 21. For all three systems the parameters used in the calculation are given in Table 1.

In order to justify our polaronic approach (Eq. 2), we investigate the effect of polaron formation on the lattice. As already mentioned above, the phonon energies undergo a rigid oscillator shift due to polaron formation, where this shift is proportional to the coupling constant γ and the band energies, eq 2. This provides a substantial temperature dependent renormalization of the local phonon mode frequency which in turn leads to a pronounced effect on the mean square Fe-As displacement. Since the renormalized frequency experiences pronounced softening, the related displacement inversely increases and is maximal when the softening is complete [24]. These effects have been observed in cuprate superconductors and have been related to the onset of the stripe phase [24]. Actually, in cuprates not only one divergence in the mean-square Cu-O displacement has been observed but two: one at the onset temperature T^* of the pseudogap phase, and another one either just at or slightly below T_c depending on which displacement is tested [see Refs. 8, 24 and references therein]. These results have been interpreted as signatures of polaron coherence at T^* and persistence into the superconducting phase, T

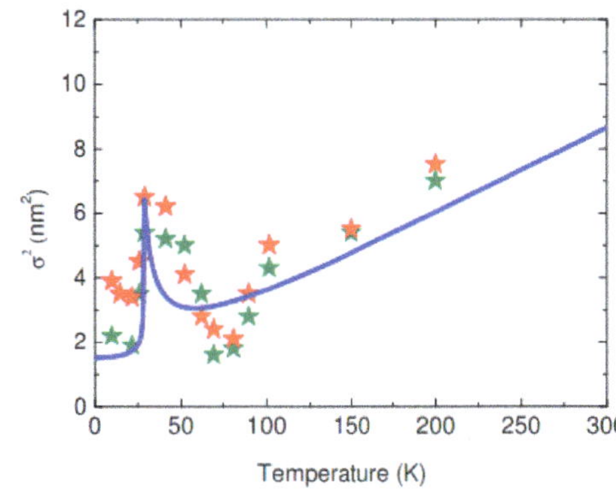

Figure 3: Temperature dependence of the mean square Fe-As displacement. Full stars correspond to experimental data from Ref. 7 for $LaFeAsO_{0.93}F_{0.07}$; the full line is theoretically derived using the procedure described in Ref. 24.

The comparison between the EXAFS data [7] and the numerical results is shown in Fig. 3. Note, that in order to avoid the use of too many parameters, no damping in the theoretical curve is included. In spite of this restriction, good agreement between theory and experiment is achieved. The polaronic coupling which has been used, is substantially smaller than that obtained at optimum doping in cuprates (γ = 75.0 − 3.0 in cuprates [24], γ = 2.0 in Fe based compounds), indicating that lattice effects in the Fe based systems are less prevalent than in cuprates. However, in order to obtain the divergence observed experimentally, polaron formation has to play a significant role also in these superconductors. This should be noticeable in an isotope effect on the penetration depth which is predicted to be small due to the rather small (of order a few %) value of γ .

Table 1: Parameters used to calculate the results shown in Figs. 1 a, b, and those of Ref. 21, not shown. The first column refers to the superconducting transition temperature Tc, the second column is the intraband interaction V multiplied by the density of states at the Fermi energy in the small gap, the third is the same but for the large gap, the fourth column is the interband interaction between both bands multiplied by the interband density of states at the Fermi energy*. The energies for the small and the large superconducting gaps (exp. values are given in parenthesis) and the corresponding ratios of the superfluid density contributions used in the calculations are displayed in columns 5 to 7.

T_c (K)	$V_l N_l(0)$	$V_s N_s(0)$	$V_{sl} N_{sl}(0)$	Δ_s (meV)*	Δ_l (meV)	$\omega/(1-\omega)$
9	0.208	0.018	0.06	0.71 0.38(1)[a]	2.42 1.6(2)[a]	0.25
32	0.297	0.018	0.06	1.71 1.5(2)[b]	8.32 9.1(2)[b]	0.54
39	0.317	0.018	0.06	1.95 2.0(3)[c]	10.09 8.9(4)[c]	2.03

[a]Ref. [33]
[b]Ref. [28]
[c]Ref. [31]
*Note that $N_s V_s(0)$ and $N_{sl} V_{sl}(0)$ are assumed to be constant for all values of T_c in order to minimize the number of parameters. In spite of this simplification, Δ_s increases systematically with increasing T_c, emphasizing the gap coupling effect through the interband channel. The difference between calculated and experimental gap values arises from the fact that the gap coupling is not included in the experimentally derived values

In conclusion, we have shown that Fe-based superconductors exhibit features common to both MgB_2 and cuprates. The link is the existence of multiple gaps, with the distinction from cuprates [31] but in analogy to MgB_2, of all being isotropic s-wave gaps. In contrast to MgB_2 local lattice effects play an important role for superconductivity, however being less effective than in cuprates. It thus seems, that pnictide superconductors lie in between these two materials and consequently have limited values of T_c which will not reach the high values observed in cuprates. It should be pointed out, however, that the multiband scenario used here and supported by many experiments [8 – 22] is questioned in Ref. 32, where a single d-wave gap has been reported. Similarly, the role of the lattice for superconductivity in Fe based compounds has been questioned in Refs. 3 – 6. Recent isotope experiments [33], on the other hand, support this approach. Since we make specific predictions, our model can be tested experimentally.

ACKNOWLEDGEMENT

It is a pleasure to acknowledge stimulating discussions with and a critical reading of the manuscript by K. A. Müller.

REFERENCES

1. Kamihara Y. et al., J. Am. Chem. Soc., 130 (2008) 3296.
2. Chen G. F. et al., Phys. Rev. Lett., 100 (2008) 247002; Zhu X. Y. et al., Supercond. Sci. Technol., 21 (2008) 105001; Chen X. H. et al., Nature (London), 453 (2008) 761; Ren Z. A. et al., Chin. Phys. Lett., 25 (2008) 2215; Johrendt D. and Pöttgen R., Angew. Chem. Int. Ed., 47 (2008) 4782; Tegel M., Bichler D., and Johrendt D., Solid State Sci., 10 (2008) 193.
3. Xu G. et al., arXiv: 0803.1282 ; Mazin I. I., Singh D. J., Johannes M. D., and Du M. H., arXiv: 0803.2740 ; Kuroki K. et al., arXiv: 0803.3325 ; Dai X., Fang Z., Zhou Y., and Zhang F., arXiv: 0803.3982 ; Han Q., Chen Y., and Wang Z. D., arXiv: 0803.4346.
4. Boeri L., Dolgov O. V., and Golubov A. A., Phys. Rev. Lett., 101 (2008) 026403.
5. Mittal R. et al., Phys. Rev. B, 78 (2008) 224518; Mittal R. et al., Phys. Rev. B, 78 (2008) 104514.
6. Higashitaniguchi S. et al., Phys. Rev. B, 78 (2008) 174507.
7. Zhang C. J. et al., Phys. Rev. B, 78 (2008) 214513.
8. Oyanagi H., Zhang C., Tsukuda A., and Naito M., J. Phys. Conf. Ser. (LEHTSC 2007), 108 (2008) 012038.
9. Cvetkovic V. et al., arXiv : 0804.4678.
10. Liu C. et al., Phys. Rev. Lett., 101 (2008) 177005.
11. Gonnelli R. S. et al., arXiv : 0902.3441 ; Boyer M. C. et al., arXiv : 0806.4400 ; Samuely P. et al., arXiv : 0902.2667.
12. Li G. et al., Phys. Rev. Lett., 101 (2008) 107004.
13. Mu Gang et al., arXiv : 0808.2941.
14. Ren Cong et al., arXiv: 0808.0805.
15. Szabó P. et al., Phys. Rev. B, 79 (2009) 012503.
16. Ren C. et al., Phys. Rev. Lett., 101 (2008) 257006.
17. Gordon R. T. et al., arXiv : 0810.2295.
18. Mallone L. et al., arXiv : 0806.3908.
19. Khasanov R. et al., arXiv : 0901.2329.
20. Martin C. et al., arXiv : 0902.1804.
21. Khasanov R. et al., Phys. Rev. B, 78 (2008) 220510(R).
22. Weyeneth S. et al., J. Supercond. Nov. Mag. online 24. 12. 2008 ; ibid online 13.02.2009.
23. Bussmann-Holder A. et al., Europhys. Lett., 77 (2007) 27002.
24. Bussmann-Holder A. et al., J. Supercond. Nov. Mag., 21 (2008) 1557.
25. Bos J. W. G., et al., Chem. Comm., (2008) 3634; Attfield J. P, private communication.
26. Bussmann-Holder A., Micnas R., and Bishop A. R., Europ. Phys. J. B, 37 (2003) 1434.
27. Mittal R. et al., arXiv: 0902.3181.
28. Lang S. G. and Firsov Yu. A., Sov. Phys. JETP, 16 (1963) 1302.
29. Micnas R. and Tobijaszewska B., J. Phys.: Cond. Mat., 14 (2002) 9631.
30. Bussmann-Holder A. and Bianconi A., Phys. Rev. B, 67 (2003) 132509.
31. Khasanov R. et al., Phys. Rev. Lett., 98 (2007) 057007; Khasanov R. et al., Phys. Rev. Lett., 99 (2007) 237601; Khasanov R. et al., J. Supercond. Nov. Mag., 21, (2008) 81.
32. Millo O. et al., Phys. Rev. B, 78 (2008) 092505.
33. R. H. Liu et al., arXiv : 0810.2694.

CHAPTER 16

SUPERCONDUCTIVITY OF GOLD ALLOYS

H. R. Khan and Ch. J. Raub

The success of Kammerlingh-Onnes in liquefying helium in 1908 led to a workable fluid with a normal boiling point of 4.2 K. In this liquid he measured the temperature dependence of electrical resistivity of gold, mercury and platinum. Both gold and platinum exhibited a temperature independent resistivity at 4.2 K, but in the case of mercury the electrical resistivity vanished completely below this temperature. Kammerlingh-Onnes called the state of matter with no electrical resistance "superconducting state" and the temperature at which matter enters into this state "critical temperature". After 66 years of active research in this field, several hundred superconducting binary and ternary alloys have been discovered with critical temperatures as high as 23.4 K. Silver, copper and gold, the best electrical conductors at room temperature, do not show any sign of superconductivity above a few millidegrees. In general elements which enter into the superconducting state easily are poorer metallic conductors, such as niobium, tantalum and lead.

To understand this peculiar mechanism of superconductivity, experimentalists measured the critical temperatures of elements and alloys and related them to other solid state properties. They found that the electronic structure of the material which gives a high electron density of states at the Fermi level is a parameter of primary importance for the critical temperature. The most favourable structure to superconductivity seems to be the A-15 or ß-W structure. B. T. Matthias (1) discovered an empirical rule which predicts that alloys with average numbers of valence electrons per atom on the lower side of valences 5 and 7 will have high critical temperatures. At the same time theoreticians tried to understand superconductivity, and Fröhlich (2) first related critical temperatures of different isotopes of an element by the expression

$$T_c \propto M^{-1/2},$$

where M is the atomic weight. The experimental confirmation of this formula led to the discovery that superconductivity is due to some kind of electronlattice interaction. Finally, Bardeen, Cooper and Schrieffer (3) gave a model to explain the phenomena of superconductivity. Their model shows that in the superconducting state the condensation of pairs of conduction electrons into a lower energy state takes place. This condensation process involves the interaction between the conduction electrons and the crystal lattice—a process which is also responsible for the electrical resistance in the normal (nonsuperconducting) state. This model could explain why superconductivity occurs in poorer metallic conductors rather than in good electrical conductors. The critical temperature is related to the Debye temperature θ_D and the electron density of states N(0) and the electron-phonon interaction parameter V by the formula

$$T_c \propto \theta_D \exp. - \left(\frac{1}{N(O)V}\right).$$

In spite of the success in explaining the superconducting behaviour, it is difficult to calculate the critical temperature of the material or even predict if a phase will superconduct or not, because the interaction energy parameter N(0)V, which is in the exponential term, is affected by several solid state parameters like lattice constant, electron density of states, and ionic mass, in a complicated way.

Gold itself does not become a superconductor - above the millidegree range even if it is extremely pure, while none of the gold-rich solid solutions so far studied have proved to be superconducting. In forming solid solutions with them in general, gold lowers the T. of other elements, for example those of niobium and vanadium. Therefore the interest of researchers has been concentrated on the intermetallic compounds. The very first intermetallic compound to be discovered was the C15-type phase Au 2Bi (4) by de Haas in 1929. As a matter of fact its existence was concluded first from superconducting measurements. Since this time superconductivity and other solid state properties of gold alloys have been studied extensively and new superconducting gold-phases have

Citation: H. R. Khan and Ch. J. Raub, Superconductivity of Gold Alloys, DOI: 10.1007/BF03216575.

emerged. The present article gives a general survey of the superconductivity of gold alloys.

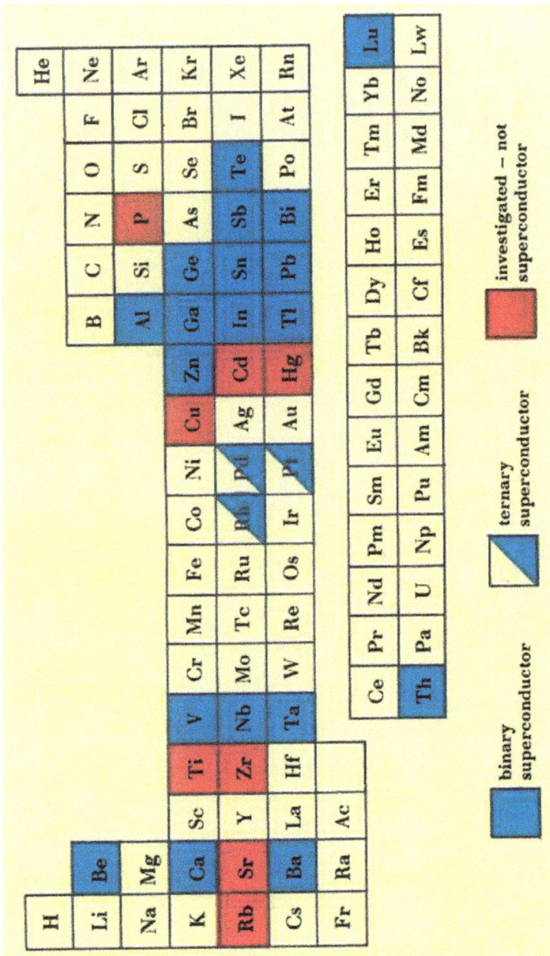

Figure 1: The distribution of the superconducting compounds of gold in the Periodic Table.

The periodic chart in Figure 1 shows the distribution of gold-containing binary and ternary superconducting compounds along with those which have been checked and are found not to be superconductors. None of the compounds formed by elements of Group IA and IB showed superconductivity. Phases checked have been Au_2Na (5) and $AuNa_2$ (6), Au_5K (5) and Au_5Rb (5), as well as AuCu and Cu_3Au. In Group IIA and IIB beryllium, calcium, barium and zinc compounds are superconducting, their transition temperatures ranging between 0.34 and 2.64 K, as shown in Table I.

Table I: Superconducting Compounds of Gold with the Elements of Group II A and II B.

Compound	T_c (K)	Crystal Structure	Ref.
$Au_{(1-x)}Be_x$	0.91–2.79	hex.	(7)
Au_5Ca	0.34–0.38	C15	(5, 8)
Au_5Ba	0.4–0.7	D_2d	(5, 11)
$AuZn_3$	1.21	P cubic	(9)
$Au_{(1-x)}Zn_x$	0.5–0.845	—	(10)

The compound Au_5Ba (11) proved to be of a special importance to superconductivity since it exists in a homogeneity range between $BaAu_{3.5}$ and $BaAu_{6.5}$. The relationship between lattice parameters, superconducting transition temperature and composition of this phase can be seen in Figure 2. On the basis of these results the authors predicted a T_c of gold in the millidegree range.

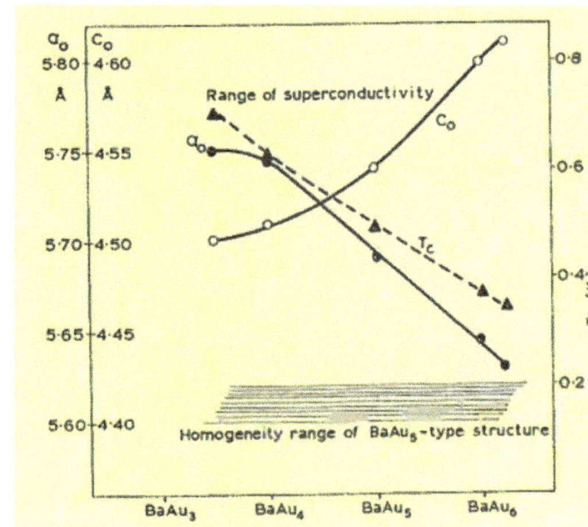

Figure 2: Relationship between lattice parameters, composition and superconductivity in the BaAa5-type structure at 700°C.

Table II: Superconducting Compounds of Gold with the Elements of Group III A and III B.

Compound	T_c (K)	Crystal Structure	Ref.
AuGa	1.2	B31	(5)
$AuGa_2$	0.34	C1	(12)
$AuIn_2$	.096–.093	—	(12)
AuIn	0.4–0.6	complex	(5)
$AlAu_4$	0.4–0.7	as A13	(5)
AuTl	1.92	—	(9)

Table II lists the superconducting compounds of gold with the elements of Group IIIA and IIIB, their crystal structures and transition temperatures. The highest critical temperature is that of AuTI (T_c= 1.92). Partial substitution of gold by palladium in a binary compound $AuGa_2$ (12) raises the T_c from .34 to 1.79 K. This shows the general trend that the addition of an unfilled d-band element often raises the T, in ternary compounds.

Of the Group IVB transition metals titanium, zirconium and hafnium, none forms superconducting phases with gold. This is even true for the A-15-type Ti_3Au. Due to stabilisation of the b.c.c. β -phase, gold enhances the T_c of zirconium (13) to about 2.80 K, but decreases the T_c of Ti-Rh (14) alloys. Germanium (15), tin (16) and lead (17) form superconducting compounds with gold. With tin and lead phases it has to be remembered that free tin or lead might have caused the superconductivity.

The composition and structure of superconducting Au-Ge (T_c=2.7 K) are not established for sure, since it is only formed by rapid quenching. Compounds AuSn and $AuSn_4$ have hexagonal and primitive cubic structures respectively. Their T_c values are between 2.0 and 3.8 K. Normally the addition of gold to lead lowers its T, yet there is a compound Au_x,Pb_y of unknown crystal structure with a T_c of 7.2 K. Since this value is identical with the one for pure lead (T_c= 7.193 K) the cause for superconductivity might be the preserves of filaments of lead.

Of the Group VA elements only antimony (18) and bismuth

react to form a superconductor. Au_2P_3 is not superconducting above 10 mK, $AuSb_2$ is at 0.58 K and Au_2Bi as mentioned at 1.84 K.

Much work has been done on the low temperature properties of the intermetallic compounds of gold with the Group VB elements vanadium, niobium and tantalum, especially since all form A15 phases. Nb_3Au (19) has the highest T, of 10.5 and Ta_3Au (20) the lowest at 0.5 K. Au_4V (21) is one of the few ferromagnets, where none of the constituent elements are ferromagnetic.

Substitution of rhodium (unfilled 4d band) and platinum (unfilled 5d band) for gold in Nb_3Au raise the T_c to 10.9 for $Nb_3Au_{0.98}Rh_{0.02}$ (22) and to 12.5 K for $Nb_3Au_{0.7}$. $7Pt_{0.3}$ (19). A further increase in rhodium concentration lowers the T_c. Metallurgical conditions have important influence on the superconducting behaviour of these ternary alloys. As a rule of thumb one can say that in general heterogeneous alloys exhibit higher T_c's than homogeneous phases of the same composition. The classic examples of inhomogeneous systems having high transition temperatures are Nb_3 (A1,Ge) (23) and Nb_3Ga (24).

The Nb_3Au (25) phase occurs up to 1560°C. Its maximum homogeneity range is between 75 and 81 atomic per cent niobium. If Nb_3Au is annealed at 1000, 1200, 1400 and 1550 °C and is afterwards quenched the homogeneity of the samples increases with the annealing temperature, yet T_c, decreases with improving homogeneity (Figures 3 and 4). A reason for this could be the atomic disordering due to annealing at higher temperatures. Recently we investigated the superconducting behaviour of the alloy $Nb_4Au_{0.25}Pt_{0.75}$ (26). In the as-cast state the T_c is 10.0 K but annealing for 14 days at 1200°C decreased the transition temperature down to 5.0 K. In contrast, for $Nb_3Pt_{(1-X)}Au_x$. alloys, annealing increased the superconducting transition temperature as shown in Figure 5. The story of the relationship of multiple phases to the transition temperature is not clear yet; once completely understood it would be of great help in raising the T_c.

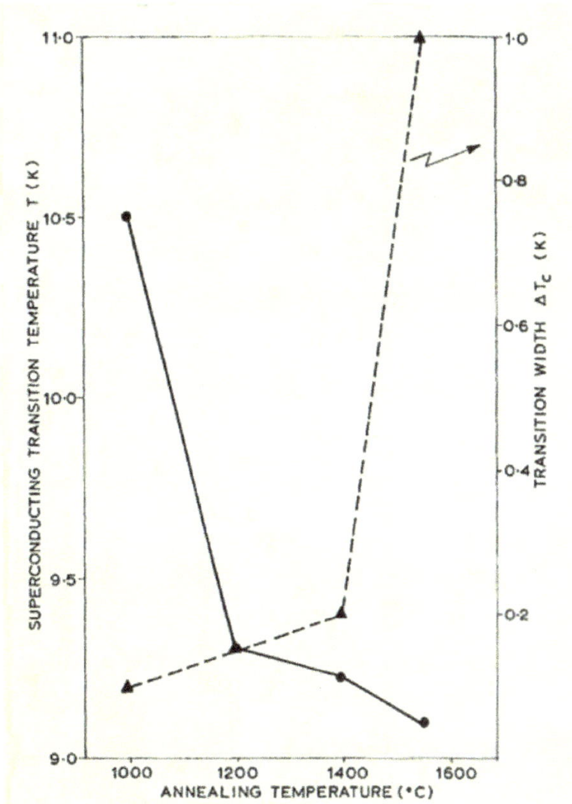

Figure 3: Superconducting transition temperatures and transition widths of Nb3Au alloys annealed at different temperatures.

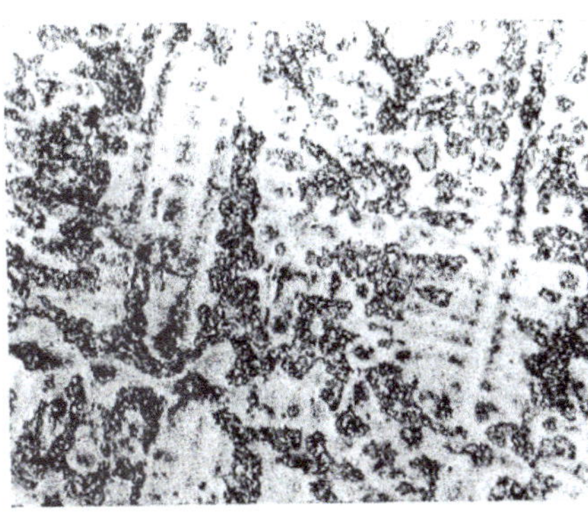

A Annealed at 1000°C, 4 weeks, fast quenched in water

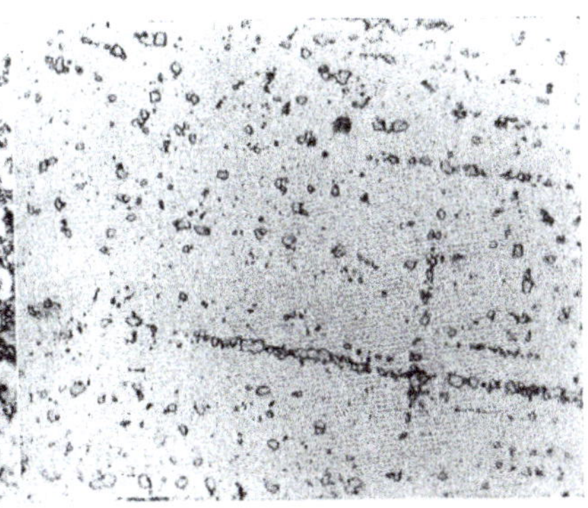

B Annealed at 1200°C, 1 week, fast quenched in oil

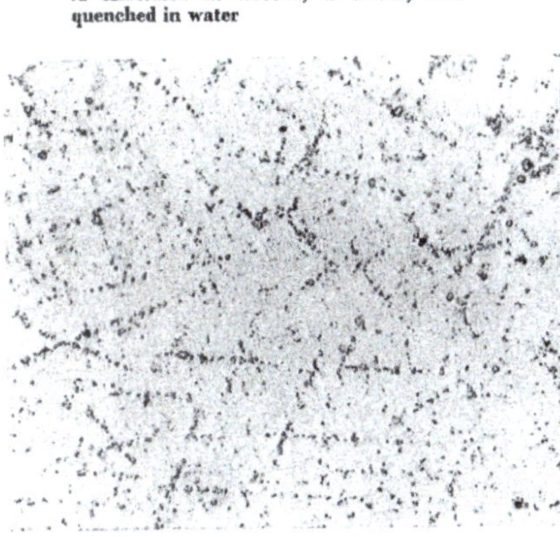

C Annealed at 1400°C, 7 hours, fast quenched in oil

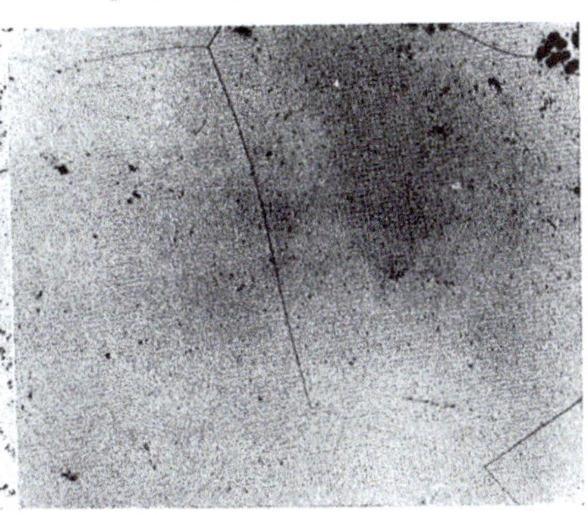

D Annealed at 1550°C, 2 hours, fast quenched in oil

Figure 4: Microstructures of Nh_3Au alloys. The homogeneity of the alloys increases with the annealing temperature but the critical temperature decreases with increasing homogeneity.

V_3Au also crystallises in the A-15 structure. It has the same valence electron concentration as Nb_3Au, but its critical temperature is only 0.74 K. Van Reuth and others (27) were able to raise its T_c to 2.98 K by low temperature annealing. They also measured the order parameter by X-ray diffraction technique and found that atomic ordering increased. This ordering raises the electron density of states and subsequently the critical temperature. From their measurements they introduced an ordering parameter in superconductivity. From a technological point of view this increase in T_c, by 2 K does not sound very interesting, but its physical implications are important. Once the ordering mechanism is fully understood and could be utilised, superconductors having a T_c value of at least 30 K can be imagined. This would have an enormous impact on the technical applications.

Among the Group VI A elements only tellurium forms a superconducting phase Au_3Te_5 (28) (T_c= 1.62). None of the gold compounds with elements of the V1 B to VIII B groups superconducts. Other phases worth mentioning are AuTh 2 (6) (T_c=3.65) and AuLu (29) (T_c=0.35). A pseudobinary carbide $AuC_{1.3}Y_{0.9}$ (30) was stated to have a T, of 10.1 K, but again the possibility of filaments of the binary carbide being present cannot be excluded.

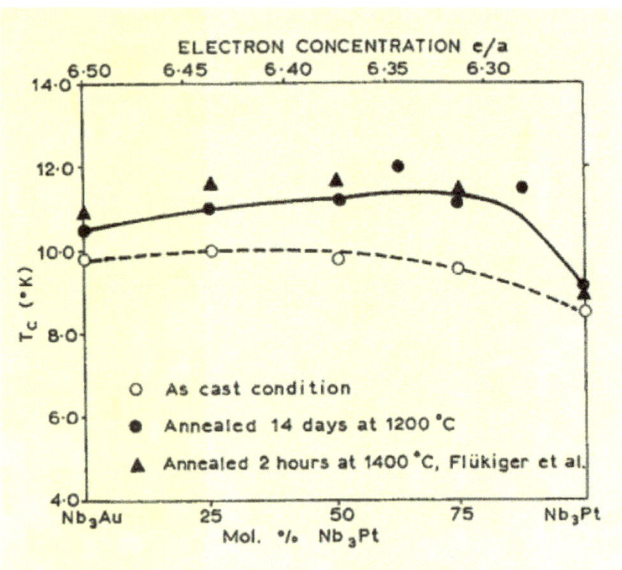

Figure 5: Variation of superconducting transition temperature of $Nb_3Pt_{(1-x)}Au_x$ alloys in the as-cast and annealed conditions as a function of mol. % Nb_3Pt.

The technological applications of superconductivity are emerging, but for materials research and solid state physics it has already become of increasing importance. Scientists involved have two things in mind, to synthesise a superconducting material having mechanical properties suitable to industrial needs and at the same time to understand the phenomenon in order to help in synthesising superconductors of high critical temperatures. The investigation of gold compounds has helped modern solid state physicists to understand more about superconductivity. Whether a gold compound will ever show a T_c, above 20 K one cannot predict. In no other field of science does the number of theoretical explanations of experimentally found effects surpass by so much the number of experimentally confirmed theoretical predictions as in the field of superconductivity.

REFERENCES

1. B. T. Matthias, Phys. Rev., 1955, 97, 74
2. H. Fröhlich, Phys. Rev., 1950, 79, 845
3. Bardeen, L. N. Cooper and J. R. Schrieffer, Phys. Rev., 1957, 108, 1175
4. W. J. de Haas and J. Turriaanse, Proc. Acad. Sci. Amsterdam, 1932, 35, 748
5. D. C. Hamilton, Ch. J. Raub, B. T. Matthias, E. Corenzwit and G. W. Hull, Jr., Y. Phys. Chem. Solids, 1965, 26, 665
6. E. E. Havinga, H. Damsma and P. Hokkeling, J. Less-Common Metals, 1972, 27, 281
7. C. E. Olsen, B. T. Matthias and H. H. Hill, Z. Phys., 1967, 200, 7
8. Ch. J. Raub and D. C. Hamilton, Y. Less-Common Metals, 1964, 6, 486
9. B. W. Roberts, Progress in Cryogenics, Vol. 4, Heywood, London, 1964, p. 160
10. D. Farrell, J. G. Park and B. R. Coles., Phys. Rev. Letters, 1964, 13, 328
11. G. Arrhenius, Ch. J. Raub, D. C. Hamilton and B. T. Matthias, Phys. Rev. Letters, 1963, 11, 313
12. J. H. Wernick, A. Meuth, T. H. Geballe, G. Hull and J. P. Maita, J. Phys. Chem. Solids, 1969, 30, 1949
13. B. W. Roberts, Progress in Cryogenics, Vol. 4, Heywood, London, 1964, p. 160
14. Ch. J. Raub., Z. Phys. 1964, 178, 216
15. B. Stritsker, and H. Wuhl, Z. Phys., 1971, 243, 361; H. L. Luo, M. F. Merriam and D. C. Hamilton, Science, 1964, 145, 581
16. E. Klokholm, and C. Chion, Acta Met., 1966, 14, 565; H. L. Luo, M. F. Merriam, and D. C. Hamilton, Science 1964, 145, 581
17. Ref. 6 and V. Sadogopan, E. Pollard and H. C. Gatos, Solid State Commun., 1965, 3, 97; H. L. Caswell, Phys. Letters, 1964, 10, 44
18. B. T. Matthias, T. H. Geballe, and V. B. Compton, Rev. Mod. Phys., 1963, 35, 1
19. P. Spitzli, R. Flükiger, F. Heininger and J. Muller, Phys. Letters, 1969, 30 A, 170
20. H. L. Luo, E. Vielhaber and E. Corenzwit, Z. Phys., 1970, 230, 443
21. L. Creveling, H. L. Luo and G. S. Knapp, Phys. Rev. Letters, 1967, 18, 851
22. S. T. Zegler, Phys. Rev., 1965, 137 A, 1438
23. G. Arrhenius, E. Corenzwit, R. Fitzgerald, G. W. Hull, Jr., H. L. Luo, B. T. Matthias and W. H. Zachariasen, Proc. Natl. Acad. Sciences, 1968, 61, 621
24. G. W. Webb, L. J. Vieland, R. E. Miller and A. Wicklund, Solid State Commun., 1971, 9, 1769
25. E. Röschel, O. Loebich, Jr. and Ch. J. Raub, Z. Metall 64, 359 -kunde, 1973,
26. H. R. Khan, E. Röschel and Ch. J. Raub, Z. Phys., 1973, 262, 279
27. E. C. van Reuth, R. M. Waterstrat, R. D. Blaugher, R. A. Hein and J. E. Cox, Proc. 10th Internat. Conference Low

Temp. Physics Vol. IIB, p. 137; R. A. Hein, J. E. Cox, R. D. Blaugher, R. M. Waterstrat and E. C. van Reuth, Proc. Internat. Conference Science of Superconductivity, Stanford University, Stanford, California, 1969; E. C. van Reuth and R. M. Waterstrat, Acta Cryst., 1968, 24, 187; R. A. Hein, J. E. Cox, R. D. Blaugher, R. M. Waterstrat and E. C. van Reuth, Physica, 1971, 55, 523

28. B. W. Roberts, Progress in Cryogenics, Vol. 4, Heywood, London, 1964, p. 160

29. T. F. Smith and H. L. Luo,. Phys. Chem. Solids 1967, 28, 569

30. M. C. Krupka, A. L. Giorgi, N. H. Krikorian and E. G. Szklarz,,. Less-Common Metals, 1969, 19, 113

CHAPTER 17

HIGH-TEMPERATURE SUPERCONDUCTING FIBER

Daniel Homa, Yongxuan Liang, Gary Pickrell

Keywords: Superconductivity, Superconductor cables, Fiber optics, Optical fiber, High-temperature superconductors

The promise of superconductivity continues to cultivate a sustained and intense interest from the scientific community because of the inverse relationship between finite sources of energy and the demands of a growing industrialized global community. Superconductivity is an inherent solution to the challenges of electric power transmission, transportation, and magnetic energy storage [1]. Since its discovery in 1911, the extremely low transition temperatures often relegated the use of superconductivity to the laboratory because of the expense and difficulties associated with the use of liquid helium [2]. The trajectory of research and developments in the field of superconductivity appeared to change in 1987 upon the discovery of "high-temperature" superconductivity in YBCO because of the higher transition temperatures, 93 K, that allowed the use of a much more cost effective cryogen: liquid nitrogen [3, 4] Although the pursuit of "room-temperature" superconductivity continues, significant effort has been dedicated to improved cable designs and manufacturing processes for these basic high-temperature superconductors [5, 6]. Although the pioneering work over the past few decades has made superconductivity relevant in an increasing number of applications, superconductivity has yet to experience the revolution seen with semiconductors and fiber optics [7–9]. The lack of traditional cable structures has, in part, impeded mass commercialization.

To meet this need, superconductivity was demonstrated in a lead core fiber via a multidisciplinary approach akin to current research dedicated to the integration of semiconductor devices into optical fibers [10–12]. Unfortunately, the extremely low transition temperature, $T_c \approx 7.2$ K, of lead necessitated the use of liquid helium, which is both expensive and difficult to handle. Thus, a superconducting fiber with a high-temperature superconducting core material such as yttrium barium copper oxide is preferable for the majority of applications.

In this letter, we demonstrate a high-temperature Type II superconducting fiber, as seen in Fig. 1. The yttrium barium copper oxide core of these types of fibers exhibited zero resistance at temperatures of approximately 93 K.

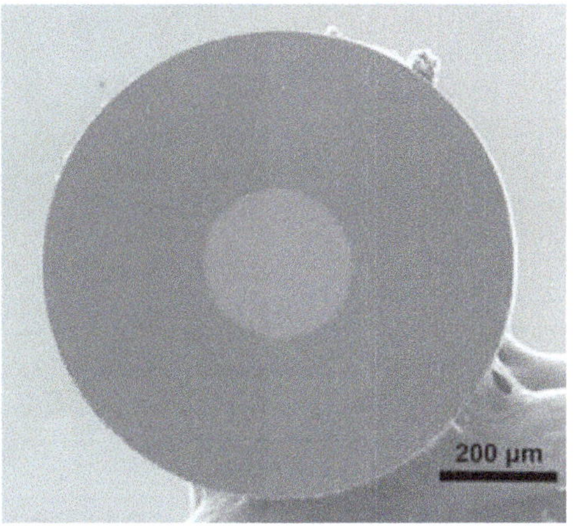

Figure 1: SEM image of superconducting fiber with a 200 μm yttrium barium oxide copper oxide core and fused silica cladding with an overall diameter of 800 μm.

The superconducting core fibers were prepared by the melt-draw technique on a traditional glass working lathe, as seen in Fig. 2 [10, 13]. First, a fused silica substrate tube (GE214, OD=8 mm, ID=3 mm) was fused to a processing tube (GE214, OD=9.5 mm, ID=7 mm). A smaller diameter fused silica rod (GE214, 8 mm) was used to push the yttrium barium copper oxide melt into the substrate tube forming a preform with a YBCO core. Finally, the preform was drawn into a fiber via a technique similar to the Taylor process [14, 15].

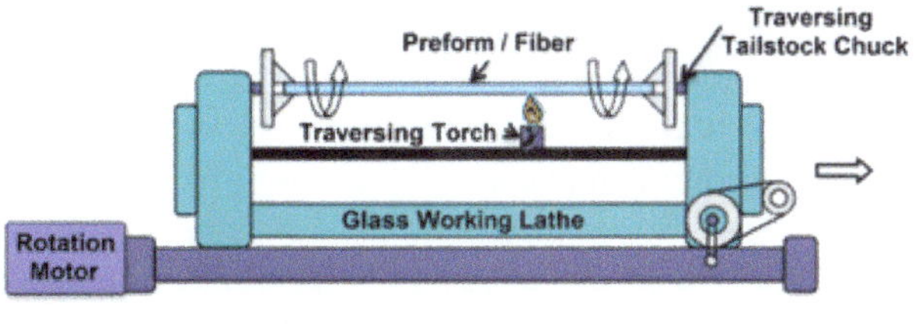

Figure 2: Schematic of the custom fiber drawing system utilized to fabricate the high-temperature superconducting core fibers

Citation: Daniel Homa, Yongxuan Liang, Gary Pickrell, High-Temperature Superconducting Fiber, DOI: 10.1007/s10948-014-2483-6.

The drawing temperature required for fused silica based optical fibers was on the order of 2000–2100 °C, which is much higher than the melting point of YBCO (~1010 °C) [16]. As shown in Fig. 3, the rapid consolidation of the core melt upon fiberization produced a YBCO core with minimal porosity. However, the as-drawn YBCO core fibers were essentially nonsuperconducting due high-temperature phase separation, loss of oxygen upon relatively rapid cooling, possible silicon diffusion from the cladding, and thermal decomposition [17]. Thus, the as-drawn fiber was annealed in an oxygen rich environment to recover it to superconductive state. Generally, the YBCO core fibers were heated to 950 °C, at a rate of 5 °C/min, and held at this temperature for a period of for 12 hours, cooled to 500 °C at a rate of 1 °C/min and held for 12 hours, and then allowed to naturally cool to room temperature [18,19].

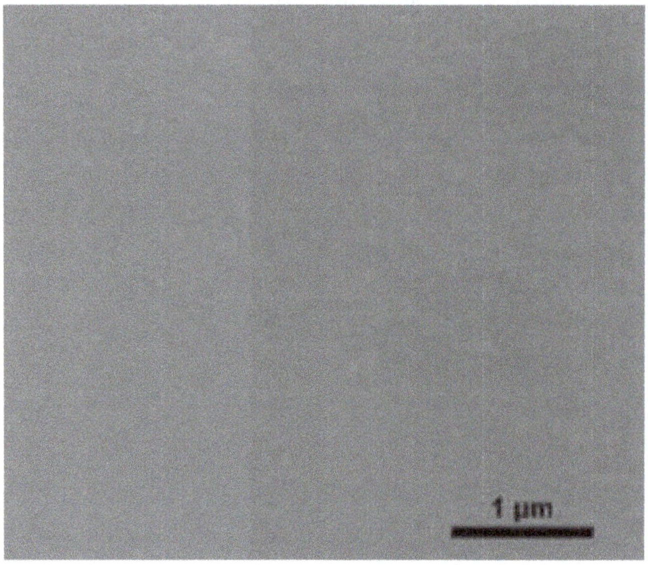

Figure 3: SEM image of the yttrium barium copper oxide core with minimal or no evident porosity (magnification bar of 1 μm for reference).

The fibers were fabricated on a glass working lathe, Litton Model HSJ143. The maximum achievable fiber lengths were limited by the working distance, approximately 1.2 meters, between the chuck faces, the preform length and draw rod at the tailstock end. Fiber diameters in the range of 100–900 microns were routinely fabricated to maximum lengths of 1.2 meters. Fiberization via the glass working lathe allows for the preliminary determination of feasibility prior to process implementation on a more traditional fiber drawing tower. The fundamental techniques are based on the Taylor and Taylor–Ulitovsky methods, which are employed to manufacture fiber lengths on the order of kilometers [20–24].

Polished fiber cross sections were characterized with a scanning electron microscope (SEM, LEO 1550). As shown previously in Fig. 1, the superconducting fiber possessed impressive core/clad concentricity and a distinct interface between the YBCO core and the fused silica glass cladding. The fused silica cladding maintained a diameter of approximately 800 μm and YBCO core diameter of 250 μm, respectively. Energy dispersive spectroscopy (EDS) chemical composition and mapping were performed with an attached IXRF system, Inc Iridium Microanalysis System, at an accelerating voltage of 20.0 kV. The EDS mapping images at the core-cladding interface can be seen in Fig. 4. A distinct transition between the cladding, SiO_2, and the core, YBCO, exists at the interface, although cross diffusion of both yttrium and barium is evident. Preliminary EDS chemical composition analysis suggested a Y:Ba:Cu elemental ratio of approximately 1:1.96:2.57, but further investigation is required to minimize the potential measurement artifacts caused by silicon and silicon dioxide contamination of the core material upon sample preparation and analysis and the characterization methods.

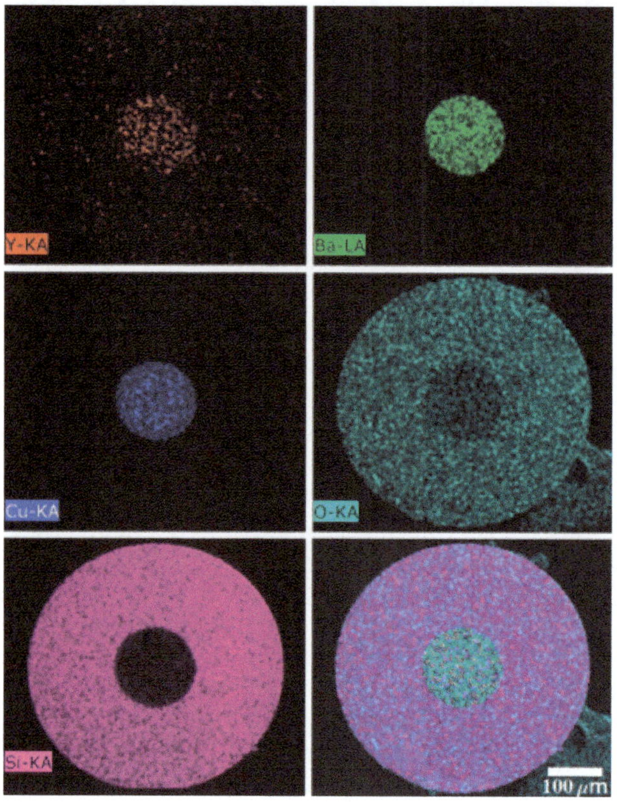

Figure 4: X-ray dot mapping of yttrium (*red*), barium (*green*), copper (*blue*), oxygen (*turquoise*), and silicon (*pink*). A composite X-ray dot map image is shown in the *bottom right-hand corner*. There is limited cross diffusion of the elements between the core and clad, but yttrium diffusion from the core to the clad and silicon diffusion from the clad to core is evident. A well-defined elemental transition is noted at the core-clad interface region.

The superconductivity of the YBCO core fibers was verified via the four-probe test method as shown in Fig. 5 [25]. The effects of contact resistance are eliminated via the four-probe test method. Resistance to the flow of electrical current will produce a voltage drop measured between the two voltage probes. Conversely, a superconductor will exhibit no voltage drop below the critical temperature.

The fused silica cladding could not be readily removed via hydrofluoric etching and/or mechanically because of the brittle nature of the YBCO material and necessitated the use of an alternative to expose the core. The YBCO core fiber was mounted into an epoxy resin mold and then carefully polished to expose the core in lengthwise fashion. The polished probe ends were placed in direct contact with the YBCO core with a custom test fixture to assure stable measurements. Furthermore, the probes were mounted in another mold at specific intervals of approximately 0.25 mm. Liquid nitrogen was used for the YBCO core fiber because the critical temperature is approximately 93 K, and the boiling point of liquid nitrogen is 77 K. A HP Hewlett-Packard Agilent, 6633A System was utilized as the direct current supply and ammeter. An Agilent, 34405A, 5 1/2 Digit Multi-meter was connected to the "voltage probes" to accurately measure the voltage drop. Both Type T and E thermocouples were utilized to continuously monitor the temperate of the sample.

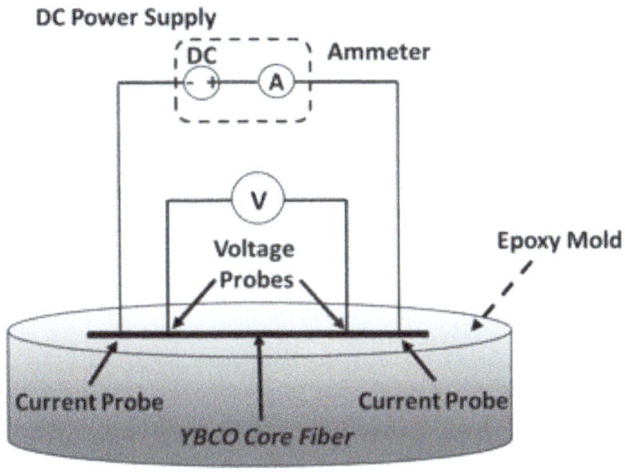

Figure 5: Schematic of the Four-Point Probe test setup. Three samples were tested: (1) "as received" commercial brass laminated high-temperature superconducting wire, (2) a commercial bulk YBCO disk, and (3) YBCO superconducting core fiber with the fused silica clad. A custom mechanical fixture was used to assure stable contact with the superconducting material. The samples were immersed in a liquid nitrogen housed in a Styrofoam™ container. A voltage drop of approximately 1.1 mV at 25 °C was set for each sample by adjusting the current.

A baseline measurement was performed with a commercially available brass laminated high-temperature superconducting wire (thickness ≈ 0.4 mm, width ≈ 4.4 mm) in a liquid nitrogen bath to confirm the validity of the test setup. As seen in Fig. 6(a), the resistance of the commercial superconducting wire disappeared at temperatures on the order of 93 K. Although, the initial electrical resistance was considerably larger, the superconducting YBCO core fibers appeared to demonstrate a slightly higher T_c of approximately 95 K. As shown in Fig. 6(b), the superconducting transition was also evaluated for a commercially bulk YBCO disk (diameter ≈ 25 mm, thickness ≈ 3.1 mm) to further confirm the performance of the superconducting fiber. The transition temperature of the bulk YBCO, $T_c \approx 93$ K, was lower than both the commercial wire and superconducting fiber. Although further investigation is required, the difference in T_c is currently attributed to measurement errors arising due the variation in thermocouple positions, thermal electromotive forces, external electromagnetic radiation, Faraday currents, and the degradation of the electrical contacts, as well as basic errors in the experimental setup [26, 27]. The critical temperature has been verified and extensively studied in the literature and is generally accepted as approximately 93 K, which is in good agreement with the measurements in this study [28].

The demonstration of a Type II superconducting fiber is one step closer to simple and cost effective superconducting cables for long haul power transmission, sensors, and devices. High-temperature superconductivity in traditional geometries such as fibers may well provide pave the way for the significant advancement of superconductivity in the landscape of emerging technologies.

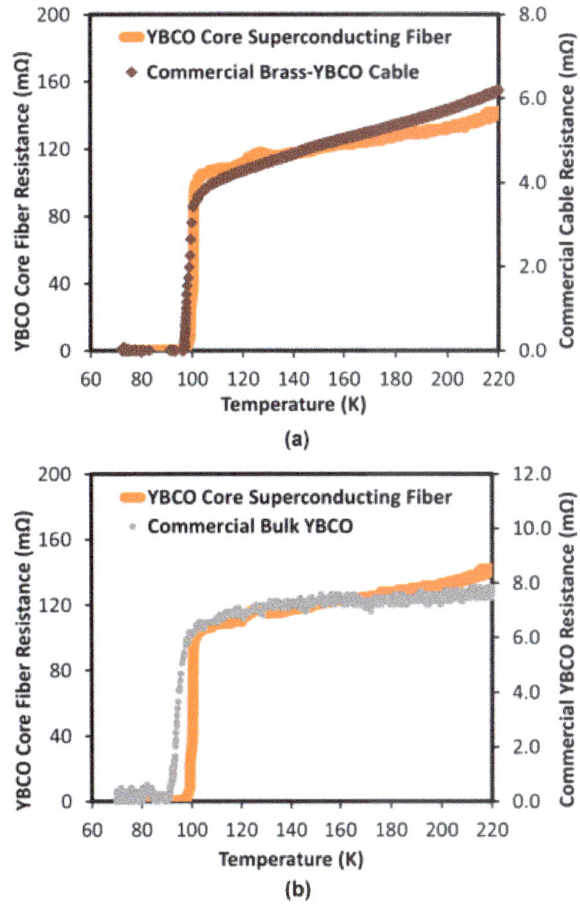

Figure 6: Electrical resistance as a function of temperature upon full immersion in liquid nitrogen for (**a**) commercial brass laminated high-temperature superconducting wire and a YBCO core superconducting fiber with a fused silica clad and (**b**) a commercial bulk YBCO disk and a YBCO core superconducting fiber with the fused silica clad. The *primary vertical axis* is the resistance of the YBCO core samples, and the *secondary vertical axis* is the resistance of the commercial high-temperature superconducting samples. All samples maintain a critical temperature of approximately 90–100 K.

REFERENCES

1. Zimmerman, G.O.: Superconductivity: the promise and reality. Int. J. Mod. Phys. B 17(18–20), 3698–3701 (2003)
2. Williams, R.: APS News 20(5), 1, 4 (2011)
3. Hikami, S., Hirai, T., Kagoshima, S.: High transition temperature superconductor: Y–Ba–Cu oxide. Jpn. J. Appl. Phys. 26, L314–L315 (1987)
4. Paranthaman, M.P., Izumi, T.: High-performance YBCO-coated superconductor wires. Mater. Res. Soc. Bull. 29(08), 533–541 (2004)
5. Maguire, J., Schmidt, F., Bratt, S., Welsh, T., Yuan, J., Allais, A., Hamber, F.: Development and demonstration of a HTS power cable to operate in the long island power authority transmission grid. IEEE Trans. Appl. Supercond. 17(2), 2034–2037 (2007)
6. Chen, M., Donzel, L., Lakner, M., Paul, W.: High temperature superconductors for power applications. J. Eur. Ceram. Soc. 24(6), 1815–1822 (2004)
7. Sazio, P.J., Amezcua-Correa, A., Finlayson, C.E., Hayes,

J.R., Scheidemantel, T.J., Baril, N.F., Jackson, B.R., Won, D.-J., Zhang, F., Margine, E.R.: Microstructured optical fibers as high-pressure microfluidic reactors. Science 311(5767), 1583–1586 (2006)

8. Ballato, J., Hawkins, T., Foy, P., Yazgan-Kokuoz, B., McMillen, C., Burka, L., Morris, S., Stolen, R., Rice, R.: Advancements in semiconductor core optical fiber. Opt. Fiber Technol. 16(6), 399–408 (2010)

9. Scott, B., Wang, K., Caluori, V., Pickrell, G.: Fabrication of silicon optical fiber. Opt. Eng. 48(10), 100501 (2009)

10. Homa, D., Liang, Y., Pickrell, G.: Superconducting fiber. Appl. Phys. Lett. 103(8), 082601 (2013)

11. Pickrell, G.R., Smirnova, E.: Novel structures in random hole optical fibers. In: IEEE Sensors (2005). 4 pp

12. Ballato, J., Hawkins, T., Foy, P., McMillen, C., Burka, L., Reppert, J., Podila, R., Rao, A., Rice, R.R.: Binary III–V semiconductor core optical fiber. Opt. Express 18(5), 4972–4979 (2010)

13. Scott, B.L., Wang, K., Pickrell, G.: Fabrication of n-type silicon optical fibers. IEEE Photonics Technol. Lett. 21(24), 1798–1800 (2009)

14. Taylor, G.: A method of drawing metallic filaments and a discussion of their properties and uses. Phys. Rev. 23(5), 655 (1924)

15. Grodkiewicz, W.: Fused silica fibers with metal cores. Mater. Res. Bull. 10(10), 1085–1090 (1975)

16. Lian, Z., Pingxiang, Z., Ping, J., Keguang, W., Jingrong, W., Xiaozu, W.: The properties of YBCO superconductors prepared by a new approach: the 'powder melting process'. Supercond. Sci. Technol. 3(10), 490 (1990)

17. Jin, S., Tiefel, T., Sherwood, R., Kammlott, G., Zahurak, S.: Fabrication of dense Ba2YCu3O7–δ superconductor wire by molten oxide processing. Appl. Phys. Lett. 51(12), 943–945 (1987)

18. Müller, P., Schubert, M., Rodig, C., Fuchs, G., Fischer, K.: Heat treatment of wires on the basis of the high T superconductor YBaCuO. Appl. Phys. Lett. 55, 917 (1989)

19. Jee, Y., Kim, C., Sung, T., Hong, G.: Top-seeded melt growth of Y–Ba–Cu–O superconductor with multiseeding. Supercond. Sci. Technol. 13(2), 195 (2000)

20. Donald, I., Metcalfe, B.: The preparation, properties and applications of some glass-coated metal filaments prepared by the Taylorwire process. J. Mater. Sci. 31(5), 1139–1149 (1996)

21. Pardoe, G., Butler, E., Gelder, D.: Rapid quenching by the Taylor wire technique. J. Mater. Sci. 13(4), 786–790 (1978)

22. Butler, I., Kurz, W., Gillot, J., Lux, B.: The production of metal fibres and wires directly from the melt. Fibre Sci. Technol. 5(4), 243–262 (1972)

23. Zhukov, A., González, J., Blanco, J., Vazquez, M., Larin, V.: Microwires coated by glass: a new family of soft and hard magnetic materials. J. Mater. Res. 15(10), 2107–2113 (2000)

24. Bayindir, M., Abouraddy, A.F., Shapira, O., Viens, J., SayginHinczewski, D., Sorin, F., Arnold, J., Joannopoulos, J.D., Fink, Y.: Kilometer-long ordered nanophotonic devices by preform-to-fiber fabrication. IEEE J. Sel. Top. Quantum Electron. 12(6), 1202– 1213 (2006)

25. Vazquez, V., Pérez-Amaro, N., Canizo-Cabrera, A.: Selected error sources in resistance measurements on superconductors. Rev. Sci. Instrum. 72(8), 3332–3339 (2001)

26. Cohen, M., Falicov, L., Phillips, J.: Superconductive tunneling. Phys. Rev. Lett. 8(8), 316–318 (1962)

27. Hua, J., Xiao, Z., Rosenmann, D., Beloborodov, I., Welp, U., Kwok, W., Crabtree, G.: Resistance anomaly in disordered superconducting films. Appl. Phys. Lett. 90(7), 072507 (2007)

28. Genoud, J.-Y., Graf, T., Junod, A., Sanchez, D., Triscone, G., Muller, J.: Preparation, resistivity, magnetic properties and specific heat of the 95 K superconductor $YBa_2Cu_{3.5}O_{7.5+x}$ ("247"). Physica C, Supercond. 177(4), 315–329 (1991)

CHAPTER 18

DIELECTRIC PROPERTIES OF (ZN)X/CUTL-1223 NANOPARTICLE-SUPERCONDUCTOR COMPOSITES

M. Mumtaz, Liaqat Ali, Shoaib Azeem, Saad Ullah, G. Hussain, M. W. Rabbani, Abdul Jabbar, K. Nadeem

Keywords: (Zn)x/CuTl-1223 nanoparticle-superconductor composites, dielectric properties, activation energy

INTRODUCTION

One of the most eminent phase of $(Cu_{1-x}Tlx)Ba_2Ca_{n-1}Cu_nO_{2n+4-\delta}$ (n = 2, 3, 4, ...) high temperature superconducting family is $(Cu_{0.5}Tl_{0.5})Ba_2Ca_2Cu_3O_{10-\delta}$ (CuTl-1223) [1]. CuTl-1223 superconducting phase has long coherence length along c-axis (ξ_c) [2], low superconducting anisotropy ($\gamma = \xi_{ab}/\xi_c$) [3], small penetration depth (l), high zero resistivity critical temperature ($T_c(0)$), high irreversibility field (H_{irr}), and high critical current density (J_c) [4,5]. The performance of bulk high temperature superconductors can be severely affected by inter-grain voids, micro-cracks, and pores [6–8]. Different strategies have been exercised to minimize the voids and pores in CuTl-1223 phase [9,10]. Here, we have implemented an easy, effective, and innovative technique of the addition of Zn nanoparticles in CuTl-1223 superconducting matrix. The places, where these nanoparticles reside are only the voids and grain boundaries or grain surfaces. The key objective of this research is the progression of inter-grain connectivity and to scrutinize its effects on superconductivity and dielectric properties of $(Zn)_x$/ CuTl-1223 nanoparticle–superconductor composites.

In CuTl-1223 phase, the charge reservoir layer $(Cu_{0.5}Tl_{0.5})$ $Ba2O_{4-\delta}$ consists of Tl3+, Cu2+, and Ba2+ ions, which can localize the charge carriers, while the conducting CuO2 plane consists of mobile charge carriers. The dielectric constant ascends from the accretion of carriers in between (Cu0.5Tl0.5) $Ba_2O_{4-\delta}$ charge reservoir layer and CuO2 plane [11,12]. The charge reservoir layer, inter-grain boundaries, and micro-cracks act as dielectric media, which can be polarized by the displacement of charge carriers with the assistance of an external electric field from their equilibrium position. The material may undergo four primary mechanisms of polarization [13–16] each of which involves a short-range motion of charges and subsidizes to the total polarization of the material:

- Electronic polarization (α_e). It arises when shifting of electronic charge clouds of atoms and ions takes place. This is observed at very high frequencies such as 1016 Hz, i.e., in ultra violet optical region.
- Lattice polarization (α_a). The origin of this polarization is the combined displacement of ion cores and their electronic charge distribution. This is observed at frequencies ranging from 1010 to 1013 Hz, i.e., in infrared optical region.
- Dipolar polarization (α_0). It originates from the permanent dipoles or from the presence of charge carriers, which is active at lower frequencies from 10^3 to 10^6 Hz, i.e., in the sub-infrared optical region due to their longer relaxation time.
- Interfacial polarization (α_i). This polarization is sensitive at low frequency range of 103 Hz and may extend to a few kHz [6,13,17].

Dielectric properties of $Cu_{0.25}Tl_{0.75}Ba_2Ca_3Cu_4O_{12-\delta}$ superconductor with addition of MgO nanoparticles were investigated in frequency range from 100 Hz to 4 MHz [7]. Volume fraction and connectivity between grains were

Citation: M. Mumtaz, Liaqat Ali, Shoaib Azeem, Saad Ullah, G. Hussain, M. W. Rabbani, Abdul Jabbar, K. Nadeem, Dielectric properties of (Zn)x/CuTl-1223 nanoparticle-superconductor composites, DOI: 10.1007/s40145-016-0185-1.

found to be increased by the addition of these nanoparticles by healing micro-cracks. Increase in ε_r' and decrease in σ_{ac} were observed up to certain optimum level of the nanoparticle content. Dielectric constants (ε_r', ε_r'') and dielectric loss (tand) of Tl-based high temperature superconductors (HTSCs) were found to depend strongly on operating temperature and frequency of external applied ac-field [6]. Negative capacitance (NC) phenomenon was observed in $Cu_{0.5}Tl_{0.5}Ba_2Ca_3Cu_{4-y}ZnyO_{12-\delta}$ (y = 0, 3) superconductor at different frequencies and temperature [13,18]. Effect of oxygen post-annealing on superconductivity and dielectric properties of $Cu_{0.5}Tl_{0.5}Ba_2Ca_3(Cu_{4-y}Cd_y)O_{12-\delta}$ was investigated [19]. A strong dielectric dispersion was observed at low frequencies and temperature, while reverse effects were observed at high temperature and frequencies. The dielectric constant determines the nature of the superconducting material, i.e., the binding forces to which the mobile carriers are tied among the conducting CuO_2 planes and $Cu_{0.5}Tl_{0.5}Ba_2O_{4-\delta}$ charge reservoir layers. The real part of the dielectric constant (e¢) shows energy stored within the material when exposed to electric field. The imaginary part of the dielectric constant (e ¢¢) indicates the absorption and the attenuation of energy across the interfaces under an external electric field. Electrical nature of materials can be described by dielectric properties. Materials with high dielectric constant are used in micro-electronic devices. In this article, we reported the frequencydependent dielectric measurements for $(Zn)_x$/CuTl- 1223 nanoparticle–superconductor composites. Different parameters related to dielectric measurements like dielectric constant (both real and imaginary parts), tangent loss, and ac-conductivity have been calculated in the frequency range of 10^{-1} to 10^7 Hz at room temperature, i.e., 293 K.

Experimental Details

$(Zn)_x$/CuTl-1223 nanoparticle–superconducting composites were synthesized by using solid-state reaction method. Initially Cu(CN), $Ba(NO_3)_2$, and $Ca(NO_3)_2$ compounds were mixed in appropriate ratios to prepare bulk $Cu_{0.5}Ba_2Ca_2Cu_3O_{10-\delta}$ precursor material. The mixed material was ground in agate mortar and pestle for 2 h. After grinding, the mixed material was loaded in quartz boat and fired at 860 °C for 24 h in pre-heated chamber furnace followed by furnace cooling to room temperature. The firing step was repeated after 2 h grinding under the same conditions.

The calculated amounts of Tl_2O_3 and Zn nanoparticles were mixed in $Cu_{0.5}Ba_2Ca_2Cu_3O_{10-\delta}$ precursor material. The mixed material obtained was pelletized under 3.8 tons/cm^2 pressure by using hydraulic press [20,21]. The pellets were wrapped in gold capsules (to avoid Tl leakage) and sintered for 10 min at 860 °C in pre-heated chamber furnace followed by quenching to room temperature in order to obtain the final product $(Zn)_x$/CuTl-1223 (x = 0–4 wt%) nanoparticle–superconductor composites.

These composites were characterized by X-ray diffraction (XRD),scanning electron microscopy (SEM), temperature-dependent resistance measurements, and dielectric properties. The structural phases of nanoparticles and composites were determined by XRD. The average size of nanoparticles was calculated by using Debye–Scherrer's formula. The cell parameters were determined by using CheckCell computersoftware. The surface morphology and chemical composition were examined by SEM and energy dispersive X-ray (EDX) analysis, respectively. Conventional four-probe method was used to measure temperature-dependent resistance of these samples. The frequency-dependent dielectric measurements were carried out by using an LCR meter at frequencies from 10^{-1} to 10^7 Hz. Typical two-probe technique was used for frequency-dependent dielectric measurements. Silver paint was applied to both the surfaces and was dried at room temperature. By measuring the conductance (G) and capacitance (C), the dielectric constant (r_e), dielectric loss (tand), and ac-conductivity (s_{ac}) of the samples were determined by using following equations [13,20]:

$$\varepsilon_r = \frac{\varepsilon_m}{\varepsilon_0} \quad (1)$$

$$\varepsilon' = \frac{Cd}{\varepsilon_0 A} \quad (2)$$

$$\varepsilon'' = \frac{Gd}{\varepsilon_0 A\omega} \quad (3)$$

$$\tan\delta = \frac{\varepsilon''}{\varepsilon'} \quad (4)$$

$$\sigma_{ac} = \omega\varepsilon'\varepsilon_0 \tan\delta \quad (5)$$

where d is the thickness of pellets, A is the area of electrode, G is the conductance, w = 2π f is the angular frequency, f is the frequency of applied ac-field, ε_m is the permittivity of material, ε_0 is the permittivity of free space, and ε_r is the relative permittivity.

RESULTS AND DISCUSSION

XRD spectrum exhibits prominent sharp diffraction peaks indexed with hexagonal structure of Zn nanoparticles as shown in Fig. 1. XRD analysis shows the distinct diffraction peaks at 2θ = 36.34°, 39.06°, 43.28°, 54.36°, which are indexed to (002), (010), (011), and (012) planes, respectively. The hexagonal structure of Zn nanoparticles matches well with the database of JCPDS, indicating that Zn nanoparticles have crystalline structure. The average size of Zn nanoparticles calculated by Debye–Scherrer's formula is 50 nm. No prominent peak of impurity is found in XRD spectrum of Zn nanoparticles. The XRD spectra of $(Zn)_x$/CuTl-1223 nanoparticle–superconducting composites with different weight percentage of Zn nanoparticles (i.e., x = 0–4 wt%) are shown in Fig. 2. Most of the diffraction peaks are well indexed in accordance to CuTl-1223 tetragonalstructure, following P4/mmm space group. These peaks correspond to CuTl-1223 phase with lattice parameters a = 4.01 Å, c = 14.95 Å for x = 0; a = 3.03 Å, c = 15.85 Å for x = 1 wt%; a = 3.77 Å, c = 15.25 Å for x = 2 wt%; a = 3.70 Å, c = 15.20 Å for x = 3 wt%; and a = 3.70 Å, c = 15.48 Å for x = 4 wt%. A few un-indexed peaks of low intensity also appear, which are possibly due to presence of impurities or some other superconducting phases.

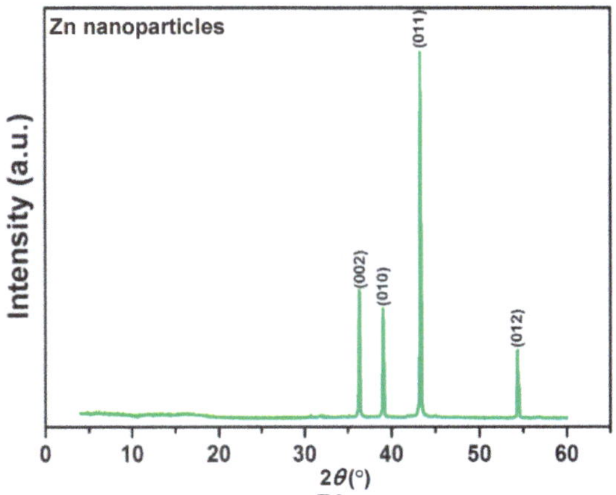

Figure 1: XRD spectrum of Zn nanoparticles.

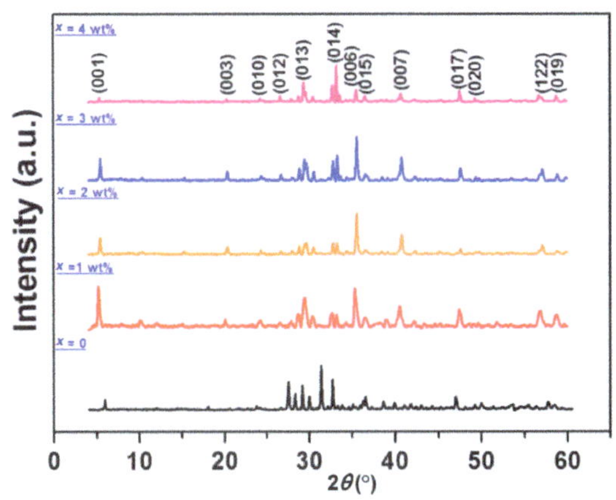

Figure 2: XRD spectra of $(Zn)_x$/CuTl-1223 nanoparticle– superconductor composites with x = 0–4 wt%.

A slight shift of the diffraction peaks to lower angles is observed, which may be due to strains produced in the material after the addition of Zn nanoparticles. Figure 3 shows SEM images of $(Zn)_x$/ CuTl-1223 nanoparticle–superconductor composites with x = 0, 2 wt%. High porosity, weak inter-grain connectivity, and disorientation of grains in pure CuTl-1223 superconducting phase can be visualized in Fig. 3(a). Inter-grain connectivity and improved grain size after addition of Zn nanoparticles in CuTl-1223 superconducting matrix can be seen in Fig. 3(b). The improvement in inter-grain connections and grain size may be due to healing up of inter-grain voids after inclusion of metallic Zn nanoparticles. EDX spectra of $(Zn)_x$/ CuTl-1223 nanoparticle– superconductor composites (x = 2, 4 wt%) are shown in Fig. 4, which show the presence of different elements in the composites. XRD, SEM, and EDX show that Zn nanoparticles occupy the interstitial spaces between the grains and do not affect the crystal structure of host CuTl-1223 phase. The temperature-dependentresistance measurements of $(Zn)_x$/ CuTl-1223 nanoparticle– superconductor composites (x = 0, 1, 3 wt%) are shown in Fig. 5. The metallic variation in resistance from room temperature down to onset of superconductivity with critical temperature T_c (0) around 92, 101, and 100 K for x = 0, 1, 3 wt% has been observed, respectively.

These measurements indicate the increase in critical temperature T_c (0) from 92 to 101 K after the addition of 1 wt% of Zn nanoparticles. The increase in T_c (0) is due to the facilitation of mobile free carriers and improvement in the inter-grain connectivity after addition of Zn nanoparticles in CuTl-1223 phase. The initial rise in T_c (0) is due to increase in superconducting volume fraction and improved weak-links after the addition of Zn nanoparticles at the grain boundaries of CuTl-1223 matrix.

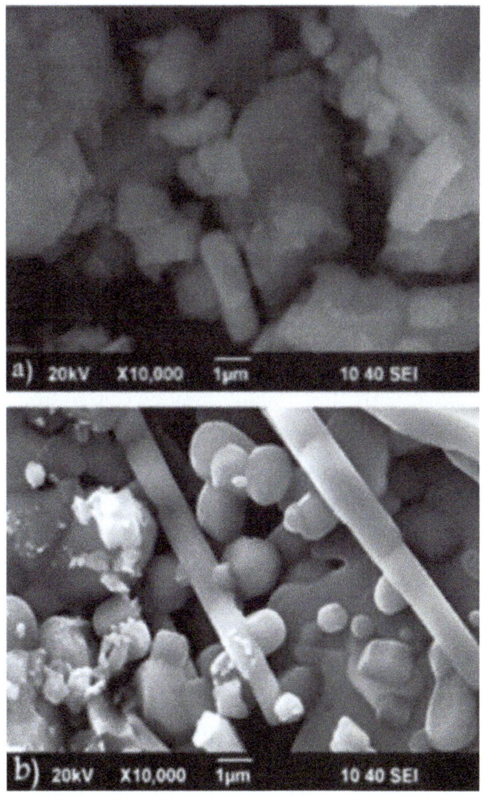

Figure 3: SEM images of $(Zn)_x$/CuTl-1223 nanoparticle– superconductor composites with (a) x = 0 and (b) x = 2 wt%.

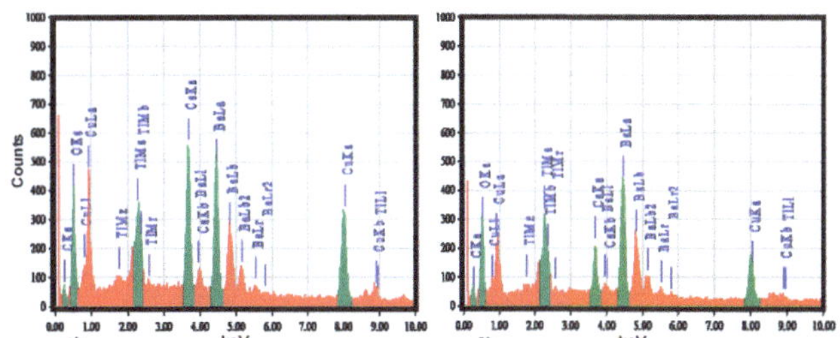

Figure 4: EDX spectra of $(Zn)_x$/CuTl-1223 nanoparticle–superconductor composites with (a) x = 2 wt% and (b) x = 4 wt%.

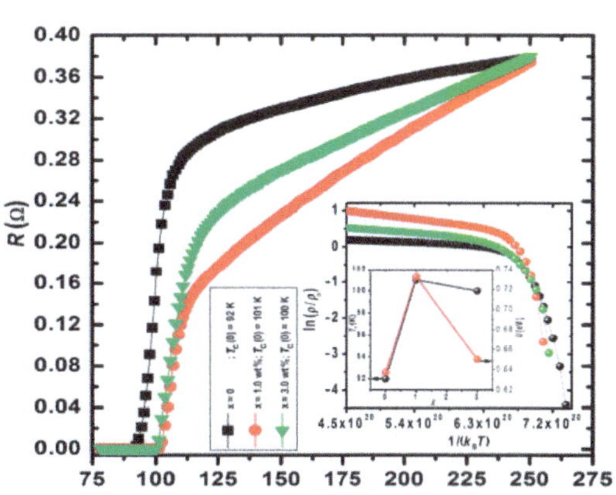

Figure 5: Resistance versus temperature measurements of $(Zn)_x$/ CuTl-1223 nanoparticle–superconductor composites with x = 0, 1, 3

wt%. In the inset, there are shown the Arrhenius plots and variation of T_c and U as a function of Zn nanoparticle content (i.e., x = 0, 1, 3 wt%).

The dipolar polarization relaxation time obeys the Arrhenius equation according to the Debye model, which is

$$\frac{\rho}{\rho_0} = \frac{U}{k_B T} \qquad (6)$$

where U is the activation energy; B_k is the Boltzmann constant; and ρ_0 is the normal state resistivity. The plot of $\ln(r/\rho_0)$ versus $1/(k_B/T)$ showing the activated behavior is given in the inset of Fig. 5. To calculate the activation energy, the region close to superconducting transition is chosen and it is observed that the activation energy is increased after the addition of Zn nanoparticles in CuTl-1223 matrix. The maximum value of activation energy is found in the sample with x = 1 wt%. The increase of activation energy may be due to interaction of carriers with the metallic Zn nanoparticles at grain boundaries.

Dielectric properties of $(Zn)_x$/CuTl-1223 nanoparticle–superconductor composites with x = 0– 4 wt% have also been examined. The variation in frequency-dependent dielectric constant (ε_r) of the samples at room temperature is shown in Fig. 6. The dielectric constant is decreased with increase of frequency. The carriers can not follow the frequency of external applied electric ac-field at high frequencies and dielectric response becomes ineffective. The maximum values of r e at frequency 10^{-1} Hz are found around 8.96×10^{12}, 1.53×10^{13}, 3.74×10^{12}, 1.181×10^{12}, and 1.22×10^{13} for x = 0, 1, 2, 3, 4 wt%, respectively. The variation of r e is non-monotonic, which can be due to heterogeneous mixing of Zn nanoparticles in CuTl-1223 matrix, but the overall suppression of dielectric properties may be due to reduction of defects after the addition of metallic Zn nanoparticles. The variation of maximum values of ε_r with Zn nanoparticle content in $(Zn)_x$/CuTl-1223 nanoparticle– superconductor composites is given in the inset of Fig. 6. The increase of maximum values of ε_r may be due to agglomeration and segregation of Zn nanoparticles in the bulk CuTl-1223 phase at higher concentration of Zn nanoparticles.

The variation in frequency-dependent real part of dielectric constant (e¢) of these samples at room temperature is shown in Fig. 7. At lower frequency, dipolar polarization is dominant but when the frequency increases, the oscillation of the applied field increases and the dipolar polarization characteristic time becomes longer than the time constant of applied field. That is why decrease in e¢ is observed. The maximum values of e¢ at frequency 10–1 Hz are found around 1.32×10^{9} , 2.18×10^{9} , 2.16×10^{8} , 4.75×10^{7} , and 1.77×10^{9} for x = 0, 1, 2, 3, 4 wt%, respectively. The variation of maximum values of e¢ with Zn nanoparticle content in $(Zn)_x$/ CuTl-1223 nanoparticle–superconductor composites is given in the inset of Fig. 7. The variation in frequency-dependent imaginary part of dielectric constant (e ¢¢) of the samples at room temperature is shown in Fig. 8. All the investigated samples exhibit dielectric dispersion where decrease in e ¢¢ is observed with the increase in frequency. The dielectric dispersion phenomenon has been explained on the basis of Maxwell–Wagner model of dielectrics [22,23], according to which a dielectric medium is assumed to be made of well conducting grains separated by grain boundaries. At lower frequencies, the grain boundaries are more effective, due to which charges hop, resulting in higher dielectric constant.

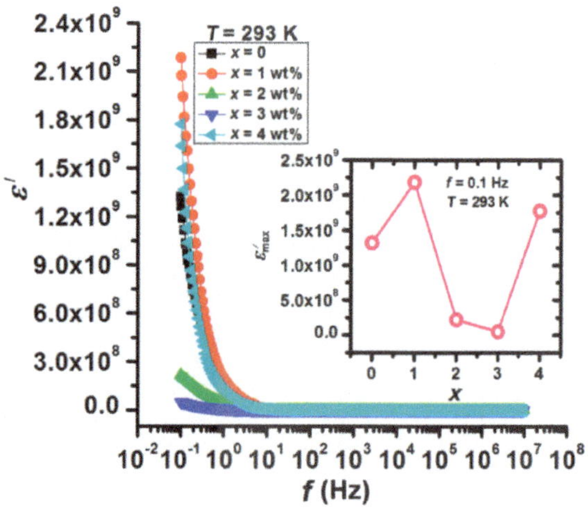

Figure 7: Variation of real part of dielectric constant (e¢) of $(Zn)_x$/ CuTl-1223 nanoparticle–superconductor composites versus frequency for x = 0–4 wt%. In the inset, there are shown the variation of maximum e¢ versus Zn nanoparticle content (i.e., x).

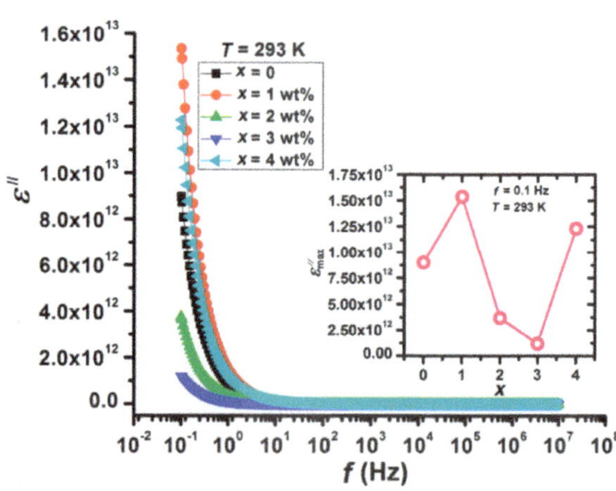

Figure 8: Variation of imaginary part of dielectric constant (e¢¢) of $(Zn)_x$/CuTl-1223 nanoparticle–superconductor composites versus frequency for x = 0–4 wt%. In the inset, there are shown the variation of maximum e¢¢ versus Zn nanoparticle content (i.e., x).

The decrease observed with increasing frequency is due to decrease in polarization as the dipoles do not have the capacity to follow the applied field at high frequencies. The maximum values of e ¢¢ at frequency 10^{-1} Hz are found around 8.96×10^{12}, 1.53×10^{13}, 3.64×10^{12}, 1.15×10^{12}, and 1.23×10^{13} for x = 0, 1, 2, 3, 4 wt%, respectively. The variation in frequency-dependent tangent loss (tand) of the samples at room temperature is shown in Fig. 9. The peaks appeared correspond to dispersion positions indicating the relaxation process. Maximum dissipation is observed because of relaxation time of electrical dipoles, which is near the applied electric field. The maximum values of tand at frequency 10^{-1} Hz are found around 6.89×10^{3} , 7.03×10^{3} , 1.73×10^{4} , 2.49×10^{4} , and 6.92×10^{3} for x = 0, 1, 2, 3, 4 wt%, respectively. A decreasing trend in tand is seen at low frequency, which is due to the effective grain boundaries.

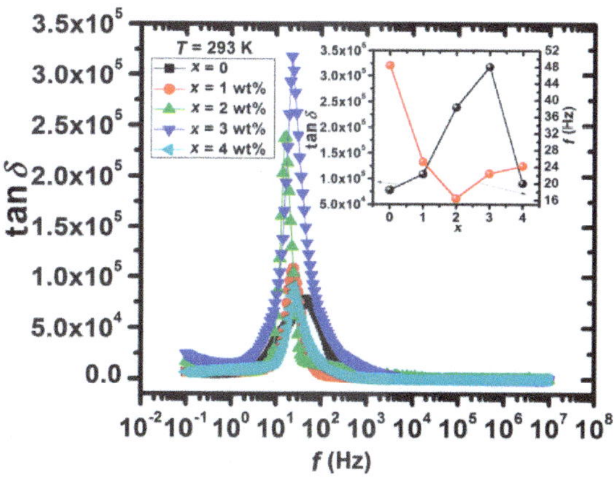

Figure 9: Variation of dielectric loss factor (tand) of (Zn)x/CuTl-1223 nanoparticle–superconductor composites with frequency for x = 0–4 wt%. In the inset, there are shown the variation of maximum tand versus Zn nanoparticle content (i.e., x).

The peak in tand can be observed at different frequencies with the addition of these Zn nanoparticles, which is 7.8×10^4 at 48.48 Hz, 1.09×10^5 at 24.25 Hz, 2.39×10^5 at 16.50 Hz, 3.18×10^5 at 22.45 Hz, and 9.08×10^4 at 24.25 Hz for the samples with x = 0, 1, 2, 3, 4 wt%, respectively.

At relaxation frequency, the most apparent relaxation time, t , is given by

$\omega t = 1$ (7)

where w is the angular frequency.

The variation in frequency-dependent acconductivity (σ_{ac}) of $(Zn)_x$/CuTl-1223 nanoparticle– superconductor composites at room temperature is shown in Fig. 10. Maximum values of σ_{ac} at frequency 10^7 Hz are found to be 0.69, 1.27, 0.21, 0.077, and 0.88 S/cm for x = 0, 1, 2, 3, 4 wt%, respectively. The value of s ac increases as a whole after the addition of Zn nanoparticles in CuTl-1223 superconducting matrix. The possible reason for this increase in s ac is due to healing up of the micro-cracks, which results in the improvement of inter-grain connectivity. The ac-conductivity (s_{ac}) of our samples increases at high frequencies, which happens due to bound carriers trapped in the samples. When ac-conductivity (s_{ac}) decreases with increasing frequency, it happens due to the free carriers. In the inset, there is shown the variation of maximum value of s_{ac} versus Zn nanoparticle content, showing a random trend in variation,

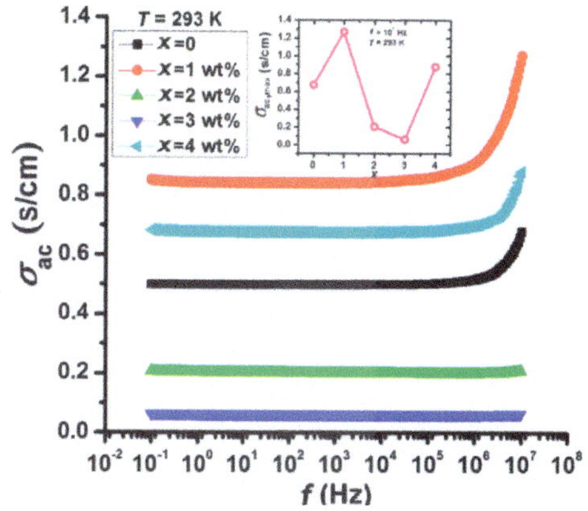

Figure 10: Variation of ac-conductivity (s_{ac}) of $(Zn)_x$/CuTl- 1223 nanoparticle–superconductor composites with frequency for x = 0–4 wt%. In the inset, there is shown the variation of maximum s_{ac} versus Zn nanoparticle content (i.e., x).

which may be due to inhomogeneous distribution of these nanoparticles in the CuTl-1223 matrix.

CONCLUSIONS

$(Zn)_x$/CuTl-1223 nanoparticle–superconductor composites were synthesized successfully by solid-state reaction method and characterized by different experimental techniques. Structural, morphological, compositional, superconducting, and dielectric properties were explored by XRD, SEM, EDX, and dielectric parameters of these composites. Granular structure with improved grain size could be observed by SEM micrographs. Tetragonal crystal structure of CuTl-1223 matrix was not altered after the addition of Zn nanoparticles, which provided evidence about their occupancy at the grain boundaries in the bulk CuTl-1223 material. The value of U was increased after the addition of Zn nanoparticles in CuTl-1223 composites. The improvement of superconducting properties could be accredited to the metallic nature of Zn nanoparticles, which could facilitate the mobility of carriers by improving the weak-links. Dielectric parameters have been decreased with the increase of frequency and became saturated at higher frequencies. The ac-conductivity (σ_{ac}) was seen to be increased non-monotonically due to heterogeneous distribution of Zn nanoparticles in CuTl-1223 matrix.

Acknowledgements

Higher Education Commission (HEC) of Pakistan is acknowledged for financial support through Project No. 20-1482/R&D/09-1472. Authors are also highly thankful to Prof. Xiang-Gang Qiu, Beijing National Laboratory of Condensed Matter Physics, Institute of Physics (IOP), Chinese Academy of Sciences (CAS), Beijing, China, for providing the characterization facilities.

REFERENCES

1. Yamamoto H, Tanaka K, Tokiwa K, *et al.* Synthesis of $Cu_{1-x}Tl_xBa_2Ca_2Cu_3O1_{1-y}$ (x~0.7) superconductor by hot press. *Physica C* 1998, **302**: 137–14
2. Awad R. Enhancement the formation of (Cu, Tl)-1223 superconducting phase by Cd-substitution. *J Alloys Compd* 2009, **474**: 517–521.
3. Ihara H, Tanaka K, Tanaka Y, *et al.* Mechanism of T_c enhancement in $Cu_{1-x}Tl_x$-1234 and -1223 system with T_c > 130 K. *Physica C* 2000, **3412-348**: 487–488.
4. Ihara H. How to achieve the best performance superconductor based on Cu-1234. *Physica C* 2001, **364–365**: 289–297.
5. Tanaka K, Iyo A, Terada N, *et al.* Tl valence change and T_c enhancement (> 130 K) in $(Cu,Tl)Ba_2Ca_2Cu_3O_y$ due to nitrogen annealing. *Phys Rev B* 2001, **63**: 064508.
6. Çavdar Ş, Koralay H, Tuğluoğlu N, *et al.* Frequency-dependent dielectric characteristics of Tl-Ba-Ca-Cu-O bulk superconductor. *Supercond Sci Technol* 2005, **18**: 1204.
7. Mohammed NH. Effect of MgO nano-oxide additions on the superconductivity and dielectric properties of

$Cu_{0.25}Tl_{0.75}Ba_2Ca_3Cu_4O_{12-\delta}$ superconducting phase. *J Supercond Nov Magn* 2012, **25**: 45–53.

8. Nkum RK, Gyekye MO, Boakye F. Normal-state dielectric and transport properties of In-doped Bi-Pb-Sr-Ca-Cu-O. *Solid State Commun* 2002, **122**: 569–573.

9. Younis A, Khan NA, Bajwa NU. Dielectric properties of $Cu_{0.5}Tl_{0.5}Ba_2Ca_3Cu_{4-y}Zn_yO_{12-\delta}$ (y=0,3) superconductors. *J Korean Phys Soc* 2010, **57**: 1437–1443.

10. Mumtaz M, Khan NA, Khurram AA. Enhanced superconducting properties of $Cu_{0.5}(Tl_{0.5-y}Hg_y)Ba_2Ca_3Cu_4O_{12-\delta}$ (y = 0, 0.15, 0.25, 0.35) superconductor. *J Alloys Compd*2008, **452**: 435–440.

11. Khan NA, Husnain G, Sabeeh K. Enhanced superconductivity in $Cu_{0.5}Tl_{0.5}Ba_2Ca_{n-1-y}Be_yCu_nO_{2n+4-\delta}$ (n = 3, 4 and y = 0. 7, 1.5, 1.7, 2.0) system with oxygen doping. *J Phys Chem Solids* 2006, **67**: 1841–1849.

12. Ihara H, Tokiwa K, Tanaka K, *et al.* $Cu_{1-x}Tl_xBa_2Ca_3Cu_4O_{12-y}$ ($Cu_{1-x}Tl_x$-1234) superconductor with T_c = 126 K. *Physica C* 1997, **2822-287**: 957–958.

13. Ben-Ishai P, Sader E, Feldman Y, *et al.* Dielectric properties of $Na_{0.7}CoO_2$ and of the superconducting $Na_{0.3}CoO_2 \cdot 1.3H_2O$. *J Supercond* 2005, **18**: 455–459.

14. Kamalasanan MN, Kumar ND, Chandra S. Dielectric and ferroelectric properties of $BaTiO_3$thin films grown by the sol-gel process. *J Appl Phys* 1993, **74**: 5679.

15. Xu X, Jiao Z, Fu M, *et al.* Dielectric studies in a layered Ba based Bi-2222 cuprate $Bi_2Ba_2Nd_{1.6}Ce_{0.4}Cu_2O_{10+\delta}$. *Physica C* 2005, **417**: 166–170.

16. Çavdar S, Koralay H, Altindal S. Effect of vanadium substitution on the dielectric properties of glass ceramic Bi-2212 superconductor. *J Low Temp Phys* 2011, **164**: 102–114.

17. Annabi M, Bouchoucha I, Azzouz FB, *et al.* Effect of ZnO and $Zn_{0.95}Mn_{0.05}O$ nano-particle inclusions on YBCO polycrystalline pinning properties. *IOP Conf Ser: Mater Sci Eng* 2010,**13**: 012009.

18. Mumtaz M, Khan NA. Dielectric properties of $Cu_{0.5}Tl_{0.5}Ba_2Ca_3Cu_4O_{12-\delta}$ bulk superconductor. *Physica C* 2009, **469**: 728–731.

19. Mumtaz M, Rahim M, Khan NA, *et al.* Dielectric properties of oxygen post-annealed $Cu_{0.5}Tl_{0.5}Ba_2Ca_3(Cu_{4-y}Cd_y)O_{12-\delta}$ bulk superconductor. *Ceram Int* 2013, **39**: 9591–9598.

20. Abdeen W, Mohammed NH, Awad R, *et al.* Influence of nano-Ag addition on phase formation and electrical properties of $(Cu_{0.5}Tl_{0.5})$-1223 superconducting phase. *J Supercond Nov Magn* 2013, **26**: 623–631.

21. Jabbar A, Mumtaz M, Nadeem K. Noble metals (Ag, Au) nanoparticles addition effects on superconducting properties of CuTl-1223 phase. *Eur Phys J Appl Phys* 2015, **69**: 30601.

22. Wagner KW. Zur Theorie der unvollkommenen Dielektrika. *Annalen der Physik* 1913, **345**: 817–855.

23. Tselev A, Brooks CM, Anlage SM, *et al.* Evidence for power-law frequency dependence of intrinsic dielectric response in the $CaCu_3Ti_4O_{12}$. *Phys Rev B* 2004, **70**: 144101.

CHAPTER 19

EXPERIMENTAL PRESENTATION OF MICROWAVE ABSORPTION DUE TO SHAKING OF JV BY AC MAGNETIC FIELD IN BI2212 AND BI2223

D. Shaltiel, H.-A. Krug von Nidda, B. Y. Shapiro, B. Rosenstein, A. Loidl, B. Bogoslavsky, I. Shapiro, T. Fujii, T. Watanabe, T. Tamegai

Keywords: High anisotropy superconductors, Bi-based cuprates, Josephson vortices and fluxons, Response to microwave and to magnetic field in superconductors, Superconductivity phase diagram

INTRODUCTION

Investigation of Josephson sine-Gordon solitons, known as fluxons, has drawn great interest in basic and applied physics [2]. Most investigations are performed in mesastructures of intrinsic Josephson junction IJJ stacks of superconducting layers coupled by Josephson effect. The linear size of sacks is usually less than 100 μm, where the Josephson solitons move along its junction [3]. The purpose of those investigations is to study the dynamic behavior of fluxons where emphasis is drawn on the collective motion of the Josephson vortices generated by electrical current and detected via the I–V curves [4]. The investigations are motivated for the possible application to generate radiation at THz frequencies [5].

In the present work the experimental method of "Induced Microwave Dissipation by AC Magnetic Field" (IMDACMF) [6] is introduced to study the dynamic behavior of solitons in bulk high-anisotropy single superconductor crystals of Bi2212 and Bi2223, via their interaction with the magnetic component of the microwave field. The interaction is attained by applying an AC magnetic field collinear with a DC magnetic field parallel to the a–b plane. The AC field varies the density of the fluxons due to their motion in and out of the superconductor, causing interaction with the microwave within its skin depth. The interaction induces a microwave dissipation caused by the collective motion of the JV; we show that the dissipation is by far larger than the dissipation due to the interaction of the microwave with the JV without the presence of the AC magnetic field [7]. Differences in the microwave power dissipation of more than two orders of magnitude were observed with or without the presence of AC field in Bi2212 crystals where a very similar experimental setup was applied. It indicates that the AC field strongly enhances the microwave dissipation in highanisotropy superconductors that results from the interaction of the JV with microwave magnetic field. The results indicate that, depending on the variables involved, the microwave response can be linear or nonlinear with the AC field. Furthermore, the signal phase with respect to the external AC phase can vary with the change in the variables involved.

Based on the present experimental results, the ACinduced microwave dissipation was recently interpreted theoretically in the framework of Josephson electrodynamics in layered superconductors [1]. It shows that the enhancement of the microwave dissipation in the presence of AC field results from the shaking of JV by the AC field. The IMDACMF technique introduces a new method to investigate the dynamics of fluxons. It is different from the IJJ method mentioned above. In the IJJ method the motion of the JV vortices is obtained by applying electric current perpendicular to the conducting planes of the superconductor and its effect is observed by DC resistivity and I–V characteristics measurements. The present method is contactless; the motion of fluxons is induced by an AC field collinear to the DC field and parallel to the conduction planes, and its effect is detected via microwave dissipation. It can be applied to investigate the effect of different treatments of the superconductor on the motion of the fluxons. A greater

Citation: D. Shaltiel, H.-A. Krug von Nidda, B. Y. Shapiro, B. Rosenstein, A. Loidl, B. Bogoslavsky, I. Shapiro, T. Fujii, T. Watanabe, T. Tamegai, Experimental Presentation of Microwave Absorption due to Shaking of JV by AC Magnetic Field in Bi2212 and Bi2223, DOI 10.1007/s10948-008-0407-z.

variety of samples, small and large single crystals, can be used with this technique.

It should be mentioned that after the discovery of high T_c superconductors, EPR spectrometers where extensively used to investigate their interaction with microwave [8]. Though interesting results where obtained, the technique was soon almost abandoned as it did not reveal additional substantial information. The present work shows a new experimental procedure to obtain results on the interaction of vortices with the microwave in high-anisotropy superconductors, using EPR spectrometers. In the following we describe the experimental set up and the measuring procedure. We then present the experimental results of the AC-induced microwave dissipation in the high anisotropy Bi2212 and Bi2223 crystals, as a function of the variables involved, namely, DC magnetic field, AC magnetic field and temperature, where similarities and differences in their behavior are exposed. The next section analyzes the experimental results. Then the experimental results are compared with the theoretical work [1]; it shows a good qualitative agreement with the experimental results. The discussion summarizes the important results and points out new possibilities to investigate the dynamics of JV by the IMDACMF technique.

EXPERIMENTAL

Experimental Setup

The experiment block diagram is shown in Fig. 1. It consists of a Bruker ELEXSYS continuous EPR spectrometer working at X-band frequency (9.36 GHz) and a digital oscilloscope. A microwave source feeds a cavity where highanisotropy Bi2212 or Bi2223 crystals ($1 \times 1 \times 0.1$ mm^3) are placed in the cavity region where only the microwave magnetic field is present. The sample, whose temperature could be varied down to helium temperature, was exposed to a collinear DC and AC (100 Hz to 100 kHz) magnetic fields. The microwave magnetic field was parallel to the a–b plane and perpendicular to the DC magnetic field. The AC magnetic field induces a signal into the microwave power reflected from the cavity. It is rectified by a diode that feeds either a lock-in detector or a digital oscilloscope. The lock-in gives the signal intensity and phase of the first AC harmonic; the oscilloscope displays also higher harmonics within its band width. The combined detectors present direct information of the microwave dissipation at the AC frequency. Unless otherwise specified, the magnetic fields were parallel to the a–b plane where only JV are formed in contrast to magnetic field parallel to the c-axis where only PV are formed (Fig. 2). The experiments were performed using the ZFC procedure, namely, the sample was cooled at zero magnetic field from above T_c to the required temperature. The AC or DC field was then set to a predetermined value and the DC or the AC magnetic field respectively was increased in steps. The signal was registered at each field. The reason for increasing the fields in steps and not continuously is that not only the signal intensity but also the AC frequency signal phase varied as a function of the variables, as is demonstrated in Fig. 3. Thus when measuring with the lock-in detector it was necessary to determine the phase at maximum signal for each measurement. Here the phase values are stated with respect of the phase of the external AC field h_{AC}. Analyzing the results obtained from the lock-in and from the digital oscilloscope gave a full picture of the signal response, including signal intensity, signal shape and signal phase. The method of aligning the crystal can be deduced from a previous work [9]. It is performed by rotating the crystal around an axis parallel to the c-axis [6, Fig. 2].

Two high-anisotropy compounds were investigated, an optimally doped Bi2212 crystal $T_c = 90$ K and several Bi2223 crystals whose T_c varied between 110 K and 105 K. First, the experimental results of the Bi2212 are presented, followed by the Bi2223 results.

Experimental Results of Bi2212

Figure 3 shows the oscilloscope signals of a Bi2212 crystal for the same AC field at two different DC fields (the upper frame), and for the same DC field at two different AC fields (the lower frame). The figure indicates that the signal can have a sinusoidal or a non-sinusoidal shape indicating linear or nonlinear response, respectively, for the same AC field or for the same DC field. The figure also shows that the AC phase of the fundamental harmonics depends on the intensities of the AC or DC field. Additional measurements have shown that the phase may depend also on the magnitude of other variables, such as temperature.

Figure 4 shows signal intensity as a function of temperature from slightly above T_c down to 4 K for magnetic field, with H_{DC} parallel to the c direction or parallel to the a–b plane. At magnetic field parallel to the c-axis, where only PV are formed (Fig. 2), the signal intensity down to 4 K is zero, except the narrow range signal just below T_c. It results from the variation of the thermal activated flux flow resistivity [10] (TAFF), and is reported in detail in reference [6]; it will not be discussed further. For H_{DC} parallel to the a–b plane where only JV are formed (Fig. 2) the signal intensity increases sharply at T_c and even further down to 4 K. The results in Fig. 4 indicate that the AC magnetic field induces strong microwave dissipation when interacting with JV but almost none when interacting with PV.

The lock-in signal intensity and signal phase as a function of DC fields at 40 K at a low-AC field and a highAC field of 0.1 mT and 1.2 mT, respectively, are shown in Fig. 5a. Different functional behavior is observed at these two AC fields. At the low-AC field the signal intensity below $H_{DC} = 4$ mT is small. An asymmetric bell-shaped curve is observed at higher fields, with a stronger slope in the low field side and a maximum at 13 mT. The signal phase is close to $-180°$ in the whole field region. The high-AC field signal intensity has a maximum at 2 mT, the intensity which is more than three orders of magnitude larger than the intensity of the low-AC field signal with the same DC field. This maximum decreases strongly, followed by a second broader maximum at 19 mT. The signal phase associated with the first maximum is around $-30°$: it decreases sharply at higher fields and tends asymptotically to $-160°$ at the highest DC field.

The digital oscilloscope signals along the DC field for the low-AC field and the high-AC intensity curves of Fig. 5a are shown in Figs. 5b and 5c respectively. Figure 5b shows that for $h_{AC} = 0.1$ mT the signal intensity at the two low-DC fields is weak and its signal shape deviates strongly from the sine, indicating a nonlinear response. The intensity increases strongly at higher DC fields and its signal shape is close to sine, indicating a linear response. Figure 5c shows that the signal shape for h_{AC} = 1.2 mT at the magnetic field close to the first maximum is close to sine, indicating a linear response. It starts to deviate from the sine already at 10 mT with maximum deviation from the sine at 15 mT. At higher fields the deviation from sine shape decreases with increasing field. Thus the signal shape in Figs. 5b and 5c clarifies the changes in the signal intensities and phases as a function of DC field, as was shown in Fig. 4a.

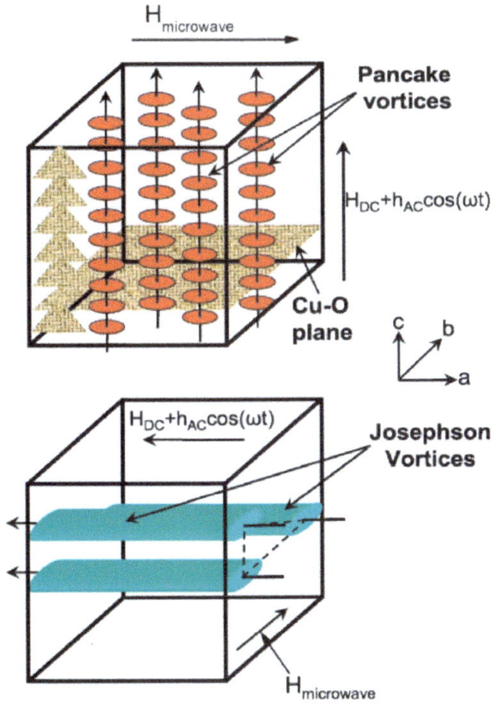

Figure 2: Vortex structures in high-anisotropy superconductors for magnetic field $H_{DC}\|_c$-axis where only PV are formed and $H_{DC}\|a-b$ plane where only JV are formed.

Figure 6a shows the lock-in signal intensity and phase as a function of AC fields at 40 K at a low-DC field of 2.5 mT and at a high-DC field of 22 mT. The two DC fields are close to the two maxima shown in Fig. 5 for h_{AC} = 1.2 mT. Different behavior is observed at these two curves. At the low-DC field's maximum curve, the lock-in signal intensity is close to zero below a threshold AC field. Above this threshold the microwave dissipation increases linearly. At the high-DC field's maximum curve the signal intensity increases linearly from zero up to about h_{AC} = 0.5 mT, then it saturates strongly up to the highest AC field. The phase of the low-DC field signal is close to zero (−25° on average). The phase of the high-DC signal is equal to −180° at low-AC fields and tends asymptotically to −140° at the highest AC field. The digital oscilloscope signals along the curves of the low-DC field maximum and the high-DC maximum are shown in Figs. 6b and 6c respectively. Figure 6b shows that at h_{AC} = 0.1 mT the low-DC signal intensity is weak and its signal shape deviates strongly from sine, indicating a nonlinear response. The intensity increases strongly at higher AC fields and its signal shape is close to the sine, indicating a linear response. Figure 6c shows that the high DC field signal shape at h_{AC} = 0.1 mT is the sine indicating a linear response. At higher AC field the signal shape deviates from the sine indicating a nonlinear response. Thus the signal shape in Fig. 6b and Fig. 6c clarifies the changes in the signal intensities and phases as a function of AC field shown in Fig. 6a

Figure 7 presents the temperature evolution of the signal intensity as a function of DC field from 80 K to 4 K for the "low-AC" field (0.1 mT) and the "high-AC" field (1.2 mT). It shows that the intensity increases with decreasing temperature as also is indicated in the insets of the figure.

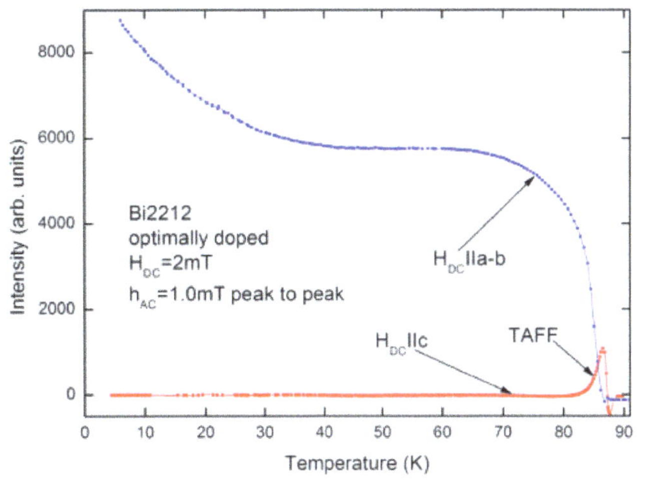

Figure 4: Intensity as a function of temperature when applying magnetic field parallel and perpendicular to the conduction planes. At H||a − b a strong signal is observed from T_c down to 4 K. The intensity as a function of temperature for H||c is zero except for the FFR signal just below T_c.

Dependence of Microwave Dissipation on the Microwave Power

The applied microwave magnetic field in the present work was 0.007 mT (6.3 mW in a cavity with Q = 3300). The interaction of the JV with the microwave for different microwave power, Fig. 8, indicates that the signal intensity for the same AC and DC fields is proportional to the square root of the microwave power implying an ohmic behavior. Similar ohmic behavior was observed by Enriquez [11] for the same experimental setup but without the application of the AC field. The unusual high intensity response of microwave dissipation by AC magnetic field in Bi2212 was not observed in the low-anisotropy optimally doped $YBaCuO_7$ superconductor with magnetic fields parallel to the a–b plane. A fundamental difference between these two compounds regarding Josephson vortices is that in the high-anisotropy Bi2212 the coupling mechanism between the layers is due to the Josephson effect. To find out if this dissipation is related to the JV inter-layered coupling, we have extended the investigated to Bi2223. Though its anisotropy is slightly lower than that of Bi2212 [12], the coupling mechanism between the layers is due to the JV.

Experimental Results of Bi2223

Several Bi2223 crystals were investigated whose T_c ranges from 105 K to 111 K. Here we present the results of a T_c = 105 K crystal the transition temperature of which though was slightly lower than in other crystals, the results show a clearer behavior attributed to smaller number of defects [13]. Figure 9a shows the signal intensity as a function of temperature for H||a − b and H||c, with H_{DC} = 1 mT and h_{AC} = 0.4 mT. At H||c the signal just below T_c results from the TAFF resistivity discussed earlier. A strong anomalous signal with a maximum at 84.5 K and a width of 4 K is observed. The signal intensity below 85

K is practically zero. At H||a–b the signal intensity increases slightly and gradually down to around 88 K, where a strong anomalous signal is observed at 83 K, similar in shape and intensity to the anomalous signal observed for H||c. Below the anomalous signal, a relatively-high intensity signal is observed down to 4 K. At high-anisotropy superconductors the TAFF signal intensity for H||a − b is almost zero [6]. The small TAFF signal observed at this orientation in Fig. 9a results from a slight deviation of the magnetic field (about 3° away the a–b plane). Indeed, when aligning the crystal almost parallel to the a–b plane, the small TAFF signal just below Tc is not observed (Fig. 9b). Figure 10 shows the lock-in signal as a function of DC field at 4 K for a low-AC and a highAC field of 0.1 mT and 1.2 mT, respectively. The low-AC field signal intensity has an asymmetric bell-shaped curve with a smaller slope at the high field side with the maximum at 8 mT. The high-AC field signal intensity has two maxima. The maximum of the low field signal is at 1 mT and its intensity is more than 2 orders of magnitude stronger than the signal intensity of the low-AC field intensity signal at the same DC field. The intensity below this maximum decreases sharply with the field and then increases towards a second maximum. Above the second maximum, the intensity decreases gradually towards zero at 100 mT. The signal phase of the low-AC field is about −180°. The signal phase of the high-AC field has a maximum of −60° followed by a sharp decrease that tends asymptotically to −160° at fields higher than 100 mT.

Figure 11a shows the lock-in signal intensity and phase at 20 K as a function of AC field for a low-DC field and a high DC field of 6 mT and 20 mT, respectively. The low and high DC field intensities were chosen to be close to the maxima observed in Fig. 10. At the low-DC field the signal intensity above a threshold AC field increases almost linearly up to about h_{AC} = 2 mT and then it starts to saturate. The signal intensity at the high-DC field increases linearly already from zero but saturates already at h_{AC} = 0.6 mT. The phase of the low-DC field is close to −180° at AC field of 0.1 mT and increases sharply at higher AC field. The phase of the high-DC field signal has an almost constant value of −170° up to h_{AC} = 0.3 mT and increases gradually to −130° at the highest h_{AC} field of 3 mT. The digital oscilloscope signals along the two curves of Fig. 11a are shown in Figs. 11b and 11c. They clarify the strong changes in the intensity and phase as a function of AC field observed in the two curves of Fig. 11a.

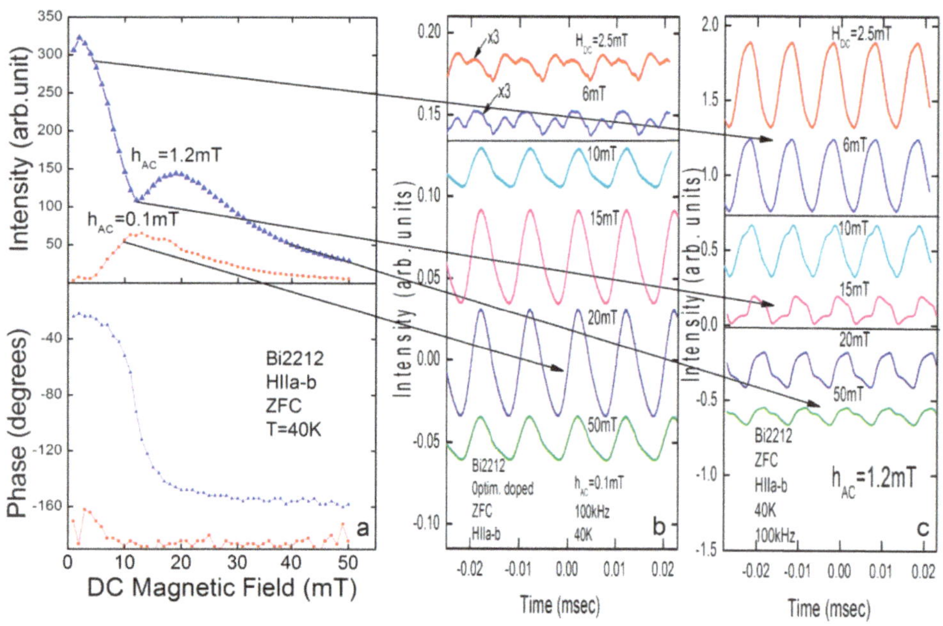

Figure 5: (a) Lock-in signal intensity and phase as a function of DC fields at 40 K at a low-AC field and at high-AC field of 0.1 mT and 1.2 mT respectively. (b) Oscilloscope signal along the curve of the low-AC field shown in (a). (c) Oscilloscope signal along the curve of the high-AC field shown in (a).

The evolution of the signal intensity as a function of DC field from 90 K to 4 K for a low-AC field of 0.3 mT and a high-AC field of 1.2 mT is shown in Fig. 12.

ANALYSIS OF THE EXPERIMENTAL RESULTS

Bi2212

Figure 4 shows that the AC field collinear with the DC field induces microwave dissipation as a function of temperature down to 4 K for magnetic field parallel to the Cu planes, where only JV are formed. There is no microwave dissipation for magnetic field parallel to the c-axis where only PV are formed. It leads to the fundamental conclusion that the induced microwave dissipation results from the AC interaction with JV but not with PV. The increase of the signal intensity with decreasing temperature for H||a − b (Fig. 4) signifies that the dissipation does not result from a static property, as decrease in the absorbed microwave power with decreasing temperature is expected due to pinning. Thus the observed signals result from dynamic interaction of the moving fluxons with the microwave field, within the superconducting microwave skin depth, whose motion is induced by the AC field. Figure 3 shows that the time dependent of the microwave dissipation depends strongly on the variables involved, such as DC magnetic field and AC magnetic field. Small changes in one parameter may change the signal shape from sinusoidal to non-sinusoidal and vary the phase more than 90°. These changes are related to the variation of the signal phase and shape as a function of DC and AC fields shown in Figs. 5 and 6 respectively.

The properties of this dynamic interaction as a function of the DC and AC magnetic fields at 40 K are manifested in the signal intensity and phase in Fig. 5a and Fig. 6a, and their corresponding signal shapes are displayed in Figs. 5b–5c and 6b–6c, respectively. The lock-in signal intensities and their corresponding phases at 40 K as a function of DC field (Fig. 5), indicate that the functional behavior at the 0.1 mT low-AC field differs substantially from the 1.2 mT high-AC field. It confirms the conclusion obtained from Fig. 3 that the AC-induced interaction with the fluxons depends strongly on the values of the variables involved. The low-AC field curve has an asym metric bell-shaped signal whose phase is close to −180° (Fig. 5a). Its signal shape is sinusoidal indicating a linear response (Fig. 5b). Thus linear response occurs at a phase values close to −180° at low-AC field. The high-AC field curve shows two maxima, at a low-DC field of 2 mT and at a high-DC field of 15 mT (Fig. 5a). The intensity of the low- field maxima is by far larger than the intensity at the low-AC field for the same DC fields. The phase of this low-field maximum is close to zero (Fig. 5a) and its line shape (Fig. 5c) is close to the sine, indicating that for phase close to zero a linear response is observed. At magnetic fields above this maximum, there is a strong decrease of the signal intensity followed by a broad maximum. The phase decreases rather sharply with the decrease of the intensity at higher DC fields and tends asymptotically toward −160°. The line shape in this field range deviates from sinusoidal shape (Fig. 5c) indicating a nonlinear response. Thus Fig. 5 shows that the signal intensity as a function of DC field is strongly affected when exposed to low-AC or to high-AC field.

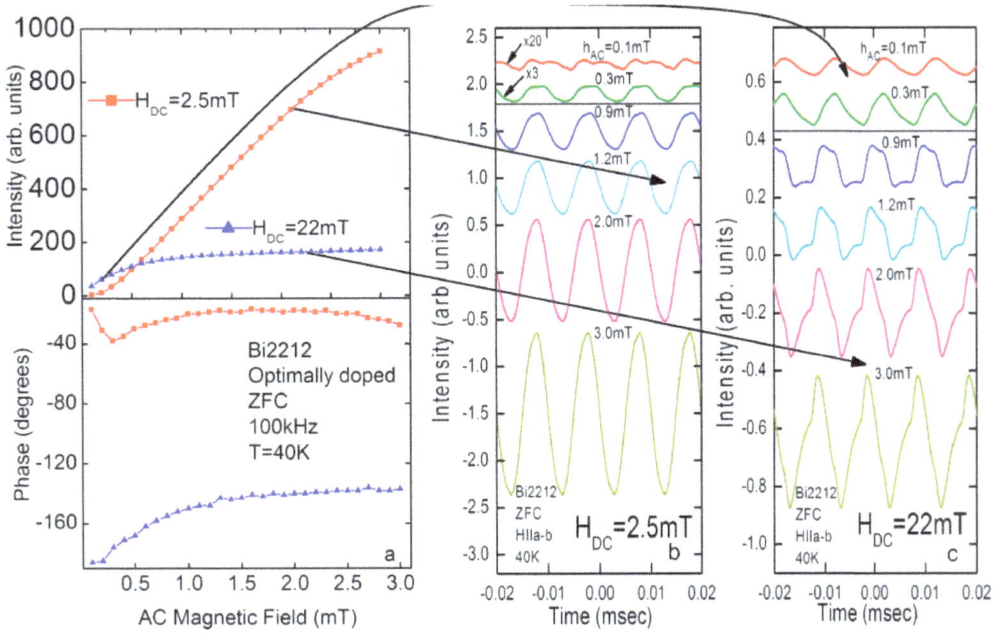

Figure 6: (a) Lock-in signal intensity and phase in Bi2212 as a function of AC fields at 40 K at H_{DC} = 2.5 mT and H_{DC} = 22 mT, corresponding to the two DC fields where the respective maxima of the signal intensity as function of DC shown in Fig. 5 were observed. (b) Oscilloscope signal along the curve for DC field of 2.5 mT shown in (a). (c) Oscilloscope signal along the curve for DC field of 22 mT shown in (a)

A clearer behavior of the effect of the AC field on the microwave dissipation is obtained from the signal intensities as a function of AC field for a 2.5mT at the low-DC field and 22 mT at a high-DC field, where the two DC fields correspond to the position at the low and at the high DC field maxima with h_{AC} = 1.2 mT of Fig. 5. Strong differences were observed in the signal intensities and signal phases as a function of AC field at 40 K at these two DC fields (Fig. 6). It implies again that the AC interaction with the solitons strongly depends on the values of the variables involved. There is a sharp contrast in the functional dependence at the two fields: The low-DC field signal intensity is very small: up to h_{AC} = 0.3 mT it increases linearly up to h_{AC} = 2.5 mT. The high-DC field signal increases linearly only up to $h_{AC} \approx$ 0.3 mT and remains almost constant up to h_{AC} = 3 mT (Fig. 6a). The signal phase value of the low-DC is close to zero. The high-DC signal phase is close to −180° up to hAC ≈ 0.3 mT then it increases with AC field, saturating at −140° at the maximum AC value of 3 mT (Fig. 6a). The signal shape of the low-DC field signal is sinusoidal indicating a linear response in the whole AC field range where the phase is close to zero (Fig. 6b), while the signal shape of the high-DC field is sinusoidal only up to h_{AC} = 0.3 mT where the phase is close to −180°, and deviates from the sine shape above this value (Fig. 6c).

The phase behavior as a function of DC field is similar to the phase behavior as a function of AC field. In both cases, a linear response is observed at phase values either close to 0° or to 180°. Nonlinear response is observed in the narrow region where the phase values shift from close to 0° toward 180°.

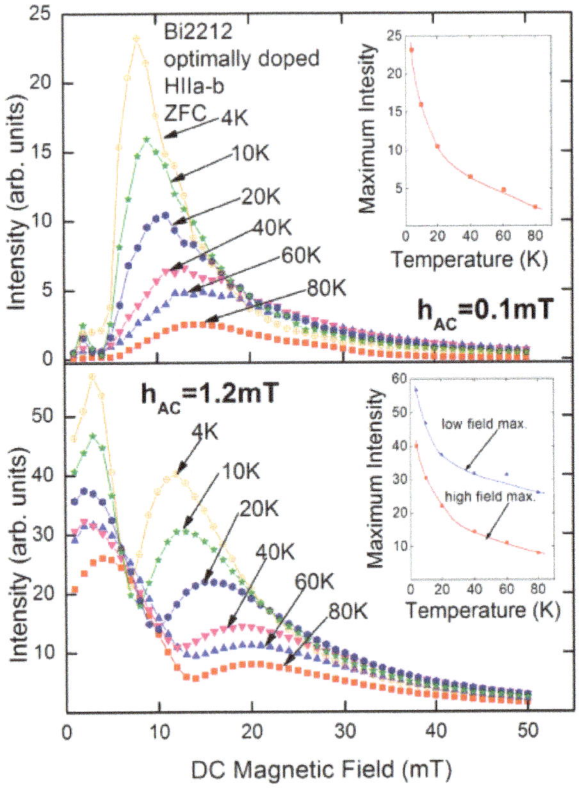

Figure 7: Temperature evolution of the signal intensity as a function of a DC field from 80 K to 4 K for a low-AC field (0.1 mT) and a high-AC field (1.2 mT). It shows that the intensity increases with decreasing temperature as also indicated in the insets.

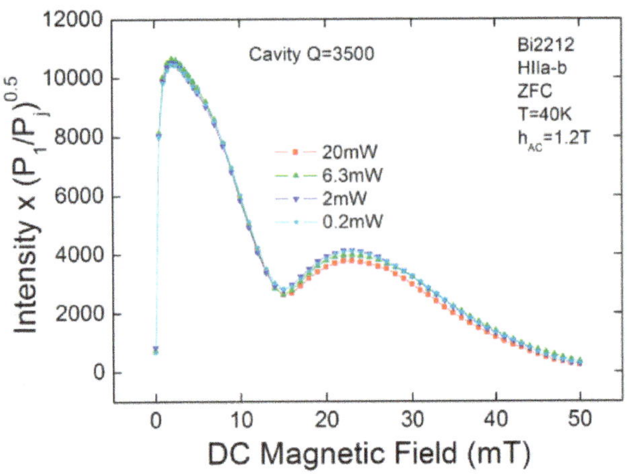

Figure 8: Normalized signal intensities as a function of DC magnetic field, for microwave power difference of up to two orders in magnitude.

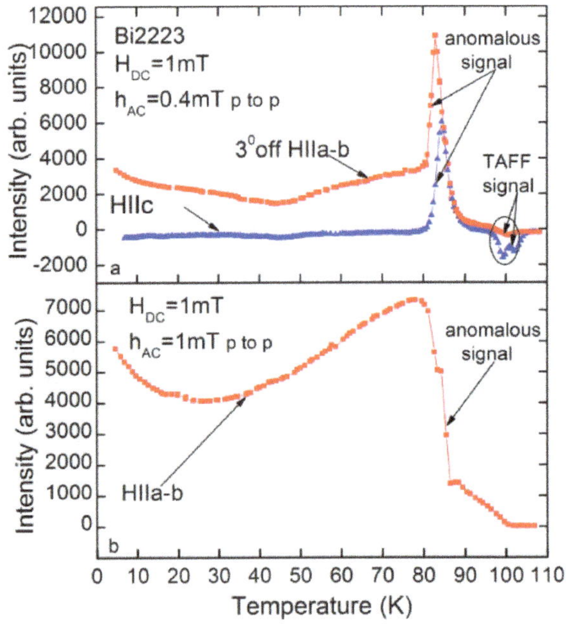

Figure 9: (a) Signal intensity as a function of temperature in Bi2223 for magnetic fields: for H almost parallel to the a–b plane (30 off) and for H parallel to the c-axis. The strong signals at 85 K for both orientations represent an anomalous signal. The small signals just below T_c result from TAFF resistivity. (b) Signal intensity as a function of temperature in Bi2223 for H parallel to the a–b plane. The absence of the small TAFF resistivity signal just below indicates that it is aligned almost parallel to the conduction planes

Comparison between Bi2212 and Bi2223

Induced microwave dissipation by AC field observed in Bi2212, where the layered superconductors are coupled via the Josephson effect, is not observed in optimally doped YBCO where there is no Josephson coupling between the layers. We deduce that coupling between the conducting layers by JV is a necessary condition to induce the microwave dissipation. If this assumption is correct the AC field should induce microwave dissipation in Bi2223 similar to that observed in Bi2212 as a function of any of the variables, namely DC magnetic field, AC magnetic field and temperature should show similar behavior. Indeed, the analysis presented below of the experimental results in the two compounds displays very similar behavior.

(a) Signal as a function of temperature, Figs. 4 and 9.

The signal intensity in both compounds as a function of temperature at magnetic fields parallel to the a–b plane shows strong signal intensities from 85 K down to 4 K with similar behavior. At magnetic fields parallel to the c-axis, a FFR signal is observed just below T_c with zero signal intensity at lower temperatures.

(b) Signal as a function of DC magnetic fields at a constant temperature, Figs. 5 and 9.

Applying a low-AC field in the two compounds induces a similar asymmetric bell-shaped signal with a signal phase of −180°. Two maxima are observed when exposed to a highAC field. The intensity at the lower DC field maximum is strong and its width is narrower compared to the width of the higher

DC field maximum. The phase at high-AC field is closer to zero for the first maximum, and it decreases sharply with DC field tending asymptotically to −160° at higher DC field.

(c) Signal as a function of AC magnetic field at a constant temperature, Figs. 4 and 10.

In both compounds the signal intensities as function of AC field for the low-DC field (above a threshold value) increases linearly with the AC field. For the "high-DC" field the intensity increases linearly already from zero AC field and saturates above a small value of the AC magnetic field. Their corresponding phases at low and high DC fields show similar behavior (Figs. 4a and 8a). Comparison of the digital oscilloscope signals in Bi2212 (Figs. 4b and 4c) with those in Bi2223 (Figs. 10b and 10c), shows similar linear and nonlinear response at low and high intensity of the AC field.

(d) Variation of the signal intensity for temperatures 80 K to 4 K as a function of DC magnetic field at low-AC and at high-AC magnetic field, Figs. 7 and 11.

The signal intensities increase with decreasing temperatures for temperatures below 80 K in Bi2212, and for temperatures below 40 K in Bi2223. The very similar behavior as a function of AC and DC field in these two high-anisotropy superconductors indicates that the origin of the induced microwave dissipation stems from interaction of AC field with fluxons that reside between conducting planes.

Although the two compounds exhibit similar behavior, there are some significant differences. The sharp increase of the signal intensity just below T_c in Bi2212 for H||a − b (Fig. 4) is not observed in the Bi2223 close and below T_c (Fig. 9). Instead, a gradual increase in the intensity with decreasing temperature is followed by a strong anomalous signal at about 85 K as shown in Figs. 9a and 9b. For H||c the signal intensity in Bi2212 at temperatures below the FFR signal is zero down to 4 K, whilst in Bi2223 a strong anomalous signal is observed at about 85 K, very similar to the anomalous signal observed at H||a − b. Our results show that the intensity of the anomalous signal decreases more than two orders of magnitude when increasing the DC field from 0.004 T to 0.03 T. Thus the interaction of the AC field with the vortex system below T_c down to 80 K in Bi2223 is different than expected assuming that the response below T_c for the two compounds should be similar. However, the two compounds show similar behavior below 80 K, including the sharp increase in the signal intensity in Bi2223 (Fig. 9b).

COMPARISON BETWEEN THE EXPERIMENTAL RESULTS AND THEORY

Based on the present experimental results, the microwave dissipation in layered highly-anisotropic superconductors subjected to a magnetic field parallel to the layers has been studied theoretically [1]. The analysis of the microwave dissipation in the presence of an AC field is based on Josephson electrodynamics in layered type II superconductors. The reader is referred to the article that discusses the microwave dissipation as a function of the DC field and temperature with and without the application of AC field. It shows that the addition of the low frequency AC magnetic field greatly enhances the microwave signal. It is explained by AC "shaking" effect that depins the Josephson vortices. Thus the theoretical article, referred here as the Theory, and the present experimental work, are complementary. In order to present to the reader a more comprehensive discussion without referring him continuously to the theoretical work, we bring below the experimental fit to the theoretical results including important conclusions.

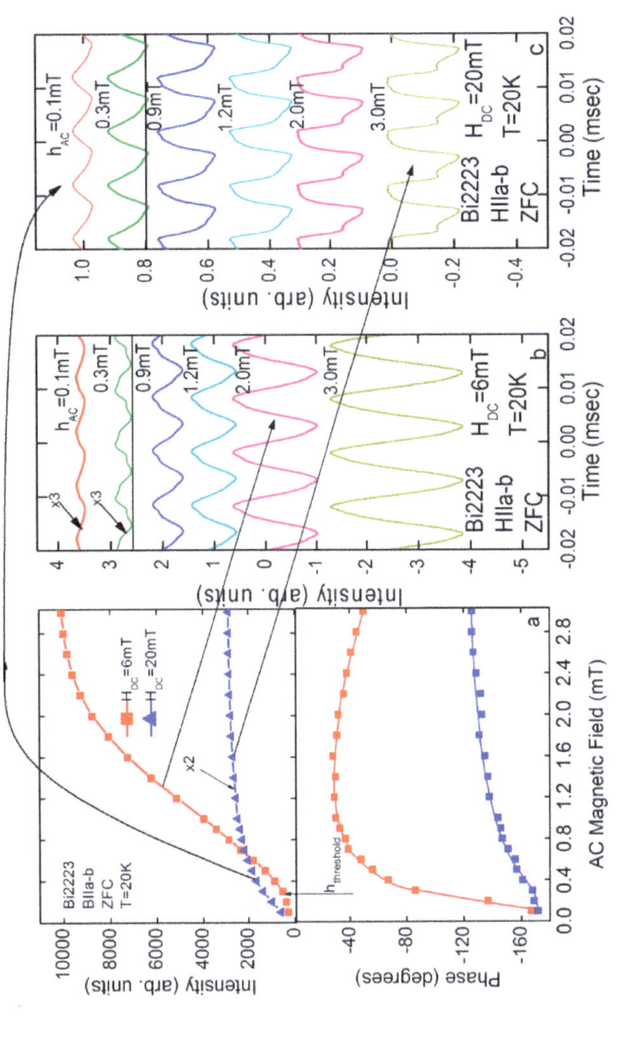

Figure 11: (a) Lock-in signal intensity and phase in Bi2223 as a function of AC field at 40 K at a low-DC field and a high-DC field of 6 mT and 20 mT, respectively. (b) Oscilloscope signal along the curve of the low-DC field shown in (a). (c) Oscilloscope signal along the curve of the high-DC field shown in (a).

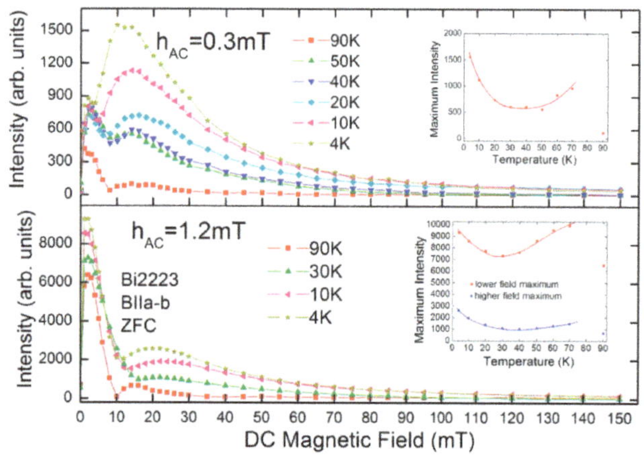

Figure 12: Evolution with temperature of the signal intensity as a function of a DC magnetic field in Bi2223 for AC fields of 0.3 mT and 1.2 mT. The inserts show the variation of the maxima of these signals with temperature.

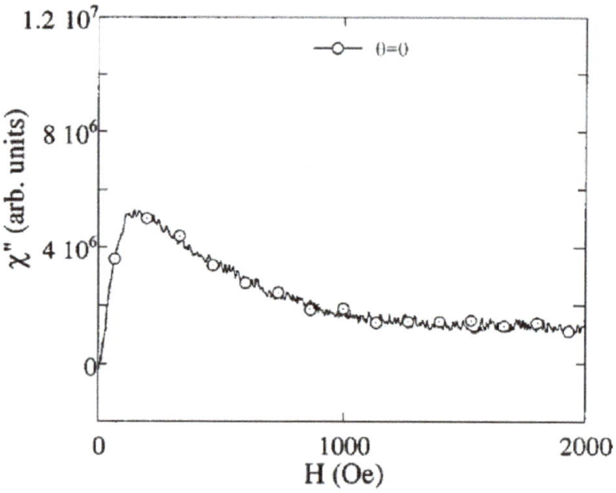

Figure 13: Bi2212 microwave susceptibility as a function of magnetic field at 70 K obtained by microwave power modulation and not by of AC magnetic field modulation for $H_{DC}||a-b$ ($\theta = 0$) [7].

The experimental results in the high-anisotropy superconductors Bi2212 and Bi2223 show, in general terms, the presence of two DC magnetic field regions: a low-DC and a high-DC region in which the AC field induces sinusoidal shape signals that are in phase with and out of phase to the AC frequency, respectively. In the small in-between DC field region the signal deviates from the sine form and the phase of the fundamental frequency varies sharply with increasing field, from close to 0° toward 180°. The microwave dissipation that results from the interaction of the microwave magnetic field with JV in Bi2212 as a function of a DC magnetic field parallel to the a–b plane, but without the presence of the AC field, has been investigated for DC magnetic field parallel to the a–b plane by Enriquez et al. [7]. The experimental setup was identical to the present experiment using microwave power modulation instead of the magnetic field modulation in the present work. Results of the power dissipation as a function of magnetic field at a temperature of 70 K without AC field [7] are shown in Fig. 13. The signal intensity in Enriquez experiment were a few orders of magnitude smaller than those obtained in the present experiment, indicating that the AC field strongly enhances the interaction of the JV with the microwaves in accordance with the shaking effect.

Figures 14, 15 and 16 correspond to Figs. 4a, 5 and 3 of the Theory article [1]. Figure 14 presents a schematic plot of the calculated signal intensity as a function of a DC magnetic field. It shows the presence of two linear regions separated by a nonlinear region in which the phase shifts from 0° at low field to 180° at high field. This result is in agreement with the results of Fig. 5 and Fig. 10 for Bi2212 and Bi2223 respectively (at the high-AC field of 1.2 mT) that show two linear regions whose phase varies from close to 0°, at low-DC fields, to close to −180° at high-DC fields, separated by a nonlinear region where the phase shifts continuously toward −180°.

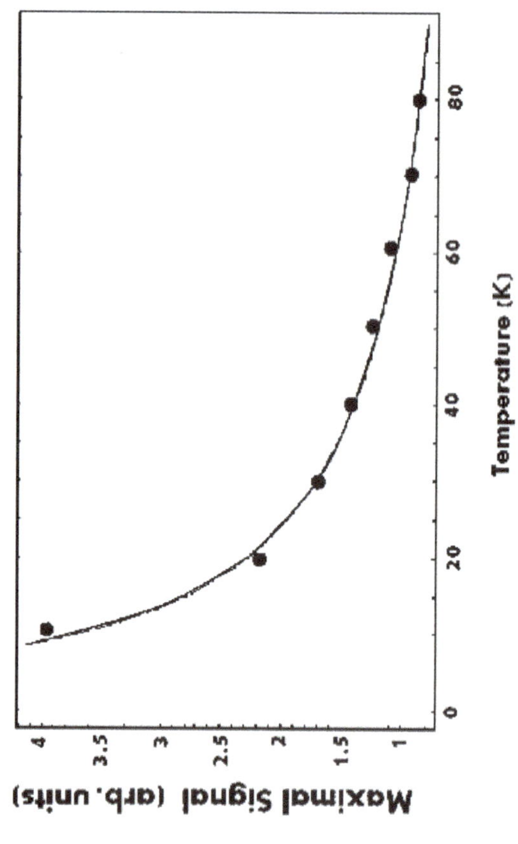

Figure 15: Fit of the calculated maximum intensity in the signal at temperatures 80 K to 10 K with the measured intensities for $h_{AC} = 1.2$ mT of Fig. 6.

Figure 15 presents the exponential fit to the maximum signal intensity, S, as a function of temperature shown in Fig 6. This exponential fit $S = T^{-\alpha}$ was derived from (42) with $\alpha = 0.7$ of the Theory paper. A remarkable result of the theoretical work was obtained when fitting the experimental results of Enriquez et al. [7] that show the microwave dissipation as a function of DC field in Bi2212 without the application of AC field, Fig. 13. The full circles were fitted in the theoretical work using (36) of the theory article [1]. That infers that the interaction of the JV with the microwave magnetic field with and without the presence of the AC field originates from the same mechanism. The AC field induces the shaking of the JV that strongly enhances the microwave dissipation.

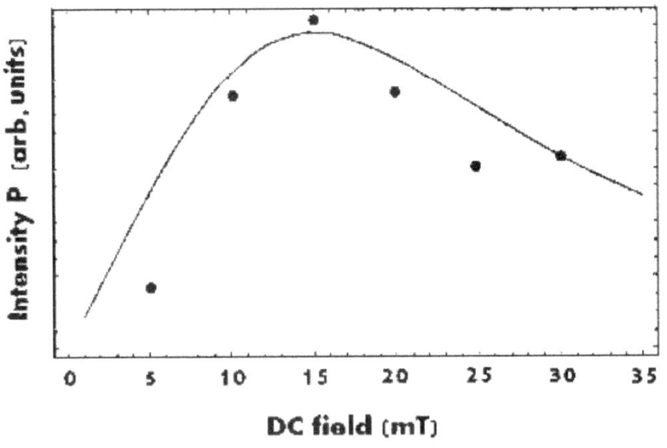

Figure 16: Theoretical fit to the microwave absorption intensity as a function of HDC, without the presence of the AC field, observed by Enriquez et al. in Bi2212 [7].

DEPENDENCE OF MICROWAVE DISSIPATION ON THE MICROWAVE POWER

The applied microwave magnetic field in the present work was 0.007 mT (6.3 mW in a cavity with Q = 3300). The interaction of the JV with the microwave for different microwave power in Fig. 8 shows that the signal intensity for the same AC and DC fields is proportional to the square root of the microwave power implying an ohmic behavior. A similar ohmic behavior was observed by Enriquez [11] using the same experimental setup but without the application of the AC field (see Fig. IV.2.1 in reference [11]). It implies that the JV interaction with the microwave field is not affected by the application of the AC field.

DISCUSSION

The present work investigates experimentally dynamic properties of JV in the highly anisotropic superconductors Bi2212 and Bi2223. The JV are generated by applying a DC magnetic field parallel to the conducting plane. Their motion is caused by AC magnetic field collinear to the DC field and detected by microwave dissipation due to interaction of the JV with the microwave magnetic field. Thus three magnetic fields, DC, AC and microwave, affect the intensity of the interaction. In addition, other variables such as temperature, frequency of the AC magnetic field and pinning center can affect the JV interaction with the microwave. Deviation of the magnetic field from the a–b plane [14] introduces PV that impedes the motion of JV. We have shown that small variations of any of the above variables can strongly influence the experimental results. The strong dependence of the observed signal on the large number of parameters made the results complicated to analyze and difficult for the reader, as it is necessary to specify the different parameters when analyzing or presenting the results. Initially the different response of the microwave dissipation as a function of a DC magnetic field with or without the application of an AC field raised difficulties in the analysis of the results. However, the theoretical work presented recently in the theory article has shown that both dissipations stem from the interaction of the JV with the microwave and that the differences in their functional behavior result from the shaking of the JV by the AC field.

Ohmic dissipation of the microwave in the power range 0.2 mW to 20 mW with and without and the application of the AC field observed in the present experiment and in Enriquez experiment, respectively, implies that the interaction of the AC magnetic field—but not the microwave magnetic field—is responsible to the observed strong power dissipation. Hence the AC modulating field has an active role in inducing strong power dissipation, in contrast to the passive modulating role observed in electron spin resonance experiments. The theory article [1] explains the results shown in Figs. 6 and 8 that indicate the presence of two regions at a low-DC and a high-DC fields, in which the dissipation signals differ by 180° and the response to the AC field is linear. Inbetween these two regions the signal phase shifts from close to 0° to close to 180° with a nonlinear signal shape. The strong increase of the microwave dissipation with decreasing temperature shown in Fig. 7 was not observed in microwave dissipation measurements by Enriquez [11] where the AC field was absent. This difference results from the AC field shaking effect, as described in the theory work and presented in Fig. 14.

The response of the signal as a function of DC field for AC frequencies 0.1 kHz to 100 kHz for the same AC field was very similar. It indicates that the response in this AC frequency range is not affected by frequency. The almost identical behavior of the signal intensity in Bi2212 and Bi2223 confirms the theoretical investigation that the microwave dissipation results from the AC-shaking of the JV in high-anisotropy superconducting compounds. It therefore predicts that similar behavior in the microwave dissipation should be observed in all highly anisotropic superconductors. A list of these compounds is discussed by Shimoya et al. [12]. Thus one would expect to observe IMDACMF signals in the under-doped high anisotropy superconductor YBCO6.6 (T_c = 60 K) where the coupling between the layers is through the Josephson effect; it has not been observed in the low-anisotropy optimally doped YBCO (T_c = 89 K) where there is no Josephson coupling between the layers.

The functional dependence of Bi2212 and Bi2223 at temperatures below 80 K is similar, in agreement with the theory. Above 80 K the response to the AC field shows different behavior, including the presence of an anomalous signal in Bi2223, observed also at magnetic fields parallel to the caxis. It may indicate the presence of a different JV lattice structure above 80 K that so far has not been observed by other methods. Figures 6 and 11 show that the signal intensity asa function of the AC field at a high-DC field is saturated above a certain AC field. The reason for this saturation is not clear. It can be due to the depinning force induced by the AC field. A further investigation is needed. No AC-induced microwave dissipation is observed for magnetic fields parallel to the c-axis where only PV are formed. The unique property where AC field induces the shaking of only JV and not PV opens a vast field of investigation where the dynamics of JV exposed to various pinning centers can be investigated. Indeed recently, by using this technique, increase of the interaction between PV and JV obtained by tilting the magnetic field away from the a–b plane has revealed a strong effect of PV on the dynamics of JV [14]. Also this technique has shown a memory effect where a magnetic field can be imprinted into the JV structure [15] and irreversibility behavior of the JV system that results from field cooling procedure [16].

In the present work the AC microwave dissipation has been investigated using an EPR spectrometer at 9.3 GHz. EPR spectrometers, whose frequency range is from a few MHz to hundreds of GHz have been reported [17]. Applying the technique using EPR spectrometer that operates a different frequencies would provide information on the interaction of fluxons with the electromagnetic field in a very large frequency range. The technique can also be used to investigate interaction of JV with different defects, including irradiation defects, by comparing the microwave dissipation results before and after introducing the defects. Intense research has been carried out on the dynamic properties of solitons in intrinsic Josephson junction IJJ stacks of highanisotropy superconductors, where the motion of the JV is generated by a DC current parallel to the c-axis and detected by I–V curves [18]. It would be interesting to compare the present results with a technique where the motion of the JV is generated by AC current parallel to the c-axis and detected via microwave dissipation [19]. It can clarify the difference between the two types of induced motion.

Acknowledgements

This work was supported by The Israel Science Foundation Grant No. 499/07, by Heinrich Hertz Minerva Center for High Temperature Superconductivity, by the German Bundesministeriu für Bildung und Forschung BMBF under Contract No. VDI/EKM 139617, by the Deutsche Forschungsgemeinschaft DFG within Sonderforschungsbereich SFB 484 Augsburg, and partly supported by Grant-in-Aid for Scientific Research from the Ministry of Education, Culture, Sports, Science, and Technology, Japan. B.R. is supported by DOE and NSC of R.O.C. Grant No. M93-2112-M009-023. One of us—D.S.—would like to thank O. Shaltiel for analyzing the experimental data.

REFERENCES

1. Shaltiel, D., Krug von Nidda, H.-A., Loidl, A., Rosenstein, B., Shapiro, B.Ya., Shapiro, I., Tamegai, T., Bogoslavski, B.: Phys. Rev. B 77, 014508 (2008)
2. Ustinov, A.V.: Physica D 123, 315 (1998)
3. Clauss, T., Oehmichen, V., Mössle, M., Müller, A., Webber, A., Koelle, D., Kleiner, R.: Supercond. Sci. Technol. 15, 1651 (2002)
4. Hechtficher, G., Kleiner, R., Schlenga, K., Walkenhorst, W., Muller, P.: Phys. Rev. B 55, 14638 (1997)
5. Filatrella, G., Rotoly, G., Grombench-Jensen, N., Pormetier, R.D., Pedersen, N.F.: J. Appl. Phys. 72, 3179 (1992)
6. Shaltiel, D.: J. Low Temp. Phys. 103, 383 (2003)
7. Enriquez, H., Bontemps, N., Foumier, P., Kapitulnik, A., Maignon, A., Ruyter, A.: Phys. Rev. B 53, R14757 (1996)
8. Bohandy, J., Kim, B.F., Adrian, F.J., Moorjani, K., Darcangelis, S., Cowan, D.O.: Phys. Rev. B 43, 3724 (1991)
9. Shaltiel, D., Ginodman, V., Golosovsky, M., Katz, U., Boasson, H., Gerhouser, W., Fisher, P.: Physica C 202, 303 (1992)
10. Tinkham, M.: Phys. Rev. Lett. 61, 1658 (1998)
11. Enriquez, H.: These de doctorat de l'Université de Paris (1997)
12. Shimoya, J., Kitazava, K., Shimizu, K., Ueda, S., Orii, S., Chikumoto, N.: J. Low Temp. Phys. 131(5/6) (2003)
13. Shaltiel, D., Krug von Nidda, H.-A., Loidl, A., Lin, C.T., Liang, B., Kurtz, T., Bogoslavsky, B.: Physica C 433, 93 (2005)
14. Shaltiel, D., Krug von Nidda, H.-A., Shapiro, B.Ya., Rosenstein, B., Loidl, A., Bogoslavsky, B., Shapiro, I., Tamegai, T.: Phys. Rev. B 77, 214522 (2008)
15. Shaltiel, D., Krug von Nida, H.-A., Loidl, A., Tamegai, T., Watanabe, T., Fujii, T., Mattsuda, A.: In: VORTEX V Conference Rhodes (2007)
16. Shaltiel, D., Ioshpe, D., Grayevsky, A., et al.: Phys. Rev. B 36, 4090 (1987)
17. Beeman, B., Fink, H.J., Shaltiel, D.: Phys. Rev. 147, 454 (1966)
18. Yurgens, A.A.: Supercond. Sci. Technol. 13, R85 (2000)
19. Madsen, S., Filatrella, G., Pedersen, N.F.: Eur. Phys. J. B 40, 209 (2004)

CHAPTER 20

3D-MODELING NUMERICAL SOLUTION OF GUIDANCE FORCE OF HTSC BULK ABOVE PERMANENT MAGNETIC GUIDEWAY

Yujie Qin and Baoluo Liu

Keywords: Permanent magnetic guideway, Guidance force, Grid multiplexing technology

INTRODUCTION

The stability of bulk superconductor which can levitate over (or below) permanent magnets makes the potential engineering application possible [1, 2], especially in a hightemperature superconducting maglev vehicle system [3]. As is known widely, the horizontal stability of the maglev vehicle arises from the guidance force. Research shows that the lateral guidance is dependent on the trapped flux lines, which are pinned by the pinning centers in the HTSC bulk. However, up to now, there has been no effective way to deal with the pinning properties of HTSC. On the other hand, the guidance force is closely related to the c-axial current; usually the critical current density along the a–b plane is about three times that along the c-axis in the bulk YBCO interior. But the proposed 2D-models, limited to relatively simple geometries under homogeneous or axially symmetric external fields [4–10] did not deal with it. In these numerical or analytical methods, a simple Bean critical-state model, a shielding-current model and a frozen-field model were used.

For a typical superconducting levitation style which is composed of a bulk superconductor suspended above a permanent magnet guideway (PMG), based on the Prigozhin model Wang Xiaorong et al. have built up a 2D model where a bulk superconductor moves horizontally along the guideway. The model assumed the bulk is isotropic, that is, the influence of the c-axial current on the magnetic force is ignored. Afterwards, Ren Zhongyou et al. proposed an isotropic model, a–b plane model and c-axial model based on it, and interpret that the c-axial current only contributes to the levitation force in an axially symmetric external field. When the applied field is non-symmetric, the c-axial current must be considered and we must deal with the anisotropic problem by weighting. But this method cannot describe the macro-current flowing in a finite-dimension superconductor. So it is necessary to be able to simulate well the guidance force properties in a 3D model. Based on the proposed 3Dmodeling numerical solutions of electromagnetic behavior of a HTS bulk sample, this paper proposed firstly grid multiplexing to deal with the pinning properties of HTSC. The resolving code of the 3D-modeling of a high-temperature maglev was developed with the FORTRAN language. Comparisons between the experimental and numerical results indicated the effectiveness of the method utilized in the paper.

Citation: Yujie Qin, Baoluo Liu, 3D-Modeling Numerical Solution of Guidance Force of HTSC Bulk Above Permanent Magnetic Guideway, DOI: 10.1007/s10948-010-0833-6.

NUMERICAL METHOD

Analytic Model

According to the 3D superconducting model, the partial equation [11] describing the magnetic intensity interior HTSC bulk is as follows:

$$\frac{\partial}{\partial t}(\mu\vec{H}) + \nabla \times \frac{1}{\sigma_{sc}}\nabla \times \vec{H} + \nabla \times \frac{1}{\sigma_{sc}}\vec{Q} = 0 \quad (1)$$

where $\vec{Q}$ Q is a vector parameter dealing with anisotropy of the HTSC bulk current. σ_{sc} is the effective conductivity of the superconductor.

For the high-T_c superconductors, we use (2) to describe their electrical behavior:

$$E_{sc} = E_{c0}\left(\frac{J_{sc}}{J_c(H_{sc})}\right)^n \quad (2)$$

Here E_{c0} is material- and temperature-dependent constant. E_{sc}, J_{sc} and H_{sc} are the electric field, current density and magnetic field of the high-T_c superconductor, respectively. J_c is the critical current density. In this paper, we choose the value n = 30. The relation between σ_{sc} and Esc is as below:

$$\frac{1}{\sigma_{sc}} = \frac{E_{sc}}{J_{sc}} = \frac{E_{c0}}{J_c}\left(\frac{J_{sc}}{J_c}\right)^n \quad (3)$$

Superconducting constitutive relation is described through the Bean model:

$$J = \begin{cases} J_c\frac{E}{|E|} & |E| \neq 0 \\ \frac{\partial J}{\partial t} & |E| = 0 \end{cases} \quad (4)$$

For the Bean critical model the critical current density J_c can be a function of magnetic field, while the temperature T is a constant.

$$J_c(|\mathbf{B}|, T) = J_{c0}(T) \quad (5)$$

For (1), we use the FEM and finite differences to solve the nonlinear problem of the bulk YBCO electromagnetic behavior in applied non-uniform external magnetic field which is offered by a permanent guideway. In this solution, we can obtain H, then we will get the guidance force through $\vec{J}(H(r)) = \nabla \times \vec{H}(r)$.

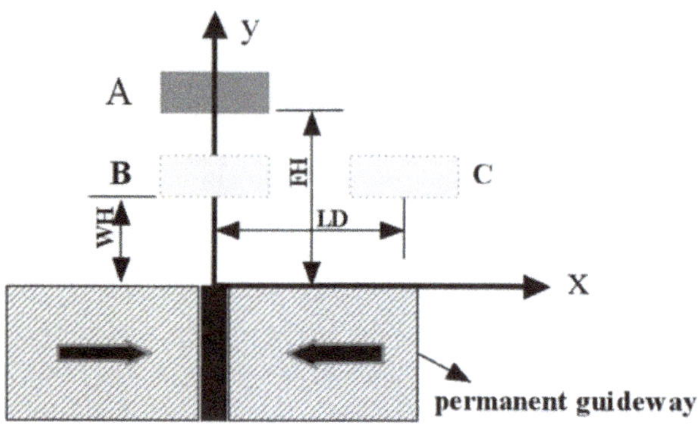

Figure 1: PMG vs. YBCO bulk levitation system.

Pinning Properties

Based on the proposed 3D-modeling numerical solutions of electromagnetic behavior of a HTS bulk sample, a grid multiplexing technology in which the meshing nodes within the HTS bulk were used to represent the centers of pinning to simulate flux pinning of HTS bulk is used. If the solution area lies in the interior of the HTS bulk and the meshing nodes is within the HTS bulk, each meshing node is defined as a pinning center to simulate pinning characteristic of flux, and an applied field value just at the position is forced to the node as an initial value in which the value is to solve at field-cooling position when the initial unit stiffness matrix is obtained. That is, the field cooling sense has been fixed. This process can be written by

$$H_i^n(x, y, z) = f(x, y, z) \quad (6)$$

where superscript number n is the nth element of the sample, and the subscript i denotes the ith node of the nth element, and f (x,y,z) is the value of the external magnetic field caused by a permanent guideway at point (x,y,z).

The merit of grid multiplexing technology lies in the facility to deal with solving codes. The nodes which simulate the pinning properties of HTSC overlap the meshing nodes which have been used to solve (1), but the shape function for interpolation is related to geometrical features of the solving areas. So it is effective to form stiffness matrices using the same shape function. When we use interior nodes to simulate the pinning centers, only dealing with a loading vector in unit stiffness matrices makes the computation simplified.

SIMULATION AND MEASUREMENT RESULT

We used the proposed methods in Sect. 2 to simulate the guidance force of the simple levitation system, composed of a HTSC bulk and a permanent magnetic guideway. This levitation system is the prototype of the man-loading maglev system [3]. Figure 1 shows a schematic diagram of the levitation system during lateral displacement of the simulation and experiment. Figure 2 shows the magnetic distribution of the guideway that the system offered. In the figures, FH is the cooling field height, MH is measurement height or calculation height, and LD is lateral displacement. Positions A and B are at the right center above the PMG. The axis line of the cylinder YBCO bulk is parallel to the y-direction.

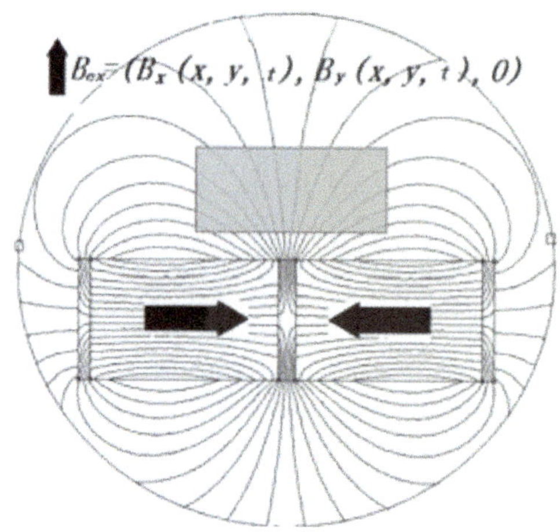

Figure 2: Magnetic distribution of PMG.

The cylindrical bulk YBCO is 30 mm in diameter and 15 mm in thickness. The PMG is made of NdFeBs of 40 mm in width, 40 mm in height and 80 mm in length and iron yokes. The cross section is parallel to the x–y plane of the PMG, and the magnetic field induced by the PMG is symmetric in the x–y plane. For the magnetic field, the usual analysis method with the equivalent surface current model is used to calculate. Figure 3 shows the calculated and measured results of magnetic flux density along the y-direction with gap 3 mm above PMGs. Figure 3 shows that the calculated values of magnetic flux density generated by the PMG are a little smaller than the measured values near the center of PMG. This is because the calculation of the magnetic field of the PMG is carried out in a perfect state which ignores the influence of magnetic field concentration effect of the iron yokes.

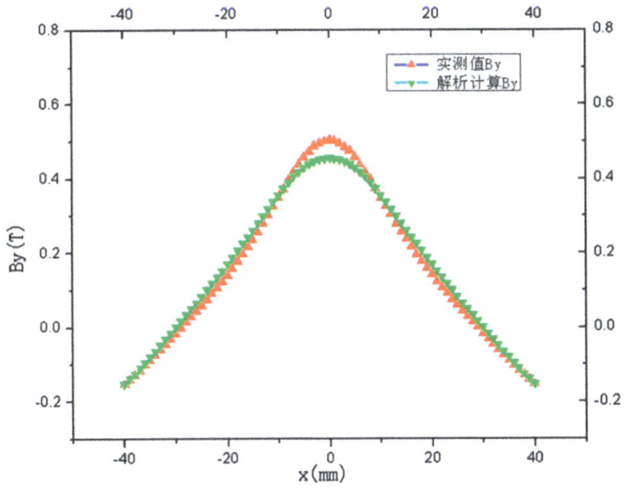

Figure 3: The distribution of the calculated and measured results of magnetic flux density along the y-direction with gap 3 mm above PMGs.

As Fig. 1 shows, during the simulation, the bulk HTSC is in field-cooling (FC) mode. A simple magnetic forcedisplacement loop is simulated in this way: firstly, the simple was moved horizontally to 20 mm from the original point B (where B overlaps A) to point C of gap 10 mm, that is, FH = 10 mm, MH = 10 mm. Then we go back and forth from 20 mm to −20 mm above the PMG with velocity 1 mm/s. Because the velocity is small, the displacement current is ignored. Physical parameters in the calculation are shown in Table 1.

In similar experiments with these instruments progress is made to measure the guidance force between the bulk YBCO and the PMG. In the experiment the sample is cooled by liquid nitrogen (77 K) in the case of FC. Measurement and calculating results are shown in Fig. 4. In order to further verify the reliability of the model, comparison between calculating results and measurement results under different FH, with lateral displacement in the range from 0 to 20 mm, is shown in Fig. 5.

Table 1: Invariant parameters of the levitation system.

Name	Value
PMG width	110 mm
PMG height	60 mm
Iron yokes width	10 mm
J_c	6.5×10^7 A/m^2
E_c	1.0×10^{-4}

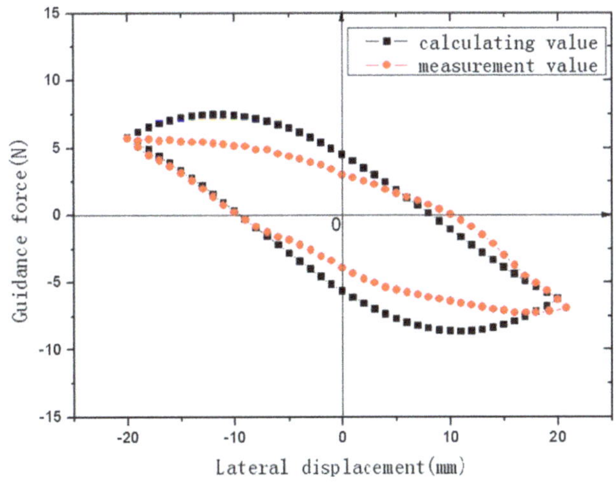

Figure 4: Experiment and calculation of guidance force–displacement loops plot.

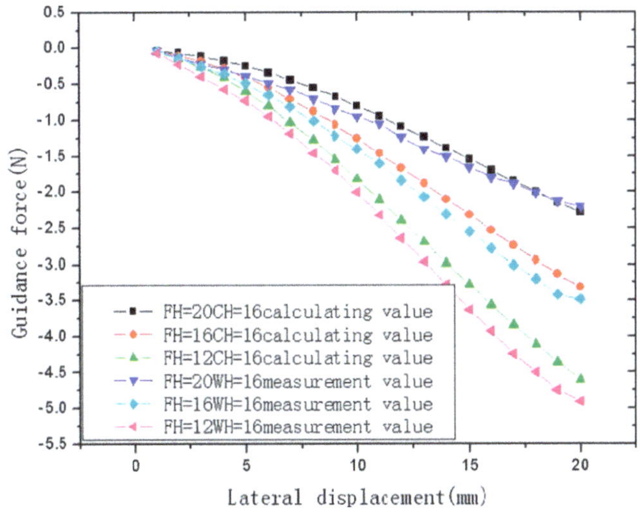

Figure 5: Experiment value and calculation value of guidance force under different FH.

From Fig. 4, we can see that the guidance force-displacement loop exhibits some small distinction between the calculation and measurement value. But the variable tendency has good consistency, as Fig. 5 shows well.

CONCLUSIONS

Based on the proposed 3D-modeling numerical solutions of electromagnetic behavior of a HTS bulk sample, a grid multiplexing method has been proposed to simulate flux pinning properties. The resolve code is successfully developed with the FORTRAN language. The guidance force of a simple cylinder shape bulk HTSC above PMG is successfully simulated by the proposed method. A similar experiment of the simulation is carried out and the measured results show that the calculated and measured values match well. It indicated the effectiveness of the method.

ACKNOWLEDGEMENTS

This work is supported by the National High Technology Research and Development Program of China

(2007AA03Z210), the National Natural Science Foundation in China (50677057).

REFERENCES

1. Hull, J.R., Murakami, M., et al.: Proc. IEEE 92(10), 1705–1718 (2004)
2. Ma, K.B., Postrekhin, Y.V., Chu, W.K.: Rev. Sci. Instrum. 74(12), 4989–5017 (2003)
3. Wang, J.S., Wang, S.Y., Zeng, Y.W.: Physica C 378–381, 809–814 (2002)
4. Davis, L.C.: J. Appl. Phys. 67(5), 2631–2636 (1990)
5. Johansen, T.H., Bratsberg, H.: J. Appl. Phys. 74(6), 4060–4065 (1993)
6. Tsuchimoto, M.: J. Mater. Processing. Technol. 108, 188–192 (2001)
7. Tsuchimoto, M., Kojima, T., et al.: IEEE Trans. Magn. 31(3), 1781–1784 (1995)
8. Kudou, Y., Ueda, H., Ishiyama, A.: IEEE Trans. Appl. Supercond. 15(2), 2230–2233 (2005)
9. Akamatsu, T., Ueda, H., et al.: IEEE Trans. Appl. Supercond. 13(2), 2161–2164 (2003)
10. Okano, M., Iwamoto, T., et al.: Physica C 386, 500–505 (2003)
11. Lu, Y.Y., Wang, J.S., Wang, S.Y., Zheng, J.: J. Supercond. Nov. Magn 21(8), 467–472 (2008)

INDEX